Laser Acceleration of Particles

(Malibu, California, 1985)

AIP Conference Proceedings
Series Editor: Rita G. Lerner
Number 130

Laser Acceleration of Particles

(Malibu, California, 1985)

Edited by
Chan Joshi and Thomas Katsouleas
University of California, Los Angeles

American Institute of Physics
New York 1985

L.C. Catalog Card No. 85-48028
ISBN 0-88318-329-3
DOE CONF-850128

TABLE OF CONTENTS

PREFACE

The second workshop on "Laser Acceleration of Particles," was held at the Norton Simon Malibu Beach Conference Center of the University of California, Los Angeles from January 7-18, 1985. The Chairman of the organizing committee was Dr. Lee Teng of Fermilab.

This book contains most of the invited presentations and contributed papers by the attendees. The executive summary by Dr. Lee Teng and the working group summaries provide overviews of the present state of the field on laser acceleration of particles and can serve as introductions for the nonexpert. In addition to the working group summaries, a basic overview of laser technology and accelerator technology is provided by the laser and accelerator tutorials. These reflect the effort to establish and unite (at least through common language) a joint scientific community with the diverse backgrounds that laser acceleration of particles involves. The tutorials should provide a valuable reference to researchers who have not specifically worked in both disciplines.

The proceedings of the first workshop on Laser Acceleration of Particles, held at Los Alamos in February 1982, were also published as AIP Conference Proceedings, No. 91, edited by P. Channell. The workshop and these proceedings were supported by the Office of Energy Research of the U.S. Department of Energy, the National Science Foundation, the American Physical Society and UCLA. Special thanks are due to Dr. David Sutter of the DOE and Dr. David Berley of the NSF for support and encouragement, and to Ms. Maria Gonzales for help with the organizational details.

Chan Joshi and Thomas C. Katsouleas
Electrical Engineering Department
University of California, Los Angeles
Los Angeles, CA 90024

BRIEF REPORT ON "THE SECOND WORKSHOP ON LASER ACCELERATION OF PARTICLES"

Lee C. Teng
Fermi National Accelerator Laboratory*, Batavia, IL. 60510

INTRODUCTION

The Second Workshop on Laser Acceleration of Particles was held at UCLA on January 7-17 1985; the first one had been held almost 3 years ago at LANL in February 1982. This workshop was attended by some 120 physicists and engineers experienced in particle-accelerator and laser technology, a 100% increase in participation over the first workshop. Between the two workshops we have seen tremendous progress in the field of laser acceleration methods. This progress is a direct result of the emphasis placed by research institutions and the strong support given by funding agencies on initiatives in this area which has made possible both theoretical and experimental efforts. Despite the title, the workshop broadened to include studies of all novel acceleration schemes, whether lasers are involved or not. To accomplish this and to cover all important work carried out since the first workshop we had to extend the duration of this workshop from the original 5 days to 9 days.

PROGRESS ON VARIOUS ACCELERATION SCHEMES

Substantial progress both theoretical and experimental has been made on all three basic types of laser accelerators: near field, far field, and plasma accelerators. Understandings gained from detailed studies of problems on laser excitation and propagation and on particle beam containment and stability have led to concentrations on those specific schemes which appear to be the most promising.

a. Near field accelerator - The possibility of forming miniature open linac structures with liquid metal micro-droplets is under close scrutiny. The accelerating gradient in such structures is limited to ~1 GeV/m by droplets turning into plasma.

b. Far field accelerator - This is now considered to be simply the inverse free-electron-laser accelerator, the principle of which has already been demonstrated in several experiments. The maximum accelerating rate obtainable is ~0.2 GeV/m and the top electron energy is limited to a few hundred GeV by synchrotron radiation loss from the electrons.

c. Plasma accelerator - The most promising scheme at this time is the beat-wave accelerator. Excitation of plasma waves by the beat-wave of two lasers has now been demonstrated experimentally to be reproducible and in good agreement with computer simulation result. An accelerating field of ~1 GeV/m was measured in the experiment. Fields higher by an order of magnitude should be attainable.

---

*Operated by the Universities Research Association, Inc., under contract with the U.S. Department of Energy.

For non-laser accelerators, the wakefield or two-beam acceleration schemes are the most promising and have received the most attention. In both schemes the electric field generated by a low energy high current beam is used to accelerate a low current second beam to high energies. The two beams can be traveling in separate cavity structures, and the field is transmitted from one cavity to the other through couplers. Or, the two beams can be traveling in the same cavity. In either case the ratio of the electric fields on the two beam trajectories must be very large. This is equivalent to having a large transformer ratio in a voltage step-up transformer. Experiments are being carried out at DESY (single cavity wakefield accelerator), and at LBL and LLNL (dual cavity two-beam accelerator). In addition, several other non-laser schemes were proposed and discussed at the workshop, but more careful and more extensive analyses are needed for these schemes.

REQUIREMENTS FOR COLLIDING BEAMS APPLICATIONS

At the first workshop the emphasis was almost exclusively on high accelerating fields and high energies. However, to gain full benefit from high energy colliding beams; i.e. to provide easily detectable rates for the occurrence of interesting events, one must also have sufficiently high luminosity, hence high beam intensity and low emittance. One set of parameters presented by Burton Richter in his opening talk for a 5 TeV on 5 TeV $e^{\pm}$ linear collider is the following:

Beam energy	5 TeV + 5 TeV
Luminosity desired	$10^{34}cm^{-2}sec^{-1}$
Normalized emittance	$1.2x10^{-7}$ m-rad
Beam radius at collision	$1.0x10^{-2}$ μm
Bunch length at collision	0.3 mm
Number of particles per bunch	$2.5x10^{9}$
Bunch repetition rate	5000 Hz
Average power for two beams	20 MW

These numbers clearly illustrate the rather extreme magnitude of the demand. A sizeable effort on the part of one of the working groups at this workshop was devoted to examining means to fulfill these requirements on repetition rate, average beampower, and beam emittance.

The requisite high average beam power calls for high efficiencies both in the production of driver power whether laser or microwave, and in the energy transfer from driver to particles. Although capable of extremely high peak electric field lasers are, at present, not energy efficient; and their construction cost per unit average power is rather high. Energy efficiency considerations point to two directions:

a. Optimization of the driver-to-particle energy transfer efficiency tends to push toward lower peak accelerating field and longer accelerating structure. This also accentuates the need for staging the accelerating structure and the driver source.

b. Similar consideration also points to the use of millimeter and centimeter waves as driver instead of lasers. Compared to lasers the energy efficiency of microwave sources is much higher, the construction cost per unit average power is much lower, although the peak field attainable is not as high. But in generating microwaves, laser also finds application. Microwaves can be induced in cavities by tightly bunched intense electron beams produced from photocathodes irradiated by pulsed laser light.

Nevertheless investigation on the possibilities of developing high efficiency, high repetition rate, short pulsed lasers should be pursued with vigor. Preliminary examinations carried out at the workshop give encouragement that these may indeed be realistic goals.

The desire for high luminosity also led to the need for very low beam emittance and very strong focusing lenses. To this end the possibility of using laser powered electric micro-quadrupoles to provide the required strong focusing was investigated. It appears possible to obtain a focal length of 5 m for a 50 GeV electron beam.

WORKING GROUP ACTIVITIES

At this workshop the three identified schemes for laser acceleration of particles: the near field, the far field and the plasma accelerators, were further studied in light of the new intensity requirement. To these three topics two others dealing with investigations of the necessary laser and accelerator technologies were added. Finally, a catch-all group was formed to look at all other schemes. In the following, we give short summaries of the activities and findings of each of these six working groups.

a. Near field accelerators

These are miniature open linac structures which support longitudinal accelerating field components (non-plane wave). Resonant structures such as arrays of liquid metal droplets ejected from micro-orifices (c.f. ink jets) were studied. Maximum accelerating fields with such structures without plasma production appear to be of the order of 1 GeV/m for 3 ps pulses and possibly 3 GeV/m at 300 fs. At higher gradients plasma production appears to rule out resonant structures. Non-resonant structures were also discussed and, in the droplet case, might provide very high gradients but probably with lower efficiency. Near field structures can also generate super strong radio-frequency transverse quadrupole focusing which could be useful for focusing the beam at the colliding point.

A possible proof-of-principle experiment using the SLAC injector is planned. The setup could also be used for experiments on other types of accelerators such as inverse free electron laser, inverse Cherenkov, etc.

b. Far field accelerators

In this scheme the beam is wiggled laterally so that the particles are synchronously accelerated by the alternating transverse electric field of a plane laser wave. The principle scheme is the inverse free electron laser (IFEL) in which the electron beam is wiggled by a series of alternating static magnetic fields (undulator). Since the first workshop, free electron laser experiments have verified the theory,and the mechanism of IFEL has been observed. The acceleration rate for an IFEL accelerator is limited to less than a few hundred MeV/m, and the maximum energy is restricted by synchrotron radiation loss to less than a few hundred GeV. The necessary wiggle amplitude, hence the synchrotron radiation loss can be reduced by modifying the resonance condition with the addition of a gaseous medium. Also, to form long sections of IFEL accelerator the laser beam must be kept focused in a wave-guide with dielectric walls. All these concepts look all right on paper, but should be confirmed experimentally.

The use of an IFEL to reduce the emittance of an electron beam was discussed. The damping effect is indeed present but rather weak. With presently available technology the damping length is too long to be useful. Further studies are required.

c. Plasma-laser accelerators

Although there are several promising plasma accelerator schemes, much of the discussion centered around the plasma beat wave accelerator. In this scheme particles are accelerated by the very high electric field in a highly modulated high-density plasma wave which is, in turn, resonantly driven by the beat-wave of two laser beams. A great deal of analytical and numerical simulation work has been done since the first workshop.

On the experimental side, the UCLA experiment, using 10.6 μm and 9.6 μm CO_2 radiation, generated and detected the fast beat wave reproducibly over a length of 1.5 mm (limited by the size of the plasma). The electron density modulation in the wave was measured to correspond to fields in the 300 MeV/m to 1 GeV/m range. A similar experiment is planned at the RAL using 1.06 μm and 1.05 μm radiations from glass lasers. An unsuccessful experiment at LANL using a single frequency and a long pulse from the HELIOS laser was analyzed by two dimensional computer simulation, and the outcome of the experiment was found to be predictable.

The transverse focusing of the particles by the plasma tends to be excessively strong making matching between stages rather difficult. This prompted people to look into very long self-focused laser channels. Such channels appear to be feasible.

The near term goals for experiments were discussed and should consist of:

1. Demonstrate significant acceleration of electrons, say, from 10 to 100 MeV.

2. Study relativistic self-focusing of the laser beam and its propagation over long distances, say, up to 100 m.

3. Investigate whether there is much emittance blow-up during acceleration.

4. Determine the efficiencies of energy transfer from laser to plasma wave and from plasma wave to particle beam.

d. Accelerator technology

A central problem is the derivation and retention of a small electron beam emittance. Several alternative methods were investigated. Damping the emittance in a damping ring looks most favorable. Single-pass damping by the use of a wiggler requires excessive length. Beam collimation by scraping is effective but will greatly reduce the beam intensity. Some emittance blow-up during acceleration is unavoidable due to misalignments and non-linear fields. The different acceleration schemes were studied and compared in this regard.

Phase space matching between acceleration stages was also studied in some detail. Longitudinally, phase locking between stages is principally a problem in laser technology and appears to be soluble. Transverse phase space matching between stages of the very strong focusing beat-wave plasma accelerator may require the use of the very strong "laser lens".

The basic scaling laws for high energy colliding beam parameters such as energy, luminosity, disruption parameter, beamstrahlung, etc. were reviewed. It was concluded that the sets of equations used by different investigators are consistent and appropriate, and can be used with confidence to consider trade-offs.

The acceleration schemes were studied also in regard to energy efficiency and cost. The overall wall-plug to beam power efficiency for all the laser schemes is, at present, one to two orders of magnitude lower than the >10% achievable with conventional radiofrequency technology. The same is true for cost per unit average power. Thus the challenge to laser technology has been clearly formulated.

e. Laser technology

The newly introduced requirements on the laser systems are the need for a large number of precision phase-locked stages, and the desire for high average power (high repetition rate) and high energy conversion efficiency. Studies made during the workshop resulted in the conclusion that:

1. Phasing between multiple stages presents no major problem, and synchronization between electron and laser beams is obtainable at repetition rates ≤ 15 Hz.

2. Wall-plug to laser energy conversion efficiency higher than 10% would be difficult. Multiple use of the gain medium is necessary to approach even this value. To obtain high efficiency for FEL's, fast switching methods are needed.

3. Repetition rates up to 100 Hz are possible for glass lasers with thin panes of glass and up to 1000 Hz are possible for gas lasers. To this extent, high average power is obtainable except it will be very costly.

4. Ultra-short, high peak power lasers with pulse length down to 1 ps are possible.

5. Side coupling, such as to an open linac structure, is easy. End coupling, such as to a beat-wave self-focused channel, is more difficult.

6. For the plasma beat-wave accelerator, detailed examinations led to the conclusions that short wavelengths (<1 μm) are best, and plasma wavelengths of 10-100 μm seem to give the best options for laser pulse length and energy. Propagation of the laser in the dispersive plasma is a concern, as is the use of long self-focused laser beam channels.

f. Other schemes

The major effort was spent on the wakefield or two-beam acceleration scheme in which the field produced by a low energy, high intensity beam is used to accelerate a second low intensity beam to a high energy. The two beams could travel in separate cavity structures with the fields transmitted from the generator structure to the driver structure through couplers. This is equivalent in principle to powering a linac structure by a klystron. In the experiment being conducted by LBL and LLNL 1-cm waves are produced by a wiggling low energy beam in a long generator structure through the FEL action. The energy of the beam is replenished in regular intervals by induction cores. The two beams could also be in the same cavity. In this case, one can prove a theorem which states that the transformer ratio (energy of beam 2 particle to that of beam 1 particle) can at most be 2 if the two beam bunches are colinear and have zero length (δ-functions). To obtain a useful transformer ratio one must, therefore, use long beam bunches or non-colinear geometries. A simple and effective arrangement due to Voss and Weiland is one for which the generating beam is in the shape of a series of rings and travels inside a tight-fitting cylindrical cavity. The wake field propagates radially inward and superimposes constructively to produce a very high longitudinal accelerating field on the axis. The axial field can be further enhanced by using irises in the wave guide to channel the propagation of field toward the axis. Several other geometries were proposed and studied. The wakefield acceleration principle can also be applied to beams traveling in a plasma.

Other schemes considered are:

1. The Inverse Cherenkov accelerator, proposed a long time ago and for which a great deal of theoretical design work has been done, was tested in a proof-of-principle experiment. The main concern for this scheme is the emittance growth due to multiple scattering in the gas medium.

2. The Laser-Focus accelerator, which was also described at the first workshop, has been tested experimentally. Ion energies greater than 100 MeV have been observed.

3. The Radial Implosion accelerator, in which radially inward electron beams are used to "squeeze" an axial magnetic field and hence, produce a high electric field (up to 3 GeV/m) on the axis.

4. The use of laser to focus, not to accelerate, an electron beam was investigated. It appears possible to develop rather strong lenses (3 m focal length for 50 GeV electrons).

Experiments are starting in DESY on the ring-beam wakefield accelerator and are in progress at LBL on the two-beam accelerator. The latter may yield, as a first step, a high intensity 1 cm microwave source. The inverse Cherenkov accelerator, the laser-focus accelerator, and the laser lens are ready for further experimental studies. Theoretical and numerical work should be pursued on all other schemes.

CURRENT AND PROPOSED PROGRAMS

A. Current programs

<u>Beat-wave plasma accelerator</u>

UCLA, LANL, RAL, NRL

Droplet/Grating accelerator
SLAC/BNL/NRCC, Cornell
FEL/IFEL accelerator
BNL, Cornell, Adelphi Tech.
Wakefield accelerator
DESY/LANL, SLAC/UCLA
Two-beam accelerator
LBL/LLNL
Inverse Cherenkov accelerator
UCSB
Photocathode/Lasertron
LANL, SLAC, Japan
Laser technology
MSNW (now Spectra Tech.), NRCC

B. Proposed additional programs
Beat-wave plasma accelerator
Wisconsin/FNAL/LANL
Wakefield accelerator
ANL/Wisconsin, LANL
Plasma wakefield accelerator
LANL
Switched pulse-field linac
CERN/LBL
Laser photodiode source
LANL
Short-pulse laser technology
LANL

CONCLUSIONS

This second workshop confirmed the promise and likely payoff of a variety of schemes. Gauged from the very large proportions of the Superconducting Super Collider (SSC), it is clear that the efforts on research and development into new methods of acceleration are the only hope for further advances in high energy accelerators and particle physics beyond the SSC. Thus, not only do we see needs for continued support of the on-going efforts, but we also visualize the exigency for additional and intensified new activities to ensure continued strong growth in this field.

REQUIREMENTS FOR VERY HIGH ENERGY ACCELERATORS*

BURTON RICHTER
Stanford Linear Accelerator Center
Stanford University, Stanford, California 94305

I. INTRODUCTION

In this introductory paper at the second Workshop on Laser Acceleration my main goal is to set what I believe to be the energy and luminosity requirements of the machines of the future. These specifications are independent of the technique of accelerations. But, before getting to these technical questions, I will briefly review where we are in particle physics, for it is the large number of unanswered questions in physics that motivates the search for effective accelerators.

The first particle accelerators were built roughly fifty years ago. These first machines had energies of the order of MeVs and were used to study a world that looked relatively simple. Matter was composed of four basic constituents: protons, neutrons, electrons, and neutrinos. These constituents interacted via four forces: the weak (to account for radioactivity); the electromagnetic (to account for the interaction between charges and currents); the strong (to bind the nucleus together); and the gravitational (to account for the interaction of masses at large distances.) All our attempts at understanding matter were guided by two dynamical principles – relativity and quantum mechanics.

In the intervening years, the energy of our accelerators has grown by six orders of magnitude to reach the TeV level. Our old view of what were the elementary constituents of matter has turned out to be wrong. The simple picture of four constituents became ever more complicated as machines of higher energy were built and more and more mesons and isobars of the nucleon were discovered. In the early '60s there were more than one hundred of the "elementary particles." All of this was swept away in the '60s to be replaced with the quark model, wherein the proton, the neutron, all of those mesons and other particles became composites of combinations of quarks and antiquarks.

In these last fifty years we seem to have lost one force. Our present picture is that the weak and the electromagnetic forces are but different manifestations of the same basic force. Our theoretical colleagues are struggling (so far unsuccessfully) with models that try to combine the strong force and perhaps even gravity into a unified picture.

* Work supported by the Department of Energy, contract DE – AC03 – 76SF00515.

Our dynamical principles remain the same. Relativity and quantum mechanics are still our guide and space is still thought to be continuous although some are questioning that, too.

Experiments and theory of the last fifty years have given rise to our present generation of models that allow us to calculate what happens at the fundamental level down to distances as short as 10^{-17} centimeters. The key to this great advance in our understanding of the fundamental structure of matter and the forces of nature has been the accelerators that have allowed experiments that probe matter to ever smaller distances. We have gone from Cockcroft–Walton generators to Van de Graaffs to cyclotrons to synchrotrons to strong focusing to linacs to colliding beams to superconductivity. The energy of our machines has gone up by six orders of magnitude while the cost per unit energy has gone down by nearly five orders of magnitude in the same period of time. To continue our study of the fundamental nature of matter we will need more powerful and cost-effective accelerators that will probe distances where we already know our present theoretical models to be inadequate. Here are a few of the problems that exist with our present framework:

1. We have no quantum theory of gravity and such a theory is clearly required to understand the things that happen at the highest energy and the smallest sizes.

2. We have no unified picture incorporating the strong interaction and indeed the first attempt to make such a theory, the SU-5 theory, failed when tested by experiment.

3. We don't understand the relation between the quark and lepton masses and our present models need 20 apparently arbitrary parameters to specify these parameters.

4. We seem to have three families of quarks and leptons which differ only in the fact that each family is heavier than the one before. Why are there three? Are there more?

5. We have what seem to be 37 elementary constituents – 18 colored quarks, 6 leptons, a photon and 3 massive vector bosons to carry the electroweak force, 8 gluons to carry the strong force, and 1 graviton to carry the gravitational force. This seems a bit much.

There are many different theories available in the literature today which purport to explain some of the unexplained and to predict what we will see when we probe still deeper into the fundamental structure of matter. All of these models predict various new phenomena at higher masses or shorter distances than are now accessible and only experiment can sort out which, if any, of the currently popular "next step" in theory is the right direction.

The experiments that will be required will need a new generation of accelerators. These machines will have to have much higher energy than is available today and will have to be built at a cost that the taxpayers of the country (or perhaps the world) will be willing to bear. In the past the scientific community has come up with new techniques of acceleration when the progress of science required it and when the cost of the old techniques, extrapolated to higher energy, became prohibitive. That is what this meeting is about. You are all here to try to see whether the enormous fields available, in principle, from focused laser beams can somehow be transformed into a mechanism for accelerating particles to very high energy in a cost-effective fashion. If progress is to be made it will take the talents of a mixture of accelerator physicists, plasma physicists, and laser physicists. All of those disciplines are represented at this workshop and I look forward to seeing how far you all get with this job in the time you are here working on it.

I now turn to the technical questions. I will review the energy and luminosity as a function of energy required for both very high energy electron and proton machines. Electron machines will turn out to be most promising, and I will review the design principles for very high energy machines.

II. LUMINOSITY AND ENERGY REQUIREMENTS

A. Proton Machines

Protons are composite particles. Their constituents are three valence quarks (u, u, d); gluons that are exchanged between the quarks to bind the system together; and the so-called "Sea" quarks which are virtual quark–antiquark pairs generated by the interaction of the gluons and the valence quarks. This multitude of constituents (partons) within the proton share the proton's energy.

A proton–proton collision is like two bags, each containing many constituents, hurtling at each other. The hard collisions, the ones that lead to the production of large mass phenomena, are collisions of one of the constituents in one of the bags with a constituent in the other bag. These hard collisions are relatively improbable, and when they occur tend to produce final state particles with large transverse momentum and leave behind a collection of excited debris in the bags. The individual partons tend to have low energy fractions and so the center of mass energy in the parton–parton collision is, on the average, much smaller than the center of mass energy of the proton–proton system.

Figure 1 shows the momentum distribution within the proton of the valence quarks, the Sea quarks and the gluons.[1] The quantity x is the fraction of the proton momentum carried by a given constituent. The momentum distribution is itself a function of the momentum transfer in the hard collision of the constituents. For example, the valence quark momentum distribution is shown

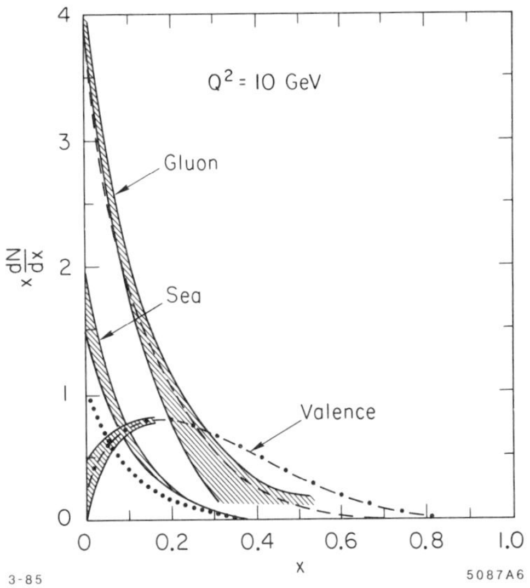

Fig. 1. The gluon, valence quark, and Sea quark distributions at a momentum transfer of 10 GeV^2 from Ref. 1.

schematically at several momentum transfers in Fig. 2. The higher the momentum transfer the smaller is the average fraction of the momentum of the proton carried by a particular constituent.

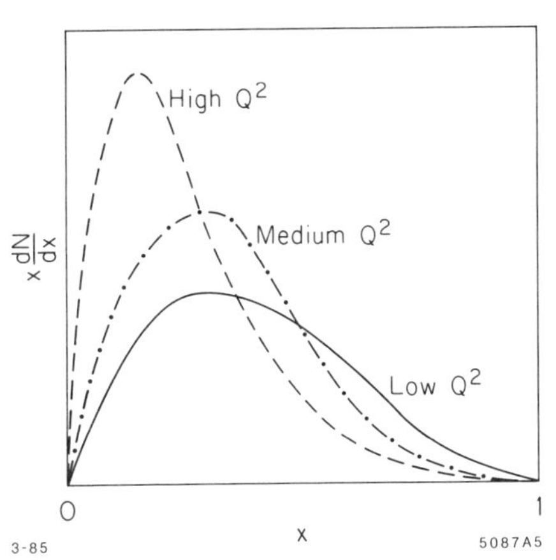

Fig. 2. The evolution of the valence quark distribution as Q^2 increases. At higher Q^2 the distribution becomes more peaked and is shifted to lower x.

What all of this means is that while the total cross section for a proton–proton collision is very large, the partial cross section for a hard collision is very small and depends strongly on the mass of the final state produced. The cross section for the production of some final state with a mass M plus the excited proton fragments X has an energy and mass dependence given by

$$\sigma(M + X) \propto \frac{1}{M^2} \quad f\left(\frac{M^2}{E^{*2}}\right) \tag{1}$$

where E^* is the center of mass energy of the proton–proton system. An example of the energy and mass dependence of the cross section is given in Fig. 3. It shows the cross section for the production of a Higgs boson as a function of Higgs mass for various proton–proton center of mass energies. This cross section decreases rapidly with increasing mass at a fixed center of mass energy and decreases rapidly with decreasing center of mass energy at a fixed boson mass.

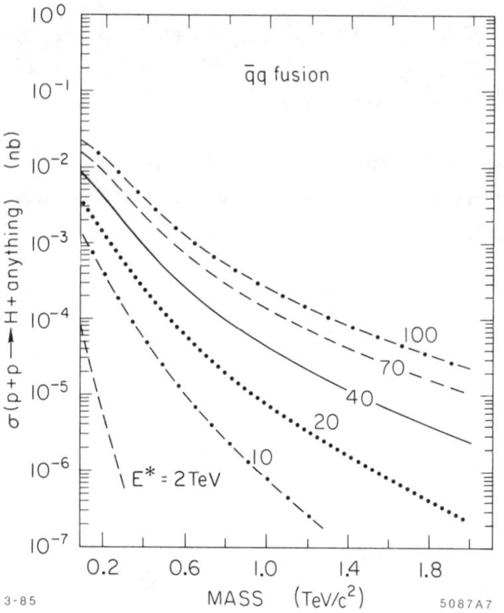

Fig. 3. The total cross section for Higgs boson production by quark–antiquark fusion in proton-proton collisions as a function of Higgs boson mass for various center of mass energies from Ref. 1.

One can do this kind of analysis for any process one cares to study, and from this kind of analysis can define the “discovery limit” of a given machine. This discovery limit relates the mass of the phenomena that can be studied to both the center of mass energy and the luminosity of a proton–proton collider. Figure 4 shows an example for the case of the Higgs boson. This analysis tells us that if the Higgs boson mass is 1 TeV, then a machine with 40 TeV in the center of mass of the proton–proton system will have to deliver an integrated luminosity of 10^{40} cm^{-2} to produce a handful of Higgs events above the expected background. Thus a 40 TeV cm machine has to have an instantaneous luminosity of about 10^{33} if one is to get this handful of events in one year of running.

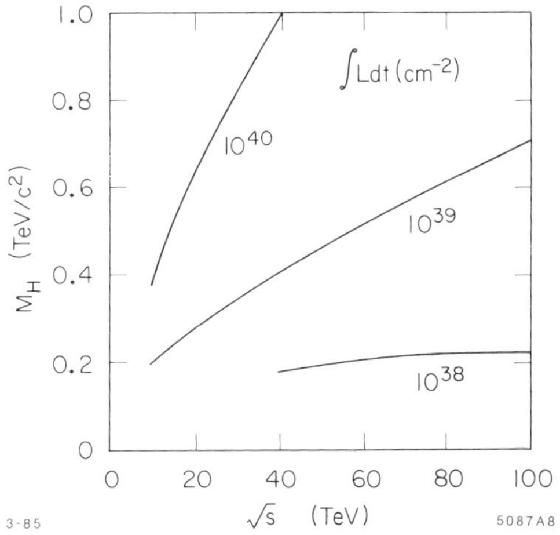

Fig. 4. The "discovery limit" for the Higgs boson relating the maximum mass of the Higgs which can be seen and the center of mass energy of the proton–proton collision for different integrated luminosities.

The SSC, now in the preliminary design phase, has a design center of mass energy of 40 TeV and a luminosity of 10^{33} cm^{-2} s^{-1}. I list in the table below the upper limit on the mass detectable in various kinds of phenomenon where this upper limit is set at that mass that results in a handful of events in a running year.

The SSC thus has a discovery limit that depends on the process studied and ranges from 0.4 TeV for new lepton pairs, to 8 TeV for jet pair formation. A crude mean for the mass reach of the SSC is about 3 TeV. However, it should be noted that because of the energy and mass dependence of the cross section for a given process (Eq. (1)), the SSC is a "discovery machine" at this TeV mass region, and is a precision machine giving very high event rates at a few hundred GeV mass.

Final State	Mass Limit (TeV)
Jet pairs	8.0
lepton pairs	0.4
W'	3.6
Z'	1.6
η_T	3.2
$\tilde{g}$	4.8
$Q\bar{Q}$	4.8
H	1.0
Mean Limit	3.0

We now have to look at the requirements for a proton machine going beyond the SSC. Suppose we want to move up a decade in mass. To move the "discovery" limit up by a factor of ten we have to increase the energy or the luminosity or both. Equation (1) shows that raising the center of mass proton–proton collision energy by a factor of ten and the luminosity by a factor of a hundred over those of the SSC moves this discovery limit up by the required factor of ten. Can one build such a machine using storage ring technology, and could one use such a machine if one could build it? I think the answer is no in both cases.

An obvious problem with the machine will be the luminosity lifetime. Particles will be lost from the circulating beams by proton–proton interactions at the collision point. This is already a significant problem at the SSC, where the luminosity lifetime for the presently favored design is about 20 hours. In our super SSC with an energy ten times higher than the SSC, we would probably get our luminosity up by a factor of a hundred by increasing the number of bunches circulating in the machine by a factor of ten, and getting the other factor of ten from the decreased size of the colliding bunches resulting from the adiabatic damping that occurs in acceleration to the higher energy. If one gets the luminosity up in this fashion and adds in the increase of the total cross section expected from the increase in the center of mass energy, the luminosity lifetime goes down by a factor of fifteen from the SSC value to roughly 1.5 hours. This is probably too short a lifetime to allow for injection and ramping up to energy in a storage ring design.

As far as the experimental detectors are concerned, the problems are probably overwhelming. There are approximately 100 proton–proton interactions per beam–beam collision, and I don't believe that you can make detection apparatus to stand that kind of rate. There are some who argue now that the 10^{33} luminosity of the SSC will be very hard to use with "real world" detectors; and I doubt that anyone can demonstrate a usable detector technology at 10^{35}.

B. Electron–Positron Machines

In contrast to protons, from what we know now electrons and positrons are elementary particles. There are no "partons" to share the momentum of the primary electron and proton and thus reduce the effective collision energy. The energy you build is what you get. However, cross sections for particular processes are small and thus large luminosities are required. The cross section for a given process is given by

$$\sigma_i \approx 10^{-37} E^{*^{-2}} R_i \quad (cm^2) \tag{2}$$

where R_i is the ratio of the cross section for process i divided by the cross section for mu pair production through the electromagnetic interaction only. Some typical values of R_i are listed in the table below.

Final State	R
$\mu^+\mu^-$	1.2
$Q\bar{Q}$ (charge $\frac{2}{3}$)	2.0
$Q\bar{Q}$ (charge $\frac{1}{3}$)	1.2
W^+W^-	25
Z^0Z^0	25
$Z^0\gamma$	25
Z^0H	0.2
Z'	1000
ρ_T	7
$\widetilde{\nu}\,\widetilde{\bar{\nu}}$	0.6

We can define "discovery" limits for the electron–positron machines, too. I will set the required yield as 100 events per 10^7 seconds. The table below gives the center of mass energy at which one would get 100 events in an integrated luminosity of 10^{40} cm^{-2} s^{-1}.

Channel	$E^*(TeV)$ at $L = 10^{33}$
$Q\bar{Q}$ (charge $\frac{2}{3}$)	4.5
Jet- -Jet (old quarks)	10.0
Z^0H	1.4
$\widetilde{W}^+\widetilde{W}^-$	4.5
$\widetilde{\nu}\,\widetilde{\bar{\nu}}$	2.5

As in the case of the proton machines, one spans quite a range of masses as one looks at different processes. Here an integrated luminosity of 10^{40} is enough to study jet–jet phenomena up to 10 TeV mass or to study Z^0 plus

Higgs production to 1.4 TeV mass. I will interpret this table as implying that very roughly a machine with 3 TeV in the center of mass requires a luminosity of 10^{33}. The luminosity required for machine of other energies is given by

$$\mathcal{L} = 10^{33} \left(\frac{E^*}{3} \right)^2 \; cm^{-2} \; s^{-1} \tag{3}$$

where the center of mass energy E^* is in units of TeV.

There are background processes in electron–positron collisions which will eventually give multiple events per beam crossing for sufficiently high luminosity. The dominant background is the so-called two photon process. However, the total cross section for this process is much smaller than the background generating cross section in proton–proton collisions and there is no problem with the two photon process until luminosities are much higher than 10^{35} cm^{-2} s^{-1}.

C. A Quick Summary of Proton and Electron Colliders

For proton Colliders:

1. The effective center of mass energy is much lower than the proton–proton center of mass energy.
2. Cross sections are proportional to $M^{-2} f\left(\frac{M^2}{E^{*2}}\right)$
3. The SSC has an effective discovery limit of 3 TeV if its luminosity is 10^{33} cm^{-2} s^{-1}. To go to higher energy, the energy, the luminosity or both have to be increased.
4. If the luminosity is held fixed, the machine energy must be scaled roughly as the square of the mass limit.

For electron–positron Colliders:

1. The energy built is what you get.
2. The cross section is proportional to ${E^*}^{-2}$.
3. The luminosity required is proportional to the square of the cm energy and is roughly given by

$$\mathcal{L} = 10^{33} \left(\frac{E^*(TeV)}{3} \right)^2 \tag{4}$$

4. Background is not a problem until the luminosities are much larger than 10^{35} cm^{-2} s^{-1}.

III. THE BASIC DESIGN OF HIGH ENERGY LINEAR ELECTRON COLLIDERS

The technique in use up to now for electron–positron colliders is that of the colliding beam storage ring. This technology is well understood and is being used to construct the 27 km. circumference LEP storage ring at CERN. However, the cost of storage rings at fixed luminosity scales as the square of the center of mass energy and so runs into "fiscal feasibility" problems at energies much higher than LEP's. A technique with different scaling laws is required and I believe that that technique is the linear collider.

The basic design of high energy linear colliders is much more complicated than that of high energy electron storage rings. In colliding beam storage rings the technology is well known and the limits on performance are well understood. It is possible to write a few simple equations that define the parameters of an optimized storage ring and determine its costs for any choice of energy and luminosity. However, linear electron colliders are new and we are still learning to understand them. In this section I will summarize some of the basic design equations and constraints and give a few examples of parameters for very high energy machines. My aim is to introduce some realism into the discussion of new technologies for acceleration.

The beam–beam interaction can be much stronger in a linear collider than in a storage ring. In an electron–positron collider the collective fields of one beam will focus a single particle in the other beam, as illustrated in Fig. 5. The strength of the interaction is measured by a dimensionless parameter D (the disruption parameter) which is the ratio of the bunch length to the focal length of an equivalent lens. For round trigaussian beams D is given by

$$D = \frac{\sigma_z}{F} = \frac{r_e \sigma_z N}{\gamma \sigma_{r_o}^2} \tag{5}$$

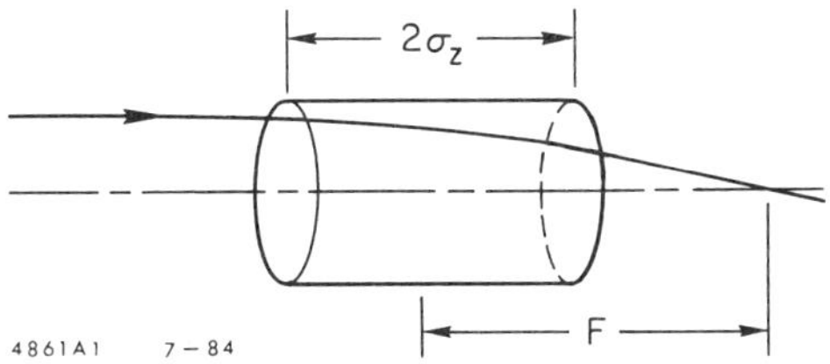

Fig. 5. The effect on a particle in one beam of the macroscopic fields from all of the particles in the other beam in a linear collider.

where the bunch has a longitudinal standard deviation σ_z, a radial standard deviation σ_{r_o}, a number of particles N and an energy γ in rest mass units; r_e is the classical electron radius; and F is the small amplitude focal length of an equivalent thin lens. The effective fields in a linear collider tend to be very large and the focal lengths tend to be small. For example, in the SLC project now under construction at SLAC, the fields are on the order of megagauss, F is on the order of millimeters, and D is about 1.

The luminosity equation of a linear collider is given by

$$\mathcal{L} = \frac{N^2 f}{4\pi} \left\langle \frac{1}{\sigma_r^2} \right\rangle \equiv \frac{N^2 f}{4\pi\sigma_{r_o}^2} H \qquad (6)$$

where the charge in the two bunches is assumed equal, f is the collision frequency, σ_{r_o} is the radial standard deviation of the charge distribution before the collision, and H is an enhancement factor which measures the effect of the beam–beam interaction on the transverse dimension of the beams during the collision. The beam–beam interaction in linear colliders can be so strong that a kind of mutual pinch occurs, reducing the radius of both beams during the collision period and hence enhancing the luminosity. H has been calculated by means of a computer simulation by Hollebeek,[2] and his results for a round gaussian beam are shown in Fig. 6. H is by definition 1 at small values of the disruption parameter and rises to an asymptotic value of around 6 for disruption parameters greater than 2.

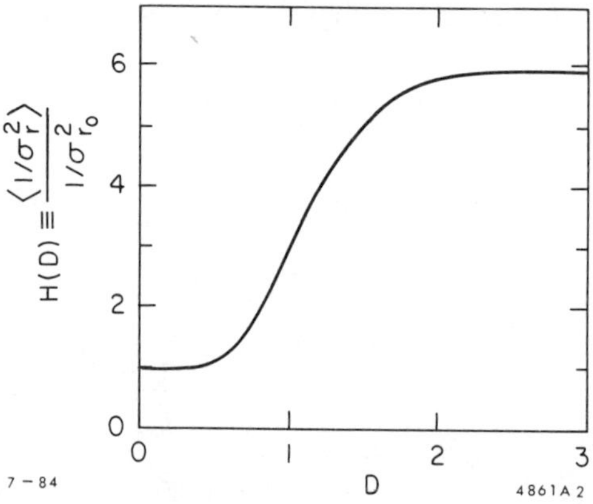

Fig. 6. The luminosity enhancement factor, H, as is a function of the disuption parameter, D.

The large effective fields in the collision region can generate very intense synchrotron radiation. At high luminosity the synchrotron radiation, called "beamstrahlung", dominates the energy spread in the beams. Classically, the synchrotron radiation spectrum is a universal function of the photon energy divided by a parameter E_c called the critical energy.

$$E_c = 3\hbar c \frac{\gamma^3}{2\rho} \tag{7}$$

In this equation h is Planck's constant, c is the velocity of light, γ is the energy in rest mass units, and ρ is the bending radius of the particle in the field of the other beam. Classically, if the beamstrahlung photon energy is measured in units of the critical energy, the spectrum is like that shown by the heavy line in Fig. 7, rising to a maximum at $x = 1$ and decreasing exponentially for $x > 1$. This classical spectrum is good as long as the beam energy divided by the critical energy is much greater than one.

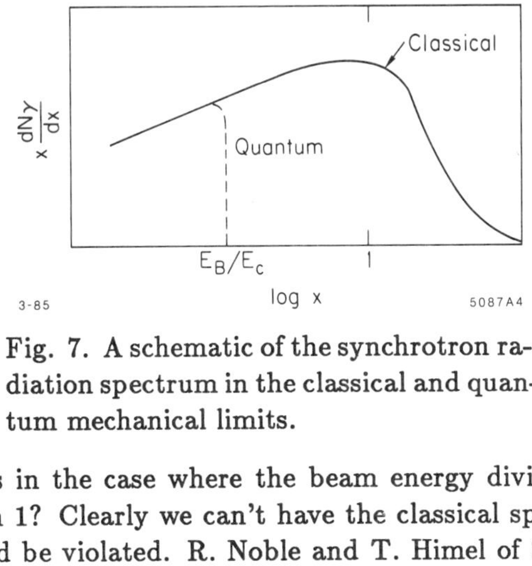

Fig. 7. A schematic of the synchrotron radiation spectrum in the classical and quantum mechanical limits.

What happens in the case where the beam energy divided by the critical energy is less than 1? Clearly we can't have the classical spectrum, for energy conservation would be violated. R. Noble and T. Himel of SLAC have worked out this problem and the results are shown by the dashed line in Fig. 7. If effect, the beamstrahlung spectrum follows the classical spectrum up to $x = E_b/E_c$ and then drops rapidly to zero. In this case, less beamstrahlung is emitted than the classical equations imply.

The ratio of beam energy to critical energy is given by

$$\frac{E_b}{E_c} = \frac{D f^{\frac{1}{2}} P}{3\hbar c \gamma r_e^2 (4\pi \mathcal{L})^{\frac{3}{2}}} \tag{8}$$

where P is the power in one beam and $\mathcal{L}$ is the luminosity. A useful approximation is

$$\frac{E_b}{E_C} \approx 5 \times 10^{-4} \frac{f^{\frac{1}{2}} DP}{EL^{3/2}} \tag{9}$$

where P is measured in megawatts, E is in TeV, and L is in units of 10^{33} cm^{-2} s^{-1}.

It turns out that all low energy machines like the SLC are in the classical regime and all interesting very high energy machines are in the quantum mechanical regime. For the SLC, E_b/E_c is 15 and safely classical. A high energy machine which might be of interest could have a beam energy of 1.5 TeV, a luminosity of 10^{33}, a frequency of 1,000 hertz, a disruption parameter of 1, and a beam power of 1 megawatt. Such a machine would have E_b/E_c of 0.01 and would be very definitely in the quantum mechanical regime.

The fractional energy loss δ of a particle in one beam in passing through the other beam is given by

$$\begin{aligned} \delta_{QM} &\approx \delta_{\text{classical}} \times \left(\frac{E}{E_c}\right)^{4/3} \\ &\approx \frac{2mc^2}{3} \left(\frac{r_e}{3\hbar c}\right)^{4/3} \left(\frac{DP}{\gamma f}\right)^{1/3} \\ &\approx 4 \left(\frac{D\ P(MW)}{f\ E(TeV)}\right)^{1/3} \end{aligned} \tag{10}$$

This is all very new, and hence I would not be surprised if I had lost a factor of 2 here or there in the above equations. I hope they will be checked shortly.

A parameter of importance for the high energy physics experiments to be done with the machine is the center of mass energy spread which is given by

$$\sigma_{E/E} \approx \delta_{QM}/2\sqrt{3} \tag{11}$$

For the high energy example given above δ is about 0.3 and $\sigma_{E/E}$ is about equal to 8.5%.

What does all this mean for very high energy machines? I cannot claim to have fully digested the implications of the quantum mechanical beamstrahlung regime on machine design. Rather than trying to develop an optimized set of parameters, I will give several sets of consistent parameters for a machine of sufficiently high energy and luminosity to be interesting. I will take a center of mass energy of 10 TeV; a luminosity of 10^{34} cm^{-2} s^{-1}; an interaction region β function of 1 centimeter (though I have no idea if the magnets can be made strong enough to realize such a small beta); a disruption parameter of 2, which implies a H of 5; and a center of mass energy spread of 10%. Four sets of consistent parameters are given in the table below. In the table ϵ_N is the invariant emittance defined as $\gamma\sigma_x\sigma_{x'}$.

CONSISTENT NON-OPTIMIZED SETS OF PARAMETERS

P (MW)	1	3	10	30
f(HZ)	500	1500	5000	15,000
$N(e^+$ or $e^-)$	2.5×10^9	2.5×10^9	2.5×10^9	2.5×10^9
σ_z(mm)	0.03	0.1	0.3	1.0
$\epsilon_N(M)$	1.2×10^{-8}	3.6×10^{-8}	1.2×10^{-7}	3.6×10^{-7}
σ_{r_o}(microns)	3.5×10^{-3}	6.0×10^{-3}	1.0×10^{-2}	1.8×10^{-2}

In all of the cases the energy delivered to the collision region per bunch of electrons or positrons is constant. As the total power in the beam increases, the invariant emittance, and hence the radius at the collision point also increases. In all of these cases the invariant emittance is considerably smaller than that of the SLC and the beam radii are tiny indeed. I emphasize again that these parameter sets are not meant to be taken as optimized sets – they are only consistent sets.

IV. CONCLUSIONS

My conclusions are relatively simple, but represent a considerable challenge to the machine builder.

High luminosity is essential. We may in the future discover some new kind of high cross section physics, but all we know now indicates that the luminosity has to increase as the square of the center of mass energy. A reasonable luminosity to scale from would be 10^{33} cm^{-2} s^{-1} at a center of mass energy of 3 TeV.

The required emittances in very high energy machines are small. It will be a real challenge to produce these small emittances and to maintain them during acceleration. The small emittances probably make acceleration by laser techniques easier, if such techniques will be practical at all.

The beam spot sizes are very small indeed. It will be a challenge to design beam transport systems with the necessary freedom from aberration required for these small spot sizes. It would of course help if the beta functions at the collision points could be reduced.

Beam power will be large – to paraphrase the old saying, "power is money" – and efficient acceleration systems will be required.

REFERENCES

1. This figure as well as Figs. 3 and 4 are reprinted from "Supercollider Physics," E. Eichten *et al.*, Rev. Mod. Phys. 56, 579 (1984), and Fermilab Pub-84/17-T, February, 1984.

2. R. J. Hollebeek, Nucl. Inst. and Methods, 184, 333 (1981).

Fundamental Physics During Violent Accelerations

KIRK T. MCDONALD

Joseph Henry Laboratories
Princeton University
Princeton, New Jersey 08544

ABSTRACT

When a powerful laser beam is focussed on a free electron the acceleration of the latter is so violent that the interaction is non-linear. We review the prospects for experimental studies of non-linear electrodynamics of a single electron, with emphasis on the most accessible effect, non-linear Thomson scattering. We also speculate on the possibility of laboratory studies of a novel effect related to the Hawking radiation of a black hole.

1. Introduction.

It is widely appreciated that the greatest laboratory acceleration obtainable with present technology is that of an electron in a high intensity laser beam. In this paper we concern ourselves with the intrinsic interest of the acceleration, rather than considering it as a step in the production of high energy particle beams. In particular we wish to draw attention to the various non-linear radiation effects which can be explored with strongly accelerated electrons.

An important qualitative feature distinguishes our understanding of the electromagnetic interaction from that of the strong, weak and gravitational interactions. Namely that the latter are fundamentally non-linear, meaning that the bosonic quanta which mediate these interactions can couple to themselves. However experimental application of our understanding is largely limited to situations in which these self-couplings can be ignored, so that the interaction is effectively linear. Yet a complete understanding of the fundamental interactions will require mastery of their non-linear aspects as well. As a step towards the elucidation of fundamental non-linear phenomena we consider the unusual case of very strong electromagnetic fields, in which non-linear effects can be induced.

We will discuss three types of non-linear electrodynamic processes. In Section 2 we speculate on the interaction of a highly accelerated electron with the zero-point energy of the electromagnetic field, in close analogy to Hawking radiation of black holes. Sections 3 and 4 review various possible effects of vacuum polarisation. In Section 5 we note that the most accessible non-linear effect is the production of higher harmonic radiation in Thomson scattering. A sketch of an experiment to detect the latter is given in Section 6.

2. A Conjecture for Future Experimental Study.

In this section we discuss a speculative effect which first aroused the author's interest in non-linear electrodynamics. This example may serve to indicate the possibilities for phenomena beyond the more standard features to be reviewed in Sections 3-5.

An appropriate point of departure is the work of Hawking (1974,1975) in which he associates a temperature with a black hole:

$$T = \frac{\hbar g}{2\pi c k}.$$

Here g is the acceleration due to gravity measured by an observer at rest with respect to the black hole, and k is Boltzmann's constant. The significance of this temperature is that the observer will find that (s)he is immersed in a bath of black-body radiation of characteristic temperature T. This is in some way due to the effect of the strong gravitational field on the ordinarily unobservable zero-point energy structure of the vacuum.

Contemporaneous with the work of Hawking several people considered quantum field theory according to accelerated observers. By the equivalence principle we might expect accelerated observers to experience much the same thermal bath as Hawking's observer at rest near a black hole. The efforts of Fulling (1972,1973), Davies (1975), and Unruh (1976) indicate that this may well be so. If a is the acceleration as measured in the instantaneous rest frame of an observer, then (s)he is surrounded by an apparent bath of radiation of temperature

$$T = \frac{\hbar a}{2\pi c k}.$$

Additional discussions of this claim are given by Sciama, Candelas and Deutch (1981), and by Birrell and Davies (1982). 'Elementary' discussions are given by Boyer (1980), and by Donoghue and Holstein (1984).

Of experimental interest is the case when the observer is an electron. Then the electron can scatter off the bath of radiation producing photons which can be detected by inertial observers in the laboratory. This new form of radiation, which we will call *Unruh radiation*, is to be distinguished from the ordinary radiation of an accelerated electron. In particular, the intensity of radiation in the thermal bath varies as T^4. Hence we expect the intensity of the Unruh radiation to vary as $T^4 \sim a^4$. This result contrasts with the a^2 dependence of the intensity of Larmor radiation.

We illustrate this further with a semi-classical argument. The power of the Unruh radiation is given by

$$\frac{dU_{\text{Unruh}}}{dt} = \text{energy flux of thermal radiation} \times$$
$$\times \text{ scattering cross section.}$$

For the scattering cross section we take

$$\sigma_{\text{Thomson}} = \frac{8\pi}{3} r_o^2$$

where r_o is the classical electron radius. The energy density of thermal radiation is given by the usual Planck expression :

$$\frac{dU}{d\nu} = \frac{8\pi}{c^3} \frac{h\nu^3}{e^{h\nu/kT} - 1},$$

where ν is the frequency. The flux of the isotropic radiation on the electron is just c times the energy density. Note that these relations hold in the instantaneous rest frame of the electron. Then

$$\frac{dU_{\text{Unruh}}}{dt d\nu} = \frac{8\pi}{c^2} \frac{h\nu^3}{e^{h\nu/kT} - 1} \frac{8\pi}{3} r_o^2.$$

On integrating over ν we find

$$\frac{dU_{\text{Unruh}}}{dt} = \frac{8\pi^3\hbar r_o^2}{45c^2}\left(\frac{kT}{\hbar}\right)^4 = \frac{\hbar r_o^2 a^4}{90\pi c^6},$$

using the Hawking-Davies relation $kT = \hbar a/2\pi c$.

A variation of the preceding argument has been given by Gerlach (1982). Again the key idea, taken as an assumption, is that an accelerated observer can have a non-trivial interaction with the vacuum fluctuations of the electromagnetic field. We suppose that these fluctuations, $\Delta\mathbf{E}$, lead to an additional acceleration of the observer, which for an electron yields $\langle\Delta a^2\rangle = e^2\langle\Delta\mathbf{E}^2\rangle/m^2$. The apparent strength of the vacuum fluctuations depends on the acceleration of the electron, Planck's constant, and the speed of light, but not the external electric field. By dimensional arguments, $\langle\Delta\mathbf{E}^2\rangle \sim \hbar a^4/c^7$. The Unruh radiation rate can now be estimated from the Larmor formula by inserting the fluctuation acceleration:

$$\frac{dU_{\text{Unruh}}}{dt} = \frac{2}{3}\frac{e^2\langle\Delta a^2\rangle}{c^3} \sim \frac{\hbar r_o^2 a^4}{c^6}.$$

A numerical comparison with Larmor radiation is instructive:

$$\frac{dU_{\text{Unruh}}}{dt} \sim 4.1 \times 10^{-118} a^4 \qquad \text{(in c.g.s. units)}$$

$$\frac{dU_{\text{Larmor}}}{dt} \sim 5.7 \times 10^{-51} a^2.$$

The two radiation effects are comparable for $a \sim 3\times10^{33}\text{cm/sec}^2 \sim 3\times10^{30}g$, where g is the acceleration due to gravity at the surface of the earth. If this acceleration is to be provided by an electric field we then need

$$E \sim 2\times10^{17}\ \text{volts/cm}.$$

This is about one order of magnitude larger than the 'critical' field $m^2c^3/e\hbar$ to be discussed in Section 4b. The consequence is that the Unruh radiation effect will manifest itself only in the context of other non-linear electrodynamic phenomena. Untangling the various features of radiation when $E > 10^{17}$ volts/cm will be a formidable challenge. For example, the radiation will include e^+e^- pairs as well as photons.

Note that E_{lab} need not be $\sim 3 \times 10^{17}$ volts/cm if a relativistic electron probes the field. For a 50 GeV electron $\gamma \sim 10^5$ so that $E_{\text{lab}} \sim 3 \times 10^{12}$ volts/cm will suffice. This is still large by present standards but far short of the fundamental limitation for laboratory field strengths to be discussed in Section 4d.

The Unruh radiation effect is purely electrodynamic in origin, and hence should be contained in a complete theory of quantum electrodynamics. This is claimed to be proven true in the work of Myhrvold (1983), but the form of his argument does not shed immediate light on experimental considerations.

The possibility of observing the effect of Unruh radiation in another context has been considered by Bell and Leinaas (1983). Salam and Strathdee (1977), Barshay and Troost (1977), and Hosoya (1978) have considered the possible relevance of the Hawking-Davies temperature to thermodynamic models of the strong interaction.

3. Light by Light Scattering.

A fundamental non-linearity in electrodynamics is light by light scattering. It has not yet been observed directly. This process is expected to occur in higher order of perturbation theory due to the production of a virtual e^+e^- pair. A related non-linear process is two photon production of an electron-positron pair,† $\omega\omega \to e^+e^-$.

† We use ω to symbolize a photon, as well as its energy. The symbol γ will be reserved for Lorentz transformations.

The cross sections for these processes are not particularly small‡ :

$$\sigma_{\omega\omega\to\omega\omega} \sim \frac{\alpha^4}{\omega_1\omega_2} \qquad \text{if } \omega_1\omega_2 \gtrsim m^2 \,(\text{Achieser, 1937})$$
$$\sim \frac{10^{-30}}{\omega_1\omega_2}\ \text{cm}^2 \qquad \text{for } \omega_1, \omega_2 \text{ in MeV}.$$

Thus we might expect one light by light scattering if two beams of 10^{15} photons of energy 1 MeV are brought into collision. As laboratory beams in the MeV range have typically $\lesssim\ 10^{10}$ photons it has not been practical to search directly for light by light scattering.

Mikaelian (1982) has suggested colliding 50 Gev photons against a laser beam of 5 eV photons. Then $\omega_1\omega_2 \sim m^2$ so that $\sigma \sim 10^{-30}$ cm^2. A one joule laser pulse would contain $\sim 10^{18}$ photons of 5 eV, so we might expect one scattering for every 10^{12} high energy photons. While the prospects for this experiment are not immediate it is conceivable it will become possible within the next decade.

Indirect observation of light by light scattering effects has been made at the PEP and PETRA electron-positron storage rings via the reaction $e^+e^- \to e^+e^- + X$. Each electron radiates a photon which then collide yielding final state X.

The cross section for pair production by two photons, $\omega\omega \to e^+e^-$, is larger than than for $\omega\omega \to \omega\omega$ by $(1/\alpha)^2$ (Breit and

‡ In making estimates of cross sections and reaction rates it is convenient to use units in which $\hbar = c = 1$. Then in dimensional arguments we have

cross section $= \sigma \sim$ length$^2 \sim 1/$energy^2
rate $\sim 1/$time $\sim 1/$length $\sim$ energy
rate/volume $\sim$ energy^4.

To obtain numerical estimates we insert appropriate powers of $\hbar$ and c, noting

$\hbar c \sim 2 \times 10^{-11}$MeV cm; $\qquad \hbar \sim 6.6 \times 10^{-22}$MeV sec.

Of course, $\alpha = e^2$ $(= e^2/\hbar c) \sim 1/137$, and the electron rest energy $=$ m $(= mc^2) \sim 0.5$ Mev.

Wheeler, 1934):

$$\sigma_{\omega\omega\to e^+e^-} \sim \frac{\alpha^2}{\omega_1\omega_2} \qquad \text{if } \omega_1\omega_2 \gtrsim m^2.$$

This process could be observed with present techniques, as discussed in Section 4c.

4. Multiphoton Effects.

The development of the laser has fostered fruitful growth of the field of non-linear optics. The basic effect was first observed by Franken et al. (1961). The motion of atomic electrons in an intense laser beam must be described by an anharmonic potential. The forced oscillations in this potential include higher harmonic components. Equivalently, two or more photons are absorbed by the electron before the absorbed energy is radiated away as a single higher frequency photon. In the last 20 years a rich variety of related phenomena has emerged, little anticipated at the time of the initial explorations into non-linear optics.

In the 1960's considerations of non-linear effects were extended to the case of a free electron illuminated by a laser beam. The rest of this paper arises out of these considerations.

a. Relativistic effects.

Multiphoton effects become important in the electrodynamics of a single electron when the fields are strong. There are two different measures of how strong the field must be, applicable to two different classes of phenomena. The first criterion is classical and applies to wave fields. The field is strong when the potential difference across one wavelength is greater than the rest energy of the electron. We define

$$\eta = \frac{eE/k}{mc^2} = \frac{eE}{m\omega c} \left(= \frac{eE}{m\omega} \right)$$

where E is the electric field strength, ω is the wave frequency, and

k is the wave number. This is a relativistic invariant†:

$$\eta = \frac{e}{mc^2}\sqrt{-A_\mu A^\mu} \qquad \text{where } A_\mu = 4 - \text{vector potential.}$$

When $\eta \gtrsim 1$ an electron in the wave field takes on velocities comparable to c during its oscillatory motion. In this case higher multipole radiation is emitted with comparable intensity to dipole radiation. In other words, several photons are absorbed before a higher frequency photon is emitted. The higher harmonic radiation rate is non-linear in the field strength. This phenomenon should be readily observable with present technology. We defer detailed discussion until Sections 5 and 6. The remainder of this section concerns non-linear effects associated with the production of e^+e^- pairs.

b. Spontaneous pair creation.

Another criterion is that the field is strong when the potential difference across one Compton wavelength of an electron is greater than the rest energy of the electron. We define

$$\xi = \frac{eE \cdot \hbar/mc}{mc^2} = \frac{\hbar eE}{m^2c^3}\left(= \frac{eE}{m^2}\right) = \frac{E}{E_{\text{critical}}}.$$

It is interesting to note the value of the field strength when $\xi = 1$:

$$E_{\text{critical}} = \frac{m^2c^3}{e\hbar}\left(= \frac{m^2}{e}\right) \sim 1.3 \times 10^{16} \text{ volts/cm.}$$

Note that ξ is not a relativistic invariant so that it must be invoked with care. However for a static, purely electric field ξ is well defined. Then if $\xi \gtrsim 1$ the static field is unstable against spontaneous production of e^+e^- pairs, certainly a non-linear effect (Schwinger, 1951, 1954; Brezin and Itzykson, 1970). We estimate the rate of spontaneous pair production per cm^3 as:

---

† We use a metric such that $A_\mu A^\mu = A_0^2 - \mathbf{A}^2$.

rate/cm$^3 \sim \alpha^2\times$ field energy density $\times$ 'barrier penetration factor'.

For the 'barrier penetration factor' we adopt the picture that virtual e^+e^- pairs are bound in the vacuum by energy $2m$. In a field E the e^+ and e^- must be moved distance $\pm m/eE$ to overcome the 'binding'. As a measure of the probability of this occurence we take the ratio of the squares of the (non-relativistic) wave function at distances 0 and $2m/eE$ in the potential eEx:

$$\text{factor} = \exp\left(-2\int_0^{\frac{2m}{eE}} \sqrt{2m(2m - eEx)dx}\right) = e^{-\frac{16m^2}{3eE}} = e^{-\frac{16}{3\xi}}.$$

(See, for example, Landau and Lifshitz, Quantum Mechanics, secs. 46 and 77.) It turns out the the coefficient 16/3 should be more like 8/3. Then

$$\begin{aligned}\text{rate/cm}^3 &\sim \alpha^2E^2\ e^{-\frac{8}{3\xi}} \sim \alpha m^4\xi^2\ e^{-\frac{8}{3\xi}}\\ &\sim 10^{50}\xi^2\ e^{-\frac{8}{3\xi}}\ \text{pairs/sec/cm}^3.\end{aligned}$$

To have a rate of 1/sec/cm^3 requires $\xi \sim 1/40$ or $E \sim 3\times10^{14}$ volts/cm. The strongest static electric fields produced in the laboratory are perhaps 10^5 volts/cm. Thus spontaneous pair production rates will be low.

c. Stimulated pair creation.

Spontaneous pair production is not possible in a plane wave field as energy-momentum conservation would not be satisfied. However if the wave field is probed by another wave (or particle) stimulated emission of an e^+e^- pair can occur. For example, consider the reaction

$$n\omega + \omega' \to e^+e^-.$$

Here n laser photons collide with a high energy photon ω' to produce the pair. This may be thought of as a multiphoton version of the Breit-Wheeler process, and was anticipated by Toll (1952). To estimate the rate we first consider the process in the rest frame of the e^+e^- pair. In this frame the probe photon has energy m, so the Lorentz boost from the lab frame is $\gamma \sim \omega'/2m$, assuming $\omega' \gg m$. Hence the electric field strength of the laser beam in the e^+e^- rest frame is $\omega' E/m$, supposing the photons ω and ω' move in opposite directions. Comparing with the case of spontaneous emission we infer that parameter

$$\chi = \frac{\omega' E/m}{E_{\text{critical}}} = \frac{\omega\omega'}{m^2}\frac{eE}{m\omega} = \frac{\omega^\mu \omega'_\mu}{2m^2}\eta$$

will govern the pair production rate. Note that χ is a Lorentz invariant. Furthermore in the e^+e^- rest frame we estimate

$$\text{rate} \sim \alpha m^4 \chi^2 \; e^{-\frac{8}{3\chi}} \times \text{effective volume probed by photon } \omega'.$$

We estimate the probed volume as only $\lambda^3_{\text{probe}} \sim 1/m^3$, evaluating λ_{probe} in the e^+e^- rest frame. Then

$$\text{rate in rest frame} \sim \alpha m \chi^2 \; e^{-\frac{8}{3\chi}}$$

and hence

$$\begin{aligned} \text{rate in lab} &\sim \frac{1}{\gamma} \times \text{rate in rest frame} \\ &\sim \alpha \frac{m^2}{\omega'} \chi^2 \; e^{-\frac{8}{3\chi}} \quad \text{pairs/sec/probe photon.} \end{aligned}$$

The more exact result is

$$\begin{aligned} \text{rate} &\sim \frac{3}{8}\alpha \frac{m^2}{\omega'} \left(\frac{\chi}{2\pi}\right)^{3/2} \; e^{-\frac{8}{3\chi}} \\ &\sim 10^{19} \frac{m}{\omega'} \left(\frac{\chi}{2\pi}\right)^{3/2} \; e^{-\frac{8}{3\chi}} \quad \text{pairs/sec/probe photon.} \end{aligned}$$

If we use a probe photon of 50 GeV then $\omega'/m \sim 10^5$. The Omega Nd:glass laser of the U. of Rochester (Seka *et al.* 1980) is quoted as achieving fields equivalent to $\eta \sim 3$. The 1 μm photon energy is $\omega/m \sim 10^{-5}$. If such a laser could be brought to a 50 GeV photon beam then we would have $\chi \sim 3$, and the pair production rate would be $\sim 10^{23}$ per sec per probe photon. That is, each probe photon would have essentially unit probability of producing an e^+e^- pair.

Thus stimulated pair creation is observable with present technology. One must merely overcome the logistical difficulty of bringing a high intensity laser beam near a high energy photon beam

For $\chi \gg 1$ our simple rate estimate should be modified according to the detailed QED calculations of Reiss (1962); Nikishov and Ritus (1964, 1965, 1967); and Narozhny, Nikishov and Ritus (1965).

It is instructive to continue our considerations of the process $n\omega + \omega' \rightarrow e^+e^-$. We can calculate the number n of photons absorbed from the laser beam. In the e^+e^- rest frame the total energy of these photons must also be m[†] : $n\omega^\star = m$. Now

$$\omega^\star \sim 2\gamma\omega = \frac{\omega\omega'}{m} \qquad \text{and hence} \qquad n = \frac{m^2}{\omega\omega'}.$$

Thus we have

$$\chi = \frac{\omega\omega'}{m^2}\eta = \frac{\eta}{n}.$$

For a significant pair production rate the laser field strength must satisfy $\eta \gtrsim n = \omega\omega'/m^2$. The weakest laser field that gives signficant pair production is such that $\eta \gtrsim 1$ (provided also that $\omega\omega' \gtrsim m^2$). In this case only one photon is absorbed from the laser beam, exactly as in the process studied by Breit and Wheeler. It is notable that the classical criterion for strong fields, $\eta \gtrsim 1$, must be satisfied in practice before the quantum non-linear effects will be observable.

---

† We use the superscript * to indicate quantities evaluated in the rest frame.

We can verify that our present rate estimate is consistent with the Breit-Wheeler cross section discussed in section 1. When $\chi \sim 1$ we have

$$\text{rate} \sim \frac{\alpha m^2}{\omega'} \sim \alpha\omega\eta \text{ pairs/sec/probe photon.}$$

But

$$\text{rate} = \sigma \times \text{flux of laser photons} = \sigma\frac{E^2 c}{8\pi\hbar\omega} \sim \sigma\frac{E^2}{\omega}.$$

Thus

$$\sigma \sim \frac{\alpha\omega^2\eta}{E^2} = \frac{\alpha^2}{m^2\eta}.$$

If in addition $\eta \sim 1$, which implies $n = 1$ also, then

$$\sigma \sim \frac{\alpha^2}{\omega\omega'} \sim \sigma_{\text{Breit-Wheeler}}.$$

Another measure of the reaction rate is the mean free path of the probe photon before pair production occurs in the laser beam. Again we consider the case when $\chi \sim 1$, and again

$$\text{rate} \sim \frac{\alpha m^2}{\omega'} \sim \alpha\omega\eta \text{ pairs/sec.}$$

Then

$$\text{mean free path} \sim \frac{c}{\text{rate}} \sim \frac{c}{\alpha\omega\eta} \sim \frac{\lambda}{\alpha\eta}.$$

If $\eta \sim 1$ also, then

$$\text{mean free path} \sim 137\lambda.$$

Stating this yet another way, when the pair production rate becomes significant ($\chi \sim 1$, $\eta \sim 1$) then the probability of pair production is roughly α per wavelength of laser beam traversed by the probe.

d. A fundamental limit to laser field strength.

Consider a laser beam focussed by a lens. The field is no longer a plane wave and now ‘spontaneous’ pair production is possible in the absence of a probe particle. Photons which are converging on the focus can be transformed to a frame in which they collide head on, and so pair production is kinematically allowed.

If $\theta =$ half angle of convergence at the focus, then we find that the critcal field strength for pair production is

$$E_{\text{critical}} \sim \frac{m^2}{e \sin\theta}.$$

If the focus is diffraction limited the converging photons cannot be effectively in collision outside a volume much larger than λ^3. Nor will the strong field region be much larger than this. To estimate the fraction of the beam energy lost in spontaneous pair creation we return to the rate expression for static fields. If $E \sim E_{\text{critical}}$ at the focus,

$$\text{rate} \sim \alpha^2 E^2 \text{ pairs/sec/cm}^3.$$

Then

$$\Delta U_{\text{lost}} \sim m \times \text{rate} \times \frac{\lambda}{c} \times \lambda^3 \sim \frac{\alpha^2 m}{\omega} E^2_{\text{crit}} \lambda^3 \sim \frac{\alpha^2 m}{\omega} U_{\text{beam}},$$

and

$$\frac{\Delta U}{U} \sim \frac{\alpha^2 m}{\omega}.$$

For laser photons of energy less than 25 eV, $\Delta U/U \rightarrow 100\%$. In this sense

$$E_{\text{critical}} \sim \frac{m^2}{e \sin\theta} \sim \frac{1.3 \times 10^{16}}{\sin\theta} \text{ volts/cm}$$

represents a fundamental limit to the field strength obtainable in focussed laser beams.

e. Restrictions on stimulated pair creation.

In principle the preceding analysis could be applied to production of pairs of any charged particle, not just the electron. Consider the attempt to produce a pair of particles of mass M in a strong field by probing with a photon ω'. For a significant rate we need, according to the preceding,

$$\chi = \frac{\omega' E/M}{M^2/e} \sim \frac{\omega' m^2}{M^3} \sim 1,$$

noting that $E \sim m^2/e$ will be the strongest achievable electric field. Thus

$$M^3 \sim \omega' m^2.$$

Even with $\omega' = 500$ GeV $\sim 10^6 m$ we can produce only $M \sim 100m$. However the next lightest particle after the electron is the muon with $M_\mu \sim 206m$.

A similar limitation obtains if we collide two laser beams against each other. The number of photons which can be extracted from a beam to produce a particle of mass M is roughly $\eta = eE/M\omega$. The energy of these photons is then eE/M, so we must have field strength $E \sim M^2/e$. But as we are limited to $E \lesssim m^2/e \sin\theta$, the only possibility is $M \sim m_{\text{electron}}$.

Thus it seems likely that future studies of pair production by intense laser beams will be limited to electron-positron pairs only.

5. Multiphoton Effects in Thomson Scattering.

As suggested in Section 4a the most accessible non-linear effect in the interaction of electrons and photons involves higher harmonic radiation in Thomson scattering:

$$n\omega + e \to \omega' + e'.$$

a. Classical analysis.

For electric field strengths of the incident wave less than the critical field m^2/e (in the rest frame of the electron) the interaction is essentially classical. It is useful to think of the scattering process in terms of multipole radiation. The motion of the electron in the wave field generates all orders of multipole moments, but the radiation due to the nth order moment is smaller than dipole radiation by a factor of order $(v/c)^{2n-2}$. [See Dyson (1979) for an interesting application of this memorable fact.] As the electron's velocity in the wave field approaches c the intensity of multipole radiation at higher harmonic frequencies becomes comparable to that of dipole radiation.

The case of a circularly polarised incident wave is simplest to analyze. In the frame in which the electron has no motion in the direction of the wave, the electron moves in a circle with angular velocity ω equal that of the wave frequency. Thus $eE = m\omega^2 r = m\omega v$ and

$$\frac{v}{c} = \frac{eE}{m\omega c} = \eta.$$

Strictly speaking, for $\eta \gtrsim 1$ we should write

$$eE = \gamma m\omega v \qquad \text{where} \qquad \gamma = 1/\sqrt{1-\beta^2} \qquad \text{and} \qquad \beta = v/c.$$

Then

$$\gamma\beta = \eta \qquad \text{so that} \qquad \gamma = \sqrt{1+\eta^2} \qquad \text{and} \qquad \beta = \eta/\sqrt{1+\eta^2}.$$

Also, the radius of the circle is

$$r = \frac{eE}{\gamma m\omega^2} = \frac{\eta}{\sqrt{1+\eta^2}}\frac{\lambda}{2\pi} \leq \frac{\lambda}{2\pi}.$$

For waves with other than circular polarisation it is convenient to define

$$\eta^2 = \frac{e^2\langle E^2\rangle}{m^2\omega^2c^2},$$

where the average is with respect to time. Then expressions involving η deduced for the case of circular polarisation will hold for all types of polarisation.

In any case we see that the condition $v/c \sim 1$ for a significant rate of higher harmonic radiation becomes $\eta \gtrsim 1$. Then the cross section for scattering into frequency $n\omega$ varies approximately as

$$\sigma_{n\omega} \sim \sigma_{\text{Thomson}} \eta^{2n-2} \sim r_o^2 \eta^{2n-2} \qquad \text{for } \eta \lesssim 1.$$

These arguments can be expressed in terms of photons by noting that radiation at frequency $n\omega$ corresponds to absorption of n photons by the electron followed by emission of one photon at frequency $n\omega$. A simple QED estimate of the cross section would be

$$\sigma_{n\omega} \sim \frac{\alpha^{n+1}}{m^2} = \alpha^{n-1} r_o^2$$

counting one power of α for each external photon, and noting $r_o = \alpha/m$. On comparison with our classical argument we reach an important conclusion. In strong fields where a QED approach is necessary $(E > m^2/e)$, the appropriate dimensionless QED expansion parameter which depends on e^2 is not α but rather η^2. Thus we anticipate that a detailed QED analysis of higher harmonic radiation in the regime $\eta \gg 1, E > m^2/e$ cannot be done via Feynman diagrams, but must involve 'non-perturbative' techniques.

b. The mass shift effect.

If we prefer to think of the non-linear Thomson scattering in terms of photons we must take account of an effect which may seem surprising, although it is readily understood classically. When illuminated by the wave field, the electron undergoes oscillatory motion. The frequency of oscillation is that of the incident photons, and the amplitude of the oscillation is less than a wavelength of the photons. Thus the incident wave cannot resolve the details of the oscillatory motion (although a higher frequency probe could indeed trace the classical path of the electron). The scattering process can be affected by this motion only in an average way. The physical effect can be summarized by saying that the effective mass of an electron in a circularly polarised wave is

$$\overline{m} = \gamma m = m\sqrt{1+\eta^2}.$$

Recall that we are in a frame in which the average velocity of the electron is zero.

By effective mass we mean that the rapid oscillatory motion which causes the mass increase cannot be resolved by the scattering process, so that in effect we are dealing with a heavy electron which doesn't oscillate. In kinematic calculations we must use an effective 4-vector for the electron with invariant mass $\overline{m}$ rather than m, as described in the next section.

c. The drift velocity.

An important subtlety should be noted (Brown and Kibble, 1964; this is overlooked in such treatments as Landau and Lifshitz, Classical Theory of Fields, secs. 47-48). If the electromagnetic wave is incident on a free electron the subsequent motion of the electron is not purely oscillatory. The electron also takes on a 'drift' velocity along the direction of propagation of the wave, which can be substantial for $\eta \gtrsim 1$. The drift velocity is the result of transient effects when the electron first encounters the field. During this time the electron motion is not perfectly in phase with the wave and the $\mathbf{v} \times \mathbf{B}$ force has a component in the direction of the wave propagation.

We can avoid detailed consideration of the transient behavior and calculate the drift velocity of the electron once steady motion is established by thinking of the reaction in terms of photons. The argument which led to the relation $\overline{m} = m\sqrt{1+\eta^2}$ was made in the frame in which the drift velocity of the electron vanishes. However this frame is not necessarily the lab frame. In the general case suppose p_μ is the 4-momentum vector of the electron in the lab frame before entering the wave, and ω_μ is the 4-vector of a photon in the lab. Then $\overline{p}_\mu$, the effective 4-vector of the electron in the wave, must have the form

$$\bar{p}_\mu = p_\mu + \epsilon\omega_\mu .$$

We know that

$$\bar{p}^2 = \overline{m}^2 \equiv m^2 + \eta^2 m^2$$

Hence

$$\epsilon = \frac{\eta^2 m^2}{2 p_\mu \omega^\mu} .$$

For example, if the electron is initially at rest,

$$p_\mu = (m, 0, 0, 0), \qquad \omega_\mu = (\omega, 0, 0, \omega),$$

so that

$$p_\mu \omega^\mu = m\omega \qquad \text{and} \qquad \epsilon = \frac{\eta^2 m}{2\omega} .$$

Then

$$\bar{p}_\mu = \left(m\left(1 + \frac{\eta^2}{2}\right), 0, 0, \frac{\eta^2 m}{2} \right) .$$

The drift velocity is thus

$$\beta_{\parallel} = \frac{\eta^2}{2 + \eta^2} .$$

d. The frequency of the scattered light.

For $\eta \gtrsim 1$ the laboratory motion of an electron initially at rest becomes relativistic, which causes further changes in the appearance of the scattering process. To quantify this change we calculate the kinematics of the multiphoton reaction

$$n\omega + e \rightarrow \omega' + e'.$$

In terms of 4-vectors this is

$$n\omega_\mu + \overline{p}_\mu = \omega'_\mu + \overline{p}'_\mu.$$

Noting that $\overline{p}'^2 = \overline{m}^2$ we find

$$n\omega'_\mu \omega^\mu + \omega'_\mu \overline{p}^\mu = n\omega_\mu \overline{p}^\mu.$$

For example, if the electron is initially at rest and we write

$$\omega'_\mu = (\omega',\ \omega' \sin\theta,\ 0,\ \omega' \cos\theta)$$

then we find

$$\omega' = \frac{n\omega}{1 + \left(\frac{n\omega}{m} + \frac{\eta^2}{2}\right)(1 - \cos\theta)}.$$

This is the appropriate form of the Compton scattering relation for high field strengths.

If we scatter optical photons off electrons then $\omega/m \lesssim 10^{-5}$ which is negligible as usual. But η^2 is large when there is significant probability for higher harmonic radiation. In this case there is substantial variation of the frequency ω' with scattering angle. This effect occurs for the fundamental harmonic ($n = 1$) as well. Measurement of the frequency shift with angle allows a direct determination of the parameter η^2, independent of the intensities of the various higher harmonics. This feature will aid greatly in the interpretation of experimental results.

In the mid 1960's a small controversy arose as to the observability of the electron mass shift and of the frequency shift of the scattered photon. Although these are essentially classical effects, there is some difficulty in demonstrating them in a QED approach. This arises because QED calculations for strong fields can only be made at present for plane waves of infinite extent and duration. Reasonably convincing arguments show that the QED calculations indeed have the proper classical limit in the case of pulsed fields (Kibble, 1966b; Eberly and Sleeper, 1968). See also Eberly (1969) for a review with extensive references. The conclusion remains that the photon frequency shift should be detectable, even if its value is not exactly that given by arguments based on plane waves.

e. The possibility for experiment.

The program of a possible experiment is now reasonably clear. The key is the production of a laser beam for which the field strength satisfies $\eta = eE/m\omega c \gtrsim 1$. We anticipate that in practice this can be obtained only by focussing the beam, and that for a diffraction limited focus the condition $\eta \sim \eta_{\max}$ can be maintained only over a volume $\sim \lambda^3$. The beam then scatters off any electrons in this volume leading to a discrete spectrum of radiation at any fixed angle. The fundamental scattered frequency will be lower than the laser frequency. The experiment should detect this frequency shift, as well as measure the intensity of the various harmonics of the scattered light.

A rate estimate for the non-linear Thomson scattering experiment can be made by recalling the Larmor formula,

$$\frac{dU}{dt} \sim \frac{e^4E^2}{m^2}.$$

The energy radiated in one cycle of the wave is

$$dU \sim \frac{e^4E^2}{m^2\omega} \text{ per cycle,}$$

and the number of photons radiated is

$$dN = \frac{dU}{\omega} \sim \frac{e^4E^2}{m^2\omega^2} \sim \alpha\eta^2 \text{ photons/cycle.}$$

This rate is of course to be multiplied by the number of electrons in volume λ^3. Thus once fields with $\eta \sim 1$ have been achieved the scattering rate is quite substantial. [Strictly speaking the preceeding argument holds in the average rest frame of the electron. The conclusion is clearly invariant of the choice of frame.]

Non-linear effects have been observed in the scattering of light from a CO_2 laser or a Nd:glass laser off an electron plasma. See Burnett *et al.* (1977) and McLean *et al.* (1977). Up to the 46th harmonic has been observed by Carman *et al.* (1981). In these experiments the laser beam induces a sharp density gradient on the plasma which then supports non-linear plasma oscillations leading to higher harmonic radiation. This effect is to be distinguished from present concerns of scattering off a single electron. In particular the electron mass shift and attendant frequency shift of the scattered photon have not been observed in the plasma experiments.

f. The need for relativistic electrons.

In practice it is doubtful that a free electron which is initially at rest can occupy the focus of an intense laser beam. The electron will be expelled from the strong field region by the 'ponderomotive' or 'field-gradient' force (Kibble, 1966a,b). A sense of this force can be gotten from a non-relativistic argument. The effective mass $\overline{m}$ can be thought of as describing an effective potential for the electron inside the wave field.

$$U_{\text{eff}} = \overline{m}c^2 = mc^2\sqrt{1+\eta^2} \sim mc^2 + \frac{1}{2}mc^2\eta^2.$$

Hence

$$\mathbf{F} = -\nabla U \sim -\frac{1}{2}mc^2\nabla\eta^2 = -\frac{4\pi e^2}{m\omega^2 c}\nabla I$$

where I is the intensity of the wave. In a focussed laser beam the intensity has a strong transverse gradient which will push an electron away from the optical axis.

Another view of this effect is obtained by analogy to the reflection of low frequency light off of an electron plasma. From the dispersion relation for light in a plasma,

$$\omega^2 = k^2c^2 + \omega_p^2 \qquad \text{where } \omega_p = \text{plasma frequency},$$

we infer that a photon inside the plasma has an effective mass given by

$$m_{\text{eff}}^2 = \left(\hbar\omega_p/c^2\right)^2 .$$

For an electron in an intense photon beam we found a mass shift of $\Delta m^2 = \eta^2 m^2$. If we consider the laser beam as a kind of 'plasma' of photons then we see that the quantity which plays the role of the plasma frequency is $\eta mc^2/\hbar$. Just as photons with $\omega < \omega_p$ can't penetrate an electron plasma, we infer that electrons with momentum less that ηmc can't penetrate a photon beam. Thus electrons with initial velocities such that $\gamma\beta < \eta$ will be expelled from the strong field region of the laser beam.

The conclusion is that intense laser beams can only be probed by relativistic electrons.

g. Very strong fields.

We remark briefly on the case when the wave field is so strong that $\eta \gg 1$. Consider a circularly polarised wave of frequency ω incident head on with an electron with Lorentz factor γ. If the electron is not to be badly deflected by the wave then we need $\gamma \gg \eta$ according to the preceding argument. Following an analysis like that of Section 5c we find that the energy of the electron once inside the wave remains γm to first order in η/γ. Of course much of this energy is in the transverse motion of the electron, so its longitudinal velocity has been decreased. This is described by the effective mass according to $\gamma m = \gamma_{\|}\overline{m}$. Since $\overline{m} = m\sqrt{1+\eta^2} \sim m\eta$ we have that $\gamma_{\|} \sim \gamma/\eta$. On transforming to the average rest frame of the electron the wave has frequency $\omega^\star \sim 2\gamma_{\|}\omega \sim 2\gamma\omega/\eta$. In this frame the electron is not at rest but moves in a circle at relativistic velocities, described by the Lorentz factor $\gamma_\perp = \overline{m}/m \sim \eta$. The radiation of the circling electron can be thought of as synchrotron radiation. From a classical analysis of the latter we know that the spectrum peaks at frequency $\omega^{\star\prime} \sim \gamma_{\perp}^{3}\omega^{\star} \sim \eta^{3}\omega^{\star}$, corresponding to maximum strength for the harmonic with $n \sim \eta^{3}$. This radiation is strong only in the plane of the circular motion (in the average rest frame of the electron). On transforming back to the lab frame, the radiation has characteristic frequency $\omega' = \gamma_{\|}\omega^{\star\prime} \sim \gamma\eta^{2}\omega$, and lies in a narrow cone of half angle $\sim \eta/\gamma$.

For strong enough wave fields the resulting radiation will include $e^{+}e^{-}$ pairs and a classical analysis no longer suffices. The threshold for this QED correction is that the energy of the radiated photon as viewed in the average rest frame of the electron is equal to $\overline{m}$. The critical energy is $\overline{m}$ and not m as any pairs produced must have the energy of the transverse oscillations of electrons in the strong field. The condition $\omega^{\star\prime} \gtrsim \overline{m}$ transforms to lab frame quantities as $\gamma\eta\omega \gtrsim m$ or $\gamma E \gtrsim m^{2}/e$ recalling that $\eta = eE/m\omega$. Not surprisingly, this is just the condition that the electric field in the true rest frame of the electron be stronger than the critical field introduced in Section 4b. Then the electron probes the wave field so as to stimulate pair creation as discussed in Section 4c.

There is of course another limit in which QED corrections are important. Namely when $\gamma\omega \gtrsim m$ the Thomson scattering of a single photon off an electron goes over to Compton scattering. This possibility is unrelated to the intensity of the photon beam and is not a non-linear effect.

Detailed QED calculations of the higher harmonic spectrum are available for the case of a plane wave of arbitrary intensity incident on an electron. However, only when $\gamma E \gtrsim m^{2}/e$ or $\gamma\omega \gtrsim m$ will the results depart significantly from those of a classical calculation. The results are given in a rather complicated form and will require numerical integration in order to be compared with a specific experiment. The case of circular polarisation is somewhat simpler, and has been considered by Goldman (1964); Narozhny, Nikishov and Ritus (1965); Nikishov and Ritus (1965,1967) (see also Landau and Lifshitz, Relativistic Quantum Theory, sec. 98). The case of linear polarisation is treated by Brown and Kibble (1964), and Nikishov and Ritus (1964). Earlier considerations include Sengupta (1949,1952), Fried (1961,1963), and Vachaspati (1962).

The calculations are based on the exact solution of the Dirac equation for an electron in a plane wave (Volkov, 1935; Schwinger, 1951,1954). The radiated photon is treated as a perturbation, leading to only one power of α in the results. Some of the authors give series expansions suitable for the limits $\eta \ll 1$ or $\eta \gg 1$, which are not of present interest. In comparison of these results with experiment one must also consider how well a focussed laser beam approximates a plane wave.

g. Comparison with radiation in an undulator.

The phenomenon of non-linear Thomson scattering is closely related to the production of higher harmonic radiation in the passage of an electron through a static but spatially oscillating magnetic field (Motz, 1951). In the analysis of an 'undulator' or 'wiggler' magnet of periodicity λ_o one introduces the dimensionless parameter $\eta = eB\lambda_o/2\pi mc^2$, where B is the r.m.s. spatial average magnetic field. For example, the radiation emitted in the forward direction has wavelength $\lambda \sim 2\lambda_o/\gamma_{\|}^2 = 2\lambda_o(1+\eta^2)/\gamma^2$ where γ is the Lorentz factor of the incident electron. Although the total velocity of the electron is unaffected by the undulator, its longitudinal component is reduced to compensate for the transverse oscillations. This leads to the appearance of η in the expression for λ. For $\eta \gtrsim 1$ the oscillations are strong enough that higher harmonic radiation is probable. The higher harmonic radiation has been observed (Billardon *et al.*, 1983), and is considered a background to the operation of undulators as free electron lasers, where it is desired to amplify only one frequency. Coherent production of higher harmonic radiation has been observed by Girard *et al.* (1984) from a bunched electron beam obtained in an 'optical klystron.'

The magnetic field of the undulator can be thought of as consisting of 'virtual photons.' In the lab frame these carry momentum h/λ_o but no energy. However in the average rest frame of a relativistic electron inside the undulator the virtual photons appear very much like real photons. If we wish to describe the radiation process as a kind of Thomson scattering off the virtual photons we must note that the electron takes on an effective mass $\overline{m} = m\sqrt{1+\eta^2}$ inside the undulator, so that $\gamma m = \gamma_{\parallel}\overline{m}$. As for the case of an electron in a laser beam, the 'mass shift' effect is an artifice of a description which emphasizes the longitudinal motion of the electron and averages over the transverse oscillations.

6. Sketch of an Experiment.

We sketch an experiment which might be the beginning of a program to demonstrate the various non-linear electrodynamic effects discussed in this paper. The general technique is to bring a very intense laser beam into collision with a highly relativistic electron beam. The eventual goal is that $\gamma E > 10^{17}$ volts/cm, where γ is the Lorentz factor for the electron beam and E is the (focussed) laser electric field strength. A study of non-linear Thomson scattering would provide a good first opportunity to combine the needed laser and particle accelerator technologies.

a. An experiment with MeV electrons.

To produce the non-linear Thomson effect the laser field strength should satisfy $\eta = eE/m\omega c \gtrsim 1$. This field strength is presently available in CO_2 lasers such as the Antares facility at Los Alamos, and the 1 p-sec laser of Corkum (1983), and in Nd:glass lasers such as the Omega facility at the U. of Rochester (Seka *et al.*, 1980). The electron beam should have $\gamma\beta \gg \eta$ in order not to be badly deflected by the laser beam as noted in Section 5g. To match lasers with field strength $\eta \sim 1$ the electron beam energy need only be a few MeV. Thus it seems suitable to bring a small electron accelerator to an existing laser for the initial experiment.

When the laser beam scatters off the relativistic electrons the light will be doppler shifted to higher frequencies. For an electron beam with $\gamma \sim 5 - 10$ it can be arranged that the light scattered from a CO_2 laser will attain optical frequencies, which is excellent for detection of weak signals. In case of the higher frequency Nd:glass lasers the result of the doppler shift is to place most of the scattered light in the far ultraviolet, which is less suitable. Hence we will presume the use of a CO_2 laser for the non-linear Thomson scattering experiment.

We give some details of the doppler shift effect. Suppose the laser beam and the electron beam meet and angle α, where $\alpha = 0^\circ$ for a head-on collision. The scattered photon, ω', is emitted at spherical angles (θ, ϕ) with respect to the electron direction, taking the laser beam to lie in the $x - z$ plane. Then following the argument given in Section 5d we find the frequency of the nth harmonic scattering to be

$$\omega' = n\omega \frac{1 + \beta \cos \alpha}{1 - \beta \cos \theta + \Delta}$$

where

$$\Delta = \frac{\eta^2}{2\gamma^2(1 + \beta \cos \alpha)}(1 + \sin \alpha \sin \theta \cos \phi + \cos \alpha \cos \theta).$$

We have neglected a term in $n\omega/\gamma m$ in the above. If $\eta^2 \ll \gamma^2$ then the scattered photons of maximum frequency are emitted at angles $(\theta_{\max}, \phi = -180^\circ)$ where

$$\tan \theta_{\max} \sim \frac{\eta^2 \sin \alpha}{2\gamma^2(1 + \cos \alpha)}$$

supposing $\gamma \gg 1$. The maximum frequency is

$$\omega'_{\max} \sim \frac{2\gamma^2 n\omega(1 + \cos \alpha)}{1 + \eta^2}.$$

The non-zero value of $\theta_{\max}$ is due to the deflection of the electron beam by the laser beam. Note again the substantial frequency shift for radiation emitted by electrons inside the strong field.

For a laser beam with $\eta \sim 1$ it is appropriate to detect the first four harmonics of the scattered radiation. Say that the $n = 4$ harmonic is desired at 250 nm wavelength, so the fundamental should be placed at 1 μm. For a CO_2 laser the wavelength is 10 μm, so the doppler shift expression tells us that we need

$$\frac{2\gamma^2(1+\cos\alpha)}{1+\eta^2} = 10.$$

For a laser beam with $\eta = 1$ and a 5 MeV electron beam ($\gamma = 10$) the appropriate angle for the laser beam is $\alpha = 154°$. That is, the laser beam and the electron beam should converge on each other with 26° between them. In this example the deflection of the electron beam is 1.3°.

The frequency of the scattered light decreases as we move away from $\theta = 0$. To get a sense of this effect is is useful to consider the transformation of the scattered photon from the average rest frame of the electron into the lab. From this we find

$$\frac{\omega'}{\omega'_{\text{max}}} \sim \frac{1+\cos\theta^\star}{2}$$

and

$$\sin\theta \sim \frac{\sin\theta^\star}{\gamma_{\|}(1+\cos\theta^\star)} = \frac{\sqrt{1+\eta^2}\sin\theta^\star}{\gamma(1+\cos\theta^\star)}$$

again supposing $\gamma \gg 1$. If we collect light over the range $0° \leq \theta^\star \leq 60°$ then we obtain about 25% of the scattered light with a resulting spread in frequencies of 25%. In the laboratory this light appears in a cone of half angle 4.7° about the electron direction, for $\gamma = 10$ and $\eta = 1$.

The electron beam must be deflected by a magnetic field after passing through the laser beam so that the forward scattered light may be detected. In the geometry under discussion we collect the scattered light down to 20° from the laser beam. If the laser beam is not to hit the collection optics it cannot be focussed with a lens stronger than f/d = 1.4.

To obtain a field strength of $\eta \sim 1$ the laser beam must be focussed down to a spot size only a few wavelengths in diameter. The intersection of this spot with the electron beam provides a well localized source of scattered light, very suitable for analysis in a spectrograph such as that used by Carman *et al.* (1981).

The electron beam must be of sufficient quality not to blur the spectrum of the scattered photons, while being intense enough to yield a sizable counting rate. The energy spread of the electron beam should be small compared to the 25% spread in photon frequencies expected from mono-energetic electrons. The angular divergence of the electron beam should be small compared to 5°.

Only those electrons which intersect the focussed laser beam are useful. As an example, suppose the CO_2 laser pulse is 1 n-sec long and has a beam waist of 5 wavelengths, or 50 μm. We estimate that the electron beam could be brought to a 250 μm spot size with only 1° divergence. Then approximately 1/16 of the electrons would be useful. If the electron linac can deliver 500 mA current for 1 n-sec then $\sim 10^9$ electrons cross the laser focus. As noted in Section 5e about 1% of the electrons will scatter photons. Hence we expect a signal of about 10^7 scattered photons per laser pulse, spread over the various harmonics. This should be sufficient to demonstrate the non-linear Thomson effect.

b. Experiments with GeV electrons.

If a high power Nd:glass or CO_2 laser could be brought to a highly relativistic electron beam such as that at SLAC interesting experiments could be performed. The non-linear Thomson effect could be readily demonstrated, with the scattered photons taking on GeV energies. Also, the large Lorentz boost to the electron's rest frame makes the apparent field strength, γE, approach the critical value m^2/e for stimulated pair creation. Likewise $\gamma\omega \sim m$ so that direct observation of light-by-light scattering would be feasible (using a high energy photon beam derived from the electron beam). Finally, with advances in laser technology we anticipate achieving field strengths $\gamma E \sim 10^{17}$ volts/cm which would allow demonstration of the Unruh radiation effect. All of these considerations indicate that the future of experimental studies of non-linear quantum electrodynamics lies in the development of high powered lasers at the Stanford Linear Accelerator Laboratory.

References

Achieser, A. (1937) Physik Z. Sowietunion **11**, 263.

Barshay, S. and W. Troost (1978) Phys. Lett. **73B**, 437.

Bell, J.S. and J.M. Leinaas (1983) Nucl. Phys. **B212**, 131.

Billardon, M. *et al.* (1983) J. de Phys., Colloq. **44**, C1-29.

Birrell, W.D., and P.C.W. Davies (1982) Quantum Fields in Curved Spacetime (Cambridge U.P.).

Boyer, T.H. (1980) Phys. Rev. **D21**, 2317.

Breit, G. and J.A. Wheeler (1934) Phys. Rev. **46**, 1087.

Brezin, E. and C. Itzykson (1970) Phys. Rev. **D2**, 1191.

Brown, L.S. and T.W.B. Kibble (1964) Phys. Rev. **133**, A705.

Burnett, N.H. *et al.* (1977) Appl. Phys. Lett. **31**, 172.

Carman, R.L. *et al.* (1981) Phys. Rev. **A24**, 2649.

Corkum, P.A. (1983) Optics Lett. **8**, 514.

Davies, P.C.W. (1975) J. Phys. **A8**, 609.

Donoghue, J.F. and B.R. Holstein (1984) Am. J. Phys. **52**, 730.

Dyson, F.J. (1979) Rev. Mod. Phys. **51**, 447.

Eberly, J.H. (1969) in Progress in Optics, Volume VII, ed. by E. Wolf (North-Holland Publishing Company).

Eberly, J.H. and A. Sleeper (1968) Phys. Rev. **176**, 1570.

Franken, P.A. *et al.* (1961) Phys. Rev. Lett. **7**, 118.

Fried, Z. (1961) Nuovo Cimento **22**, 1303.

Fried, Z. (1963) Phys. Lett. **3**, 349.

Fulling, S.A. (1972) Ph.D. dissertation, Princeton Univ. (unpublished).

Fulling, S.A. (1973) Phys. Rev. **D7**, 2850.

Gerlach, U. (1982) Ohio State Univ. preprint (unpublished).

Girard, B. *et al.* (1984) Phys. Rev. Lett. **53**, 2405.

Goldman, I.I. (1964) Phys. Lett. **8**, 103.

Hawking, S.W. (1974) Nature **248**, 30.

Hawking, S.W. (1975) Commun. Math. Phys. **43**, 199.

Hosoya, A. (1979) Prog. Theor. Phys. **61**, 280.

Kibble, T.W.B. (1966a) Phys. Rev. Lett. **16**, 1054.

Kibble, T.W.B. (1966b) Phys. Rev. **150**, 1060.

McLean, E.A. *et al.* (1977) Appl. Phys. Lett. **31**, 825.

Mikaelian, K.O. (1982) Phys. Lett. **115B**, 267.
Motz, H. (1951) J. Appl. Phys. **22**, 527.
Myhrvold, N.P. (1983) Ph.D dissertation, Princeton Univ. (unpublished).
Narozhny, N.B., A.I. Nikishov and V.I. Ritus (1965) Sov. Phys. JETP **20**, 757.
Nikishov, A.I. and V.I. Ritus (1964) Sov. Phys. JETP **19**, 529.
Nikishov, A.I. and V.I. Ritus (1965) Sov. Phys. JETP **20**, 622.
Nikishov, A.I. and V.I. Ritus (1967) Sov. Phys. JETP **25**, 1135.
Reiss, H.R. (1962) J. Math. Phys. **3**, 59.
Salam, A. and J. Strathdee (1977) Phys. Lett. **66B**, 143.
Schwinger, J. (1951) Phys. Rev. **82**, 664.
Schwinger, J. (1954) Phys. Rev. **93**, 615.
Sciama, D.W., P. Candelas and D. Deutch (1981) Adv. in Phys. **30**, 327.
Seka, W. *et al.* (1980) Appl. Optics **19**, 409.
Sengupta, N.D. (1949) Bull. Math. Soc. (Calcutta) **41**, 187.
Sengupta, N.D. (1952) Bull. Math. Soc. (Calcutta) **44**, 175.
Toll, J.S. (1952) Ph.D dissertation, Princeton Univ. (unpublished).
Unruh, W.G. (1976) Phys. Rev. **D14**, 870.
Vachaspati (1962) Phys. Rev. **128**, 664; *erratum* **130**, 2598.
Volkov, D.M. (1935) Z. f. Physik **94**, 250.

HIGHLIGHTS OF THE WORKING GROUP ON PLASMA ACCELERATORS

John M. Dawson
Department of Physics, University of California,
Los Angeles, California 90024

INTRODUCTION

The working group on plasma accelerators took up a number of topics on the use of plasmas in an accelerating medium. In order to summarize these I have broken the material into three main topics:

(I) The beat wave accelerator;
(II) Other plasma accelerators;
(III) Miscellaneous work on plasma accelerators

Topic (I) can be further broken down into:

A Experiments on the beat wave accelerator;
B Theory and simulation of the beat wave accelerator;
C Important goals for the Beat Wave accelerator experimental program over roughly the next three years;
D Important problems for theory and simulation (and ultimately for experiments) on the beat wave accelerator.

Topic (II) is broken down into the three subtopics:

A Plasma grating accelerators;
B Plasma wave guide accelerators;
C Plasma wake field accelerators.

(I) The Beat Wave Accelerator

A Experiments on the Beat Wave Accelerator

(1) Experiments at UCLA - Probably the most definitive experimental results have been obtained on the experiment at UCLA. The working group discussed this experiment in detail. The highlights of what this experiment has accomplished are listed below.

(a) Experiments using CO_2 have been carried out at high power, operating simultaneously on 10.6 μ and 9.6 μ. There was roughly equal power in both beams; the total energy was roughly 20 joules in 1 n sec.

(b) The experiments have demonstrated that plasma of the right density ($n_e \cong 10^{17}$) so that $\omega_1 - \omega_2 = \omega_p$ can be pro-duced reproduceably (for periods of many hours) at least in cm size.

(c) A ·7 μ diagnostic laser was developed and synchronized with the CO_2 pulse. A technique for detection of scattering of the ·7 μ beam from the fast (forward) plasma wave was developed (angle of deflection, 7 m rad) and demonstrated to work.

(d) Using the .7 μ diagnostic it was demonstrated that the fast plasma wave was generated over a length of ~ 1.5 mm (15 plasma wave lengths and 150 laser wave lengths).

(e) The amplitude of the fast wave was 0.01 to 0.03 of the wave breaking limit and about half the predicted amplitude from the theory of Rosenbluth and Liu.(1)

(f) The electric fields generated by the measured fast waves are in the range of 330 MV/M to 10^9 V/M.

(g) The experiment demonstrated that a rich variety of plasma waves can be generated (a plasma wave zoo). Although a complete analysis of all experiments has not been carried out, it appears that the results are understandable using theory and simulation. Thus these seem to be reliable guides to experiments. The theory indicates that the unwanted waves can be eliminated by going to the short laser pulses in the range of 10 ~ 50 p sec.

(2) The Experimental Program at the Rutherford Appleton Laboratory - At the Rutherford Appleton Laboratory an experiment rather similar to that carried out at UCLA is being planned; the big difference between this experiment and the UCLA experiment is that it will use 1 μ solid state lasers rather than CO_2 lasers to set up the beat wave. As a consequence the phase velocity of the beat wave will be much closer to C. The planned experiment will have the following characteristics:

(a) This will be the first significant experiment using 1 μ laser light. The experiment is designed to operate at 1.053 and 1.064 μ. The higher frequency with the same resonant plasma frequency as the UCLA experiment implies that the focussed intensity must be 100 times larger than in the UCLA experiment; this is expected to be the case. The γ for the plasma wave $[(1 - V_p{}^2/c^2)^{-1/2}]$ will be 100 in this experiment rather than 10 as in the UCLA experiment.

(b) The plasma source will be a Z pinch which may be more appropriate for generating long columns of plasmas for acceleration.

(c) Some subtle problems with operating a solid state laser on two lines so close together may be encountered. This experiment should give valuable experience with 1 μ lasers.

(3) The Los Alamos Experiment - Los Alamos carried out an experiment using a single CO_2 frequency. This experiment requires that the stimulated forward Raman instability play a dominant role in the laser plasma interaction and produces the accelerating wave. The following briefly summarizes that experiment.

(a) A single frequency CO_2 laser experiment was carried out using one arm of the Helios Laser. The laser delivered 600 joules to the plasma in 1 n sec.

(b) The plasma target was formed by a laser pulse on a solid sheet of CH material 4 n sec before the accelerating pulse arrives. The plasma blows off perpendicular to the surface of the CH target; its density is a function of the distance from the target.

(c) The accelerating pulse comes in parallel to the surface of the CH sheet. By varying its height above the sheet, the density of the plasma it intersects can be adjusted to give the best interaction.

(c) They looked for both very energetic electrons, $E \geq 15$ MeV, and for sub-MeV temperature electrons (250 keV). No electrons with energies greater than 15 MeV were detected. An electron component with a temperature of 250 keV was detected.

(d) Two-dimensional computer simulations of the experiment were carried out. The results of these simulations agree reasonably well with the experiment and can be summarized as follows:

SUMMARY OF LOS ALAMOS RESULTS OF
2-D FINITE LASER BEAM BEAT WAVE SIMULATIONS

* A regime of single particle electron acceleration has been identified over a few psec time scale
  - Time to saturation and the saturation amplitude E_x, of plasma wave agrees with Rosenbluth-Liu
  - RBS, RSS, RFS and direct laser heating do not disrupt the beat wave process
  - Relativistic self-focusing enhances the on axis intensity
  - Absorption ~ 10%-15% over computational box
* Eventually the plasma wave begins to lose coherence and the acceleration is disrupted due to many competing effects
  - Rosenbluth-Liu relativistic mismatch
  - Self-focussing and filamentation
  - Ion turbulence
* Further study and optimization are required

B <u>Theory and Simulation of the Beat Wave Accelerator</u>

Theory and simulations have made a great deal of progress since the last laser accelerator meeting. Some of the important advances have already been given under the summary of the Los Alamos experiment. Here I list the most important advances; there is some overlap with the Los Alamos summary.

(1) The theory of Rosenbluth and Liu for the beat wave saturation due to the relativistic mass change predicts the observed saturations in the simulations.

(2) Two-dimensional simulations have been carried out for a finite width laser beam and for propagation distances which amount to about 0.5 mm for CO_2 (50 laser wavelength). Relativistic self focusing of the laser beam has been observed. Theory predicts that such self focusing should take place if the laser power exceeeds P_c.

$$P_c \approx \frac{2\pi}{3} nmc^2 \frac{c^2}{\omega_p^2} \frac{\omega^2}{\omega_p^2}$$

For CO_2 and $n_e = 10^{17}$ the required power is 10^{12} watts. The UCLA power is only about 2×10^{10} watts so it is not enough to produce self focusing.

In the simulation the self focused beam appears to be stable if the radius ρ satisfies the following relation

$$\rho \approx \text{Few } (c/\omega_p) \gg c/\omega_0$$

Theoretical arguments also indicate that this should be so. For channels of the above width no filamentation took place; for much wider channels filamentation was observed.

(3) Pump depletion occurs at roughly the expected rate in one-dimensional simulations. That is, it follows the relation

$$\frac{d}{dx} \frac{E_0^2 \ell_0}{4\pi} \approx \frac{E_p^2}{4\pi}$$

where E_0 is the pump electric field, ℓ_0 is the length of the driving pulse and E_p is electric field of the plasma wave.

(4) Detailed theoretical calculations showed that:

(a) Coulomb scattering of the accelerated electrons by the plasma is unimportant

(b) Synchrotron radiation is unimportant.

(c) The focusing forces of the laser and plasma wave are very strong.

C Important Goals for the Beat Wave Accelerator Experimental Program Over Roughly the Next Three Years

Some important goals for the experimental BWA for the next three years or so are:

(1) Accelerate some electrons to a significant energy (10~100 MeV).

(2) Extend the length of the region where the coherent forward beat wave exists to about 10 cm. This will be necessary to achieve significant acceleration.

(3) Demonstrate relativistic self focusing, i.e., demonstrate narrow beam propagation over many Rayleigh lengths. Included in this should be a demonstration that the proper width beam does not filament.

(4) Demonstrate that a short intense laser pulse (~50 psec) can propagate for a significant distance through a plasma (~30 cm) without generating turbulence in the neighborhood of the light pulse.

D Important Problems for Theory and Simulation

Some important problems for theory and simulation to address are the following:

(1) Can we stably propagate an intense short laser pulse for 10 to 100 meters through a plasma? What happens to the laser pulse as it undergoes pump depletion? What fraction of the laser energy can be converted to useful plasma wave energy; what fraction can be transferred to accelerated particles? Can the overall efficiency be made sufficiently high?

(2) Can we produce the quality electron (positron) beams needed by high energy physics? How much help does the phase locking in the surfatron give? Can we produce the desired quality by reflecting a preaccelerated bunch of electrons off the wake of a short laser pulse? The last situation looks promising with preliminary calculations indication that for 0.25 μ light in a plasma of 10^{16} overtaking a 10 cm long bunch of electrons preaccelerated to 5 MeV would give an energy spread of 10 GeV out of 1 TeV in transversing 100M. The emerging bunch of electrons would be of the order of microns in length.

(3) Are there good methods for rejuvenating the laser beam?

(4) How uniform must the plasma be; what level of fluctuations are allowed and how does the allowed level depend on fluctuation wavelength and frequency?

(5) Are there electron beam plasma instabilities that we must worry about?

(II) Other Plasma Accelerators

A The Plasma Grating Accelerator

Investigation has begun on another type of laser plasma accelerator; it might be called a plasma grating accelerator. The figure below illustrates this device.

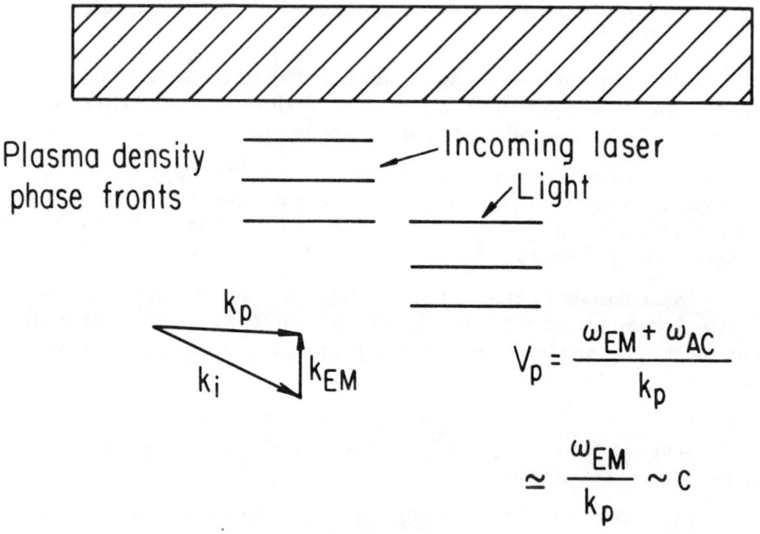

- Light brought in from side works well as a Surfatron
- 1-D simulation indicates good acceleration can be obtained

FIG. 1

A preexisting plasma column is created with an ion density wave in it (an ion acoustic wave). Let the wave number of this acoustic wave be k_i. The incident laser pulses are brought in from the side so that fresh power is brought in all along the plasma column. The laser light would be brought in as a series of pulses timed to coincide with a small packet of electrons being accelerated down the column; this is to insure that the electrons are moving in a well ordered field rather than the turbulent one which builds up after the laser has been interacting with the plasma for a few instability growth times (typical times for turbulence to develop might be 10 psec). The light interacts with the ion waves to produce an electron disturbance with wave number $k_{plasma} = k_i \pm k_{EM}$ and frequency $\omega = \omega_{EM} \pm \omega_{AC}$. Since the acoustic frequency, ω_{AC}, is very small it can generally be neglected. The phase velocity of the disturbence is

$$V_p = \frac{\omega_{EM}}{k_{plasma}}$$

and by proper choice of k_i and ω_{EM} can be made equal to c. Theory predicts the amplitude of the electric field associated with this disturbence is

$$E_{plasma} = E_{EM} \frac{\delta n_{AC}}{n_0} \frac{\omega_p^2}{\omega^2 - \omega_p^2 - 3k^2_{plasma} V_T^2}$$

where V_T is the electron thermal velocity and δn_{AC} is the density variation associated with the ion acoustic wave. At UCLA a 1-2/2 dimensional (x, v_x, v_y, v_z) model was used to investigate this process and it has been found that the electric field generated is in good agreement with the above value. In our simulations it was demonstrated that one could efficiently accelerate electrons; some electrons reached a γ of 50 with the value limited by the length of the run and the onset of turbulence. It was also demonstrated that the Surfatron concept works well in this geometry producing a relatively high quality beam of accelerated electrons (energy spread of 10% when accelerated out of a thermal background).

B <u>Plasma Wave Guide Accelerator</u>

Another approach to a laser plasma accelerator was presented by Tajima. He proposes to use a plasma wave guide; that is he would create a plasma in which the central part of the column is at lower density than the edge. Such a plasma configuration creates a kind of light pipe for light. Two plasma beams can be guided down such a column and if their difference in frequency is the central plasma frequency their beat will generate a plasma wave there. The axial wave length of the beat wave will be longer than precisely colinear propagating beam (the beams can be viewed as bouncing back and forth in the channel). Thus the phase velocity of the plasma wave can be made equal to c (or even higher).

C <u>The Plasma Wake Field Accelerator</u>

A quite different approach to generating the accelerating plasma wave is the plasma wake field accelerator. This is rather analogous to the electromagnetic wake field accelerators already being considered; it uses a longitudinal space charge wave in the plasma rather than an electromagnetic cavity mode.

This method would use the bunched beam coming from a linear accelerator to generate the wave in the plasma. Theory and one-dimensional simulation has shown that this method can generate strong plasma waves with large electric fields (10^9 volts/M is typical). A bunch of electrons which is phased right so as to be accelerated by this wave can be boosted in energy.

A disadvantage of this method is that the driving bunches are acted on and slowed down by the wave. For a series of uniform driving bunches and the accelerated bunch moving colinear with it there is a limit on the energy that can be gained which is $\Delta\gamma = 2\gamma_0$; i.e., electrons can be accelerated from γ_0 to $3\gamma_0$. To do better one must resort to some kind of trick. One possibility is to use nonuniform driving bunches. A second possibility is to use non-colinear driving and accelerated bunches so that the driving beams leave the system once they have generated the wave. In this last case a series of driving beams would be injected into the plasma at various axial positions so as to have new beams generating the accelerating wave as the accelerated electrons move down the plasma column. Both methods show some interesting promise and are being investigated.

(III) Miscellaneous

The Plasma Accelerator group met with the Laser Group and with the Accelerator Physics group to ascertain what might be expected in the nature of laser developments and what criteria an accelerator must meet. Most of the details in these areas are contained in the reports of those groups. I will only briefly summarize my impressions.

A Laser Builders

From this group the situation sounded encouraging with regard to producing the pico-second pulses we require with the required energy (100 to 10^4 joules per pulse). Projected efficiencies while probably not great seemed adequate. New high frequency gas lasers seem to hold a lot of promise for meeting pulse repetition rates and efficiencies required by the high energy physicists.

B Accelerator Physicists

The requirements set out by the accelerator physicists are tough to meet, but I have not seen any requirement that rules out laser plasma accelerators. There are many questions still to be answered regarding the quality of beams that can be produced, the efficiency of laser energy transfer to accelerated electrons, the efficiencies of lasers, etc. which we do not yet know the answers to and which only further research will reveal.

REFERENCES

1. M. Rosenbluth and C.S. Liu, Phys. Rev. Lett. 29, 701 (1972).

PLASMA ACCELERATORS

T. Katsouleas, C. Joshi, J.M. Dawson, F.F. Chen, C. Clayton, W.B. Mori, C. Darrow, D. Umstadter
University of California, Los Angeles, CA 90024

ABSTRACT

We review the progress made on several schemes for acceleration in a plasma medium. The beat wave accelerator is becoming fairly well developed with recent advances made on theoretical, computational, and experimental fronts. Progress on three new concepts, the surfatron, the plasma wakefield, and the plasma grating or rippled plasma accelerator is also described.

INTRODUCTION

The first question one might ask when considering a plasma as a medium for supporting large amplitude electric fields for particle acceleration is "Why use a plasma?" In light of their history of unexpected instabilities and complexity, plasmas may seem an unlikely path toward the goal of high gradient particle accelerators. In this paper, the results of investigations of plasma as a medium for high-gradient acceleration are presented. Recent work demonstrates that plasma properties are somewhat desirable and often unavoidable when contemplating advanced accelerator concepts.

As a first step toward addressing the question of why accelerate in plasma, let us briefly examine some alternatives. Conventional accelerators can be expected to produce accelerating gradients of the order 20-100 MeV/m. At field strengths not too much higher, the problem of breakdown in the accelerator walls arises. Similarly, many of the near field laser schemes considered at this conference will unintentionally become plasma schemes if the laser intensity is too high. This suggests the first advantage of a plasma accelerator: it is already fully ionized and cannot be destroyed any more than it already is.

Next, one might look to high powered lasers with fields as high as 5 GeV/cm[1] to directly accelerate particles. However, the laser fields are transversely polarized and coupling to these fields (the far field schemes) necessarily leads to transverse particle acceleration and radiation losses. Some direct acceleration in the longitudinal direction is possible from the radiation pressure of the light[2]. However, for a relativistic particle this force decreases as one over the Lorentz factor γ of the particle[3].

On the other hand, the accelerating force of a longitudinal electric field is invariant under Lorentz transformations to any frame moving along the field direction. Hence, the second advantage of a plasma accelerator: plasmas are capable of supporting large, longitudinal electric field waves (i.e., space charge waves for which the wavenumber k is parallel to the electric field E).

We can estimate just how large the plasma wave fields can be from Poisson's equation:

$$\nabla \cdot \vec{E} = 4\pi e\, \delta n_e \tag{1}$$

where δn_e is the perturbed electron density of the plasma and the ion background is assumed uniform and immobile. Now the largest density compression or rarefaction that can occur is when all of the plasma electrons are removed. In that case $\delta n_e \simeq n_o$, the equilibrium density. Assuming a plasma wave with phase velocity near c and frequency near the plasma frequency $\omega_p = (4\pi n_o e^2/m_e)^{1/2}$ so that $k \approx \omega_p/c$ and approximating $\nabla \cdot E \sim ikE$, we obtain from (1) the maximum electric field amplitude[4]:

$$eE_{MAX} \sim \frac{4\pi n_o e^2}{(\omega_p/c)} = m_e c\omega_p \approx .97\sqrt{n_o}\ \text{eV/cm} \tag{2}$$

This is sometimes referred to as the cold wavebreaking field because it is the amplitude at which a cold plasma wave steepens and becomes double valued so that the crest begins to fall into the trough[5]. This field can be quite large. For example, in a plasma of density $n_o = 10^{18}\text{cm}^{-3}$ the cold wavebreaking field is of the order 1GeV/cm, or about one thousand times the gradient of conventional linacs.

One might be concerned that a medium such as a dense plasma would interfere with the acceleration process via Coulomb scattering or Cerenkov radiation. The dominant energy loss mechanism for electrons above a few MeV is due to multiple scattering from plasma nuclei. Since the radiation length $(-U/[\partial U/\partial x])$ in a hydrogen plasma of density 10^{20}cm^{-3} is roughly 2km[6] and the scattering cross section decreases as energy squared, this should not pose a problem. Furthermore, collisional damping of the laser scales as one over laser intensity to the three halves and is minimal for intense lasers.

Having illustrated two principal advantages of plasma accelerators--immunity from breakdown and potential for large longitudinal electric fields--we turn to describing the progress that has been made toward realizing this potential.

In section I we review the state of understanding of the most developed of the plasma wave excitation schemes: plasma wave generation by the beating of two lasers[7]. Since the first workshop[8], advances have been made on theoretical modelling and two-dimensional computer modelling, and a conclusive experimental demonstration of the plasma beat wave has been performed.

In Section II we examine two mechanisms for particle acceleration in the fields of high-phase velocity plasma waves: the simple beat wave acceleration mechanism and the Surfatron mechanism (using an imposed DC magnetic field).

In Section III we explore two recently proposed plasma wave excitation schemes: the plasma wakefield accelerator[9] and a plasma grating accelerator[10]. The plasma wakefield scheme employs the electron bunches of a conventional accelerator as the free energy source to drive plasma waves which can then be used to further accelerate a trailing bunch of electrons. The plasma grating scheme employs a single laser injected from the side and couples energy to a longitudinal plasma wave via a plasma density ripple (or ion acoustic wave).

In Section IV we summarize the recent progress and discuss plans for a proof of principal 100MeV beat wave accelerator experiment.

I. PLASMA WAVE EXCITATION BY BEATING LASERS

A. THEORETICAL MODELS

The basic mechanism for excitation for a longitudinal plasma wave by laser light is illustrated in Fig. 1.

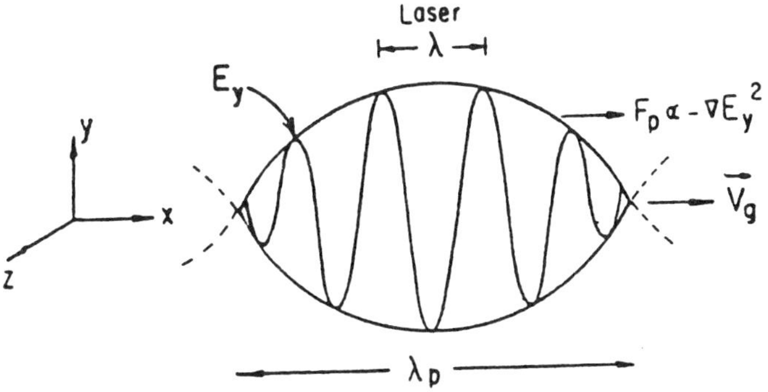

Fig. 1 The radiation pressure or pondermotive force of a packet of light drives longitudinal oscillations of plasma electrons.

Consider a packet of incident electromagnetic radiation which might be created in one of the following two ways: (1) a very short pulse from a single laser (wake plasmon excitation) or (2) by the beat envelope of two lasers of slightly different frequency (beat wave excitation). The packet of radiation exerts a force per unit area on the plasma which is given by the gradient of the stress tensor[11]:

$$F_p \approx \nabla\cdot\overleftrightarrow{T}_p - \nabla\cdot\overleftrightarrow{T}_v \approx \frac{\varepsilon-1}{8\pi}\frac{\partial}{\partial x}E_y^{\,2} \tag{3}$$

where $\overleftrightarrow{T}_p$, $\overleftrightarrow{T}_v$ refer to the stress tensor in plasma and vacuum, respectively, ε is the dielectric function of the plasma ($\varepsilon = 1 - \omega_p^2/\omega^2$, ω is laser frequency and $\omega \gg \omega_p$), $\partial/\partial x$ describes the gradient of the envelope of the packet and E_y is the laser electric field. Since we want just the force on the plasma, we subtracted $\overleftrightarrow{T}_v$, the component $\overleftrightarrow{T}$ which would exist even if there were no plasma. This so called pondermotive force displaces plasma electrons and creates a temporary charge imbalance in the plasma (the plasma ions are too massive to respond with the electrons). After the packet passes, the space charge force acts to restore the plasma electrons leading to oscillations in the $\hat{x}$ direction and and hence a longitudinal electric field wave.

In the first scenario of a very short pulse, known as the wake plasmon excitation scheme, there is no resonant interaction with the plasma and the plasma essentially receives only a single kick. As a result, the electric force of the plasma wave can never exceed the pondermotive force of the laser. Furthermore, all of the energy required for a full stage of the accelerator must be contained in a single pulse shorter than a plasma period. For a $10^{16}cm^{-3}$ density plasma, the pulse must be 1 picosecond or less. Nevertheless, a short pulse laser with normalized electric field amplitude $V_{osc}/c \equiv eE/m\omega_p c$ of order one has sufficient pondermotive force to create a plasma wave near the cold wavebreaking amplitude (Eq. 2).

In the second scenario involving the beat of two lasers, the plasma wave can be built up resonantly over many plasma periods if the beat frequency of the lasers coincides with the natural frequency at which the plasma electrons oscillate (i.e., $\omega_1 - \omega_2 \approx \omega_p$). In this way, even relatively small laser fields can lead to very large longitudinal plasma wave fields.

Consider the resonant beat wave excitation scheme. By energy and momentum conservation of the laser photons and the plasmon, we have that

$$\begin{aligned} \omega_1 - \omega_2 &= \omega_p \\ k_1 - k_2 &= k_p \end{aligned} \qquad (4)$$

If the plasma is very underdense, such that $\omega_1 \approx \omega_2 \gg \omega_p$, we find that the phase velocity of the plama waves is

$$V_{ph} \equiv \frac{\omega_p}{k_p} = \frac{\omega_1 - \omega_2}{k_1 - k_2} = \frac{\Delta\omega}{\Delta k} \approx \frac{\partial\omega}{\partial k} = V_g^{light} = c\,(1 - \omega_p{}^2/\omega^2)^{1/2} \qquad (5)$$

Thus, the phase velocity of the plasma waves equals the group velocity of the light. The light pulse and the wake of excited plasma waves move as a unit into undisturbed plasma as depicted in Fig. 2.

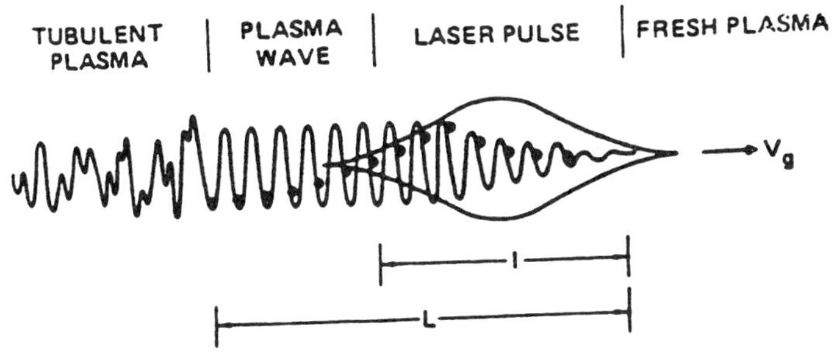

Fig. 2 The laser pulse, the wake of plama waves and the accelerating particles move together at Vg ≈ c into fresh plasma in the beat wave excitation scheme.

If the laser frequencies are much higher than ω_p (i.e., a very underdense plasma), then V_{ph} is very close to c and injected relativistic particles may stay in phase with the electric field of the plasma wave for a sufficient distance to be accelerated to high energy.

As illustrated in the figure, the plasma waves well behind the light pulse become turbulent. This can be due to competing instabilities, such as parametric decay (the decay of an electron plasma wave into a secondary plasma wave and an ion acoustic wave), or as we will discuss shortly, due to relativistic effects which detune the plasma wave frequency from the laser beat frequency. The onset of turbulence does not affect the acceleration of the particles trapped in plasma waves nearer to the light pulse since these continually move into fresh plasma. The onset of turbulence does limit the number of plasma waves that can be used for particle acceleration. For example, if turbulence first onsets due to ion instabilities which grow on a time scale characterized by the ion plasma period ($\tau_i = 2\pi/\omega_{pi}$, $\omega_{pi} = (4\pi ne^2/M_i)^{1/2}$, where M_i is the ion mass), then plasma waves can be created for a time $2\pi/\omega_{pi}$ or a distance $2\pi c/\omega_{pi}$ behind the laser pulse. For a hydrogen plasma this corresponds to 43 ($=\sqrt{M_i/m_e}$) electron plasma wave periods. Plasma waves further than 43 plasma wavelengths behind the laser will begin to become degraded and will probably not be suitable for acceleration.

The growth of the plasma waves behind the laser pulse can be quantified by considering the oscillation of a single background plasma electron driven by the pondermotive force F_p (from Eq. 3 divided by n_0 to get the force on a single electron) of the beating lasers:

$$\frac{d}{dt}\left(\gamma V_x\right) + \omega_p{}^2 x = \frac{e^2}{2m\omega^2{}_{1,2}} \Delta k E_1 E_2 \sin\left(\Delta\omega t - \Delta kx\right) \qquad (6)$$

where $E_{1,2}$ are the electric field amplitudes of each laser, $\Delta\omega$ and Δk are the beat frequency and beat wave number of the lasers (i.e., $\Delta\omega = \omega_1 - \omega_2 \approx \omega_p$), $\gamma = (1 - V_x^2/C^2)^{-1/2}$, and x is the displacement of the electron from its equilibrium position (i.e., the Lagrangian coordinate). Neglecting the factor γ, this is a simple harmonic oscillator equation with a resonant driver if the beat frequency $\Delta\omega = \omega_p$.

Now the amplitude of the electron's oscillation is proportional to the amplitude E_p of the plasma wave field that its motion supports (the coefficients of proportionality are[12] $\Delta kx = eE_p/m\omega pc \equiv \varepsilon$). Thus, renormalizing the equation, applying the chain rule to γV and expanding γ for small V/c, we obtain the equation for the beat driven plasma wave field normalized to the wavebreaking plama wave amplitude:

$$\ddot{\varepsilon} + \omega_p^2(1 - 3/2\dot{\varepsilon}^2)\varepsilon \approx \frac{\alpha_1\alpha_2}{2} \sin \Delta\omega t \qquad (7)$$

Here $\alpha_{1,2} \equiv eE_{1,2}/m\omega_{1,2}c$. This is the model of Rosenbluth and Liu[13] which includes a correction of the plasma frequency due to the relativistic mass increase of the oscillating plasma electrons. Initially, if $\Delta\omega = \omega_p$, the plasma wave amplitude exhibits secular growth

$$\varepsilon(t) = \frac{\alpha_1\alpha_2}{4} \omega_p t \qquad (8)$$

where ε is now understood to represent the amplitude of the plasma wave. As ε grows, the effective plasma frequency becomes[13] $\omega_{ef} \approx \omega_p(1 - 3\varepsilon^2/16)$. At a time t such that $\int^t(\Delta\omega - \omega_{ef})\, dt' = \pi/2 \approx \omega_p\int 3/16\, \varepsilon^2(t')dt'$, the driver has become $\pi/2$ out of phase with the oscillator and actually begins to drive the oscillation back down. Solving for the value of ε from (8) at this time gives $\varepsilon_{max} = (2\pi\alpha_1\alpha_2)^{1/3}$. A more careful solution of the model equation (7) gives the maximum value of ε as[13]

$$\varepsilon_{max} = \left(\frac{16}{3} \alpha_1\alpha_2\right)^{1/3} \qquad (9)$$

In principal, Eq. (9) is only valid for $\alpha,\varepsilon \ll 1$, but simulations indicate that it is a good approximation for values of ε as high as 0.8.

In order to model the rise time of the lasers, we modify the Rosebluth and Liu model (7) slightly by allowing the normalized laser amplitudes $\alpha_{1,2}$ to be functions of time. This leads to modified expressions for growth and saturation:

$$\varepsilon(t) = \int_0^t \frac{\alpha_1(t)\ \alpha_2(t)}{4}\, d\omega_p t \qquad (8b)$$

$$\varepsilon_{max} = \left(\frac{16}{3}\alpha_1(\tau)\alpha_2(\tau)\right)^{1/3} \quad (9b)$$

where τ is the value of t at which (8b) and (9b) become equal. Fig. 3 shows a numerical solution of the modified model equation (7) with the analytic expressions corresponding to Eqs. (8b) and (9b) plotted. Given the details of the laser pulse, Eqs. (8b) and (9b) enable predictions for the time for growth and the saturated amplitude of the resulting plasma space charge wave.

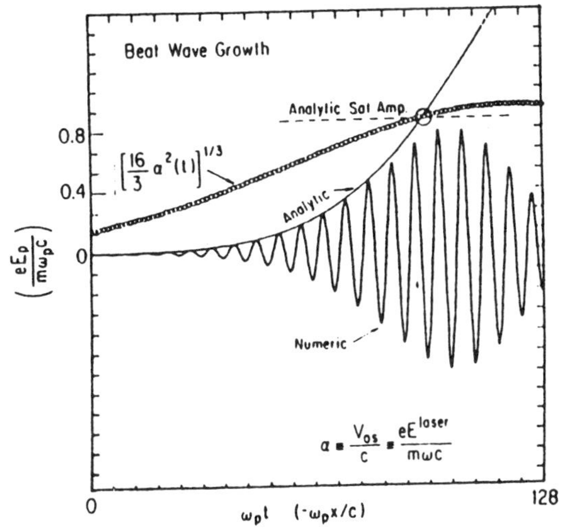

Fig. 3 Numerical solution of the modified plasma beat wave model equation (7) and the corresponding analytic growth and saturation for (8b) and (9b).

In order to minimize the effect of the detuning between the driving frequency $\Delta\omega$ and the effective plasma frequency ω_{ef}, C.M. Tang, et al.[14] have considered altering the beat frequency. They find that the beat wave growth is optimal for

$$\Delta\omega = \omega_p \left[1 - \frac{1}{2}(9\alpha_1\alpha_2/8)^{2/3}\right] \quad (10)$$

with corresponding maximum amplitude

$$\varepsilon_{max} \approx 4\ (\alpha_1\alpha_2/3)^{1/3} \quad (11)$$

an increase of about 50%. For lasers of $V_{osc}/c \equiv a = .1$, Eq. (10) corresponds to a 2% shift in the laser beat frequency below the natural plasma frequency.

The results of Tang, et al., also provide a guideline for plasma homogeneity requirements. Eq. (10) suggests not only the optimal detuning, but also a scaling law for the fractional density change ($\delta n/n \sim \delta\omega^2{}_p/\omega^2{}_p \sim 2\delta\Delta\omega/\omega_p$) that will not have a significant effect on beat wave growth. We might expect to have to maintain the plasma homogeneity to within

$$\frac{\delta n}{n} \sim (\alpha_1\alpha_2)^{2/3} \qquad (12)$$

or about 5% for lasers of $V_{osc}/c = .1$. The effect of noise in the plasma density is considered by Horton and Tajima[15], who give conditions for the resonant noise components moving with the beat wave to limit growth of the plasma wave.

B. COMPUTER MODELLING

There are a multitude of mechanisms which might cause the behavior of a real plasma to deviate from the simple model of Eq. (7). Some of these are non-linear wave steepening, thermal corrections, and ion effects. Fully self-consistent particle simulations have been invaluable in separating out the most important effects. In Fig. 4, we show the results of a 1-D simulation in which beating lasers were injected from the right[16]. In (a), the beat pattern of the rising lasers is visible and in (b), we see the characteristic growth and saturation of the plasma space charge wave. Remarkably, Fig. (4b) agrees almost identically with the numerical solution to the model Eq. (7) shown below it. This is true despite the fact that the saturation amplitude is more than 70% of the cold wavebreaking maximum. Many of the non-linear effects associated with particle trapping one normally expects at such amplitudes simply do not appear because the plasma wave phase velocity is too high to allow trapping of the background particles. Only particles with velocity V such that $V/c \gtrsim 1 - \varepsilon - 1/\gamma_{ph}$ can be trapped (see Eq. 19).

Recently, the computer models have been extended to two dimensions[17,18], and a sampling of these results are shown in Figs. 5-7. The 2-D simulations are consistent with the 1-D results and provide insight into transverse properties, such as filamentation and self-focusing.

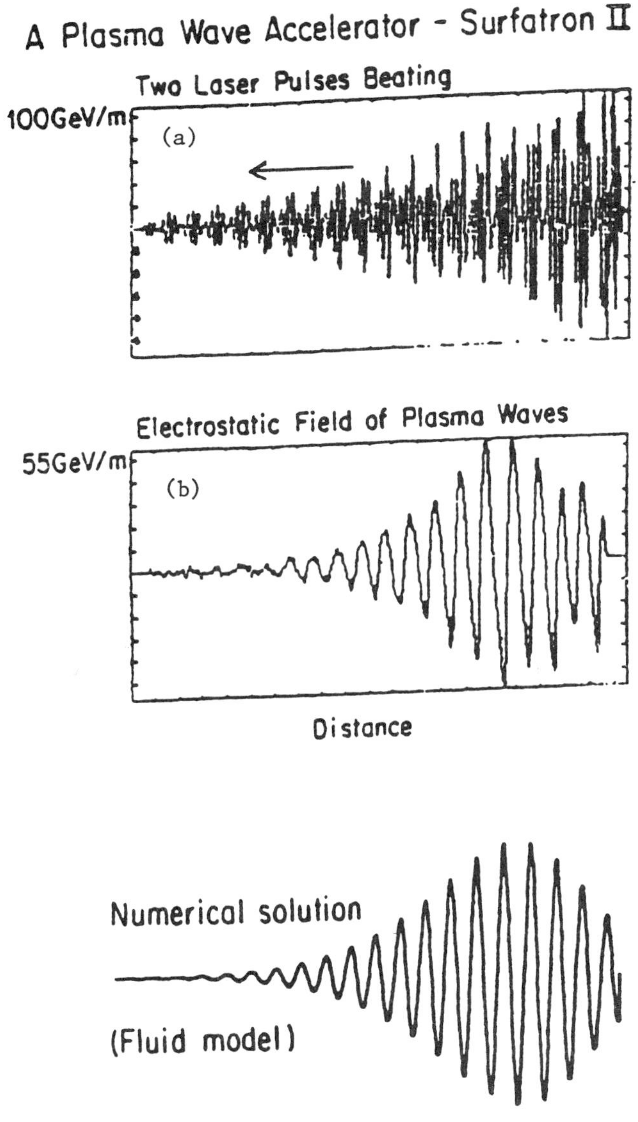

Fig. 4 1-D simulation showing (a) the laser beat pattern, (b) the space charge plasma wave. The numerical solution of the modified fluid model is shown below. $\omega_1=4\omega_p, \omega_2=5\omega_p, \alpha_1=\alpha_2$.

The 2-D contour plot in Fig. 5 shows the generation of coherent plane wave fronts (moving left to right). A slice down the axis, Fig. 5b, shows the growth of the beat wave to be in nice agreement with both the model of section A and the 1-D simulations.

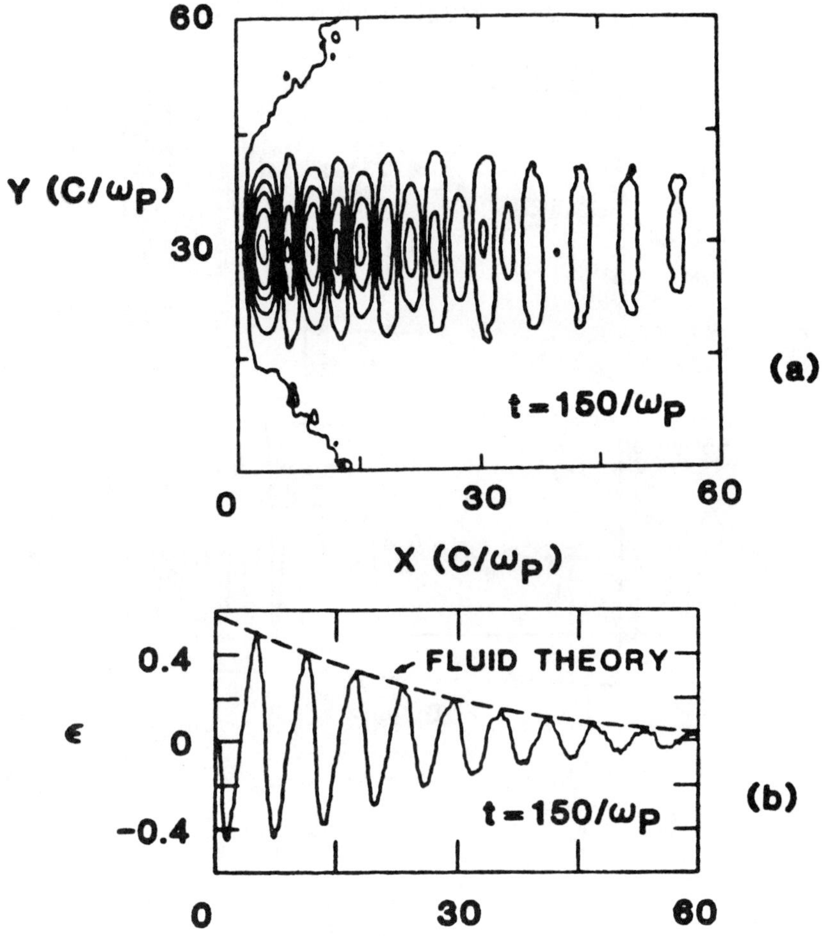

Fig. 5 2-D simulation (a) contour plot of the beat wave and (b) a slice down the y-30 axis.

Laser filamentation and self-focusing are phenomena familiar in laser fusion and can be seen clearly in Figs. 6 and 7. In Fig. 6, intense lasers ($V_{osc}/c \approx .5$) of width 20 c/ω_p incident from the left at first begin to filament into two narrower beams, then the strong radial gradient provides a pondermotive force which blows plasma out of the laser beam channels[19,20]. Pondermotive blowout is thought to be one of the major factors in the lack of accelerated electrons from a recent Los Alamos experiment. For very intense lasers, blowout can take place on a time scale as fast as the ion plasma period, necessitating the use of laser pulses shorter than this.

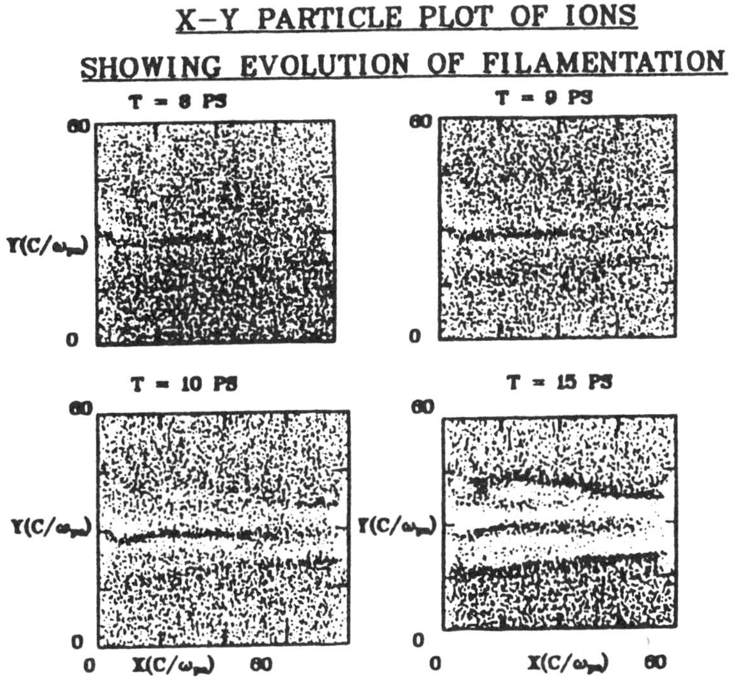

Fig. 6.

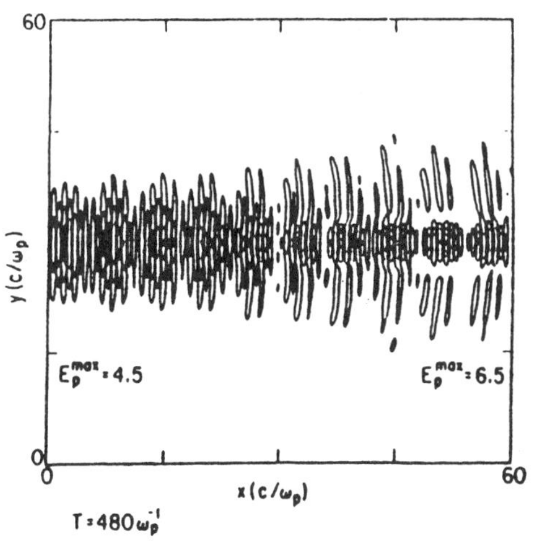

Fig. 7 2-D simulation contour plot of laser fields showing self-focusing (l to r) and increasing laser intenstiy on axis.

Laser self-focusing is a potentially beneficial phenomenon which may enable propagation of the lasers through the plasma over distances much longer than a Rayleigh length. In addition, as in Fig. 7, it may compensate for the depletion of pump intensity on axis (see Section II for a discussion of pump depletion). The mechanism for self-focusing can be understood quite simply by noting that the refractive index of a plasma is $n = (1-\omega_p{}^2/\omega^2)^{1/2}$. In a region of lower $\omega_p{}^2$, the index of refraction is higher. Inside the plasma channel where the laser is, $\omega_p{}^2$ $(=4\pi n_o e^2/m_e)$ is lower so the channel act like an optical fiber to confine the light. Inside the channel $\omega_p{}^2$ is lower for two reasons. First, the plasma density can be slightly lower due to pondermotive blowout, and second, the oscillatory motion of the background electrons in the intense laser fields gives them a relativistic mass increase. The latter effect is probably more important because it may occur faster than an ion time scale. The time scale for self-focusing can be estimated from the inverse of the growth rate ν obtained by C. E. Max and J. Arons[21]:

$$\nu = \frac{\omega_p{}^2}{8\omega}\left(\frac{V_{osc}}{c}\right)^2 \qquad (13)$$

The power threshold for self-focusing to overcome Rayleigh diffraction is approximately[22]

$$P \geq \frac{9}{2}\left(\frac{\omega}{\omega_p}\right)^2 \times 10^9 \text{ Watts} \qquad (14)$$

Some question remains as to whether the final beam radius will asymptote or oscillate about a value of the order c/ω_p.

C. EXPERIMENTAL RESULTS

The first experimental verification of the theoretical and simulation models was performed recently at UCLA by C. Joshi, C. Clayton, C. Darrow, and D. Umstadter. A more detailed description of the experiment can be found in the paper by C. Joshi, et al., in these proceedings. The 9.6 and 10.6 micron lines of a CO_2 laser ($\alpha_{10.6} \approx .03$, $\alpha_{9.6} \approx .015$) were used to resonantly drive the beat wave in a 10^{17}/cm^{-3} density plasma. Modelling the lasers rise to be linear over 1 nsec, the model equations (8b) and (9b) predict a maximum normalized wave amplitude $\varepsilon = .08$ (or E_L = 2.8GeV/m) at a time of roughly 500 picoseconds.

In order to conclusively diagnose the beat wave, one needs to know the phase velocity and amplitude of the wave. The phase velocity was obtained by measuring the frequency shift of the light from a ruby laser which is scattered by the density modulations of the beat wave. A sample of the frequency spectrum of the scattered light is shown in Fig. 8. The frequency shift of the scattered light shows that the plasma wave frequency is centered at the beat frequency of 9.6 and 10.6μ light. By varying the angle of the Thomson scattering, the k-spectrum of the plasma wave was obtained (Fig. 9). Figs. 8 and 9 together confirm that the phase velocity of the plasma wave was near c as predicted by Eq. (5).

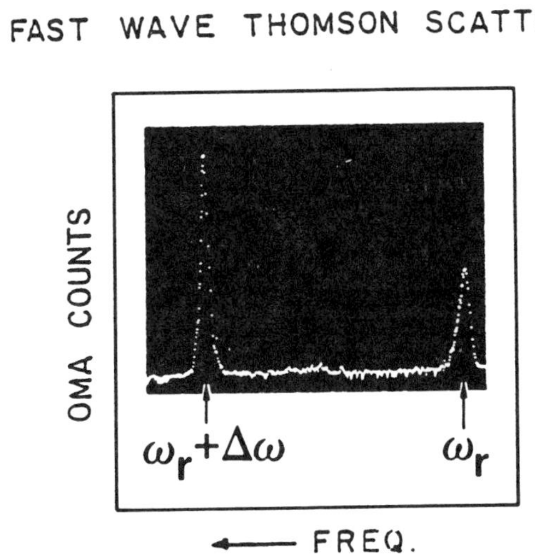

Fig. 8. Experimental small angle Thomson scattering indicating a plasma wave at the beat frequency ($\Delta\omega$) of the driving lasers (ω_r is the frequency of the diagnostic ruby laser).

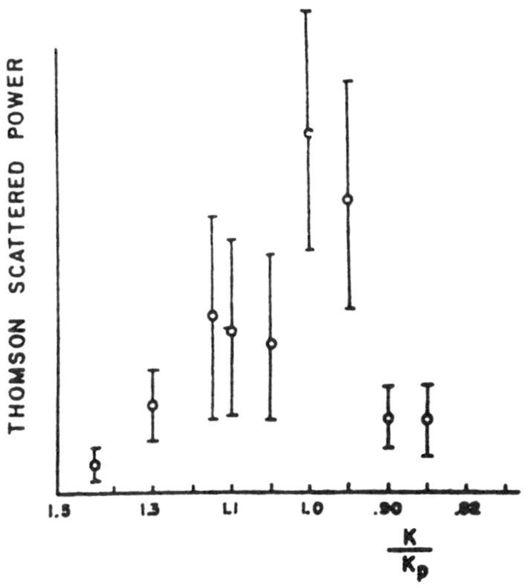

9. K-spectrum of the beat wave normalized to the theoretical beat wave wavenumber.

The time averaged plasma wave amplitude was obtained by measuring the intensity of the scattered light and calculating (assuming Bragg scattering from a moving grating) the corresponding n_1/n_0 (=ε) of the plasma wave. The resulting beat wave amplitude was n_1/n_0 = 1-3% or E_p = 330Mev/m -1GeV/m. This was the time-averaged field; the peak field may have been larger.

II ACCELERATION MECHANISMS

We now treat the acceleration of charged particles in the wave fields described in the previous section. Although it is not rigorous, we neglect the damping and distortion of the plasma waves by the accelerated particles. This is a fair approximation for moderate beams in which the number of beam particles per trapping potential is much less than the number of background plasma electrons in a plasma wavelength. Furthermore, we neglect the effect of the lasers on the accelerated particles. This is partially justified by the fact that in the beat wave frame, the laser electric field E_0 is reduced by the Lorentz transformation to E_0/γ_{ph} and the magnetic field transforms away completely. The longitudinal plasma wave field on the other hand is invariant to transformations along the direction of particle acceleration, so we consider it alone in our treatment of charged particle motion. We rely on simulations to validate these approximations and to identify the important self-consistent effects.

A particle in the plasma cannot be picked up or trapped by the accelerating wave unless it exceeds a minimum velocity in the direction of the wave, much as a surfer must paddle to catch an ocean wave.

For this reason, the accelerated particles either must be injected externally or picked up out of the high energy tail of the background plasma distribution. Before proceeding to the beat wave and surfatron acceleration mechanisms, we first calculate the minimum velocity or injection threshold for particle trapping. This calculation will also provide an estimate of the maximum plasma wave amplitude ε before trapping of background particles becomes significant (catastrophically damping the wave).

A. INJECTION/TRAPPING THRESHOLD[23]

Consider a particle of velocity V and momentum γmV in the lab frame moving in the direction of a wave of high phase velocity, $V_{ph} \lesssim c$. Its energy γ' in the wave frame is given by the Lorentz transformation to be

$$\gamma' = \gamma_{ph} \quad \gamma - \beta_{ph} \frac{P}{mc} \quad \approx \gamma_{ph}\gamma(1 - \beta_{ph}\, V/c) \qquad (15)$$

where $\beta_{ph} = V_{ph}/c$ and $\gamma_{ph} = (1 - \beta^2{}_{ph})^{-1/2}$. A particle is just trapped by the wave when the wave' s potential (in the wave frame) can overcome the particle's kinetic energy in the wave frame. That is, when

$$e\phi \geq (\gamma' - 1)mc^2 \tag{16}$$

the wave potential in the lab frame is simply

$$\phi = \phi'/\gamma_{ph} \tag{17}$$

since $E = k\phi$ is an invariant ($k'\phi' = k\phi$) and $k = \gamma_{ph}k'$. Combining Eqs. (15)--(17) gives the trapping potential

$$e\phi = mc^2\ [\gamma_{ph}\gamma(1 - \beta_{ph}V/c) - 1]/\gamma_{ph} \tag{18}$$

For $V = V_{ph}$, the minimum ϕ is zero as expected. Eq. (18) is valid for all particle and wave velocities. For $V/c \ll 1$ and $\gamma_{ph} \gg 1$, we find that

$$\phi/\phi_{cold} \approx 1 - V/c - 1/\gamma_{ph} \tag{19}$$

where ϕ_{cold} is the cold trapping threshold $\approx mc^2/e$ (from [18] with $V = 0$).

Interestingly, for electrons in the beat-wave example, cold trapping and cold wavebreaking (Eq. 2) are approximately the same threshold as can be verified from the above expression for ϕ_{cold} and $k = \omega_p/c$:

$$E_{\substack{\text{cold}\\ \text{trapping}}} \approx k\ \phi_{cold} = kmc^2/e \approx mc\omega_p/e = E_{\substack{\text{cold}\\ \text{wavebreaking}}}$$

Thus, to the extent that $|E| \approx |k\phi|$ for the plasma wave, $\phi/\phi_{cold} \approx \varepsilon \approx E/E_{max}$ (for $\gamma_{ph} \gg 1$). We comment that for ions, the cold trapping threshold would be larger than the cold wavebreaking threshold by the ion to electron mass ratio. Thus, trapping of ions will be much more difficult and will require the ions' velocity to be essentially V_{ph}.

We expect to trap few background particles if (19) is satisfied for V = a few (e.g., 3) times the thermal velocity (V_t). Thus, one should take ε less than $1 - 3V_t/c$ ($V_t \ll c$) to assure a minimum of background plasma trapping.

Expression (18) may be used to estimate what injection velocity or injection energy is required for injected particles to be picked up by a wave of given amplitude. Inverting (18) and solving for the injection energy $(\gamma - 1)$ in terms of $\phi/\phi_{cold} \equiv e\phi/mc^2 \approx \varepsilon$ yields

$$\gamma - 1 \approx \gamma^2_{ph}\ \{\varepsilon + 1/\gamma_{ph} - \beta_{ph}\ [(\varepsilon + 2/\gamma_{ph})\varepsilon]^{1/2}\} - 1 \tag{20}$$

Thus, for example if $\phi/\phi_{cold} = .1$ and $\gamma_{ph} = 10$, then the injection energy must be about 900 keV.

The injection energies versus normalized wave amplitude ε from Eqs. (18) or (20) are plotted for several values of γ_{ph} in Fig. 10.

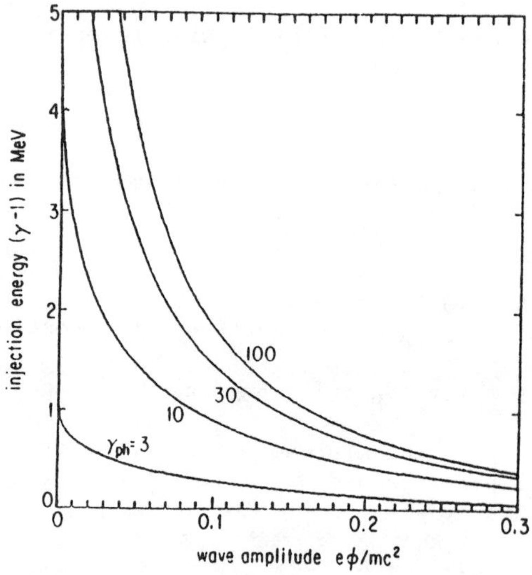

Fig. 10 Injection energy thresholds vs. normalized wave amplitude for various wave phase velocities (from Eq. 20).

B. SIMPLE BEAT WAVE ACCELERATION

In the original beat wave accelerator (BWA) concept, a particle accelerates by simply riding from the top to the bottom of the potential well of a plasma wave (i.e., a half of the plasma wavelength λ_p). Since the plasma wave moves at $V_{ph} = c\,(1 - \omega_p{}^2/\omega^2)^{1/2} \approx c(1 - \omega_p{}^2/2\omega^2)$ [see Eq. (5)] and the particle moves at nearly c,[24] the particle outruns the wave (reaches the bottom of the potential well) in a distance

$$\ell = (\lambda_p/2)\ c/(c - V_{ph}) \approx (\omega^2/\omega_p{}^2)\lambda_p \qquad (21)$$

The maximum energy gain of the particles is of order $eE_p \cdot \ell$ or

$$W_{max} = 2\varepsilon \frac{\omega^2}{\omega_p{}^2}\ mc^2 \qquad (22)$$

where we have substituted $\lambda_p = 2\pi c/\omega_p$, ε is the plasma wave amplitude normalized to wavebreaking and we have taken the average value of E_p to be its amplitude over π in order to give a result consistent with other derivations[24].

The energy gain[22] can be quite large. For example, for 1μ laser radiation incident on a 10^{17}cm^{-3} plasma, $\omega/\omega_p = 100$ and the maximum energy gain is 20GeV in a distance less than a meter (for ε near 1).

We comment that the actual region of the plasma wave useable for acceleration is likely to be less than $\lambda_p/2$. Radial fields due to the finite width of the plasma wave are defocusing to particles over part (5/16) of the accelerating portion of the plasma wave[25]. If the width of the plasma wave is given by the laser beam width and this in turn self-focuses to a few times c/ω_p as discussed in Section I, then the radial fields will be quite large and it will be necessary to avoid the regions of defocusing.

1-D simulations performed prior to the first laser accelerator workshop verified electron acceleration consistent with Eq. (22). Recent 2-D simulations[18] have substantiated the 1-D results and added information about the transverse particle dynamics. The phase space plots of Fig. 11 illustrate both electron acceleration in the plasma wave troughs and the focusing of the accelerated particles in the transverse direction. The peak γ of particles in Fig. 11 is around 25, consistent with Eq. (22) and the simulation parameters ($\varepsilon \sim .5$, $\omega_1 = 4\omega_p$, $\omega_2 = 5\omega_p$).

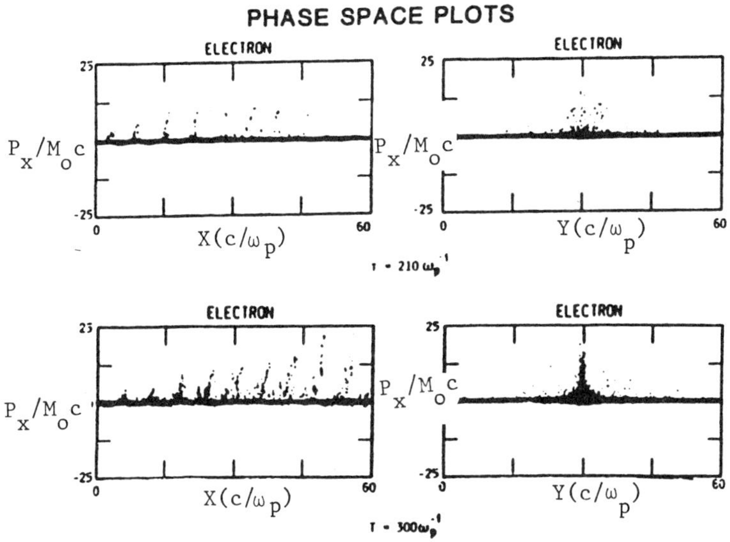

Fig. 11 2-D simulations showing acceleration in plasma wave troughs (left) and particle focusing (right) at two different times.

Fig. 12 Contour plot of azimuthal magnetic field (B_Z) in a 2-D simulation. The field near the bunches of accelerated electrons corresponds to Megagauss for CO_2 laser parameters.

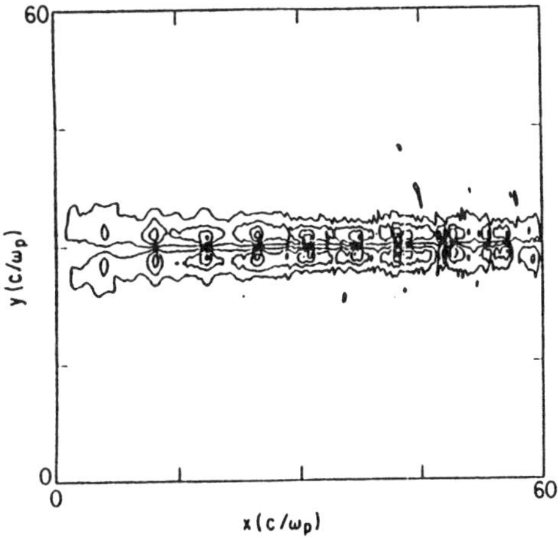

Particle focusing is important to the emittance quality of the accelerated beam. We believe that the particle focusing in Fig. 11 is caused by the self-generated magnetic fields of the accelerated electron bunches. A contour plot of the azimuthal magnetic field is shown in Fig. 12. The magnetic field concentrated near each bunch would correspond roughly to 2 megagauss for CO_2 laser parameters. Although this field cannot confine a beam in vacuum because the space charge repulsion of the beam electrons cause them to diverge, in a plasma the space charge of the beam is partly neutralized by the background plasma[26]. Thus the $\vec{V} \times \vec{B}_\theta$ force on beam particles off axis pinches them toward the center.

C. THE SURFATRON ACCELERATION MECHANISM

Expression (22) for the maximum energy gain of the BWA suggests that the lower is $\omega_p{}^2$ (i.e., the lower the plasma density), the higher is the final energy that can be obtained. Then why not eliminate the plasma so that the maximum energy can become infinite? The answer of course is contained in Eq. (2) which shows that the acceleration gradient ($\Delta W/\Delta x$) goes to zero. These two equations illustrate the beat wave dilemma:

$$W_{MAX} \sim \frac{1}{n_o}$$

$$\Delta W/\Delta x \sim \sqrt{n_o}$$

The higher the final energy required of the beat wave accelerator, the slower the acceleration gradient. For example, the maximum gradient possible in a plasma of density $10^{18}cm^{-3}$ is an impressive 1GeV/cm, but the effective acceleration length (for CO_2 parameters) is only .01 cm. If the accelerator is allowed to be any longer than this, the particle will pass the bottom of the plasma wave potential well and begin giving energy back to the plasma wave as it proceeds up the next potential hill.

The surfatron[27] is proposed as a mechanism to overcome this limitation of the BWA by phase locking the particles in the plasma wave, thereby allowing them to gain energy for as far as the plasma wave can be maintained. The basic idea, illustrated in Fig. 13, is to impose a DC magnetic field (B_z) perpendicular to the beat wave. The $-e\vec{V}_{ph} \times \vec{B}/c$ force on a trapped particle deflects the particle across the wave fronts similar to the way a surfer cuts across the face of an ocean wave. Once the particle acquires y-velocity the $-e\vec{V}_y \times \vec{B}/c$ force presses the particle up against the side of the wave where it remains phase locked.

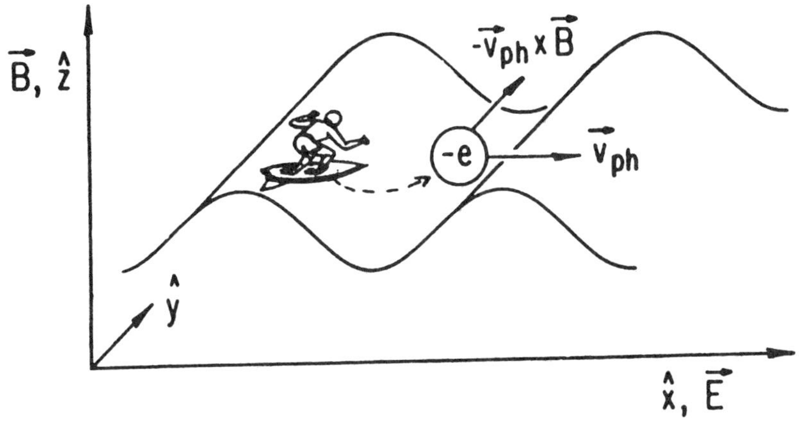

Fig. 13 A DC magnetic field deflects particles across the waves in the surfatron, preventing them from outrunning the waves.

In the wave frame (Fig. 14), the accelerated particle comes to a phase stable point in the potential well where the electric force ($E_p \sin k'x_1'$, primes denote wave frame quantities) of the plasma wave balances the magnetic force ($V_y'B_z'/C \approx \gamma_{ph}B_z$) due to motion across the wave.

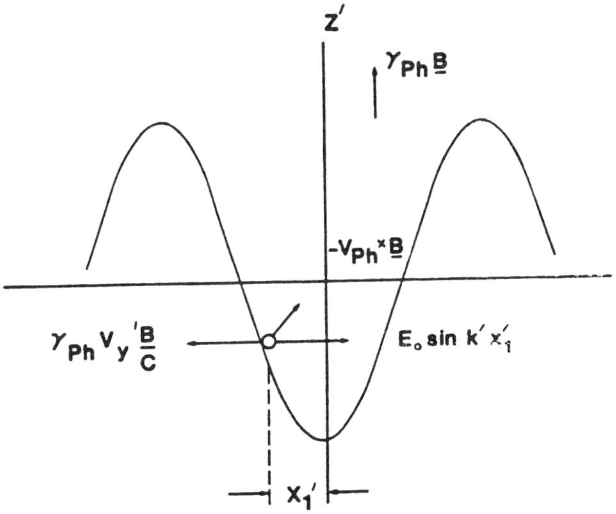

Fig. 14 In the wave frame, a surfatron particle comes to equilibrium against the side of the potential well.

Since $\sin k'x_1' \leq 1$, we must have

$$E_p > \gamma_{ph} B_z \tag{23}$$

in order to have a phase stable point. Too large a magnetic field will cause the particle to detrap and exhibit cyclotron motion.

If the wave phase velocity is near c, the particle travels a nearly linear trajectory at a small angle (θ) to the direction of the plasma wave. This can easily be seen from the fact that for a trapped relativistic particle

$$V_x \approx V_{ph}, \quad V_x^2 + V_y^2 \approx c^2 \tag{24}$$

Thus

$$\tan\theta = V_y/V_x \approx 1/\gamma_{ph}. \tag{25}$$

That the trajectory is linear assures that the radiation from surfatron acceleration is not large. Although it was first thought that radiation is less from surfatron acceleration than linear acceleration, it has been pointed out[28] that for the same $d\gamma/dt$, radiation from a linac and surfatron are the same.

The rate of energy gain for a surfatron particle is most easily obtained from the $\hat{y}$ equation of motion in the lab frame:

$$\frac{d}{dt}(\gamma V_y) = \omega_c V_x \tag{26}$$

where $\omega_c = eB_z/mc$, the non-relativistic cyclotron frequency. Substituting for V_x and V_y from (24) and integrating once gives the energy gain versus time or distance.

$$\gamma = \gamma_{ph}\,(V_{ph}/c)\,\omega_c t = \gamma_{ph}\,\omega_c x/c \tag{27}$$

Numerical solutions of surfatron motion in a sinusoidal plane wave electric field and uniform magnetic field are shown in Fig. 15b. The particles' energy increases linearly according to (27), and the fractional energy spread $\Delta\gamma/\gamma$ of two particles decreases as they both approach the same phase stable point. The BWA particles of Fig. 15a on the other hand spread in energy as they asymptote to the BWA limit (22). In Fig. 16, the phase space plot from a 1-D simulation shows surfatron motion in agreement with Eq. (26).

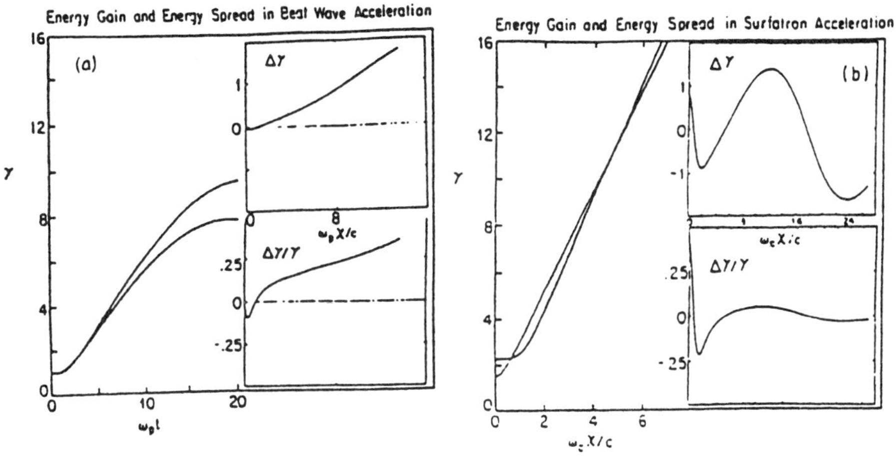

Fig. 15 Energy gain and spread for (a) two BWA particles and (b) two surfatron particles.

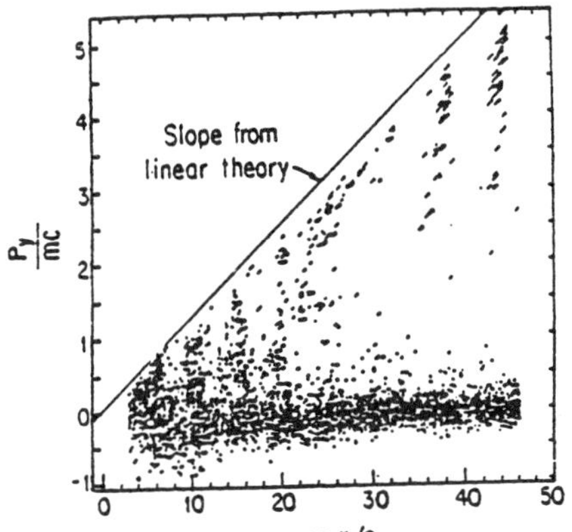

Fig. 16 Electron momentum across the wave P_y vs. distance x in a 1-D surfatron simulation.

The accelerating gradient, angle θ and trapping inequality for the surfatron can be summarized by the following formulae, valid for either electrons or protons:

$$\frac{\Delta W}{\Delta x} = \sqrt{n_{16}}\ \left(\frac{B_{kG}}{n_{16}\lambda\mu}\right)\ .1\ \text{GeV/cm} \qquad (28)$$

$$\theta \approx \lambda\mu \sqrt{n_{16}}/300 \qquad (29)$$

$$B_{kG}/n_{16}\lambda_\mu < \varepsilon = E_p/E_{wavebreak} \qquad (30)$$

where n_{16} denotes the plasma density in units of $10^{16}cm^{-3}$, B_{kG} is B_z in kilogauss, λ_μ is laser wavelength in microns and we have substituted $\gamma_{ph} = \omega/\omega_p$ from Eq. (5) and the definition of γ_{ph}. The accelerating gradient is roughly that of the BWA if B_z is chosen to make the trapping inequality (30) nearly an equality.

As an example, let us compare rough design parameters (Table I) for a 1TeV single stage BWA and surfatron accelerator using 1 micron laser light and assuming $\varepsilon \approx .5$. For the surfatron, we take $n = 10^{18}cm^{-3}$ and B = 50kG; for the BWA the density will have to be lower (from Eq. 22).

TABLE I COMPARISON OF DEVICE SIZES TO REACH 1 TeV

Device	Energy	Length	Width
Conventional	1 TeV	10-50 km	--
BWA	1 TeV	600 m	--
Surfatron	1 TeV	20 m	.6 m
Surfatron w/ optical mixing	1 TeV	20 m	.1 m

OPTICAL MIXING

As the above table indicates, a pure surfatron accelerator requires a fairly wide wavefront (.6m in this example). However, the particle trajectory is nearly linear, so much of the wave energy is wasted (see Fig. 17). C. Joshi and F.F. Chen have suggested that by optically mixing the beating lasers at a slight angle the energy can be confined along the particle path while preserving the proper wavefront geometry[29,30]. The electrons follow the path of the low frequency laser (ω_1, $\vec{k}_1$), while the high frequency laser (ω_0, $\vec{k}_0$) propagates at a small angle ϕ to the electrons.

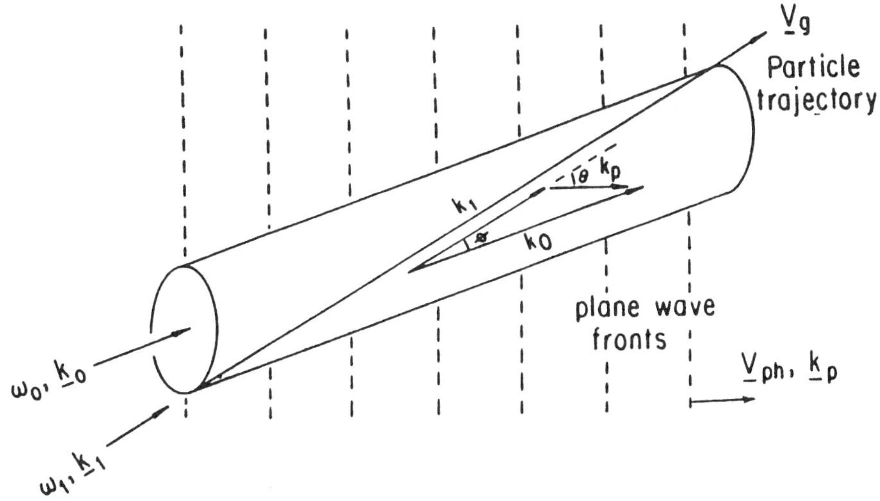

Fig. 17 By optically mixing the two lasers at small angle ϕ, the surfatron wave geometry can be created (wave fronts at angle θ to particle trajectory) without the need for a wide wavefront.

Optical mixing has two advantages. First, the wave width and laser energy requirements are greatly reduced. Second, the lasers are no longer co-linear with the particle beam, thereby allowing much easier staging.

From Fig. 17 and the law of sines

$$\sin\phi = k_p/k_0 \sin\theta \ll \sin\theta \tag{31}$$

so that the width of surfatron of length L need only be $L\sin\phi$ rather than $L\sin\theta$. From the fluid model of Section I, we expect the beat wave growth to be proportional to the product E_0E_1 of the two laser fields. Thus, we can let the lower frequency narrower laser E_1 be the more intense beam and reduce the intensity of the wider beam, while maintaining the beat wave amplitude. This lowers the energy requirement by more than just the width ratio of the devices.

The small mixing angle ϕ can be calculated from Fig. 17, Eq. (31) and the combination of the law of cosines, the dispersion relations ($\omega_i{}^2 = \omega_p{}^2 + c^2k_i{}^2$, i = 0,1), and the condition that the component of V_{ph} along k_1 ($\omega_p/k_p\cos\theta$) be c. The result is[30]

$$\phi \approx \left(\frac{\omega_p}{\omega_0}\right)^{3/2}$$

The generation of a small angle optically mixed beat wave has been supported by the 2-D simulations shown in Fig. 18. The first figure shows the contour plot of the beating between a narrow laser on axis and a wide laser at a small angle. The second figure shows that the corresponding plasma wave fronts are confined along the axis but tilted at the desired angle.

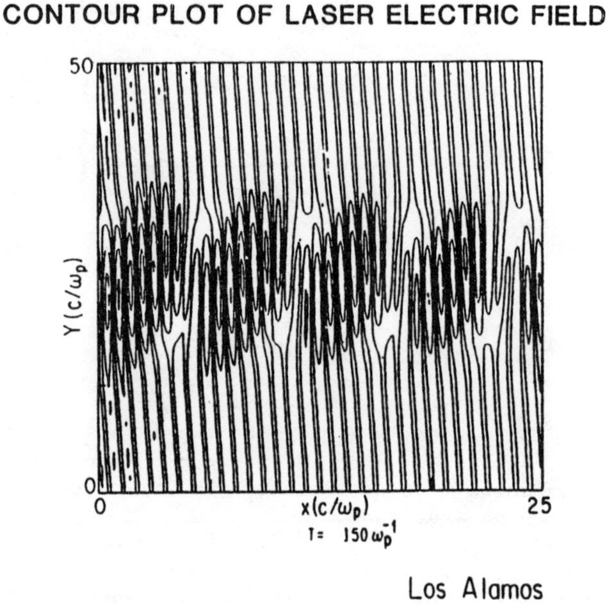

Fig. 18 2-d simulation contour plots of (a) the small angle optically mixed laser fields and (b) the resulting plasma waves.

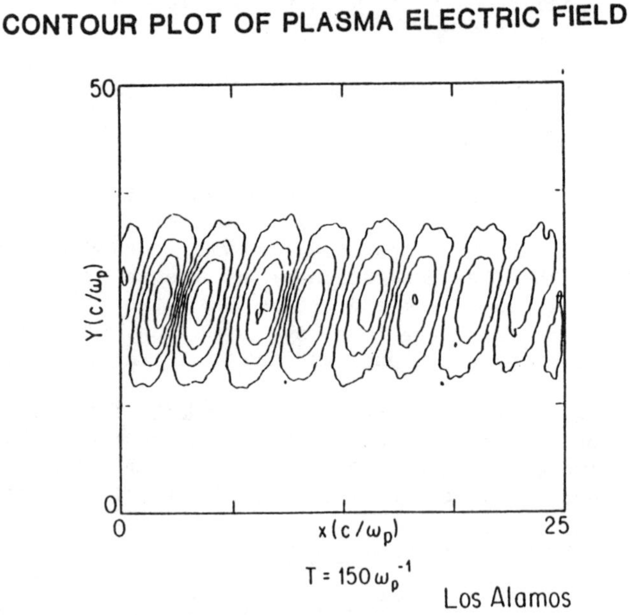

Since the particles in optical mixing move at c nearly parallel to the light pulse at V_g, the particles will eventually overtake the light pulse. If the particles are injected several (N) wavelengths behind the pusle (see Fig. 2), the distance for this overtaking to occur will be $\ell = N\lambda_p c/(c - V_g) \approx 2N\lambda_p \omega^2/\omega_p{}^2$. This is much longer than the limiting distance for the BWA (Eq. 21). As we shall see in the next section, pump depletion limits the length of acceleration before this velocity mismatch ever becomes a problem.

PUMP DEPLETION

Unfortunately, simulations of surfatron acceleration failed to show energy more than about 30% higher than corresponding BWA runs. Energy gain in the surfatron scheme is not limited by particle dynamics; it is limited instead by wave dynamics. That is, the laser pulse/plasma wake package of Fig. 2 cannot be propagated indefinitely through the plasma. The laser pulse is continually feeding energy into plasma waves. Since the plasma wave energy convects at the plasma wave group velocity ($\sqrt{3}\, V^2{}_{th}/V_{ph}$) which is effectively zero, all of the plasma wave energy is left behind. When the energy left in plasma waves becomes comparable to the energy originally in the laser pulse, the laser pulse will have become completely depleted and no more plasma waves can be produced. We can estimate this pump depletion length L_d from the simple energy balance argument that the energy density of the plasma waves times L_d must be less than the energy per unit area in the laser pulse:

$$\frac{E_p{}^2}{8\pi} \cdot L_d \leq 2 \int_0^{\tau} \frac{E_0{}^2(t)\ cdt}{8\pi}$$

where we assume both lasers to have amplitude E_0, $\ell/c = \tau$ the length of the laser pulse, E_p is the amplitude of the plasma waves and we have assumed equal cross-sectional areas for laser and plasma waves. Adopting our previous normalizations, we have

$$L_d \leq \frac{\omega_0{}^2}{\omega_p{}^2} \frac{\int^{\tau} \alpha_0^2\ cdt}{\varepsilon^2}$$

If we assume the maximum effective length τ of our laser pulse to be determined by relativistic detuning (Eqs. 8b and 9b) and substitute one factor of ε from Eq. (8b), we find that

$$L_d \leq \frac{\omega_0{}^2}{\omega_p{}^2} \frac{c}{\omega_p} \frac{4}{\varepsilon} \qquad (32)$$

independent of pulse shape. Since L_d is the distance that the laser can propagate into the plasma before becoming depleted, it is also the maximum acceleration length for the surfatron. Surprisingly, the maximum surfatron particle energy is independent of laser intensity since the energy gain is proportional to εL_d. In fact, for α greater than about .4, the plasma wave can reach wavebreaking before relativistic detuning occurs, further shortening τ and L_d.

Comparison to Eq. (21) for the dephasing length ℓ of particles in the BWA shows that L_d exceeds ℓ by only the factor $1/\varepsilon$ (neglecting a numerical factor which is approximately 1). We expect the energy gain in the surfatron and BWA to scale as

$$\frac{\Delta\gamma\ (\text{surf.})}{\Delta\gamma\ (\text{BWA})} \approx 1/\varepsilon.$$

THis is somewhat optimistic since not all of the laser pulse can be effectively converted to plasma waves. On the other hand, if the laser beats could be detuned to avoid the relativistic detuning discussed in Section I, then the effective laser pulse length τ, and consequently L_d, could be increased. Presumably τ would then be limited to about $2\pi/\omega_{pi}$ (or a few times this), the time for competing ion instabilities to degrade the plasma wave.

A final problem we have observed in surfatron simulations is the more rapid onset of turbulence in the plasma wave train. The predicted surfatron acceleration is observed for particles in the first few plasma troughs. The degradation of the later troughs seems to coincide with the development of the large self-induced magnetic fields (see section IB) of the accelerated particles. It is not clear why the plasma wave dynamics are so different for the surfatron than the BWA since the applied DC field is relatively small ($\omega_c/\omega_p < \varepsilon/\gamma_{ph} << 1$ by the trapping inequality [30]).

III NEW IDEAS FOR PLASMA ACCELERATORS

A. RIPPLED PLASMA OR PLASMA GRATING ACCELERATOR

It may be possible to overcome the pump depletion problem of the previous section by employing a non-colinear laser geometry to excite the plasma waves. Fig. 19 illustrates a mechanism for a side-injected accelerator in which a single incident laser is polarized along the direction of a static density ripple in the plasma[31]

$$n(x) = n_0 + \delta n \sin k_r x$$

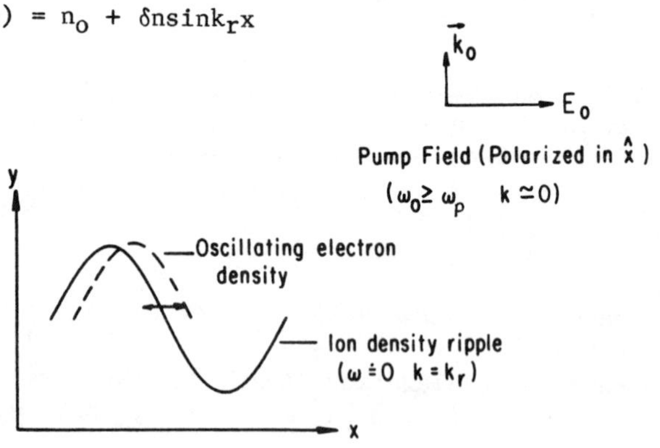

Fig. 19 A single laser incident from the side oscillates the electrons of a density rippled plasma to drive a longitudinal plasma wave in the direction of laser polarization.

Such a ripple might be produced by an ion acoustic wave or by ionizing a grating.

The laser field wiggles the electrons in the ripple by an amount

$$\delta x = (eE_0/m\omega_0^2)\cos\omega_0 t$$

while the ions are too massive to respond. This produces a longitudinal electric field distrubance given by Poisson's equation:

$$\frac{\partial E_p}{\partial x} = 4\pi e\,\delta n_e = 4\pi e\,\frac{\partial n}{\partial x}\,\delta x \approx \frac{\omega_p^2}{\omega_0^2}\,E_0\,\frac{\delta n}{n_0}\,k_r\cos k_r x\cos\omega_0 t$$

The right hand side represents a driver for the plasma wave which is bilinear in $\delta n/n_0$ and E_0 and is analogous to the beat wave ponder-motive force which is bilinear in E_1 and E_2 of the two lasers. In fact, for $\omega_0 \gtrsim \omega_p$, the initial plasma wave growth is

$$\varepsilon(t) = \frac{\alpha_0(\delta n/n_0)}{4}\,\omega_p t \qquad\qquad \varepsilon \equiv eE_p/m\omega_p c,\ \delta n/n_0 \ll 1$$

which is identical to Eq. (8) for the beat wave if $\delta n/n_0$ replaces α of the second laser. Since the interaction is only quasi-resonant ($\omega_0 > \omega_p$ to allow the laser to propagate into the plasma), the plasma wave saturation is determined by the detuning of the laser and plasma frequencies:

$$\varepsilon_{max} \approx \frac{\alpha_0\,\delta n/n_0}{2(\omega-\omega_p)} \qquad (\text{for } \varepsilon_{max} < 1)$$

The phase velocity of the plasma wave is determined by the coupling between the laser (ω_0, $\vec{k}_0$) and the ripple or acoustic wave (ω_r, $\vec{k}_r$). By choosing ω_0 near ω_p, the dispersion relation for the laser $\omega_0^2 = \omega_p^2 + k_0^2c^2$ indicates that k_0 will be small. Thus

$$V_{ph} = \frac{\omega_p}{k_p} = \frac{\omega_0 + \omega_r}{|\vec{k}_0 \pm \vec{k}_r|} \approx \frac{\omega_0}{k_r}$$

The phase velocity of the plasma wave can be controlled by varying the ripple wavenumber. The phase velocity might be increased along the accelerator to keep up with the accelerated particles or, alternatively, the particles might be surfed. Preliminary 1-D simulations of surfatron motion with an applied magnetic field show acceleration well past the unmagnetized limit [obtained from Eq. (22) with the replacement $\omega^2/\omega_p^2 \rightarrow \gamma_{ph}^2 = (1 - \omega_p^2/k_r^2c^2)^{-1}$] without the pump depletion problem. Figure 20 shows the background ripple and accelerated particles in the corresponding plasma waves of a 1-D simulation.

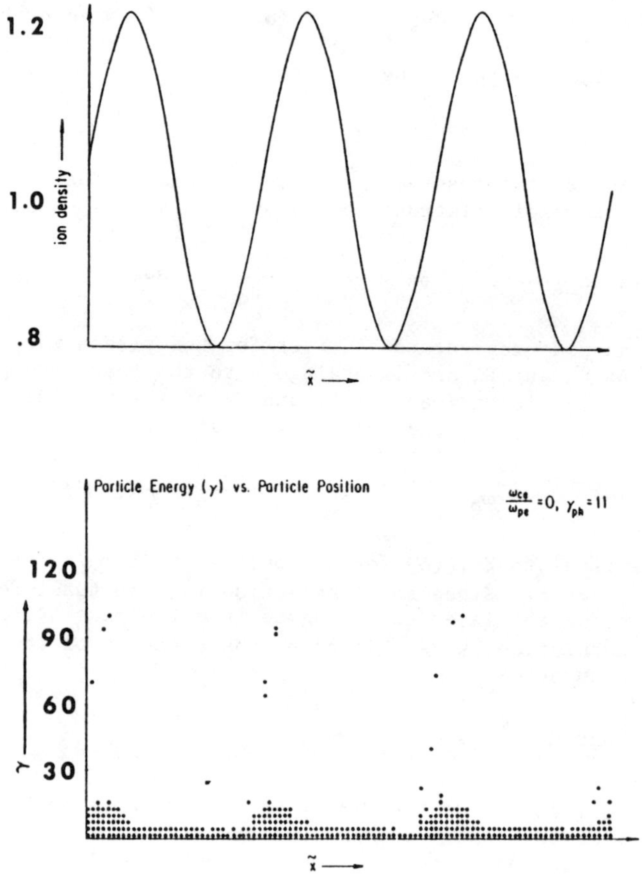

Fig. 20 1-D simulations showing (a) the initial density ripple and (b) the electron energy gained in the resulting plasma wave.

Figure 21 shows a 2-D simulation performed by W. Mori, K. Lee, D. Forslund, and J. Kindel. In this case, an intense laser is incident on a periodic array of neutral droplets rather than a density ripple. The original intention was to demonstrate near field acceleration in the normal modes of the droplet structure. Instead, the droplets were quickly ionized, forming a rippled plasma with density peaks centered at each ball. Preferred acceleration was then observed in the direction of laser polarization as in the present mechanism. At high laser intensity many schemes may unintentionally become plasma schemes. In this case, the plasma properties may actually be used to advantage.

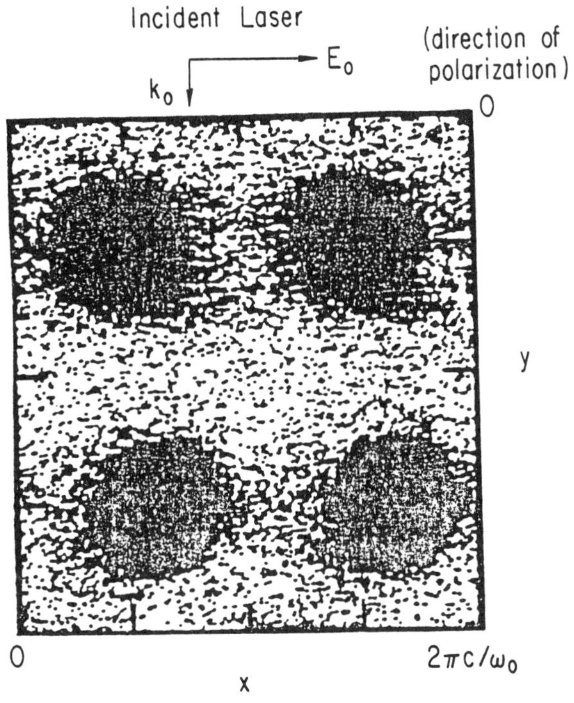

Fig. 21 2-d simulation of lasers incident on neutral droplets showing rapid formation of a (density-rippled) plasma or plasma grating.

B. CONVERGING PLASMA WAVES

Another scheme which employs a non-colinear geometry is illustrated in Fig. 22[32]. Two sets of plasma waves converge at a slight angle and particles are accelerated down the axis of symmetry between them in the superimposed fields. The phase velocity down the axis is higher than the phase velocity of the individual plasma waves (see Fig. 22b) and can be controlled by the angle θ at which the waves converge. This avoids the phase slip problem of the BWA, and the non-colinear injection facilitates resupplying the driving energy. Energy gain per stage will exceed that of the BWA if

$$\omega_p a/c > \omega/\omega_p$$

where a is wave width.

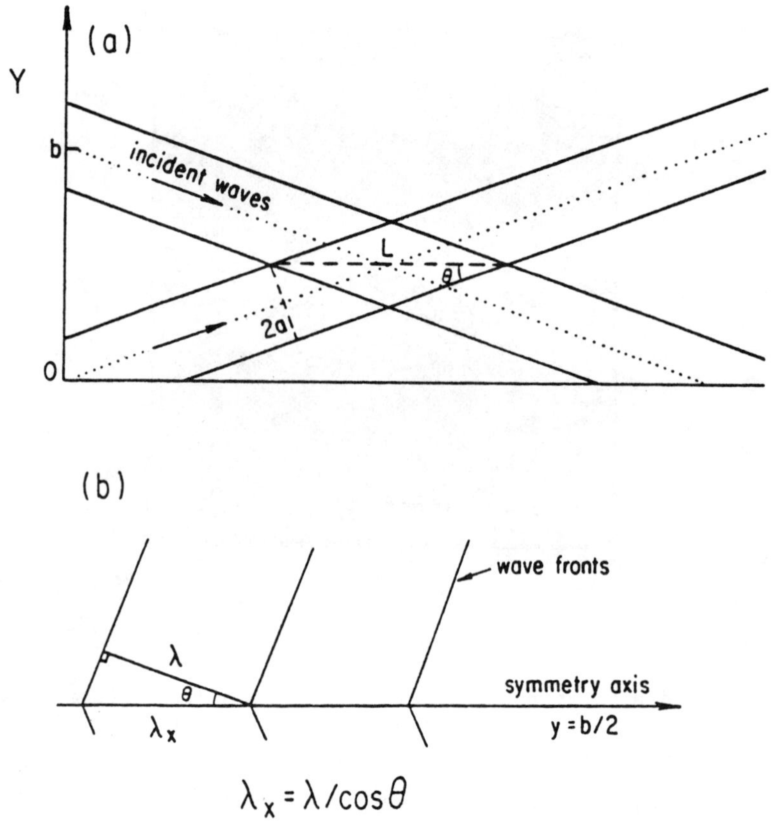

Fig. 22 A converging wave geometry (a) facilitates re-supplying the driving energy and (b) enables a higher phase velocity down the symmetry axis.

Figure 23 shows numerical solutions of (a) particle trajectories and (b) their corresponding energy gain in two sets of converging logitudinal plane wave electric fields of gaussian width a. The particles gain energy in each of the interaction regions and are either focused or defocused depending on their phase. If the focusing properties can be taken advantage of, they might offset the factor of two (or more) waste of plasma wave energy in regions where particles are not being accelerated.

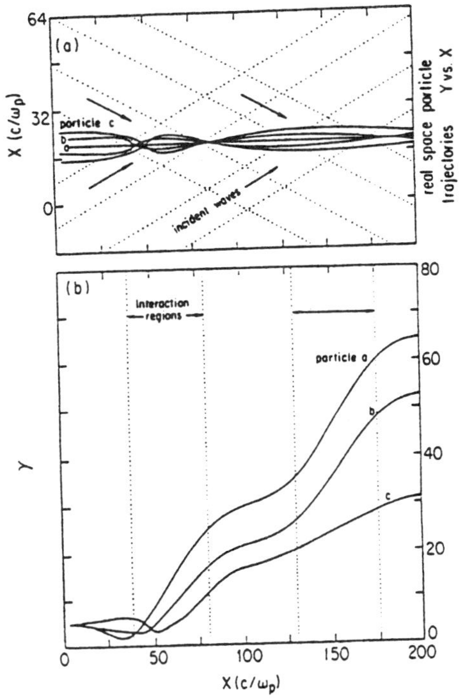

Fig. 23 Numerical solutions of (a) particle trajectories and (b) energy gain in the field geometry of Fig. 22.

In order to create the desired geometry, two pairs of beating lasers might be used. In this case, we must have $\omega_1 - \omega_2 \cong \omega_p \cong \omega_3 - \omega_4$, but $\omega_1 - \omega_4 \neq \omega_p$ to avoid unwanted couplings in the interaction region. Another way to create the waves might be to use the non-laser plasma wakefield scheme.

C. THE PLASMA WAKEFIELD ACCELERATOR

The Plasma Wakefield Accelerator[9] is a non-laser accelerator scheme which utilizes the electron bunches of a conventional linac as the free energy source to drive the plasma waves. If a later arriving, less dense, bunch of electrons is properly phased, it can ride the plasma waves or wakes of the earlier bunches and be accelerated.

The mechanism for exciting the plasma waves is similar to the plasma two stream instability except that the electron beam is made up of discrete bunches rather than a continuous stream. It is hoped that by controlling the spacing between bunches that the plasma wavelength and hence phase velocity can be controlled. The phase velocity of the pure two stream instability on the other hand is always slightly less than the stream[33] velocity with deleterious consequences for maximum energy gain.

The electric field gradients (i.e., wakefield amplitudes) possible in this scheme scale as the Coulomb Field of a driving bunch at a distance of a plasma wavelength[9]:

$$eE \sim \frac{Ne^2}{(c/\omega_p)^2}$$

where N = the number of electrons per bunch. For example, a bunch of $N = 5 \times 10^{10}$ electrons can produce a Wakefield which is near the wavebreaking amplitude in a $10^{15}cm^{-3}$ density plasma. However, the bunch length must be less than $c/\omega_p \approx .2$ mm.

In a colinear geometry, the energy transferred to a driven electron is limited by the wakefield theorem of Ruth, et al.[9], to be less than two times the energy per driving electron. To overcome this limit may require a non-colinear scheme such as that suggested in the preceeding section. Since this workshop, the use of a non-symmetric driving bunch has been proposed[34]. This overcomes the wakefield theorem and enables bunch lengths longer than c/ω_p.

IV SUMMARY

A great deal of progress has been made on plasma accelerators since the first workshop in 1982. The beat wave accelerator is becoming well developed. The growth and saturation of the beat waves is now predictable from a modified fluid model which includes the laser rise time and frequency detuning. This model gives excellent quantitative agreement with one and two dimensional computer simulations. The extension of the simulation models to 2-D has guided our understanding of effects such as laser and particle self-focusing and competing instabilities.

The theoretical and computational progress has now been complimented by the recent UCLA experiment. The UCLA results demonstrated the first conclusive experimental evidence of a beat excited fast plasma wave with an electric field on the order of 1GeV/m.

The surfatron has been proposed as a means of overcoming the energy limitation of the beat wave accelerator. A number of problems and solutions associated with the surfatron phase locking scheme have been addressed. Small angle optical mixing appears promising for reducing the width requirement of the accelerating wave. Laser detuning or "chirping" may be necessary to increase the pump depletion length and enable the surfatron to exceed the BWA in energy. Finally, if problems associated with the self-generated magnetic fields of the accelerated beam can be overcome, the surfatron enables acceleration of higher quality beams ($\Delta\gamma/\gamma \rightarrow 0$) of electrons or protons.

New ideas for acceleration in a plasma medium are constantly being explored, each with its own advantages. The plasma grating scheme described here is attractive because (a) the laser energy is injected from the side, thereby avoiding pump depletion problems, (b) only a single laser frequency is required and (c) the plasma wave phase velocity can be controlled with the density ripple or ionized grating separation. The plasma wakefield accelerator takes advantage of the efficiency and high power that are available from existing accelerator technology. Both schemes are at an early stage of development.

With the understanding gained from the simulation models and the UCLA experiment, the prospects are good for a successful proof-of-principle beat wave experiment in the next three years, contingent upon funding. Equations (8b) and (9b) provide a design guide for the laser and plasma waves. Equation (20) predicts the required injection energy and Eq. (22) gives the expected final energy. Based on these equations, table II provides a reasonable reference design for a near term experiment.

Table II LASER AND PLASMA PARAMETERS FOR A BWA EXPERIMENT

CO_2 Laser	10.6 μm, 10.3 μm
Pressure	5 atmospheres
Pulse length	30-50 ps
Intensity	5×10^{13} W/cm^2 (10J)
Focal Spot Radius	400 μm
Rayleigh Length	10 cm
Plasma Density	10^{16} cm^{-3}
Acceleration Gradient	~ 1 GeV/m
Acceleration Length	~ 10 cm
Initial Beam Energy	~ 5 MeV
Final Beam Energy	~ 100 MeV

Work continues toward realizing the high gradients, immunity from breakdown, and focusing properties that plasma accelerators offer.

ACKNOWLEDGMENT

We are grateful to Drs. J. Kindel, D. Forslund, and K. Lee for their collaboration on most of the simulations; to Drs. P. Chen and R. W. Huff for sharing their plasma wakefield work; to D. Sultana for her rippled plasma simulation; and to Dr. W. Horton for helpful comments.

Thanks to J. Rose and J. Payne for help in preparing the final manuscript.

Work supported by US DOE contract DE-AT03-83ER 40120, NSF grant ECS 83-10972, NSF grant PHY 83-04641, and Lawrence Livermore Laboratory's University Research Program.

REFERENCES

1. D. Lowenthal, see these proceedings.

2. J.D. Jackson, Classical Electrodynamics, 2nd ed., Section 6.8 (John Wiley & Sons, New York, 1975).

3. The radiation or pondermotive force on a single electron at rest in a light wave $\vec{E} = \hat{y}E_y\sin(kx-\omega t)$ is $F_p' = k'e^2E_y'^2/2\omega'^2$ (see Eq. 6 of the text), where primes denote quantities in the particle's rest frame. If the particle were moving relativistically along $\hat{x}$, then $E_y' = \gamma(E_y-\beta B_z) \approx E_y/2\gamma$, $\omega' \sim \omega/2\gamma$ $k' \sim k/2\gamma$, $Fp = \frac{dp}{dt} = 2\frac{dp'}{dt'} = 2Fp'$, so that $Fp = ke^2E_y{}^2/2\gamma\omega^2 \propto 1/\gamma$. This derivation was suggested to me by G. Schmidt; a Hamiltonian-Jacobi treatment is in Landau and Lifshitz, Classical Theory of Fields, Section 47 (Addison-Wesley, Reading, Mass., 1971), 3rd ed.

4. An exact non-linear solution to the relativistic cold fluid equations [Akhiezer and Polovin, Sol. Phys., JETP 3, 696 (1956)] shows that the maximum electric field of a highly relativistic wave can exceed this value. Their result is $eE \approx m\omega_p c(\gamma_{ph}-1)^{1/2}$. Additionally, a waterbag model of a warm plasma can be used to estimate the thermal corrections (See Ref. 23) with the result that $e\phi \approx mc^2(1-\sqrt{3}\overline{V}_{th}/c-1/_{ph})$. Note that ϕ and E are no longer simply related by the ratio c/ω_p.

5. J.M. Dawson, Phys. Rev. 113, 383 (1959).

6. Berkeley Particle Data Group, Physics Letters 111B, "Review of Particle Properties", p. 39-41 (22 Apr. 1982); see also B. Montague and W. Schnell in these proceedings for discussions of small angle scattering in plasma.

7. T. Tajima and J.M. Dawson, Phys. Rev. Lett 43, 267 (1979).

8. Laser Acceleration of Particles (Los Alamos, 1982), AIP Conference Proc. No. 91, P. Channel, Ed. (New York, 1982).

9. P. Chen, R.W. Huff, J.M. Dawson, UCLA PPG-802 (1984); P. Chen, J.M. Dawson, R.W. Huff, T. Katsouleas, Phys. Rev. Lett., 54, 693 (1985).

10. T. Katsouleas, J.M. Dawson, D. Sultana, Y.T. Yan to appear in Proc. of 13th Conference on High Energy Accelerators (TRIUMF, Canada) IEEE Trans. on Nucl. Sci (Oct 1985).

11. Landau and Lifshitz, Electrodynamics of Continous Media, Section 15, (Pergamon Press, Oxford, 1960).

12. This can be shown from a 1-D plasma model in which the electrons are treated as a series of parallel negatively charged plates in a uniform background of positive charge density n_0e. In displacing an electron (or plate) by an amount x, it passes over an amount of positive charge n_0ex giving an electric field $E=4\pi n_0 ex$ at the plate. But from Poisson's Eq., $\nabla\cdot E=4\pi en_1\sim k\cdot E$ from which $kx\approx n_1/n_0$. From the considerations in the introduction $n_1/n_0\approx eE/m\omega_p c$ for a plasma wave at c, so $kx\approx eE/m\omega_p c$.

13. M. Rosenbluth and C.S. Liu, Phys. Rev. Lett., 29, 701 (1972); to obtain the frequency shift $3/16\ \varepsilon^2\omega_p$ from Eq. (7), one must note that if $\varepsilon\sim|\varepsilon|\cos\phi$, then $\dot{\varepsilon}^2\varepsilon\sim\sin^2\phi\cos\phi = 1/4\cos\phi$ + higher harmonics of ϕ.

14. C.M. Tang, P. Sprangle, R. Sudan, appl. Phys. Lett, 45 (1984).

15. W. Horton and T. Tajima, "Laser Beat-Wave Accelerator and Plasma Noise", submitted to Phys. Rev. A (1985).

16. W. Mori, C. Joshi, J.M. Dawson, IEEE Trans. on Nucl. Sci., NS-30, 3244 (1983).

17. C. Joshi, W.B. Mori, T. Katsouleas, J.M. Dawson, J.M. Kindel, and D.W. Forslund, Nature 311, 525 (1984).

18. D. W. Forslund, J.M. Kindel, W. Mori, C. Joshi, and J.M. Dawson, Phys. Rev. Lett. 54, 558 (1985).

19. W. Mori, et al., private communication.

20. C. Joshi, C.E. Clayton, F.F. Chen, Phys. Rev. Lett. 48, 874 (1982).

21. C.E. Max and J. Arons, Phys. Rev. Lett. 33, 209 (1974).

22. G. Schmidt and W. Horton, see the proceedings of this conference.

23. T. Katsouleas, Ph.D. Thesis, Appendix C, UCLA PPG-769 (1984).

24. The derivation of maximum energy gain given in Ref. 7 does not make the simplifying assumption that the particle velocity is roughly c. Also, the factor 2 in Eq. (22) is obtained by assuming the particle gains energy by the amount of the wave's potential amplitude $e\phi'$ in the wave frame. Since the particle can actually gain $2e\phi'$ in going from top to bottom of the potential well, the result (22) should actually be $4\varepsilon(\omega^2/\omega_p^2)mc^2$. The latter expression is closer to the maximum particle energy observed in simulations.

25. J. Lawson, J. Allen, R. Bingham, J. Butteworth, F. Close, R. Evans, G. Rees, R. Ruth, Rutherford Appleton Laboratory Report RL 83057 (1983).

26. The question was raised by R. Ruth at the workshop as to whether or not the space charge neutralization is sufficient for focusing when accelerating injected electrons rather than accelerating particles from the tail of the plasma distribution as was the case in this simulation. The answer is probably yes because the fraction of electrons accelerated by the simulation was small and did not greatly affect the charge neutrality of the plasma, but this point still needs verification.

27. T. Katsouleas and J.M. Dawson, Phys. Rev. Lett. 51, 392 (1983).

28. D. Neuffer, Phys. Rev. Letters 53, 1026 (1984). It is also pointed out in this reference that radiation from the zeroth order surfatron motion will likely be insignificant compared to that due to fluctuations in E or B.

29. T. Katsouleas, C. Joshi, W. Mori, J.M. Dawson, F.F. Chen, Proc. 12th Int. Conf. on High Energy Accelerators, Fermilab (1983).

30. F.F. Chen, Proc. 12 SPIG, Sibenik, Yugoslavia (1984).

31. T. Katsouleas, J.M. Dawson, D. Sultana, Yi Ton Yan, to appear in Proc. 13th Conf. on High Energy Accererators, TRIUMF, Canada (1985).

32. T. Katsouleas, J.M. Dawson, and P. Chen, UCLA PPG-827 (1984).

33. T. Katsouleas, UCLA PPG 828 (1984).

34. P. Chen, J.M. Dawson (see these proceedings).

EXPERIMENTAL STUDY OF BEAT WAVE EXCITATION OF HIGH PHASE VELOCITY SPACE CHARGE WAVES IN A PLASMA FOR PARTICLE ACCELERATION

C. Joshi, C. E. Clayton, C. Darrow, and D. Umstadter
University of California, Los Angeles CA 90024

ABSTRACT

In this paper we present the first detailed study of the excitation of high phase velocity plasma waves by colinear optical mixing. The plasma wave frequency and wavenumber, hence the phase velocity, are directly measured using Thomson scattering. The amplitude of the plasma wave and therefore the longitudinal electric field is inferred from the magnitude of the Thomson scattered light. For a modest laser intensity of $\sim 10^{13}$ W/cm^2 in each frequency, the excited space-charge plasma wave has measured longitudinal electric fields of the order 1 GeV/m, in reasonable agreement with the fluid theory.

INTRODUCTION

Recently there has been a considerable amount of interest in exploring whether longitudinal electric field associated with space-charge plasma density waves propagating close to the speed of light can be used to accelerate particles to very high energies in a very short distance[1]. The motivation for this approach comes from the fact that the maximum electric field associated with such a wave scales as $\sqrt{n_e}$(cm^{-3}) Volts/cm. Thus for a plasma density of 10^{16} electrons/cm^3 the maximum electric field is 10 GeV/m, orders of magnitude greater than the accelerating gradients used in the current accelerators. It has been proposed that such a large amplitude, relativistic (fast) space-charge plasma wave can be driven by beating two lasers (ω_o, k_o) and (ω_1, k_1) in a plasma such that

$$\omega_o - \omega_1 = \omega_p$$
$$\underline{k}_o - \underline{k}_1 = \underline{k}_p \ . \qquad (1)$$

Here ω_p is the plasma wave frequency and $\underline{k}_p$ is the plasma wave wavenumber. Considerable theoretical work and numerous simulations have been carried out on this scheme.[2, 3] However, to-date there has been no conclusive evidence experimentally that such high longitudinal fields can be generated in a plasma by beating two colinearly propagating laser beams.[4] In this paper we report the first detailed study of the excitation and detection of high phase velocity plasma waves by colinear optical mixing; this work was carried out at UCLA as part of the program on the Plasma Beat Wave Accelerator.

EXPERIMENTAL PARAMETERS

A nanosecond risetime, 2 ns (FWHM) CO_2 laser pulse containing two frequencies, corresponding to wavelengths 9.56 μm and 10.59 μm, is used to excite a beat wave in a hydrogen plasma. Both laser wavelengths, generated from a single mode-locked TEA oscillator are amplified through the same gain medium and are finally focused using a single ZnSe, f/7.5 lens to ensure spatial overlap and temporal coincidence. The typical line intensities used, expressed as the normalized parameter v_o/c, where $v_{o(0,1)} = eE_{0,1}/m\omega_{0,1}$ is the oscillating velocity in the laser field, are $(v_o/c)_{10.6\ \mu m} \sim 0.03$ and $(v_o/c)_{9.6\ \mu m} \sim 0.015$. The resonant density where the plasma frequency is equal to the difference frequency between the two lasers is 1.15×10^{17} cm^{-3}. To produce this density we use a high current arc discharge to create a pre-ionized plasma using H_2 as the fill gas and then use the laser beams to accomplish full ionization. For our laser parameters, the plasma wave should grow to a saturated amplitude $\tilde{n}/n_o \sim 0.08$ in roughly 500 ps according to fluid theory assuming relativistic saturation.[3] This implies a longitudinal field of 2.8 GeV/m. Since the time to saturation is relatively long, it is in principle possible to time resolve the growth of the plasma wave, thereby making comparison with theory possible. On the other hand, one must hold the plasma density at a constant value for 1000's/ω_p which is a very difficult thing to accomplish.

The experimental set-up is shown in Fig. 1. Incident laser parameters are monitored by splitting a fraction of the incident energy using the beamsplitter B1. This beam is split again using the beamsplitter B2 and is then focused onto a calorimeter which monitors the incident pulse energy. The rest of the beam is incident onto a normal incidence 50 ℓ/mm grating which disperses the two wavelengths. Using a concave mirror M_1 the two wavelengths are then focused onto two separate photon drag detectors which monitor the pulse shape and the relative amplitudes of the 10.59 μm and 9.56 μm laser lines. The backscattered radiation from the plasma is also sampled using the beamsplitter B1. Since the backscattered radiation generally consists of signal from stimulated Brillouin scattering (SBS) and stimulated Raman scattering (SRS), another beamsplitter B3 is used to split the backscattered radiation. Reflected signal from B3 is sent to a scanning Fabry-Perot interferometer. The output of the F-P interferometer is detected by a Ge:Hg liquid He cooled detector. By scanning the plate separation of the F-P, a spectrum of the Brillouin backscattered light is obtained. Plasma temperature can be deduced from the red-shift of the backscattered light compared to the incident light. The backscattered radiation transmitted by the beamsplitter B3 is sent through a gas cell containing SF_6 and ethanol which strongly absorbs radiation in the vicinity of 10.6 μm and 9.6 μm, respectively. Any Raman scattered light is typically of a wavelength > 11.5 μm for our plasma conditions and is therefore transmitted by the gas cell. Infrared bandwidth filters ($\Delta\lambda \sim 2$ μm) are used in conjunction with a Ge:Cu cold detector to determine the wavelength range of scattered radiation. The Raman scattered radiation is not intense enough to be dispersed using a grating, so that the scattered

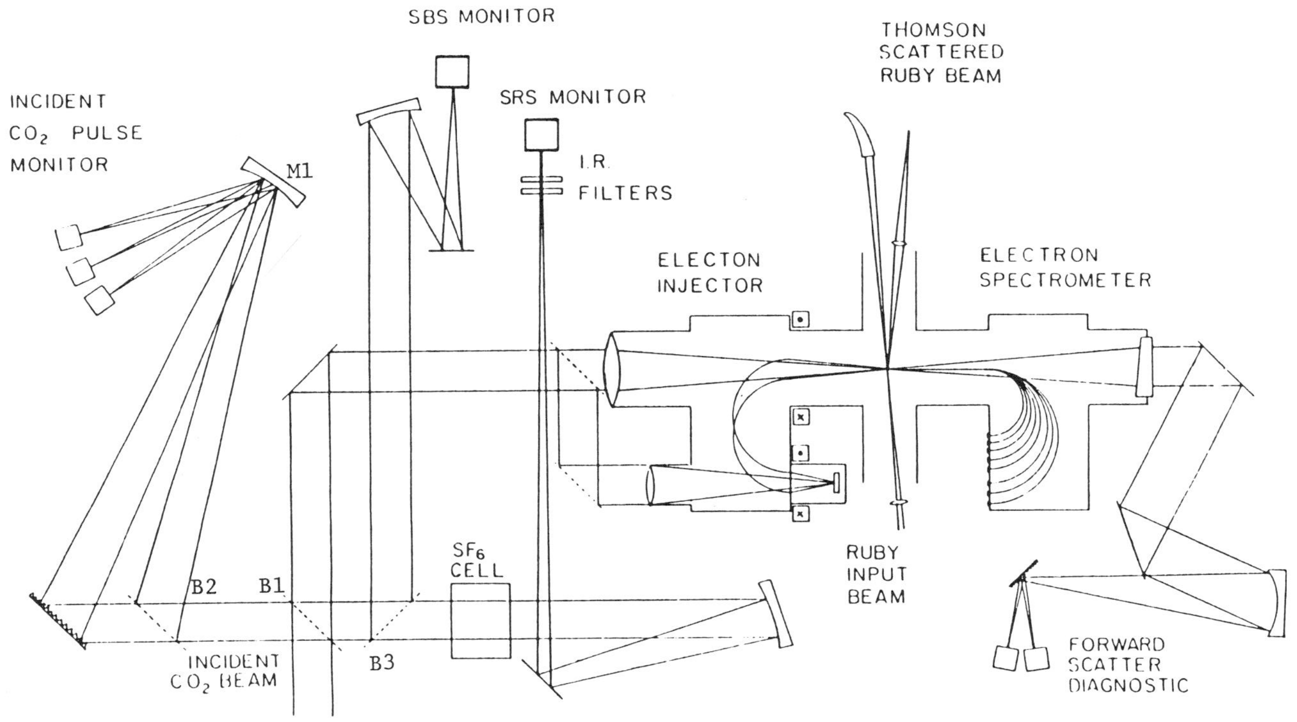

Fig. 1 Schematic of the experimental set-up.

light wavelength and hence the plasma density cannot be measured directly. Instead Ruby laser Thomson scattering is used to detect the frequency of the plasma wave excited during Raman scattering directly. By changing the angle at which the scattered radiation is observed, one can observe plasma and ion modes of different wavenumber.

The CO_2 laser beam transmitted by the plasma is collimated, split-off and its wavelength content analyzed using an IR spectrograph. Using a grating and bandpass filters, signals as small as 10^{-8} of the incident can be measured. This spectrometer is used to detect the forward Stokes and anti-Stokes radiation generated in the plasma.

OBJECT OF CURRENT EXPERIMENTS

The object of the current experiments is to quantitatively understand the physics of the beat-wave excitation and competing plasma processes. Some of the physics issues are: (a) Does the frequency difference between the two lasers, $\omega_0 - \omega_1$, exactly have to be ω_p?[5] If not, what is the width of the resonance? To answer this question one must have an extremely well diagnosed (better than 1%) plasma source. (b) How rapidly does the plasma wave build-up? Particle simulations suggest that the plasma-wave build-up is consistent with the growth rate given by the fluid theory. Ideally the growth time should be on the order of the laser pulse risetime. On the other hand, the build-up should occur faster than any instabilities. (c) What is the amplitude of the plasma wave at saturation and how does it depend upon the laser beam intensities and pulselength? (d) What is the saturation mechanism? Rosenbluth and Liu's model predicts that as the plasma wave grows, the relativistic effect on the frequency mismatch becomes important leading to the saturation of the plasma wave at an amplitude much smaller than that expected at wavebreaking. Their theory, however, still predicts large plasma waves (0(10%)) for intense laser fields ($v_o/c \sim 0.1$). An important question is whether other effects that are not included in their model such as collisions, plasma-wave convection or wave-wave coupling saturate the plasma wave to an amplitude that is lower than that due to relativistic mismatch? (e) What are the competing processes and how deleterious are they? Some of the competing processes are stimulated Brillouin scattering, Raman back and side scattering, self-focusing, filamentation and the parametric decay instability. Our current experiments are aimed at addressing these issues.

THOMSON SCATTERING

The theory for Thomson scattering of a probe laser beam from a single ion or plasma wave is analogous to Bragg scattering of x-rays from periodic electron distributions in crystals. In Thomson scattering, as in Bragg scattering, the radiation field of each electron oscillating in the probe beam's electric field is calculated and the total field is found by integrating over all the electrons in the scattering volume. In an ion wave both the electron and ion density is modulated at a certain wavenumber, whereas in a plasma wave only the plasma electrons are bunched at a given wavenumber with the ions forming a uniform background. It is found that the scattered intensity is maximized when $\underline{\Delta k} = \underline{k_i} - \underline{k_s}$ where $\underline{\Delta k}$ is the wavevector of the ion or plasma wave, $\underline{k_i}$ is the wavevector of the incident probe wave and $\underline{k_s}$ is the wavevector of the scattered probe wave. The Bragg condition is satisfied when

$$\underline{\Delta k} = 2\underline{k_i} \sin (\theta_s/2) \qquad (2)$$

where θ_s is the scatter angle. Since in an ion wave, the phasing of the electron and ion bunches travels at the ion acoustic velocity v_{ac}, the scattered light is either frequency up- or down-shifted by $\omega_{ac} = \Delta k v_{ac} = 2k_0 v_{ac}$. The phasing of the electron bunches in a plasma wave propagate much faster at a speed $\omega_p/\Delta k$ and the scattered signal is either frequency up- or down-shifted by ω_p.

The scattered light will have a small angular spread about θ_s because of diffraction. If all the light is collected, the Thomson scattered power P_{TS} is given by

$$P_{TS} = P_i \left(\frac{\pi}{2} \frac{\tilde{n}}{n_o} \frac{n_o}{n_{ci}} \frac{d}{\lambda_i} \right)^2 \qquad (3)$$

where P_i is the power of the probe beam <u>intercepted</u> by the ion/plasma wave, d is the thickness of the ion/plasma wave along the probe beam direction, n_{ci} is the critical density of the probe beam and λ_i is the wavelength of the probe beam.

In order to get a small spread of scattered light about the scattering angle θ_s and hence get a good k resolution, it is necessary to have many periods of the wave in the scattering volume. In an underdense plasma, $\omega_p << \omega_o$, the wavenumbers of the ion wave due to SBS and the plasma wave due to SRS in the backscatter direction are almost the same and $\simeq 2k_o$. The scatter angle for both these waves thus conveniently turns out to be ~7.5° when a ruby laser is used as a probe. Since the Δk is large and the scatter angle is reasonably large, spherical optics can be used to both focus the probe beam so that it covers many wavelengths and collect the scattered light. However, plasma waves produced by Raman forward scattering or colinear optical mixing have a wavelength that is (ω_o/ω_p) times the incident laser wavelength. For instance for a 10.6 μm laser propagating into a 10^{17} cm^{-3} plasma, $\omega_o = 10\ \omega_p$ and the Raman forward plasma wave has a wavelength of ~100 μm. The scatter angle is therefore only 0.4°. To probe many periods of the plasma wave, therefore, one has to use cylindrical focusing of the probe ruby beam. A combination of cylindrical optics also has to be used to collect the scattered light and physically separate it from the unscattered probe light. A novel technique has been developed to do this and is discussed elsewhere.[6]

PLASMA DIAGNOSTICS

First, the laser is operated in a single frequency mode (10.6 μm) and fired into the pre-ionized arc plasma. The peak intensity of ~10^{13} W/cm^2 in the focal volume easily exceeds the Brillouin backscatter and Raman backscatter thresholds but does not exceed the Raman forwardscatter threshold. The backscatter diagnostics show that up to 10% of the incident energy is Brillouin backscattered, whereas, the amount of light being Raman backscattered is typically less than 0.1% of the incident. The infrared spectrograph in the forward direction failed to detect any downshifted (Stokes) signal greater than 10^{-8} of the incident. Therefore, Raman forwardscatter can be neglected compared to backscatter.

In order to detect the SBS driven ion wave and SRS driven plasma wave, the ruby laser scattering is set up for 7½°. In Fig. 2 we can see the Thomson scatter data at this scattering angle for four different shots. On the right hand side of each picture we see a peak corresponding to Thomson scattering from SBS. The frequency of the ion wave as measured using the Fabry-Perot is typically ~10-15 GHz. Since the grating dispersion for resolving the Thomson scattered light is 5 Å/mm, any frequency shift less than 10 GHz cannot be resolved and, therefore, the signal essentially appears at the incident ruby wavelength. Nonetheless, from extensive previous measurements we do know that this signal is from an ion wave driven by SBS[7].

To the left hand side of the SBS peak, a second peak is seen which varies in position w.r.t. the SBS peak. This peak is characteristic of Raman backscattering and the frequency shift relative to the SBS peak is a direct measure of ω_p, and hence, the plasma density. This diagnostic is a very sensitive measure of the plasma density (better than 2×10^{15} cm^{-3}). By varying the gas fill-pressure, arc-voltage and the relative time between pre-ionization and the main laser, plasmas of different density can be produced as can be seen in Fig. 2. The plasma density can be tuned close to the resonant density of 1.15×10^{17} very accurately using this technique.

If the CO_2 laser intensity is reduced by 30% the Raman peak is found to disappear. Thus we can estimate the density scalelength assuming the threshold condition for linear and parabolic density profiles to be

$$I(W/cm^2) \geq 4 \times 10^{17}/\lambda_o L \qquad \text{(linear)}$$

$$I(W/cm^2) \geq [5 \times 10^{15}/(\lambda_o^{2/3} L^{4/3})]\, T_e^{1/3} (n_c/n_o)^{1/3} \qquad \text{(Parabolic)} \qquad (4)$$

This gives 200 < L (μm) < 1000, assuming T_e = 50 eV. Here

$$L = \left[\frac{1}{n_o}\left(\frac{dn}{dx}\right)\right]^{-1} \quad \text{(linear)} \quad \text{or} \quad L = \left(\frac{-1}{2n_o}\frac{d^2n}{dx^2}\right)^{1/2} \quad \text{(parabolic)}$$

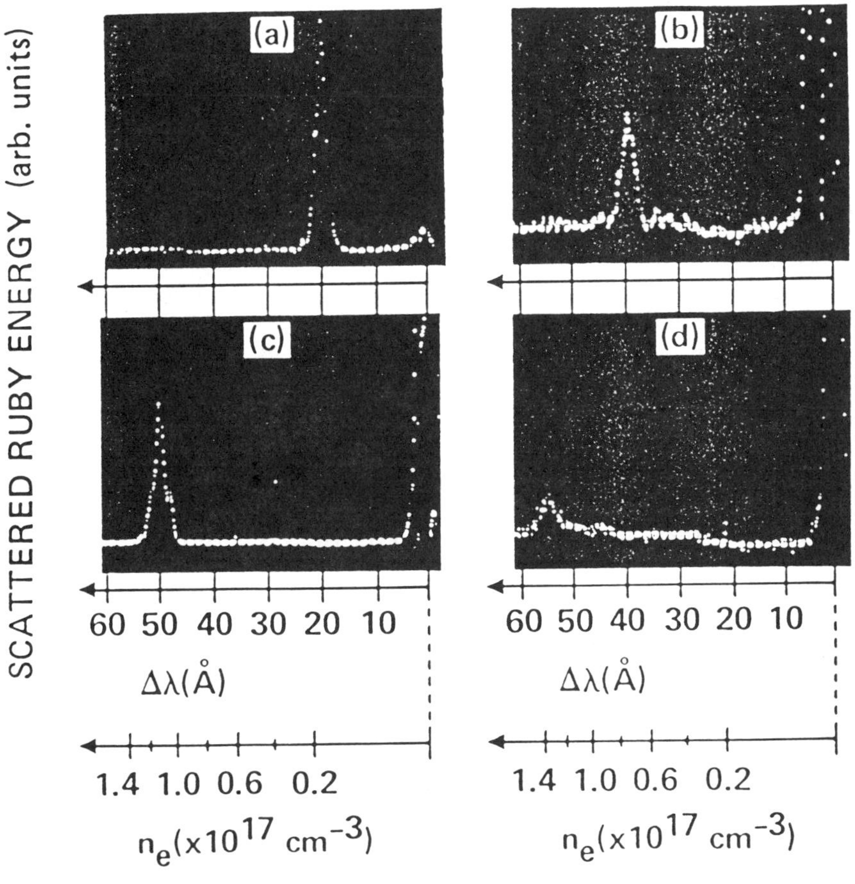

Fig. 2 Thomson scattered light spectra taken at 7½° for different plasma densities.

Apart from gross changes in density, the density profile has fine structure due to SBS. By spatially probing the plasma along the depth of focus of the CO_2 laser, the ion wave was determined to be approximately 1 cm in length. Using the Bragg formula and assuming a fully saturated ion wave we obtain a density fluctuation level of $\tilde{n}/n_o \sim 2\%$ which of course has a periodicity of 5 μm.

RESULTS

Observation of the "fast" wave: The conclusive evidence for the existance of the high phase velocity ("fast") plasma wave excited by beating of two laser beams resonantly in the plasma comes from ruby laser Thomson scattering. The scattering angle is adjusted to k match to the fast wave. Once the conditions for obtaining the correct plasma density for beat excitation have been established as discussed previously the plasma is irradiated by two laser wavelengths (10.6 μm and 9.56 μm) in a single laser pulse. Fig. 3(a) shows the frequency shift of the ruby light from the stray position to be exactly $\Delta\omega = \omega_p$. The fast plasma wave is only seen to occur under two frequency illumination and never under single frequency illumination. Thus Raman forward instability can be ruled out as the excitation mechanism. Secondly, whereas the Raman backscatter plasma wave (slow wave) is excited over a range of densities, the beat wave excited fast wave only occurs at $\Delta\omega$ ω_p. If the plasma density is tuned off resonance, the scattered signal drops (but does not shift in frequency) rapidly. We do not as yet have a quantitative measure of the width of the resonance. By moving the fiber optic which collects the scattered light on a shot-to-shot basis, the k spectrum of the fast wave has been obtained. It is shown in Fig. 3(b). In spite of the large shot-to-shot variation in the scattered light, it is clear from this figure that the scattered light spectrum is peaked about $k_p = k_o - k_1$. These direct measurements of the plasma wave frequency and the wavenumber have enabled us to unambiguously identify the wave as being the fast plasma wave ($\omega/k \sim c$). Amplitude of the fast wave: The ruby laser pulse used for Thomson scattering is 25 ns (FWHM) long and its spatial extent is 1 cm. The fraction of the ruby power intercepted by the fast wave (and therefore available for scattering) is

$$P_i = P_o \left(\frac{\tau_f}{\tau_o}\right)\left(\frac{L_f}{L_o}\right) \qquad (5)$$

where P_o, τ_o, and L_o are the total incident ruby power, pulse-length and spatial extent, respectively, and τ_f and L_f are the duration and the spatial extent of the fast wave, respectively.

The spatial profile of the fast wave was constructed by placing a 500 μm mask in the input ruby beam such that only about 5 wavelengths of the plasma wave could scatter the ruby light on a given shot. The data is shown in Fig. 4. The fast plasma wave was found to be roughly 1.8 mm (or ~18 wavelengths) long.

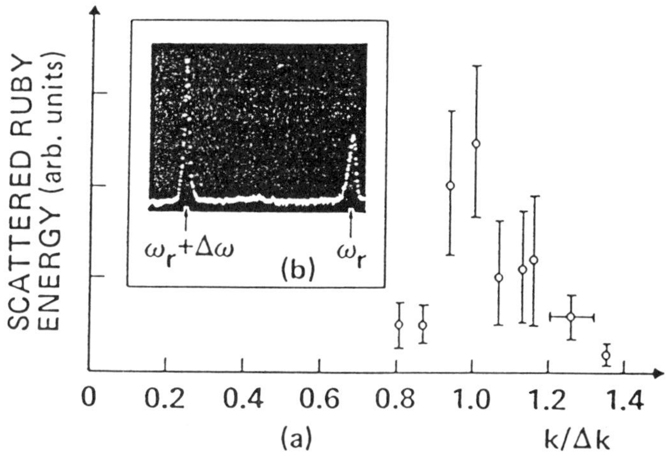

Fig. 3 (a) k spectrum of the fast wave and (b) Thomson scattering at 0.4° showing the resonant frequency shift.

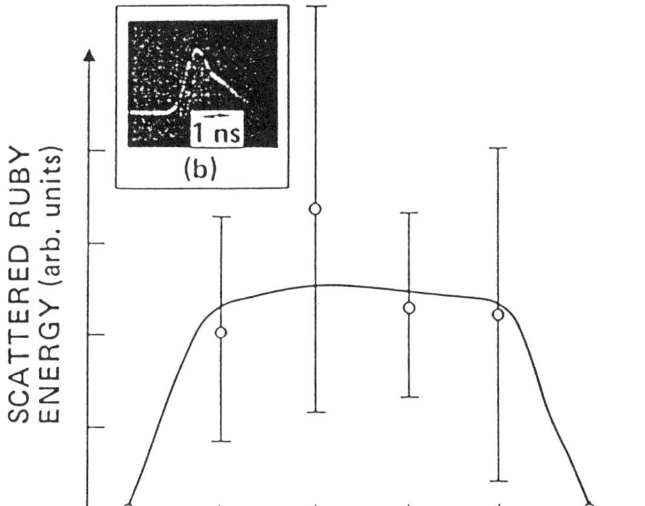

Fig. 4 (a) Axial profile of the plasma wave and (b) temporal pulse shape of Stokes radiation.

The time duration of the fast wave can in principle be determined by streaking the Thomson scattered light using a pico-second streak camera. The scattered light was typically 10^{-10} of the total incident ruby power of 40 MW, too small to streak at very high speeds. Instead, the time duration of the fast wave was inferred from the duration of the Stokes radiation, emitted in the same direction as the incident CO_2, to be approximately 1 ns as shown in Fig. 4(b).

The thickness d of the plasma wave in Eq. (3) is the thickness over which the perturbed density $\tilde{n}/n_o$ is constant. The measured transverse intensity profile of the CO_2 laser in the focal plane was approximately Gaussian with a half-energy spot size of 240 μm. Since the growth rate of the plasma wave is proportional to the product of the electric fields of the two laser beams, the plasma wave will also have a transverse profile that is approximately Gaussian with the same half width. An equivalent width d (over which the perturbed density is constant) is, therefore, approximately half the spot diameter or 120 μm.

Substituting for (P_{TS}/P_o), (τ_f/τ_f), (L_f/L_o), (d/λ_i) and (n_o/n_{ci}) we obtain the perturbed density of the fast wave $\tilde{n}/n_o$ as between 1-3%. The longitudinal electric field is simply related to the perturbed density through Poisson's equation since the plasma wave is an electrostatic wave,

$$\Delta k \, E_L = 4\pi e\tilde{n} \; . \qquad (6)$$

Substitution for Δk and $\tilde{n}$ in eq. (6) leads to longitudinal electric fields of between 300 MeV/m and 1 GeV/m. This is the first demonstration of longitudinal fields $\geq$ 1 GeV/m by any conventional or collective techniques.

<u>Observation of Stokes and anti-Stokes satellites</u>: A rather unique feature of colinear optical mixing in a plasma is the generation of satellites or electromagnetic sidebands that propagate in the same direction as the two frequency laser beam. One or more satellites each frequency downshifted by ω_p (Stokes) and frequency upshifted by ω_p (anti-Stokes) are generated in one or more ways. If the two initial pump beams are intense enough then they can undergo a cascade of decays $\omega_o \pm n\omega_p$ and $\omega_1 \pm n\omega_p$ via the stimulated Raman forward scattering instability. Pump depletion of the original laser beams is not a problem since the energy given to the forward scattered satellite is much greater than that to the plasma wave. (The latter can in principle be irreversibly lost.) This is dictated by the Manley-Rowe relation or the law of conservation of wave action which states that

$$\frac{W_o}{\omega_o} = \frac{W_1}{\omega_1} = \frac{W_p}{\omega_p} \qquad (7)$$

where $w = N\hbar\omega$. When $\omega_1, \omega_o \gg \omega_p$, multiple cascades are possible because repeated k matching (actually there is a small k mismatch)

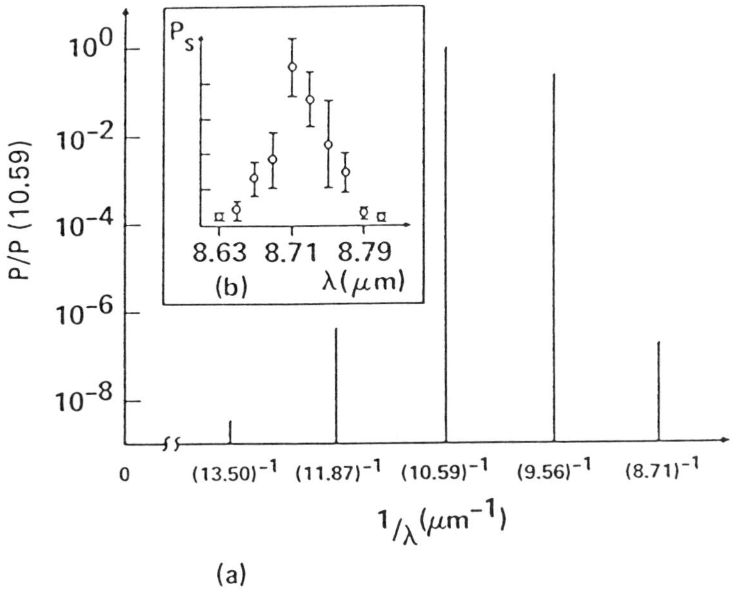

Fig. 5 (a) Forward scattered IR spectrum showing Stokes and anti-Stokes sidebands and (b) details of the anti-Stokes spectrum.

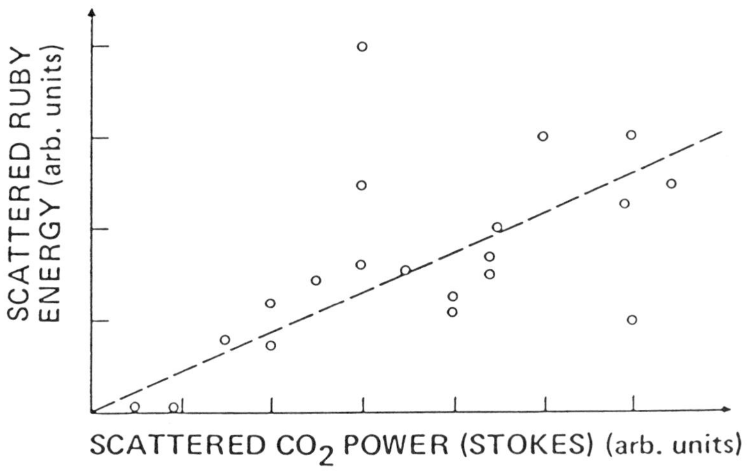

Fig. 6 (a) Correlation between Stokes and ruby Thomson scattered power at 0.4° (fast wave).

can occur until the original pump waves have cascaded down to waves with frequencies near ω_p.

If the two inital pump waves are not intense enough to produce a spectrum of satellites via the Raman forward instability, then there is the possibility of one or at most two up- and down-shifted satellites being generated by Thomson scattering. The process is analogous to the ruby Thomson scattering discussed earlier, except the pump beams themselves act as probes.

In our experiment the complete satellite spectrum was mapped out by scanning the infrared spectrometer which analyzed the forward scattered radiation. Two frequency downshifted (Stokes) and one frequency upshifted satellites were indeed observed. For the CO_2 laser intensities used in our experiment, the threshold scalelength for the Raman forward instability occuring in a parabolic density profile is 6 mm. The fact that the plasma wave is only 1.81 mm long (a length less than the theshold scalelength) and that the ratio of the Stokes to the incident 10.6 μm is only 10^{-6} leads us to believe that the Stokes (and similarly anti-Stokes) satellite is being generated by Thomson scattering. The most likely explanation of the second Stokes satellite which is frequency downshifted by $2\omega_p$ is that it is generated by the incident 10.6 μm beam Thomson scattering off the second harmonic of the plasma wave.

As in the case of ruby laser Thomson scattering, the satellites were detected only under two frequency illumination and only when ω_p was close to $\Delta\omega$. As expected, a reasonable correlation was found between the ruby Thomson scattering and the Stokes radiation. This is shown in Fig. 6. Using a Ge:Cu cold detector with a 400 MHz oscilloscope, which have a combined risetime of 600 ps, the pulse shapes of the Stokes and the anti-Stokes were resolved. When the oscilloscope risetime is deconvoluted, the satellites were found to have typical times to saturation of 360-530 ps. This is in excellent agreement with the predictions of the fluid theory.

Competing instabilites: Apart from the SBS and SRS backscatter instabilities a new and rather unexpected effect was discovered via 7.5° ruby Thomson scattering. Recall that 7.5° Thomson scattering at $k_s = \underline{k_i} + \Delta\underline{k}$ was used to probe for plasma waves generated by the Raman instability and used as a density diagnostic. The $\underline{k}$ matching assures us that a plasma wave produced by Raman backscatter decay of the incident pump always produces a blue shifted scattered ruby signal. If the plasma wave generated by SRS, which travels in the same direction as the CO_2 pump, subsequently decays into another plasma wave travelling in the opposite direction and an ion acoustic wave (this is the so called "parametric-decay instability") the present Thomson scatter geometry can give a red-shifted plasma satellite signal in addition to the blue-shifted plasma satellite. However, even when Raman scatter was large as evidenced by a large blue-shifted satellite, no red-shifted signal was ever detected. When the plasma was irradiated with two frequencies such that $\Delta\omega = \omega_p$, the scattered light showed both red- and blue-shifted satellites displaced by $\pm\omega_p$ from the SBS signal. The presence of the red and

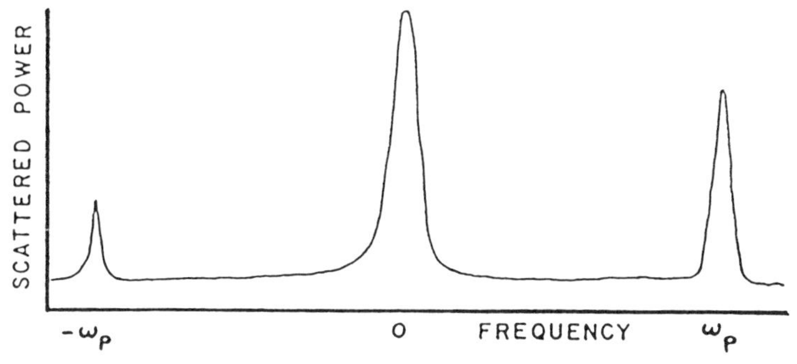

Fig. 7 Result of 7.5° Thomson scattering at $\omega_p = \Delta\omega$ and under two frequency illumination of plasma.

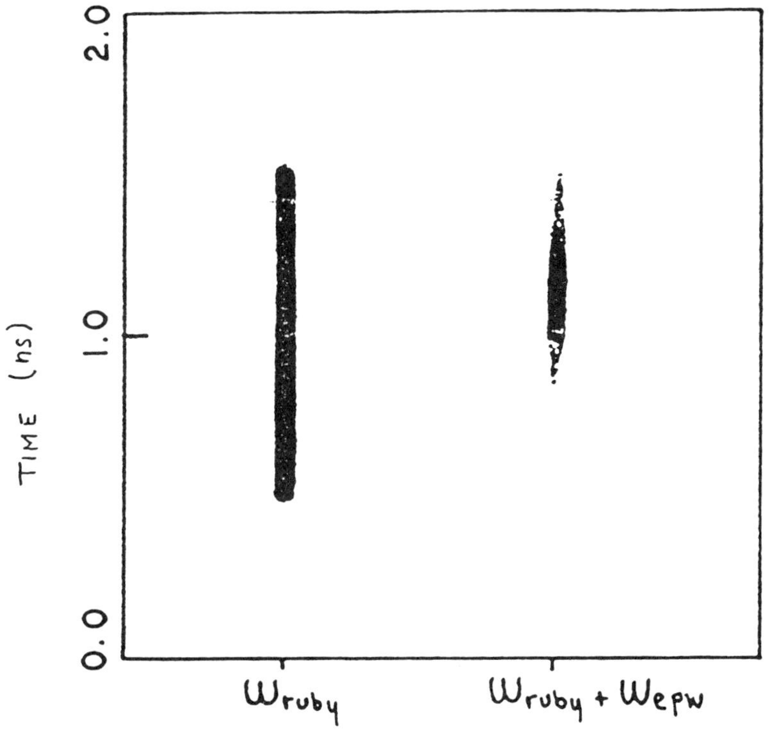

Fig. 8 Time resolved Thomson scattered light measurement at 7.5° and $\Delta\omega = \omega_p$ showing temporal coincidence of SBS and the plasma wave.

the blue satellites in Fig. 7 is indicative of two plasma waves propagating opposite one another with frequency $\sim\omega_p$ and $k \simeq 2k_o$. The SBS signal and the blue-shifted satellites were time resolved using a streak camera and were found to occur simultanously as shown in Fig. 8. Moreover, whereas single frequency Raman occurs over a range of ω_p, the two satellites were a resonant effect in that they were only displaced $\pm\omega_p = \Delta\omega$. Another qualitative feature of these satellites was that they seemed to be correlated with SBS. Large SBS produced large satellites.

There are two possible explanations for this rather curious effect. The first and the most likely is the so called induced counter-propagation optical mixing due to SBS. When the plasma is irradiated with two laser lines, SBS scatters a fraction of the energy in each line in the backward direction. The 9.56 μm incident line and the SBS induced 10.6 μm line beat and excite a slow phase velocity plasma wave propagating in the same direcion as the incident 9.56 μm line. Similarly, the incident 10.6 μm line beats with the SBS 9.56 μm line to excite a second slow plasma wave which travels in the opposite direction of the first.

The second and perhaps more intriguing possibility is the so called quasi-resonant mode coupling[8]. Simulations show that short wavelength ion fluctuations such as those produced by SBS can couple long wavelength, high phase velocity plasma oscillations (such as those produced by colinear optical mixing) to short wavelength, low phase velocity plasma modes which can subsequently be Landau damped.

Further experiments are planned to sort out which of these two (or some other) processes is occuring in the plasma. In any case, SBS plays a crucial role in these experiments, producing very rich wave-wave interaction physics. In any eventual plasma accelerator these competing effects are probably deleterious and must be avoided by going to shorter laser pulses.

CONCLUSIONS

The observation of the fast wave with the correct frequency, the expected wavenumber and hence, phase velocity, and the rather good agreement of the experimentally observed saturation time and the saturation amplitude with the predicted values from the fluid theory give us confidence that colinear optical mixing in a resonant plasma is a strong effect[9].

In spite of many competing effects, the desired phenomena, the beat wave, can be unambiguously identified and studied. The laser intensities used in the current experiment are large enough to produce a plasma wave of significant amplitude and, therefore, large longitudinal electric field. Although the beat-wave physics can be studied with nanosecond long laser pulses, much shorter pulses on the order of 50 ps are probably necessary to eliminate the competing instabilities. The current experiments have shown that this collective acceleration scheme is very rich in wave-wave interaction physics and future experiments will no doubt prove that it is equally rich in wave-particle interaction physics.

ACKNOWLEDGMENTS

We would like to thank F. F. Chen, J. M. Dawson, T. Katsouleas, W. Mori, D. W. Forslund and J. M. Kindel for many useful discussions regarding this work.

This work was supported by DOE contract DE-AT03-83ER40120, NSF grant ECS 83-10972 and LLNL University Research Program.

REFERENCES

1. C. Joshi et al., Nature 311, 525 (1984).
2. T. Tajima and J. M. Dawson, Phys. Rev. Lett. 43, 267 (1979).
   T. Katsouleas and J. M. Dawson, Phys. Rev. Lett. 51, 392 (1983).
3. D. W. Forslund et al., Phys. Rev. Lett. 54, 558 (1984).
4. N. N. Rosenbluth and C. S. Liu, Phys. Rev. Lett. 29, 701 (1972).
5. C. M. Tang et al., Appl. Phys. Lett. 45, 375 (1984).
6. C. E. Clayton et al., to be published.
7. C. E. Clayton et al., Phys. Rev. Lett. 51, 1656 (1983).
8. P. K. Kaw et al., Phys. Fluids 16, 1967 (1973).
9. C. E. Clayton et al., submitted to Phys. Rev. Letters.

ELECTROMAGNETIC EFFECTS IN RELATIVISTIC ELECTRON BEAM PLASMA INTERACTIONS

W. L. Kruer and A. B. Langdon
Physics Department
Lawrence Livermore National Laboratory
Livermore, CA 94550

ABSTRACT

Electromagnetic effects excited by intense relativistic electron beams in plasmas are investigated using a two-dimensional particle code. The simulations with dense beams show large magnetic fields excited by the Weibel instability as well as sizeable electromagnetic radiation over a significant range of frequencies. The possible relevance of beam plasma instabilities to the laser acceleration of particles is briefly discussed.

INTRODUCTION

Beam-plasma instabilities are of interest in many applications. These include plasma heating, space and astrophysical plasmas, the generation of radiation, and perhaps even some laser particle accelerators. There's been a great deal[1] of previous theory and computer simulation of beam-plasma instabilities, primarily motivated by applications to plasma heating. Our work is motivated by some recent experiments[2] by Kato, Benford and Tzach. In these experiments, the electromagnetic radiation due to a relativistic electron beam interacting with a plasma was measured. When the beam density was comparable to the plasma density, it was found that a significant fraction of the beam energy was converted to microwaves (of order 0.2%). These waves had a broad range of frequencies, extending to many times the electron plasma frequency ω_{pe}, rather than simply having frequencies of ω_{pe} and its harmonics.

To investigate the beam-plasma interaction, we have used a two-dimensional electromagnetic, relativistic particle code.[3] The beam travels in the x-direction, and the evolution of E_x, E_y and B_z is followed in the x-y plane. We started with the simplest problem-following the temporal evolution of an electron beam in a uniform plasma with no guide magnetic field. The boundary conditions were doubly-periodic, and the simulation box was rectangular with $L_x = 10.9\ c/\omega_{pe}$ and $L_y = 50.2\ c/\omega_{pe}$, where c is the velocity of light and ω_{pe} is the electron plasma frequency. The results changed by less than about 20% when both L_x and L_y were halved. The ions were

*Work performed under the auspices of the U. S. Department of Energy by the Lawrence Livermore National Laboratory under contract number W-7405-ENG-48.

usually treated as a fixed neutralizing background, and 800,000 electrons were used. Initially the beam had a kinetic energy of approximately 0.5 Mev (i.e., $\gamma = 2$) with a very narrow thermal spread of 50 eV. The background electrons had a temperature of 2.5 keV and were initialized with a drift in order to balance the initial beam current. The beam density (n_b) ranged from 0.05 to 0.5 of the total electron density (n_t).

The beam-plasma interaction in these simulations is dominated by two instabilities. The first is the well-known electron-electron beam instability, in which electrostatic waves are driven unstable. For a beam with a narrow spread in energy and angle and for $n_b/n_p\gamma_b < 1/20$, the results for the frequency and growth rate of the most unstable mode are

$$\frac{\omega}{\omega_{pe}} = 1 - 1/2\left(\frac{n_b}{2n_p\gamma_b}\right)^{1/3} f(\Theta)$$

$$\frac{\gamma}{\omega_{pe}} = \sqrt{3}/2\left(\frac{n_b}{2n_p\gamma_b}\right)^{1/3} f(\Theta), \qquad (1)$$

$$\text{where} \quad f(\Theta) = \left(\sin^2\Theta + \frac{\cos^2\Theta}{\gamma_b^2}\right)^{1/3} .$$

Here $n_b(\gamma_b)$ is the density (gamma) of the beam, n_p the plasma density, and v_b is the beam velocity. In addition, Θ is the angle between the wave vector $\underline{k}$ and $\underline{v}_b$, and $\underline{k} \cdot \underline{v}_b \simeq \omega_{pe}$. The other instability corresponds to a purely growing, spatially varying magnetic field transverse to the beam. The maximum growth rate occurs for $\Theta = 90^0$ and is

$$\frac{\gamma}{\omega_{pe}} = \left(\frac{n_b}{\gamma_b n_p}\right)^{1/2} \frac{kv_b}{(k^2c^2 + \omega_{pe}^2)^{1/2}} , \qquad (2)$$

This Weibel instability occurs since parallel currents attract. Hence beam electrons tend to bunch together or coalesce into filaments, as do also the background electrons carrying the return current. Of course, this instability also produces an electrostatic field which expells background electrons from a filament in order to preserve quasi-neutralitiy.

SIMULATION RESULTS

To illustrate the simulation results, let's consider a robust example in which $n_b = n_p$. The electrostatic field energy (EL) versus time is shown in Fig. 1.

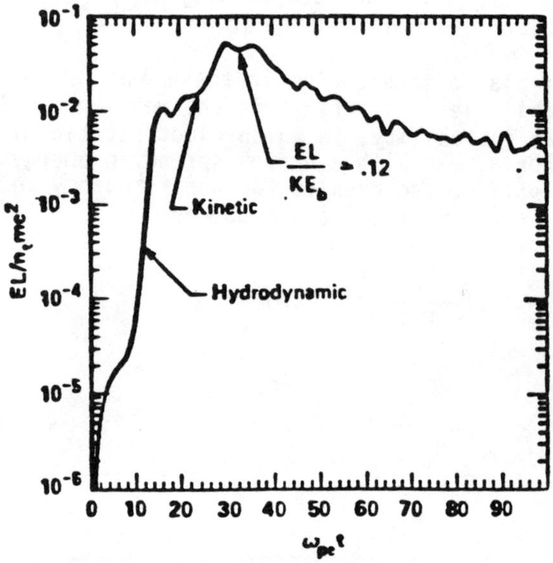

Fig. 1. Evolution of the electrostatic field energy in a 2-D simulation ($\gamma_b = 2$, $n_b = n_p$).

Note that early in time there is very rapid growth ($\gamma \simeq 1/2\ \omega_{pe}$), followed by a period of slower growth as the beam nonlinearly develops a significant spread in energy and angles. Finally, the wave energy saturates at about 12% of the initial kinetic energy of the beam as the velocity distribution becomes monotonic. The wave energy thereafter slowly decays.

The magnetic field energy ($B_z^2/8\pi$) versus time is shown in Fig. 2. The magnetic field due to the Weibel instability initially grows very rapidly ($\gamma \simeq 1/2\ \omega_{pe}$), and then grows more slowly as the beam spreads in energy and angles. Finally the magnetic field saturates at a quite large value which corresponds to $\omega_{ce}/\omega_{pe} \simeq 0.5$, where $\omega_{ce} = eB/mc$ is the electron cyclotron frequency for a nonrelativistic particle.

As the unstable waves grow and saturate, electromagnetic waves are generated as shown in Fig. 3. In this example, the transverse electric field energy (EM) reaches about 2% of the kinetic energy of the beam (KE_b). These electromagnetic waves have a significant range of frequencies, rather than just a

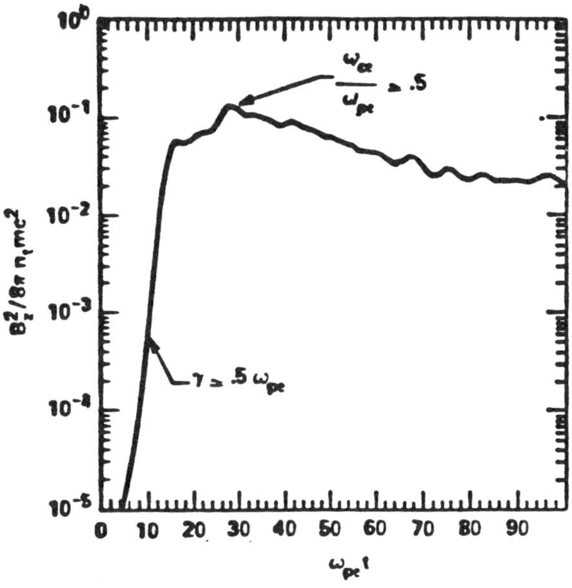

Fig. 2. Evolution of the magnetic field energy in a 2-D simulation.

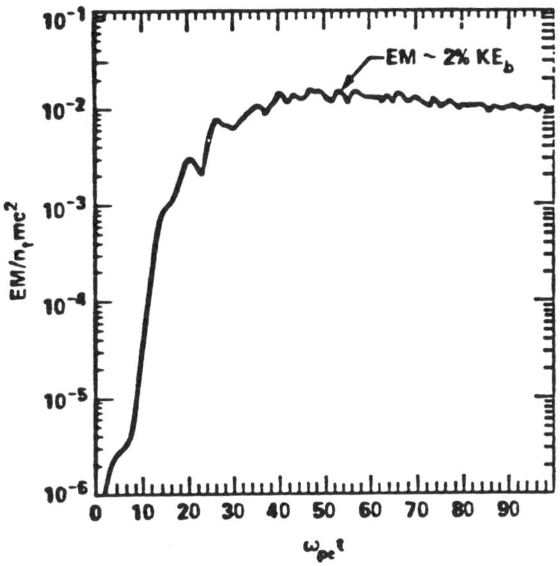

Fig. 3. Evolution of the transverse electric field energy (EM) in a 2-D simulation.

frequency of ω_{pe} and its harmonics. Figure 4 shows the Fourier transform of B_z at a given location in the plasma.

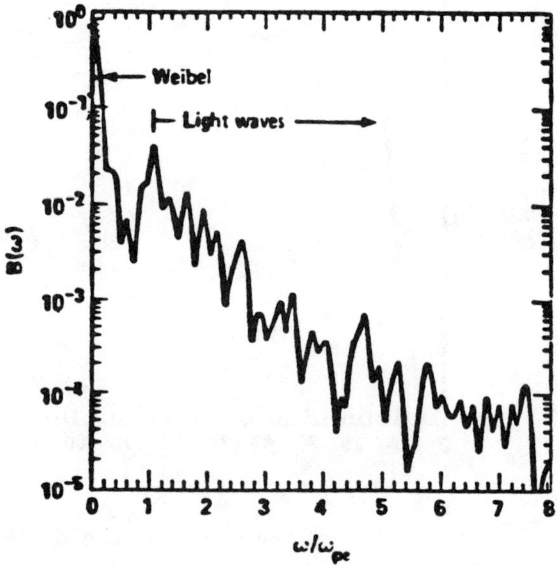

Fig. 4. Fourier transform of the magnetic field at a specific location in the plasma (γ_b = 2, $n_b = n_b$)

Note the large peak at $\omega = 0$, which is due to the Weibel instability, followed by a signal extending from ω_{pe} to 3-4 ω_{pe}, which is due to light waves. A broad range of frequencies is generated, a feature noted in the experiments. However, the amplitude of the light waves falls off more rapidly with frequency than experimentally observed.

Weaker beam densities have also been simulated. For a γ_b = 2 beam with $n_b/n_t = 0.3$, the peak magnetic field is $\omega_{ce}/\omega_{pe} \simeq 0.17$, and about 0.3% of the kinetic energy of the beam is converted to transverse electric fields. For $n_b/n_t = 0.05$, $\omega_{ce}/\omega_{pe} \simeq 0.017$ and the fraction into transverse electric fields is only about 0.06%. In all these cases, the beam loses about 40% of its energy to the plasma. The beam ends up with a spread in velocity comparable to its initial velocity as well as a broad range of angles.

Although we obtain many of the qualitative features observed in experiments, our simulations are very idealized. The transient temporal problem does not do justice to effects such as those associated with ion motion. For example, note in Fig. 1 that the electrostatic field rises and decays in a time of $\simeq$ 100 ω_{pe}^{-1}, which is too quick for the ions to respond. A spatial problem in which the beam is continually injected would be more realistic and would also admit boundary conditions closer to those in experiments. Allowance for spatial variations in the background plasma density and beam quality may also be important for details of the radiation generation and the power spectrum.

Finally our simulations indicate that beam plasma interactions (electrostatic and electromagnetic) might play a role in schemes such as the beat wave accelerator.[4] The effect of these instabilities will depend on factors such as the beam density, energy, and spatial width and structure. In experiments currently planned at UCLA,[5] it is estimated that $n_b/n_p \simeq 10^{-2}$ and $\gamma_b \simeq 20$. The growth rate of the electron-electron beam instability is then $\simeq 0.05\,\omega_{pe}$, and that of the Weibel instability is $\simeq 0.02\,\omega_{pe}$. For future applications, the beam density (and γ_b) will be significantly larger. However, it is not clear that these instabilities will have a deleterious effect. The finite radial extent of the beam can be a stabilizing influence, limiting the wave vectors of the unstable waves. In addition, the finite extent of the beam bunch may allow the beam to "outrun" the unstable fluctuations,[6] particularly since they have wave vectors at significant angles to the beam. Improved simulations can help clarify potential limitations as well as guide experiments.

ACKNOWLEDGMENTS

We are grateful for interesting discussions with J. Dawson, J. DeGroot, J. Denavit, K. Estabrook, C. Joshi, B. Lasinski, D. Meeker and R.Ziolkowski.

REFERENCES

1. B. Godrey, W. Shanahan and L. Thode, Phys. Fluids 18, 346 (1975); L. Thode, Phys Fluids 19, 305 and 316 (1976); and many references therein.
2. K.G. Kato, G. Benford and D. Tzach, Phys. Fluids 26, 3636 (1983); and references therein.
3. A. B. Langdon and B. F. Lasinski, in Methods in Computational Physics, edited by J. Killeen, B. Alder, S. Fernbach, and M. Rotenberg, Vol. 16, p. 327 (Academic, New York, 1976).
4. C. Joshi, W. Mori, T. Katsouleas, J. Dawson, J. Kindel and D. Forslund, Nature 311, 525 (1984); D. W. Forslund, J. M. Kindel, W. B. Mori, C. Joshi and J. M. Dawson, Phys. Rev. Letters 54, 558 (1985); and references therein.
5. C. Joshi, private communication.
6. J. Dawson, Private communication.

BEAT-WAVE ACCELERATOR STUDIES AT THE RUTHERFORD APPLETON LABORATORY

J D Lawson
Rutherford Appleton Laboratory, Chilton, Didcot, Oxon OX11 0QX, UK

ABSTRACT

The study carried out in 1982-83 at the Rutherford Appleton Laboratory to examine how one might use the beat-wave principle to construct a useful high energy accelerator is reviewed, and comments are made on later developments. A number of problems are evident to which solutions cannot at present be foreseen.

1 INTRODUCTION

Towards the end of 1982, following the ECFA-RAL meeting 'The Challenge of Ultra-High Energies' held in Oxford in October[1] it was decided to form a part-time study group based at the Rutherford Appleton Laboratory (RAL) for the purpose of further studying the beat-wave accelerator concept of Tajima and Dawson. The results were presented as a laboratory report in June 1983[2]. In the present report a summary is made of the findings, with comments in the light of more recent developments. Further background information on the study itself, with names of participants, are given in the acknowledgements at the end of this paper.

The idea of the beat-wave accelerator (BWA) was first described by Tajima and Dawson[3], and further papers had been given both at the Los Alamos meeting in February 1982[4] and at Oxford[1]. A study by Ruth and Chao[5] (published in ref 4 but not presented at the meeting) tackled the problem of finding a set of consistent parameters for a 5 TeV machine, based on a simplified linearized model for creating the beat-wave. The aim of the RAL study was to look in more detail at the Ruth-Chao design, and examine problems such as plasma formation, staging, gas scattering and beam focusing which had not yet been studied.

In the next section the assumptions made and parameters chosen in the Ruth-Chao study are outlined. Then follows a description of the RAL Study, and comments in the light of subsequent developments which suggest that better choices could have been made for some of the parameters. Difficulties were found that gave rise to problems for which a solution could not be forseen. Finally, some comments on the present outlook are presented.

2 THE RUTH-CHAO MODEL

There are two basic relations for the BWA, between the

accelerating field strength and the beat-wave density, and between the phase velocity of the wave and the ratio of plasma to laser frequencies. In terms of the fundamental constants and the laser and plasma frequencies ω and ω_p , the accelerating field and phase velocity of the wave are

$$E_z = \alpha m_o c\omega_p/e \tag{1}$$

$$\beta_z = \left[1 - (\omega_p/\omega)^2\right]^{\frac{1}{2}} \simeq 1 - \omega_p^2/2\omega^2 \tag{2}$$

where α is a constant of order but less than unity. It is immediately seen that large ω_p favours a high accelerating field but implies that for a relativistic particle there will be appreciable phase-slip between particle and wave, giving rise to the need for staging in very high energy machines. In addition to phase slip, laser power depletion sets a limit, which turns out to be of the same order of magnitude.

Ruth and Chao assumed Gaussian optics, with beam profiles in the 'under-dense' plasma the same as in vacuum. Although arguments for the validity of this assumption can be made, it may be that self-focusing is, in fact, a significant effect. This was, however, assumed not to occur in either the Ruth-Chao or the RAL studies. If Gaussian optics is assumed, then a short stage length allows a narrow beam, and therefore less laser power is needed to produce a given E_z . On the other hand more stages are required, implying the need for more lasers to produce the beat-waves. For stages of length limited by phase-slip or energy depletion it is found that these factors balance.

Using a simple linearized model for the build-up of the beat-wave, it is found that the laser energy required to produce a wave of given amplitude does not depend on the laser pulse length; in a real situation a very short pulse is desirable, to help combat essential non-linearities and avoid trouble from competing processes.

One of the interesting features to emerge from the Ruth-Chao analysis was that so many parameters are functions just of ω_p and ω/ω_p. This quantity can also be designated as γ_p, corresponding to the normalized total energy of a particle moving with the same velocity as the phase velocity of the wave. Some of these dependences are shown in Table 1; the symbol $\simeq$ denotes that quantities of order unity that depend on detailed assumptions are omitted. Basic assumptions are that the optics is Gaussian, particle energies are very high ($\gamma \gg \gamma_p$) and that stage lengths are limited by phase-slip.

A list of parameters proposed by Ruth and Chao for a 5 TeV machine is given in Table 2. Figures in the RAL studies were the same, except where shown in brackets. The value of α in equation (1), which can be shown to be equivalent to the square of

Table I Parameter Dependences in Ruth-Chao Analysis

Accelerating field	$E_z \simeq m_0 c\omega_p/e$
Stage length, depletion length	$L \simeq c\gamma_p^2/\omega_p$
Waist area	$\sigma_0^2 \simeq c^2\gamma_p/\omega_p^2$
Energy in laser pulse	$W\tau \simeq (m_0^2 c^5/e^2)\gamma_p^3/\omega_p$
Beat wavelength/waist radius	$\lambda_p/\sigma_0 \simeq \gamma_p^{-\frac{1}{2}}$

the ratio of the transverse oscillatory velocity of the plasma electrons to that of light, was taken as 0.5. This was derived as the criterion that electrons in the originally cold plasma should not be trapped in the wave. Injection of already relativistic particles into the accelerator was assumed.

Table II	Parameters for 5 TeV accelerator
Accelerating field	5 GeV/m
Stage length	10m (5m)
No of stages	100 (200)
Laser wavelength	1.06μ
Beat wavelength	260μ
Waist radius	1.3mm
Injection energy	>10 GeV
Laser energy per stage	17 KJ(8.5KJ)
Particles per pulse	$< 5\text{x}10^{10}$
Pulse length	140 psec (100 psec)

3 TOPICS CONSIDERED IN RAL STUDY

In the following sections brief accounts are given of the various topics considered in the RAL study, and the conclusions reached. Parameters different from those of Ruth and Chao are shown in brackets in Table II. Comments are made on these conclusions in the light of later developments.

a) Required Accelerator Parameters

It was assumed that 'conventional' machines, implying linac colliders for electron and synchrotron based storage rings for protons, might reach 0.4 and 20 TeV respectively, and that the beat-wave accelerator should aim higher than this. To be specific, an energy goal of 5 TeV for electrons was chosen. More recently there is increased confidence that 100 MeV/metre can be obtained in conventional machines, so that even 5 TeV only implies 50 km/linac; this is long, but by no means inconceivable.

For luminosity, the values considered as possibly acceptable, namely less than 10^{30} cm^{-2} sec^{-1} for protons at 10 + 10 TeV and 10^{30} for electrons at 1 TeV would now be considered too low by several orders of magnitude.

b) The Plasma Column

The central feature of the accelerator is a plasma column, the density of which must remain constant during the build-up of the beat-wave, and be everywhere uniform so that resonant conditions are maintained. For reasons given in section 3f a stage length of 5m, rather than 10m used by Ruth and Chao, was chosen; this implies twice as many stages, with 8.5 kJ laser power per stage. The tolerances depend on the build-up time of the beat-wave, which was taken as 100 psec, representing 115 cycles of the beat frequency. This value is now thought to be too long, because of the problems of relativistic detuning and the build-up of competing processes; it was chosen because the laser technology becomes more difficult as the pulse is shortened. Even with a very short pulse it is necessary to retain uniformity, and the suggested solution was to start with cold hydrogen gas. At the power levels of interest complete ionization occurs at the front end of the laser pulse, which might typically be several millimetres in length in place of the 3cm assumed earlier.

The transverse dimensions of the plasma channel depend on those of the laser pulse; it was assumed that the gas would be ionized where the power density exceeds 3 x 10^{13} watts cm^{-2}. This occurs within a radius σ_o of order 1.5σ, where the transverse power distribution is assumed to vary as exp $(-r^2/2\sigma^2)$. The column width is not constant; the stage length is assumed to be twice the Rayleigh length, and this implies that the transverse area at the waist is half that at the ends.

One obvious problem in a practical system is that of maintaining the gas column at a uniform density, and a vacuum in the region between the column and the mirror systems. No window can stand the energy deposition that would occur.

c Laser and Associated Optics

No comment was made on the laser to be used, although $\lambda = 1.06$ microns was assumed for the wavelength; the standard Nd glass laser clearly will not have adequate repetition rate.

A simple paraboloid was chosen to form the Gaussian waist. Taking 10 joules/cm^2 as the tolerable power density on the mirror surface, this requires a 16cm radius mirror with focal length about 600 metres to produce the required long and narrow waist. This distance might perhaps be considerably shortened by using more sophisticated optics with grazing incidence[6].

The mechanical arrangement of the mirrors and plasma columns requires careful consideration. It seems difficult to make the stages collinear; if they are not, then magnets are needed to deflect the particle beams. The next two paragraphs are quoted from ref 2.

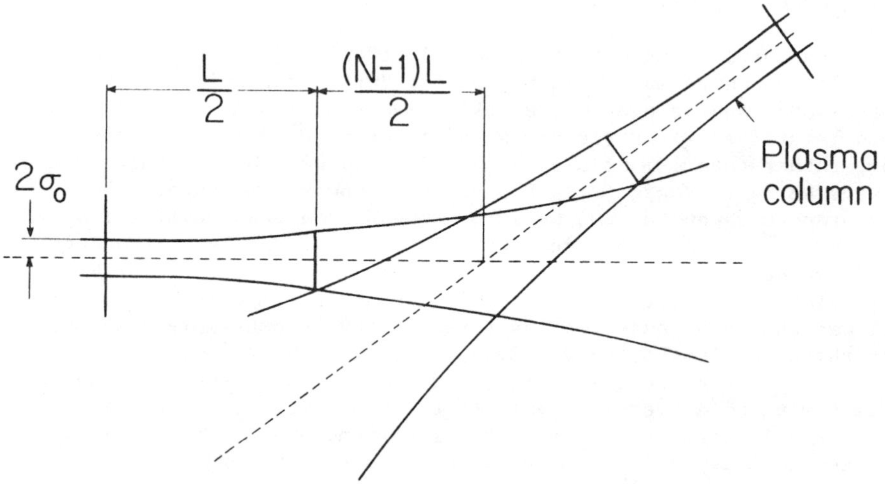

Fig 1 Beam configuration between stages

"The minimum requirement is that the mirrors should not intercept beams from other mirrors, and that only one beam should be present in each stage. Assuming mirrors with co-planar axes the deflection angle must certainly exceed the convergence angle $0.16/600 = 2.7\times10^{-4}$ radian. The ends of the stages must be separated to allow the insertion of a bending magnet system, and this further increases the minimum angle. The geometry is shown in Fig 1; it is assumed that the beams within a radius of 2σ must not intersect within the plasma region. As explained above, the stage length L=2R where R is the Rayleigh length. The beam radius σ at a distance NR from the centre of the plasma column is $2\sigma_0(1 + N^2)^{\frac{1}{2}}$. If, then, the spacing between the ends of the plasma columns is 2(N-1)R, the angle θ is $2\sigma_0/(N-1)R$. Substituting for R and σ

$$\theta = 4\sigma_0(1+N^2)^{\frac{1}{2}} / (N-1)L \tag{3}$$

If the spacing is equal to L = 2R, then N = 2 and $\theta = 4\sqrt{5}\sigma_0/L = 1.8 \times 10^{-3}$ radians. For larger values of N, θ decreases to a lower limit of $4\sigma_0/L$.

In a practical system, in the presence of a vessel to contain the gas, supports, focusing and beam bending magnets etc. it is unlikely that it will be possible to attain such a small angle."

Despite further discussions, no credible alternative layout has been devised. This is a topic that needs further consideration.

d) Bending Magnet Requirements

There are two requirements that limit the bending angle θ given in equation (3). In the first place it must be possible to design as achromatic bending system, and second, energy loss arising from synchrotron radiation must not be large. It is evident from the form of equation (3) that with the layout assumed θ cannot be decreased indefinitely by increasing the spacing between stages.

It is not possible to consider the design of the bending magnets before the beam quality, described by the emittance and $\Delta p/p$, have been specified. These questions are discussed in section 3e below. Nevertheless some simple calculations to indicate orders of magnitude are possible.

First, the fractional energy loss from synchrotron radiation in a magnet of length 5 metres is

$$\Delta\gamma/\gamma = 1.27 \times 10^{-6} B^2 WS \quad , \quad (T, GeV, m) \qquad (4)$$

Not only does the synchrotron radiation introduce loss, but there is also an energy spread of order $\Delta\gamma/\gamma$. Since this is cumulative, the permitted value of $\Delta\gamma/\gamma$ is clearly very small, especially when there are many stages.

The minimum permissible magnet length may be found from the total angle of bend

$$\theta = 0.3BS/W \qquad (5)$$

From equations (3), (4) and (5), eliminating B and θ, (set at the minimum value of $4\sigma_0/L$), we find that

$$S = 2.2 \times 10^{-4} W^3 \sigma_0^2 / L^2 (\Delta\gamma/\gamma) \qquad (6)$$

Setting σ_0=1.3mm,L=5m, $\Delta\gamma/\gamma \simeq 10^{-3}$ yields S=15m at 1 TeV. At 5 TeV this is increased by at least a factor 125, probably more since $\Delta\gamma/\gamma$ would need to be less.

This illustrative calculation demonstrates the difficulty of the staging problem. A different approach is needed.

e) Beam Quality

The required beam quality can be determined from a knowledge of the luminosity required and the power that can be afforded. These considerations have been outlined by Richter[7] and Wilson[8]. In the RAL study this topic was not adequately treated, and an emittance of 10^{-5}/ m-rad, compatible with expected performance of a SLAC type injector was assumed. The energy spread was taken as 10%, a value probably too large to allow satisfactory focusing in the collision region.

Whilst the emittance is determined by the injector, the energy spread depends on the variation of phase experienced by the particles during acceleration. Particles injected at different phases of the beat-wave acquire different energies. Phases π to 2π are decelerating, and, as shown in the next section, 0 to $\pi/2$ are radially defocusing. Only the range $\pi/2$ to π is therefore usable, as shown in Fig 2.

For a short bunch initially extending from $\phi = \pi/2$ to $\phi = \pi/2+\phi_b$, the energy spread when the bunch has slipped through an angle ϕ_s gives rise to an energy spread normalized to the mean energy gain per stage

$$\Delta\gamma/\gamma_s = \phi_b \cos(\pi/2 + \phi_s) \;/\; \left[1 - \sin(\pi/2 + \phi_s)\right] \quad (7)$$

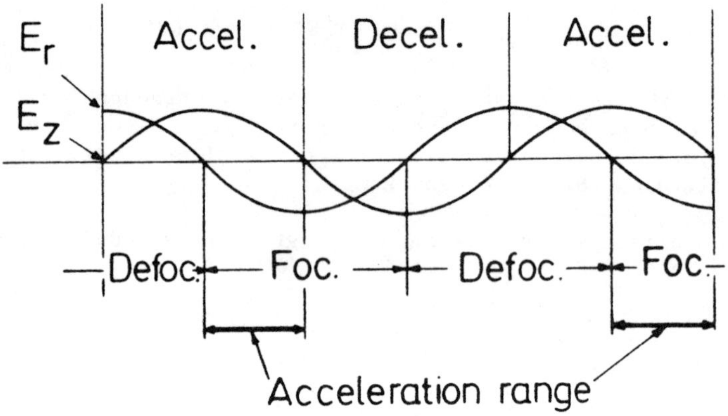

Fig 2 Accelerating and focusing fields, illustrating the phase range available for acceleration.

For ϕ = $13\pi/16$ assumed in ref 2 this is about $\Delta\gamma/\gamma_s \simeq \phi_s/2$. Thus, a 6° phase spread, corresponding to a bunch to space ratio of only 3%, corresponds to an energy spread of 5%. This is likely to be unacceptably large in the light of the requirements for final focusing.

No consideration was given in the RAL study of how these narrow bunches might be achieved.

f) Focusing of the Particle Beam in the Plasma Channel

The assumptions concerning the form of the plasma column were discussed in section 3b. The column width was determined from the breakdown strength in the gas, and was of order 2mm. Outside the column E_Z is zero, so that across the column $\partial E_Z/\partial r$, and hence $\partial E_r/\partial z$, is finite. Assuming a quadratic dependence of characteristic length r_o near the axis, it is readily shown that

$$|E_r| = \frac{\lambda r}{\pi r_o^2} |E_{zo}| \qquad (8)$$

This represents a strong field that is focusing or defocusing according to the phase, as shown in Fig 2. Only phase angles between $\pi/2$ and π can be used for acceleration. It is straightforward to calculate the focusing strength, and hence the betatron wavelength, Λ, from equation (8). For the parameters assumed in the RAL study $\Lambda|\cos\phi|$ varies from about 2-25 metres as γ increases from 35 GeV at the end of the first stage to 5 TeV. At the assumed emittance the beam diameter would be much less than that of the plasma channel.

No detailed consideration was given to the problem of achromatic focusing between stages. Rough calculations indicated that even without allowing for synchrotron radiation it would be extremely difficult.

g) Multiple Scattering of Accelerated Beam by Plasma

The effects of multiple scattering were briefly discussed in ref 2, and, although considered worthy of investigation no detailed calculations were done. It has subsequently been shown by Montague[9] that the deterioration in beam quality is negligible.

h) Beam Intensity, Pulse Length, and Repetition Rate

All these factors are important in assessing the potential of any accelerating scheme. The number of particles that can be accelerated, provided that they can be produced at the required density initially, depends on the beam loading, and the fundamental limits here are expected to be as in any other accelerator. We might hopefully assume 5% transfer of energy from laser light to accelerated particles.

The pulse length depends on the time for which the beat-wave remains coherent after the laser pulse has passed. Since energy spread is to be avoided, the amplitude must not 'droop'. Not enough is yet known about the beat-wave process to give other than hopeful guesses.

The repetition rate depends on developments in laser technology. The limitation here is probably an economic one.

i) Luminosity

With so many unknowns, it is hardly profitable at present to estimate the luminosity that might be achievable. Values consistent with those quoted by Richter[7] and Rubbia[10] are not within sight. This question was considered in ref 2. The results need re-assessment and will not be discussed here.

One point emerged, however, that is worthy of note. Owing to the fact that the bunch is split into microbunches spaced by the beat wavelength, the beamstrahlung effect is worse by a factor equal to the ratio of the pulse spacing to pulse width. The disruption parameter, on the other hand, is unaffected. This arises because of the respective squared and linear dependences of these effects on the magnetic field strength.

4 CONCLUSIONS

Not enough is known about the detailed physics of the beat wave process, nor about how one might overcome the difficulties uncovered in the RAL studies, to make any meaningful assessment of what could be achieved in practice. Some of the assumptions made in the study, such that the laser beam shape is determined by free-space Gaussian optics, may not be correct. If self-focusing is important, for example[11], much smaller laser powers might be required. There may be approaches, perhaps related to the 'surfatron', which alter the constraints arising from phase slip[12]. While there may be scope for improvement in some respects, many other factors which will cause problems have not been investigated.

5 ACKNOWLEDGEMENTS

The Rutherford Appleton Study was started in October 1982, encouraged by the European Committee for Future Accelerators (EFCA) through its chairman (Dr J H Mulvey) and by Professor A Salam. This was a part-time activity, and an attempt was made to bring together participants in the fields of High Energy Physics, Particle Accelerators, Lasers, and Plasma Physics. These are listed below in alphabetical order.

Participants	Field of Interest
J E Allen*	Plasma Physics
R Bingham	Plasma Physics
J Butterworth	Particle Beam Transport
F E Close	High Energy Physics
R G Evans	Plasma Physics and Lasers
J D Lawson	Accelerators
G H Rees	Accelerators
R D Ruth+	Accelerators

* University of Oxford

\+ Stanford Linear Accelerator Centre. At CERN during period of study group.

The Study Group as such did not continue after the publication of ref 2, but some individual members continued with basic studies of beat-wave physics. This work has led to several publications [13-15], and an experimental study in conjunction with Imperial College, London, has been approved [16].

6 REFERENCES

1 'The Challenge of Ultra-High Energies'. Proc. ECFA-RAL meeting Oxford, September 1982. ECFA 83/68. Published by Rutherford Appleton Laboratory.

2 J D Lawson 'Beat-Wave Laser Accelerators; First Report of the RAL Study Group'. RAL report RL-83-057, (1983).

3 T Tajima and J M Dawson, Phys. Rev. Lett. 43, 267 (1979).

4 P J Channell (ed) 'Laser Acceleration of Particles, Los Alamos, (1982)', AIP Conference Proceedings No 91, New York (1982).

5 R D Ruth and A W Chao, ref. 4. p. 94.

6 A Saxman, private communication at UCLA Workshop.

7 B Richter, these proceedings.

8 P Wilson, these proceedings.

9 B W Montague, these proceedings.

10 C Rubbia, presented at the Frascati Workshop on the Generation of High Fields for Particle Acceleration, September 1984.

11 J D Lawson, 'Beat-Wave Laser Accelerators, Further Comment Including Note on the 'Surfatron' Concept'. RAL report RAL-84-059, (1984).

12 T Katsouleas and J M Dawson, Phys. Rev. Lett. 51, 392 (1983).

13 R Bingham, 'Possible Instabilities in the Beat Wave Accelerator', RAL report RAL-83-058 (1983). Submitted to 'Plasma Physics'.

14 R G Evans, 'Relativistic Saturation of Beat Waves', RAL report RAL-84-086 (1984).

15 R Bingham, R A Cairns and R G Evans, 'Saturation of Plasma Beat Waves', RAL report RAL-84-122 (1984). Submitted to Phys. Rev. Lett.

16 A E Dangor, A Dymoke-Bradshaw, R Bingham, R G Evans, C B Edwards, W T Toner. These proceedings.

THE RUTHERFORD LABORATORY BEAT WAVE EXPERIMENT

A E Dangor
A Dymoke Bradshaw
Imperial College London

R Bingham
R G Evans
C B Edwards
W T Toner
Rutherford Appleton Laboratory

Following the excitement raised by the UCLA beat wave proposal[1] the Science and Engineering Research Council has funded a complementary experiment to be carried out at the Central Laser Facility. The proposal arose from the work of the Rutherford beat wave study group[2] and the interest of UK universities.

Whereas the UCLA experiment plans to use different rotational transitions of the CO2 laser to provide the different laser frequencies, the RAL experiment is based on an existing high power Nd glass laser named VULCAN. The frequency separation will be about 1% of the laser frequency so that the phase velocity of the beat wave will have a Lorentz factor $\gamma = 100$ compared with 10 in the UCLA experiment. In this sense the two experiments are largely complementary since they will provide data at different laser wavelengths, different light intensities, different pulse durations and different beat wave phase velocities.

BASIC PARAMETERS

The laser frequencies are constrained by the availability of suitable oscillator materials within the gain bandwidth of the Nd doped glass amplifiers. The crystalline materials YLF and YAG provide suitable wavelengths of 1.053 μm and 1.064 μm respectively.

The resonant density of approximately 10^{17}cm^{-3} is obtained in a coaxial Z-pinch plasma built by the Imperial College group and diagnosed as being stable, quiescent and with a very uniform central region of about 5mm diameter. The laser beams will be incident transverse to the pinch axis with a Rayleigh length of about 3mm in the focussed waist of the beams.

The laser pulse duration will be as short as is compatible with driving the beat wave to a "large" amplitude which will mean a pulse duration of about 300ps and a laser energy of 100J at each wavelength.

With a 3 metre focussing lens the spot size will be limited by the irreducible divergence of the laser beams to about 300 μm, giving a power density of 3 x 10^{14}Wcm^{-2} at each wavelength.

The time to saturation of the beat wave is expected to be about 100ps, giving the opportunity to study growth, saturation and decay with 10ps resolution optical streak cameras. If the beat wave grows to $\delta n/n = 1$ then the electric field will be $3 \times 10^8 Vcm^{-1}$.

PROPOSED MEASUREMENTS

The proposed experiment does not include injection of charged particles into the acceleration region but proposes to infer the electric fields from optical measurements. As the beat wave amplitude grows, the laser energy is degraded to longer wavelengths by Raman scattering from the beat wave. The Raman wavelengths are shifted by the difference frequency of the pump lasers and are very easily resolved by means of spectrometers in the transmitted light path. The growth of the Raman "cascade" will be measured with 10ps time resolution and the ratio of energy in successive components can be used to infer the amplitude of the beat wave. The electric field in the wave follows from its amplitude $\delta n/n$ and its wave number k.

Thomson scattering of a probe laser beam from the beat wave will be an alternative measurement of the beat wave amplitude. The scattering angle for an $0.5\mu m$ probe beam is only 0.5 degree which makes the diagnostic method difficult but the density fluctuation is measured directly by the scattered intensity, and the wavenumber of the beat wave from the scattering angle. This too can be performed with 10psec resolution.

By introducing a seed gas such as Argon into the discharge it is hoped to measure the Stark broadening of Argon spectral lines. This is a direct measurement of electric field but requires accurate computation of the atomic Stark broadening parameters. The time resolution of this diagnostic may well be inferior to the other methods depending on the amount of light detected.

Thomson scattering at a large scattering angle (incoherent scattering) will give the rate of heating of the plasma during and after the laser pulse. This shows the transition of wave energy via turbulence into thermal energy and relates to the time for which the beat wave is usable for particle acceleration.

The emittance of an accelerated beam of particles would be affected by the presence of electric fields in the plasma other than the wanted accelerating field. The most likely unwanted modes are believed to be sideways Raman scattering at an angle of about 5 degrees. This produces scattered light components which can be detected with a spectrometer and streak camera and again the level of scattered light can be used to infer the electric field amplitude.

Filamentation and Brillouin scattering are believed to be less of a problem since they require ion motion and hence have a slower response. However filamentation will be investigated by imaging the transmitted light, and Brillouin backscatter will be detected using another spectrometer and streak camera. The growth of these unwanted modes will set an upper limit on the growth time of the beat wave and hence a lower limit on the laser power.

A schematic layout of the experiment is shown in Figure 1.

LASER DETAILS

The two oscillator rods, of YLF and YAG will be in series in the same optical cavity. Each will have its own flashlamp but they will be modelocked by a common acousto optic device. This will guarantee absolute synchronism of the two modelocked pulse trains and they will have a common switchout gate. Dispersion in the oscillator cavity may cause problems with pulse durations below 200 nsec.

In order to equalise the gain of the laser glass at the two wavelengths the rod amplifier train will be a mixture of phosphate (favouring 1.053 μm) and silicate (favouring 1.064 μm) glasses. The two frequencies see an identical optical system through to the plasma and hence should accumulate exactly the same optical distortions. Even the non-linear refractive index of the glass contributes equally to the two wavelengths since the pulses propagate together through the laser system.

PLASMA SOURCE

The coaxial Z-pinch provides a stable and reproducible plasma with the appropriate density and a lifetime of microseconds. The scale length of the plasma is believed to be very long in the central 5mm but even so it may not be sufficiently uniform for the beat wave resonance. It is planned to test a multiphoton ionised plasma source which simply requires that the Z-pinch discharge should not be fired. The focussed irradiance of 3×10^{14} Wcm^{-2} is sufficiently large that multiphoton ionisation of molecular hydrogen is almost instantaneous and the resulting plasma density is as uniform as the initial gas density. The energy required to ionise the gas is a tiny fraction of the laser energy and it is hoped that this will provide the solution to uniform plasma columns of arbitrary length.

DIAGNOSTIC CHANNELS

Each diagnostic channel will consist of a coupled spectrometer and optical streak camera. It is hoped that CERN will provide support to purchase a second scattering channel so that some of the different measurements can be performed simultaneously. The intrinsic time resolution of the streak cameras is about 2psec but the uncertainty principle limits the values of temporal and spectral resolution that can be achieved at the same time.

STATUS

The Z-pinch and detector channels have been used in previous laser experiments at Rutherford Laboratory. The laser oscillator should be delivered in Spring 1985. Design work for the laser modifications are underway and it is hoped to test all parts of the laser (but not the target plasma) in July 1985. Building work at the laser facility will mean that the first experiments cannot be performed before October/November 1985.

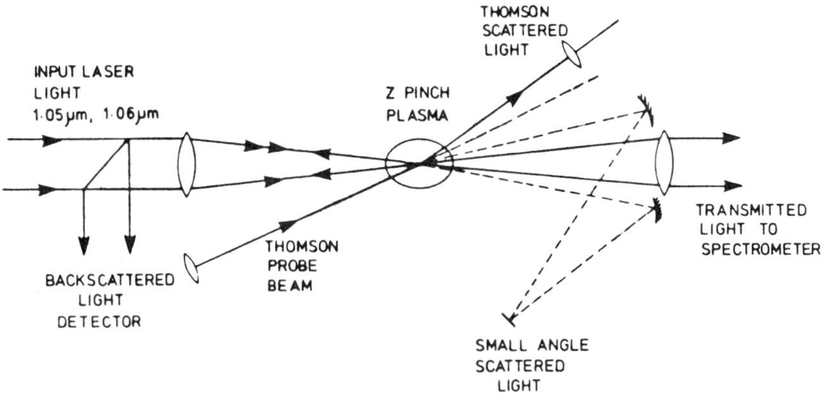

Figure 1 BEAT WAVE EXPERIMENT

REFERENCES

1. C. Joshi, Proceedings of the ECFA-RAL meeting, "The Challenge of Ultra-High Energies", Oxford, Sept. 1982.

2. J. D. Lawson, Rutherford Appleton Laboratory Report, RAL-83-057 (1983).

EFFICIENCY FACTORS IN THE BEAT WAVE ACCELERATOR

R G Evans
Rutherford Appleton Laboratory
Chilton OX11 0QZ Great Britain

ABSTRACT

The efficiency of energy transfer between the lasers and the beat wave, and the beat wave and the particle beam are examined. Considerable uncertainties are found in each case but they are amenable to further theoretical studies.

INTRODUCTION

The plasma beat wave accelerator as proposed by Tajima and Dawson[1] offers the possibility of producing accelerating fields of many GeV m^{-1}. Its ultimate usefulness as a practical particle accelerator will depend on the efficiency of the overall process including the laser efficiency, the coupling of laser energy to the beat wave, and the extraction of energy from the beat wave by the particle beam.

The fundamental parameters of the beat wave accelerator are: laser frequencies ω_1, ω_2, wavevectors k_1, k_2 plasma frequency $\omega_p = \omega_1 - \omega_2$, plasma wavevector $k_p = k_1 - k_2$, laser pump strengths $\alpha_i = \frac{eE}{m\omega c}$ The plasma wave is driven up to some fraction $\varepsilon = \delta n/n$ of the wavebreaking limit giving an electrostatic field

$$eE_p = m\,c\,\omega_p\,\varepsilon$$

The phase velocity of the beat wave is $v_{ph} = \omega_p / k_p$ corresponding to a Lorentz factor $\gamma = \omega_1 / \omega_p$. This phase velocity is generally less than the velocity of GeV electrons so that the length of one stage is limited by phase slip to $L = 2\gamma^2 c/\omega_p$ giving an energy gain per stage of $\Delta\gamma = 2\varepsilon\gamma^2$.

LASER TO BEAT WAVE COUPLING

The simplest estimate of the efficiency of energy transfer between the lasers and the plasma wave is given by Ruth and Chao[2]. The growth rate of the beat wave is $\frac{d\varepsilon}{dt} = \frac{\omega_p}{4}\alpha_1\alpha_2 = \frac{1}{4}\alpha^2\omega_p$ for equal intensity pumps. If the laser pump duration is τ then the final beat wave amplitude is $\varepsilon = \frac{1}{4}\alpha^2\tau\,\omega_p$. The total beat energy in stage length L is then $\mathcal{E}_{wave} = \varepsilon^2 mc^2 n_e.A.L$ where A is the cross-sectional area of the beat wave and the laser beams (assumed equal). The energy in the laser beams is

$$\mathcal{E}_{light} = 2.\frac{E^2}{8\pi}.c\,\tau\,A$$

giving an efficiency $\frac{\mathcal{E}_{wave}}{\mathcal{E}_{light}} = \varepsilon/_2$

If the laser frequencies can be "chirped" to avoid detuning the plasma resonance at large amplitude then at first sight it appears possible to couple up to 25% of the laser energy into the plasma before exceeding the cold plasma trapping limit $\varepsilon = \frac{1}{2}$.

If however the beat wave is viewed as composed of elementary scattering events

$$\text{photon } (\omega_1) \rightarrow \text{photon } (\omega_2) + \text{plasmon}(\omega_p)$$

then the Manley Rowe relations show that the energy transfer efficiency is ω_p/ω_1. A full treatment of the non-linear plasma equations shows that as the beat wave is generated there is a second scattering process

$$\text{photon } (\omega_2) \rightarrow \text{photon } (\omega_3 = \omega_2 - \omega_p) + \text{plasmon } (\omega_p)$$

giving rise to a new electromagnetic wave at ω_3. As the pump at ω_1 diminishes, ω_3 grows, and the process repeats giving rise to $\omega_4 = \omega_3 - \omega_p$ etc. This "Raman cascade" is limited in ultimate efficiency for the following reasons.

a) The phase velocities of the pumps at ω_2, ω_3, ω_4 are not equal giving rise to a dephasing of the pump waves.

b) If the pump waves ω_1 and ω_2 have a diffraction limited Rayleigh waist then because of the non-linearity of the generation mechanisms ω_3 etc will be spatially narrower. They will have a larger diffraction angle and will propagate out of the beat wave volume.

Both of these processes need to be analysed in detail because of the 'phase locking' inherent in all coupled wave interactions. Normally only the driven mode is phase locked but in the case of strong pump depletion the phase of the pump waves is also modified. Depending on the sign of the phase pulling this may give rise to self-focussing or defocussing. Also the phase of the beat wave will drift as the pumps cascade to lower frequencies and this will modify the effective phase velocity of the beat wave.

BEAT WAVE TO BEAM COUPLING

In a conventional RF cavity a beam of vanishingly small cross-section is able to extract energy from a large volume by virtue of its wake field partially cancelling the RF field in the cavity. In the same way the wake field of a small beam of particles in a plasma extends out to radii of at least c/ω_p by virtue of the fringe fields from the perturbed charge density in the plasma[3) 4)]. We thus obtain the extremely important result that an arbitrarily narrow beam is able to extract a large fraction of the beat wave energy over an area of order $(c/\omega_p)^2$

The energy per unit length in the beat wave is

$$\mathcal{E} = \varepsilon^2 n_e mc^2 A_{wave}$$

The energy gain per unit length by a beam containing N_b particles is $\Delta\mathcal{E} = eEN_b = \varepsilon\, mc\omega_p\, N_b$.

If the beam area is A_{beam} and occupies a length $\ell = \delta\theta\, c/\omega_p$ then the extraction efficiency is $\eta = \frac{n_b}{n_e}\frac{A_{beam}}{A_{wave}}\delta\theta\,\varepsilon^{-1}$

where n_b is the number density in the beam. If it is argued that $n_b \ll n_e$ to avoid instabilities such as the Weibel instability, and

$\delta\theta \ll 1$ to give a small energy spread then the extraction efficiency is obviously very poor. However we can have $n_b \gg n_e$ so that the plasma becomes a weak perturbation on the beam particles. The growth rate of the Weibel instability is then more usefully thought of in terms of the plasma density of the beam particles

$$\omega_{pb}^2 = \frac{4\pi n_b e^2}{m}$$

. In the rest frame of the beam the instability has a growth rate which is some fraction of ω_{pb}, but in the laboratory frame this is reduced by a factor of γ_b, which is 2×10^6 for a 1TeV beam.

In this case it is more useful to express the beam loading as the number of beam particles that extract all the beat wave energy from an area $A = \pi\ (c/\omega_p)^2$

$$N_b = \pi\varepsilon\, n_e\, (c/\omega_p)^3$$

We note that since $\omega_p \propto \sqrt{n_e}$, $N_b \propto n_e^{-\frac{1}{2}}$, ie lower densities accelerate more particles. or a typical density $n_e = 10^{17}\,cm^{-3}$ and $c/\omega_p = 2.10^{-3}$; $N_b = 2 \times 10^9\,\varepsilon$.

RAYLEIGH OPTICS

If the emittance requirements of the accelerated beam force the beam radius in the beat wave region to be much less than c/ω_p then the effective radius of the beat wave is about $2c/\omega_p$ and wave energy outside this region cannot be extracted by the particle beam. If the laser beams are focussed by diffraction limited optics to a Gaussian waist then the Rayleigh length $R = 2\pi\, a^2/\lambda$ where a is the waist radius and λ the laser wavelength. If we put $a = 2c/\omega_p$ then $R = \gamma c/\omega_p$, but the optimum stage length for energy coupling is $L = 2\gamma^2\, c/\omega_p$. If the Rayleigh length is matched to the depletion length then the beat wave is too large in cross-section for efficient extraction while if matched to the optimum beam waist the stage length is too short for efficient coupling of the laser energy to the beat wave. If $\gamma \gg 1$ then Rayleigh optics are inevitably limited to a maximum efficiency $< 1/\gamma$.

CONCLUSION

The efficiency of the beat wave accelerator is probably limited to 10%-20% in coupling laser energy to the beat wave, and a high efficiency of extraction of beat wave energy to particle beam energy would occur with 10^8 particles per bunch. The behaviour of wake fields and Weibel instabilities in this limit is unknown and needs to be calculated.

REFERENCES

1. T Tajima and J M Dawson, Phys Rev Lett 43, 267 (1979).
2. R D Ruth and A W Chao, "Laser Acceleration of Particles" Los Alamos 1982 AIP Conference Proceedings No 91 New York (1982).
3. P Chen, J M Dawson, R W Huff and T Katsouleas, SLAC report SLAC-PUB-3487, (1984) submitted to Phys Rev Lett.

4. R D Ruth, A W Chao, P L Morton and P B Wilson, Stanford University Report SLAC-PUB-3374 (1984) submitted to Particle Accelerators.

SOME NONLINEAR PROCESSES RELEVANT TO THE BEAT WAVE ACCELERATOR

R Bingham
Rutherford Appleton Laboratory, Chilton, Didcot, Oxon, UK

W B Mori and J M Dawson

University of California Los Angeles, California 90024, USA

ABSTRACT

The beat wave accelerator[1] depends on the generation of a large amplitude plasma wave with a phase velocity close to the velocity of light c. The plasma wave (ω_p, k_p) is generated by beating colinear laser beams (ω_1, k_1) and (ω_2,k_2) with $\omega_p = \omega_1 - \omega_2$, $k_p = k_1 - k_2$. Since the process involves both large amplitude transverse and longitudional waves, various nonlinear instabilities associated with either wave may occur. The object of this article is to discuss some of the processes that may compete with the beat wave generation listing their threshold and growth rate.

INTRODUCTION

The beat wave process is very similar to four-wave stimulated Raman forward scattering. Stimulated Raman scattering is normally considered as a three wave process where the incident transverse laser beam decays into another transverse wave and a Langmuir wave. In an underdense plasma where $\omega_{1,2} \gg \omega_{pe}$, ω_{pe} is the plasma frequency, the forward Raman scattering becomes important. For phase matching the wavenumber k_p is much smaller than the laser wave number $k_{1,2}$, under these conditions we must consider an up-shifted or anti-Stokes transverse component as well as the down shifted Stokes component, since both are resonant with the laser wave and the Langmuir wave. The instability then becomes a "four wave" process with the incident laser wave (ω_1,k_1) decaying into a Stokes wave ($\omega_1 - \omega_p$, $k_1 - k_p$) and an anti-Stokes wave ($\omega_1 + \omega_p$, $k_1 + k_p$) and a Langmuir wave (ω_p, k_p). Stimulated Raman backscattering or stimulated Compton scattering for $k_p \lambda_{De} \simeq 1$, where λ_{De} is the Debye length, can also occur, this instability scatters the incident laser light out of the plasma. The above instabilities operate on the electron plasma time scales ω_{pe}^{-1} On longer time scales such as the ion plasma time scale ω_{pi}^{-1} other instabilities such as stimulated Brillouin scattering, self-modulation and self-focusing or filamentation due to ponderomotive force of the incident laser beam will compete with the beat wave process. In the case of stimulated Brillouin scattering the laser light will be scattered out of the plasma. Self-modulation will cause the incident laser light to increase its bandwidth while filamentation will break the beam up into a number of smaller beams. With the intense short laser pulses proposed for the beat wave scheme[1] the electron quiver velocity or oscillating velocity υ_{osc} in the laser field is sufficiently large that the electrons become weakly relativistic. This results in relativistic nonlinearities due to the Lorentz factor producing self-focusing and filamentation as well as self-modulation of the incident beam. In regions of high intensity the laser frequency is lower due to the electron mass increase this reduces the phase velocity causing the phase front to curve such that more wave energy flows into the intense region making it even more intense.

The large amplitude Langmuir wave which constitutes the beat wave is also subject to nonlinear instabilities such as the decay instability and self-modulation as well as filamentation. The decay instability occurs on ion plasma time scales $\sim \omega_{pi}^{-1}$ this instability results in the generation of a lower frequency Langmuir wave and an ion-sound wave. Self-modulation due to the ponderomotive force also occurs on the ion plasma time scale. Self-modulation results in the collapse of the plasma wave leading to strong turbulence and the break-up of the coherent structure of the wave that is necessary for efficient acceleration. Self-modulation can also occur due to relativistic effects and takes place on a time scale faster than that due to the ponderomotive force.

STIMULATED RAMAN SCATTERING

The normal treatment of stimulated Raman scattering is a three wave process, however in the presence of a sufficiently large pump field $\upsilon_{osc}/c > (\omega_{pe}/\omega_1)^{1/2}$ the properties of the Langmuir wave is determined by the pump wave. This is regarded as the regime of strong coupling and the process is called the modified Raman scattering instability[2]. The thresholds and growth rates for these two processes are given by[2]

(a) Stimulated Raman scattering

Threshold: $(\upsilon_{osc}/c)^2 > \omega_{pe}\,\nu_{ei}^2/\omega_1^3$

Growth rate: $\gamma \simeq (\omega_{pe}\omega_1)^{1/2}\upsilon_{osc}/2c$

(b) Modified Raman scattering

Threshold: $\upsilon_{osc}/c > (\omega_{pe}/\omega_1)^{1/2}$

Growth rate: $\gamma \simeq 3(\upsilon_{osc}\omega_{pe}/4c\omega_1)^{2/3}\ \omega_1$

ν_{ei} is the electron - ion collision frequency

In the above cases $k_p = 2k_1$ for maximum growth rate.

For strongly damped plasma waves ie $k_p\lambda_{De} \geqslant 1$ the normal three wave stimulated Raman scattering becomes stimulated Compton scattering where the laser beam scatters off free electrons with a threshold and growth rate given by[3].

Threshold: $(\upsilon_{osc}/c)^2 > \nu_{ei}\ \omega_{pe}^2\ k_1^2\ \lambda_{De}^2/2\omega_1^3$

Growth rate: $\gamma \simeq 0.2\ (\omega_{pe}\ \upsilon_{osc}/\upsilon_{Te})^2/\omega_1$

The modified Raman scattering instability can also operate even if $k_p\lambda_{De} \simeq 1$.

STIMULATED RAMAN FORWARD SCATTERING

This instability occurs with $k_p \ll k_1$, with the short wavelength pump wave (ω_1, k_1) being modulated by the long wavelength electrostatic wave resulting in the generation of high frequency transverse sidebands and bunching of the plasma particles. This is the single pump analogue of the beat wave process with growth rate and threshold given by[4]

Threshold: $(\upsilon_{osc}/c)^2 \geqslant 4\nu_{ei}/\omega_1$

Growth Rate: $\gamma \simeq \omega_{pe}^2\upsilon_{osc}^2/2\sqrt{2}\omega_1 c^2$

This "four-wave" modulational instability results in the broadening of the laser frequency and generation of broadband plasma oscillations. The above instability corresponds to $k_p \ll k_1$. for intermediate cases where $\underline{k_p}.k_1 = k_p k_1\cos\theta$ the process becomes a three wave resonant side-scatter instability with the following resonance condition for forward side scatter

$$kp = k_1\cos\theta - (k_1^2\cos^2\theta - 2\omega_{pe}\omega_1/c^2)^{1/2}$$

this results in a scattered signal with a maximum growth rate at an angle $\theta_s = \sin^{-1}(2\omega_{pe}/\omega_1)^{1/2}$ the threshold and growth rate corresponding to this angle are[4]

Threshold: $\upsilon_{osc}/c \geqslant \upsilon_{ei}/2\omega_1$

Growth rate: $\gamma \simeq \omega_{pe}\, \upsilon_{osc}/\sqrt{2}c$

This side-scatter instability results in the laser light being scattered out of the interaction column, this instability has a rather fast growth rate and will effect the efficiency of the beat wave generation. One way to overcome the instability is to increase θ_s by increasing ω_{pe}/ω_1 and reduce the beam diameter this will limit the number of e-folding lengths and hence help to quench the instability.

STIMULATED BRILLOUIN SCATTERING

Stimulated Brillouin scattering results in loss of laser power by scattering off ion-modes. The instability occurs on ion plasma time scales ω_{pi}^{-1} and will be important only for long pulse lengths. As with stimulated Raman scattering, stimulated Brillouin scattering has a number of different regimes of operation, the simplest being three wave backscattering with $k_s = 2k_1$ where k_s is the wavenumber of the ion-sound wave. The threshold and growth rate for this case are given by[3]

Threshold: $\upsilon_{osc}/\upsilon_{Te} \geqslant (8\, \nu_{ei} \nu_s/\omega_1 \omega_s)^{1/2}$

Growth rate: $\gamma \simeq (\upsilon_{osc}\omega_{pe}/4\upsilon_{Te})(\omega_s/\omega_1)^{1/2}$

For sufficiently strong pumps $\upsilon_{osc}/c > (\omega_s/\omega_1)^{1/2}$ the frequency shift of the ion-sound wave can be equal to or greater than their eigenfrequency, under these conditions we have the modified stimulated Brillouin instability[3] with threshold and growth rate given by

Threshold: $\upsilon_{osc}/c \geqslant (k_s \lambda_{De})^{3/2} (\omega_{pi}/\omega_1)^{1/2}$

Growth rate: $\gamma \simeq (2\, \upsilon_{osc}\omega_{pi}/c\omega_1)^{2/3}\, \omega_1$

again this reflects the $|E|^{2/3}$ dependence of the pump wave amplitude for the growth rate for large pump values.

For a plasma with equal electron and ion temperatures ($T_e < T_i$), the ion-sound wave is heavily Landau damped and scattering of the incident transverse wave off the ions into another transverse wave occurs with a grow rate given by [5,6]

$$\gamma \simeq 0.2\, \omega_{pi}^2\, \upsilon_{osc}^2 / (\omega_1 \upsilon_{Te} k_s^2 \lambda 2_{De})^2$$

FOUR-WAVE PROCESSES: SELF-MODULATION AND FILAMENTATION

Self-modulation and filamentation of the incident laser beam can occur as a result of the ponderomotive force[7] or the relativistic term in the Lorentz factor for the electron mass[8]. The ponderomotive force effects occur on the ion plasma period time ω_{pi}^{-1} scales and will only contribute to beam break up and frequency broadening long after the acceleration phase has taken place. On the other hand relativistic effects only involve the electrons and will therefore take place on a much faster time scale $\sim\omega_{pe}^{-1}$ and could therefore disrupt the accelerating mode of the beat wave. For the ponderomotive force driving term only filamentation can occur in an underdense plasma[7], since the density perturbation (ie the ion sound-wave) has got to travel at the group velocity of the laser beams for modulation to occur. The threshold and growth rate for filamentation due to ponderomotive effects are given by[7]

Threshold: $\upsilon_{osc}/\upsilon_{Te} > 2(\nu_{ei}/\omega_1)^{1/2}$

Growth rate: $\gamma \simeq \omega_{pi}\upsilon_{osc}/\sqrt{2}c$

For the relativistic mechanism both filamentation and self-modulation will take place with a threshold and growth rate given by[8]

Threshold: $\upsilon_{osc}/c > (2\nu_{ei}/\omega_1)^{1/2}$

Growth rate: $\gamma \simeq q(\omega_{pe}\upsilon_{osc}/2\omega_1 c)^2 \omega_1$

where q is a numerical factor $\simeq 0.5$ in an underdense plasma with plane polarized waves[8].

Filamentation and self-modulation due to the relativistic term have the same threshold and growth rate but the wave number for modulation $\underline{k}$ is different. For self-modulation $\underline{k} \| \underline{k}_1$, $k = \omega_1 \upsilon_{osc}/2c^2$ for maximum growth, and for filamentation $\underline{k} = \omega_{pe}\upsilon_{osc}/\sqrt{2}c^2$ smaller by the factor ω_{pe}/ω_1 than the $\underline{k}$ for self-modulation.

NONLINEAR BEAT WAVE PROCESSES

We now turn our attention to nonlinear processes associated with the large amplitude beat wave, these include processes such as three wave decay where the large amplitude Langmuir wave decays into another Langmuir wave and an ion-sound wave and modulational type instabilities which lead to the development of strong turbulence and cavity formation. This results in complete brake-up of the plasma beat wave increasing its bandwidth, destroying the conditions required for coherent acceleration.

THE DECAY INSTABILITY

This was one of the first nonlinear wave processes to be studied, it involves the decay of a Langmuir wave into another Langmuir wave an an ion-sound wave. For Langmuir waves with large amplitude

$$\upsilon_{osc}/\upsilon_{Te} > \omega_s/\omega_{pe}$$

the instability develops into the modified decay where the frequency shift due to the pump wave exceeds the natural frequency of oscillation of the ion-sound wave. For equal electron and ion temperatures the ion-sound waves are heavily Landau damped and the process goes over to nonlinear Landau damping. For small pump waves with

$$k_p\lambda_{De} \geqslant (4/\gamma_s)(m_e/m_i)^{1/2}$$

where γ_s is the ratio of specific heats, we get the resonant backscatter de[9]ay instability with threshold and growth rates given by.

Threshold: $\upsilon_{osc}/\upsilon_{Te} \geqslant 4(\nu_{ei}\nu_s/\omega_p\omega_s)^{1/2}$

Growth rate: $\gamma \simeq \omega_{pi}\upsilon_{osc}/2\,\upsilon_{Te}$

for $k\lambda_{De} \simeq (4/\gamma_s)(m_e/m_i)^{1/2}$

where ν_s is the ion-sound damping coefficient note that υ_{osc} is the oscillating velocity of the electron in the plasma wave.

For large pump fields, the frequency of the ion-sound wave is determined by the pump, under these conditions we have the modified decay instability with threhold and growth rate given by

Threshold: $\upsilon^2_{osc}/\upsilon^2_{Te} \geqslant 0.4\, k_p \lambda_{De} m_e/m_i$

Growth rate: $\gamma \simeq \sqrt{3}\, (\upsilon^2_{osc} k^2_p \omega_{pi}/2)^{1/3}$

For $k_p\lambda_{De} < (m_e/m_i)^{1/2}/\gamma_s$, resonant decay into normal modes is no longer possible and we have instead modulational instabilities and the oscillating two stream instability[9]. These instabilities are due to the ponderomotive force and have similar thresholds and growth rates given by[10]

Threshold: $\upsilon_{osc}/\upsilon_{Te} \geqslant (4\nu_e/\omega_p)^{1/2}$.

Growth Rate: $\gamma \simeq \omega_p \upsilon^2_{osc}/8\upsilon^2_{Te}$

The above processes act on the ion-plasma time scale. However, for the beat wave relativistic effects become important and can also give rise to modulational type instabilities which act on a much faster time scale than the ponderomotive force mechanism, the threshold and growth rate for relativistic self-modulation are given by[4].

Threshold: $\upsilon_{osc}/c \geqslant (8\nu_{ei}/3\omega_p)^{1/2}$

Growthrate: $\gamma \approx 3\, \upsilon^2_{osc}\, \omega_p/32c^2$

CONCLUSIONS

In this article various nonlinear processes have been listed which could be important in the beat wave accelerator scheme. The list is by no means complete, giving only a flavour of the many processes possible. The list illustrates clearly the need to understand more fully the different processes to successfully generate a beat wave which will accelerate particles to high energies.

ACKNOWLEDGEMENTS:

I would like to thank Prof John Lawson, Dr R G Evans, Dr T Katsouleas and Dr C Joshi for comments and discussions.

REFERENCES

1. C Joshi, W.B. Mori, T Katsouleas, J.M. Dawson, J.M. Kindel and D.W. Forslund, Nat. 311, 525 (1984).

2. J.F. Drake, P.K. Kaw, Y.C. Lee, G. Schmidt, C.S. Liu and M.N. Rosenbluth, Phys. Fluids. 17, 778 (1974).

3. C.S. Liu and P.K. Kaw, Advances in Plasma Phys. Vol 6 Edited by A. Simon and W.B. THompson (1976).

4. R. Bingham "Possible Instabilities in the Beat Wave Accelerator" RAL report RAL-83-058 (1983). Submitted to "Plasma Physics".

5. A.T. Liu, J.M. Dawson, Phys. Fluids 20, 538 (1977).

6. E. Ott, W.M. Manheimer and H.H. Klein, Phys. Fluids 17, 1957 (1974).

7. R. Bingham and C.N. Lashmore-Davies Nucl. Fusion 16, 67 (1976).

8. C.E. Max, J. Arons and A.B. Langdon, Phys. Rev. Lett, 33, 209 (1974).

9. B.D. Fried, T. Ikemura, K. Nishikawa and G. Schmitt, Phys. Fluids, 19, 1975 (1976).

10. R. Bingham and C.N. Lashmore-Davies, Journal of Plasma Physics 21, 51 (1979).

MULTIPLE SCATTERING AND SYNCHROTRON RADIATION IN THE PLASMA BEAT-WAVE ACCELERATOR

B.W. Montague and W. Schnell
CERN, Geneva, Switzerland

ABSTRACT

In this paper we discuss two effects which can influence the performance of a plasma beat-wave accelerator; multiple Coulomb scattering from the plasma particles, leading to emittance growth of the accelerated beam, and synchrotron radiation due to focusing fields, which can give rise to excessive energy loss. One or both of these effects could also arise in other types of accelerator under consideration for high-energy electron linear colliders.

1. MULTIPLE SCATTERING

Scattering of accelerated particles by residual gas in a synchrotron is a familiar problem in accelerator physics and is well understood. The situation in a BWA, though similar in principle, differs in detail for two main reasons. Firstly we are dealing with an ionised plasma rather than neutral atoms or molecules, which modifies the effective range of the Coulomb force. Secondly the transverse focusing of the accelerated beam will in general be a function of energy, which influences the rate of emittance growth along the machine. A detailed derivation of the scattering formulae which takes these features into account is given in the proceedings of the Frascati Workshop[1], and we give here a summary of these results.

1.1 Small-angle scattering

Assuming a fully-ionised hydrogen plasma (unit charges) and a highly-relativistic beam of incident electrons, elastic scattering through a small angle θ by a Coulomb central force is well described by the simplified Rutherford formula:

$$\theta = \frac{\Delta p}{p} = \frac{2e^2}{pcb} \tag{1}$$

where p is the momentum of the incident particle, Δp the momentum transfer in the collision, b is the impact parameter, c the speed of light and e the unit charge. For a point Coulomb field in the absence of screening the Rutherford formula is exact quantum-mechanically for the expectation value of scattering angles.

Deviations from equation (1) occur for very close collisions (small b) and for distant collisions (large b). Outside these limits the momentum transfer Δp is less than that for a point

Coulomb field, due to saturation or screening. This is taken into account to a good approximation by imposing cut-offs b_{min} and b_{max}, corresponding to θ_{max} and θ_{min} respectively, on the range of integration of the scattering cross-section. For highly-energetic incident particles we need only consider scattering from the protons, since the plasma electrons contribute very little to the momentum transfer. Spin effects also can be neglected.

1.2 Minimum impact parameter b_{min}

For incident electrons of energy 19 GeV or more, this cut-off is determined by a quantum condition relating to diffraction of the de Broglie wavefront by the effective Coulomb radius R of the proton, which has a value of about 0.7×10^{-13} cm. This leads to

$$\theta_{max} \approx \frac{\hbar}{pR} \quad , \tag{2}$$

with the corresponding impact parameter

$$b_{min} \approx 2\alpha R \quad , \tag{3}$$

where $\hbar$ is the reduced Planck's constant and $\alpha = e^2/\hbar c$ is the fine structure constant. Since equations (2) and (3) appear only in the Coulomb logarithm we can, to a good approximation, use them even for energies down to 10 GeV, the assumed injection energy of the BWA example under consideration.

1.3 Maximum impact parameter b_{max}

In contrast to the situation in a gas composed of neutral atoms, where the Coulomb field is screened by the atomic electrons, the range of this field in a fully-ionised plasma is given by the Debye length λ_D, which is normally much greater than atomic radii. However, for relativistic incident particles the minimum scattering angle θ_{min} is constrained by the uncertainty principle, so that the momentum transfer satisfies

$$\lambda_D \, \Delta p \geqslant \hbar \tag{4}$$

Assuming that equation (1) is valid down to this cut-off angle we then have

$$b_{max} = \frac{2e^2}{pc\theta_{min}} = 2\alpha\lambda_D \quad . \tag{5}$$

1.4 Mean-square scattering angle

The differential cross-section for a cut-off Coulomb potential can be written in the form[2]

$$\frac{d\sigma}{d\Omega} = \left(\frac{2e^2}{pc}\right)^2 \cdot \frac{1}{(\theta^2+\theta^2_{min})^2} , \tag{6}$$

and, since

$$1 \gg \theta_{max} \gg \theta_{min} ,$$

the element of solid angle $d\Omega$ can be approximated by

$$d\Omega \approx 2\pi\theta d\theta .$$

The mean-square scattering angle for single collisions is then

$$\langle\theta^2\rangle = \frac{\int_0^{\theta_{max}} \theta^2 \frac{d\sigma}{d\Omega} d\Omega}{\int_0^{\theta_{max}} \frac{d\sigma}{d\Omega} d\Omega}$$

$$= 2\theta^2_{min} \left[\ln \frac{\theta_{max}}{\theta_{min}} - \frac{1}{2} \right] . \tag{7}$$

Also, since from equations (2) and (4)

$$\frac{\theta_{max}}{\theta_{min}} = \frac{\lambda_D}{R} \gg 1 ,$$

we can further approximate equation (7) to

$$\langle\theta^2\rangle \approx 2\theta^2_{min} \ln \left(\frac{\lambda_D}{R}\right) . \tag{8}$$

For a large number N of individual collisions in a distance dz the angular distribution will be approximately Gaussian with a mean square angle $\langle\Xi^2\rangle = N\langle\theta^2\rangle$. With n scattering centres per unit volume,

$$N = n\sigma dz,$$

and since the total cross-section σ, obtained by integration of equation (6), is

$$\sigma = \pi\left(\frac{2e^2}{pc\ \theta_{min}}\right)^2 , \tag{9}$$

the rate of increase in projected angular divergence on to one plane is

$$\frac{d}{dz}\langle\Xi^2\rangle_p = \pi n\left(\frac{2e^2}{pc}\right)^2 \ell n\left(\frac{\lambda_D}{R}\right) , \tag{10}$$

where we have used the fact that $\langle\Xi^2\rangle_p = \frac{1}{2}\langle\Xi^2\rangle$.

1.5 Emittance growth in a focused system

The effect of multiple scattering on a beam of particles in the presence of transverse focusing is to excite randomly betatron oscillations, leading to a growth in the invariant (or normalised) beam emittance ε, which would otherwise be independent of energy. Following the formulation of Hardt[3], it is shown in ref. 1 that the rate of emittance growth can be expressed in the form:

$$\frac{d\varepsilon}{dz} = \frac{\gamma\beta_x}{2}\frac{d}{dz}\langle\Xi^2\rangle_p , \tag{11}$$

where γ is the Lorentz energy factor, β_x is the betatron focusing function in the x-plane and the invariant emittance ε is defined at one standard deviation σ_x of the distribution by

$$\varepsilon = \frac{\gamma\sigma_x^2}{\beta_x} \tag{12}$$

It is shown in ref. 4 that, for the plasma BWA the focusing function β_x is given by

$$\beta_x = \left[\frac{-\pi\ \sigma_0^2\ mc^2\gamma}{\lambda_p\ e\ E_{zo}\cos\phi}\right]^{\frac{1}{2}} , \tag{13}$$

where σ_0 is the standard deviation of the laser beam cross-section, λ_p is the wavelength of the plasma beat-wave and ϕ is the accelerating phase angle.

The accelerating field on the axis is $E_{zo}\sin\phi$. and it is assumed that both the laser beam and the resultant beat-wave have the same Gaussian distribution transversely. Using equations (10) and (13) in equation (11), with $p = \gamma mc$, we obtain

$$\frac{d\varepsilon}{dz} = \frac{F}{\sqrt{\gamma}} , \tag{14}$$

where

$$F = 2\pi r_e^2 n \left[\frac{-\pi\sigma_0^2 mc^2}{\lambda_p e\, E_{zo}\cos\phi} \right]^{\frac{1}{2}} \ell n \left(\frac{\lambda_D}{R}\right) \tag{15}$$

and $r_e = e^2/mc^2$ is the classical electron radius.

Assuming a constant acceleration rate $eE_{zo}\sin\phi$ and writing

$$\gamma = \gamma_i + \gamma' z ,$$

where γ_i corresponds to injection energy and

$$\gamma' = \frac{d\gamma}{dz} = \frac{eE_{zo}\sin\phi}{mc^2} , \tag{16}$$

we can integrate equation (15) from γ_i to the final energy γ_f, yielding the emittance growth $\Delta\varepsilon$:

$$\Delta\varepsilon = \frac{2F}{\gamma'} \left[\sqrt{\gamma_f} - \sqrt{\gamma_i} \right] . \tag{17}$$

Rearrangement of the factors gives:

$$\frac{2F}{\gamma'} = 4\pi r_e^2 \sigma_0 n \left[\frac{mc^2}{eE_{zo}\sin\phi} \right]^{\frac{3}{2}} \left[\frac{-\pi\tan\phi}{\lambda_p} \right]^{\frac{1}{2}} \ell n \left(\frac{\lambda_D}{R}\right) \tag{18}$$

1.6 Emittance growth in a 1 TeV BWA

We can evaluate equation (18) using the following parameters adopted at the Frascati Workshop:

$n = 10^{17}\ \mathrm{cm}^{-3}$

$eZ_{zo}\sin\phi = 60\ \mathrm{MeV\ cm}^{-1}$

$\lambda_p = 10^{-2}\ \mathrm{cm}$

$\sigma_o = 2.5 \times 10^{-2}\ \mathrm{cm}$

$\lambda_D = 5.26 \times 10^{-6}$ cm (plasma electrons, kT = 5eV)

$\phi = 21\pi/32$ (average over one stage of the BWA)

and with

$r_e = 2.818 \times 10^{-13}\ \mathrm{cm}$

$mc^2 = 0.511 \times 10^6\ \mathrm{eV}$

$R = 0.7 \times 10^{-13}\ \mathrm{cm}$

we find

$$\frac{2F}{\gamma'} = 8.62 \times 10^{-10}\ \mathrm{cm}.$$

Then with $\gamma_f = 1.96 \times 10^6$ (1 TeV) and $\gamma_i = 1.96 \times 10^4$ (10 GeV), equation (17) yields

$$\Delta\varepsilon = 1.086 \times 10^{-6}\ \mathrm{cm} = 1.086 \times 10^{-8}\ \mathrm{m}.$$

This emittance growth is only about 3.6×10^{-4} of the nominal invariant emittance $\varepsilon = 3 \times 10^{-5}$ m for the SLAC Linear Collider, and therefore leaves considerable latitude for optimisation of parameters. In particular, it would permit the use of much lower-emittance beams, if these were feasible, before multiple scattering became a serious limitation.

The analysis used here can readily be adapted to other random sources of emittance growth such as plasma-density fluctuations, contamination by high-Z impurities, quantum fluctuations of the synchrotron radiation and possible random variations of the focusing forces.

2. FOCUSING

Quite generally in an electron linear accelerator of TeV energies the synchrotron radiation due to the unavoidable transverse focusing system may be of concern. The following simple expressions for the radiation loss per unit length (eqs. 23b and 32 below) should be useful for first-order estimates and scaling considerations.

2.1 Smooth focusing

In a beat-wave plasma accelerator the transverse focusing (primarily due to the transverse gradient of the accelerating force) will be continuous, i.e. the amplitude function β will not vary over an oscillation period; and even in a conventional alternating gradient system this assumption gives an interesting basis for comparison with what is treated in the next section. Thus

$$x = a \sin \frac{z}{\beta}$$

where x is the transverse displacement, a the amplitude, z the longitudinal coordinate and $2\pi\beta$ the oscillation wavelength. The curvature ρ^{-1} of any curve x(z) is given by

$$\rho^{-1} = \frac{x''}{(1+x'^2)^{\frac{3}{2}}} \tag{19}$$

where the dashes denote differentiation with respect to z. Clearly $x'^2 \ll 1$ in any high energy accelerator. Hence

$$\rho^{-2} = \frac{a^2}{\beta^4} \sin^2 \frac{z}{\beta} \tag{20}$$

and the average value over half an oscillation is given by

$$\langle\rho^{-2}\rangle = \frac{a^2}{2\beta^4} \quad . \tag{21}$$

The radiation loss per unit length is given by

$$eU' = \frac{2r_e m_0 c^2}{3} \frac{\gamma^4}{\rho^2} \tag{22}$$

where U is the particle energy over elementary charge, r_e is the classical electron radius, m_0c^2 the rest energy and γ the Lorentz factor. Substituting equation (20) for ρ^{-2} yields

$$\langle U'\rangle = 4.79\times10^{-10} \, a^2\left(\frac{\gamma}{\beta}\right)^4 \qquad [\mathrm{Vm}^{-1}] \tag{23a}$$

for the average energy loss per unit length.

A particle whose amplitude equals σ of the distribution suffers an energy loss given by

$$\langle U' \rangle_\sigma \quad = \quad 4.79\times10^{-10} \; (\gamma\varepsilon)(\frac{\gamma}{\beta})^3 \quad [\mathrm{Vm}^{-1}] \tag{23b}$$

where $\gamma\varepsilon = \gamma\sigma^2/\beta$ is the normalized emittance. At $\gamma = 2 \times 10^6$ (1 TeV) and with $\gamma\varepsilon = 3.5 \times 10^{-5}$ m (the SLC specification) one finds

Table 1, $\gamma\varepsilon = 3.5 \times 10^{-5}$ m

β	$\langle U' \rangle_\sigma$
10 m	134 V m^{-1}
1 m	134 kV m^{-1}
0.1 m	134 MV m^{-1}

Particles with amplitude $n\sigma$ will suffer n^2 times this energy loss. For higher energies the radiation loss will increase with $\gamma^{3/2}$, provided any emittance blow-up is avoided and the focusing strength is kept constant so that $\beta \propto \gamma^{1/2}$.

This result looks quite favourable but it must be remembered that equation (23b) is based on the assumption that the beam centre coincides exactly with the zero-field line of the focusing system. If, for any reason (e.g. misalignment) the beam oscillates with amplitude a around the zero-field equation (23a) has to be used. For a = 0.1 mm, say, and 1 TeV one finds

Table 2, a = 0.1 mm

β	$\langle U' \rangle$
10 m	7.7 kV m^{-1}
1 m	76.6 MV m^{-1}
0.1 m	766 GV m^{-1}

Clearly this has to be avoided by using the radiation emitted as diagnostics for aligning the system.

2.3 Alternating gradient focusing

In an alternating gradient system the radiation must be expected to be much stronger because of the uneven distribution of the deflecting forces including outward bends. Generally (in the absence of driving terms!)

$$\rho^{-2} = \left[\frac{d^2x}{dz^2}\right]^2 = [K(z)x(z)]^2 \tag{24}$$

where

$$K(z) = \frac{e\partial B}{p\partial x} \tag{25}$$

is the normalized gradient (p: particle momentum).

A thin-lens FODO structure with period length L and amplitude functions β_F, β_D in the quadrupoles may be taken for deriving an analytical formula. In the quadrupoles, of length ℓ and focal length $\pm f$ one has $K = (f\ell)^{-1}$ and hence

$$\rho^{-2} = \frac{x^2}{(f\ell)^2} \quad . \tag{26}$$

The distribution of displacements x of a particle at successive quadrupoles of same sign is sinusoidal. Hence, the average of ρ^{-2} over all F-quadrupoles (all D-quadrupoles) is

$$\langle\rho^{-2}\rangle_{F,D} = \frac{a^2_{F,D}}{2(f\ell)^2} \tag{27}$$

where $a_{F,D}$ is the peak amplitude observed at any F(D) position. For a particle at one σ of the distribution of amplitudes within the beam

$$\langle\rho^{-2}\rangle_{F,D} = \frac{\varepsilon}{2(f\ell)^2}\,\beta_{F,D} \tag{28}$$

where $\varepsilon = \sigma^2_F/\beta_F = \sigma^2_D/\beta_D$ is the emittance. As there is no radiation outside the quadrupoles the overall average is given by

$$\langle\rho^{-2}\rangle = \eta\,\frac{\varepsilon}{2(f\ell)^2}\,\beta \tag{29}$$

where $\eta = 2\ell/L$ is the filling factor and $\beta = (\beta_F+\beta_D)/2$. Introducing the phase advance per period μ_0 with

$$\frac{L}{f} = 4\sin\frac{\mu_0}{2} \tag{30}$$

$$L = \beta\sin\mu_0 \tag{31}$$

and inserting the overall average ρ^{-2} into equation (22) finally gives

$$\langle U' \rangle_\sigma = 4.79\times10^{-10} \; \frac{64}{\eta} \; \frac{(\sin \frac{\mu_0}{2})^2}{(\sin \mu_0)^4} \; (\gamma\varepsilon)(\frac{\gamma}{\beta})^3 \; [\mathrm{V\ m^{-1}}] \qquad (32)$$

to be compared with equation (23b). For $\mu_0 = 60°$, $\eta = 0.1$ the factor

$$\frac{64}{\eta} \; \frac{(\sin \frac{\mu_0}{2})^2}{(\sin \mu_0)^4}$$

by which the radiation is increased compared with smooth focusing at the same β-value amounts to 284! The factor of $64 \sin^2(\mu_0/2)\sin^{-4}\mu_0$ has a minimum at $\mu_0 = 70.5°$ where it amounts to 27.0. In spite of this the radiation loss remains quite tolerable in an a.g. focusing system for energies beyond 1 TeV and amplitude functions of the order of 10 m.

REFERENCES

1. B.W. Montague, Proc. CAS-ECFA-INFN Workshop on the Generation of High Fields, Frascati, 25 September to 1 October 1984; proceedings to be published.
2. J.D. Jackson, Classical Electrodynamics, 2nd Ed. (John Wiley and Sons, N.Y. 1975).
3. W. Hardt, Report CERN-ISR-300/GS/68-11 (1968).
4. J.D. Lawson, Report of the RAL Study Group, RL-83-057 (1983).

EVOLUTION OF THE LASER BEAM ENVELOPE IN THE BEAT WAVE ACCELERATOR

P. Sprangle and Cha-Mei Tang
Plasma Theory Branch
Plasma Physics Division
Naval Research Laboratory
Washington, DC 20375-5000

Abstract

An envelope equation is derived which describes the radial evolution of a radiation beam propagating through a plasma, which is supporting a space charge wave. The radiation envelope equation contains a defocusing term due to diffraction spreading, a focusing term due to relativistic oscillations of the plasma electrons and a rapid periodic focusing and defocusing term due to the presence of the space charge wave. In the absence of the space charge wave the condition necessary to propagate a laser beam with constant radius is found in terms of the laser power. In the presence of the space charge wave the laser envelope developes a high frequency modulation.

Introduction

The laser beat wave accelerator concept[1-13] is a collective acceleration scheme which utilizes a large amplitude plasma wave with phase velocity slightly less than the velocity of light to accelerate charged particles. The large amplitude plasma wave is generated by the nonlinear coupling of two intense laser beams propagating through the plasma. In this process the two laser beams with frequencies ω_1, ω_2 and corresponding wave numbers k_1, k_2 couple through the plasma to produce a ponderomotive wave with frequency $\omega_1 - \omega_2$ and wave number $k_1 - k_2$. If $\omega_1 - \omega_2 \approx \omega_p$, the plasma wave will initially grow linearly in time. If the laser frequencies are much greater than the ambient plasma frequency ω_p then the phase velocity of the ponderomotive wave is nearly equal to the group velocity of the laser wave. Electrons which are either injected into the plasma or part of the thermal tail of the plasma distribution can be accelerated by the large gradients associated with the plasma wave.

A potentially attractive variation of the plasma beat wave accelerator is the surfatron scheme.[14] In the surfatron configuration a transverse magnetic field is externally applied permitting the accelerated particles to effectively $\underset{\sim}{E} \times \underset{\sim}{B}$ drift in a direction transverse to the laser propagation direction. In this configuration the electrons can remain in phase with the plasma wave allowing, in principle, higher electron energies to be achieved.

In this paper the self-focusing properties of the plasma on a single intense laser beam are studied. The propagation of an intense laser beam in a plasma can lead to self-focusing.[15-16] One such self-focusing mechanism is due to the relativistic oscillations of the plasma electrons by the laser field. This effect is analyzed for a single helically polarized laser beam propagating in the z direction. The laser beam will be assumed to be axially symmetric with respect to the z axis and have a profile which is only a function of r,z and t. An equation which describes the envelope of the laser beam is derived. The laser beam envelope equation includes diffraction effects as well as relativistic plasma effects, and it has a form similar to a particle beam envelope equation.[17,18] In the present laser beam envelope equation, diffraction effects are manifested through a term which is equivalent to beam emittance in the particle beam envelope equation.

Individual Ray Equations

Consider a helically polarized laser beam propagating with a cold collisionless plasma. The vector potential of the laser field is taken to have the form

$$A_L(r,z,t) = A(r,z,t)\ \left(\cos(kz-\omega t)\hat{e}_x - \sin\ (kz-\omega t)\hat{e}_y\right), \qquad (1)$$

where the amplitude A(r,z,t) is a slowly varying function of r,z,t and the frequency ω is assumed to be much greater than the effective plasma frequency. The approximate local dispersion relation associated with the field in (1) is

$$\omega \approx ck + \left(c^2k_\perp^2 + \omega_p^2(r,z,t)\right)/2ck, \qquad (2)$$

where $k_\perp = \sqrt{k_x^2 + k_y^2}$ is the transverse wave number, $k_\perp^2 << k^2$ and $\omega_p^2(r,z,t)$ is the effective background plasma frequency. For purposes of this discussion we can write the effective plasma frequency in the form

$$\omega_p^2(r,z,t) = \frac{n(r,z,t)}{n_0}\ \frac{\omega_{po}^2}{\gamma_\perp(r,z,t)}, \qquad (3)$$

where $\omega_{po} = (4\pi|e|^2 n_o/m_o)^{1/2}$ is the ambient plasma frequency, $n(r,z,t)$ is the modified electron density of the plasma due to the excited plasma wave, $\gamma_\perp(r,z,t) = (1+a^2(r,z,t))^{1/2}$ is the relativistic mass factor and $a(r,z,t) = |e|A(r,z,t)/m_o c^2$ is the normalized laser field amplitude. The relativistic factor $\gamma_\perp$ arises from the plasma electron's relativistic mass change due to their transverse oscillations in the laser field.

Using the ray equations from geometric optics, the transverse motion of the electromagnetic rays are given by

$$\frac{d\underset{\sim}{r}_\perp}{dt} = \frac{\partial\omega}{\partial \underset{\sim}{k}_\perp} \tag{4a}$$

$$\frac{d\underset{\sim}{k}_\perp}{dt} = -\frac{\partial\omega}{\partial \underset{\sim}{r}_\perp} \tag{4b}$$

where $\underset{\sim}{r}_\perp = x(t)\hat{e}_x + y(t)\hat{e}_y$ is the transverse position of a ray, $\underset{\sim}{k}_\perp = k_x(t)\hat{e}_x + k_y(t)\hat{e}_y$ is the transverse wave number of a ray and ω is given by (2). Substituting (2) into (4a) and (4b) yields the transverse ray equations

$$\frac{dx^2}{dt^2} + \Omega^2(r,z,t)x = 0, \tag{5a}$$

$$\frac{dy^2}{dt^2} + \Omega^2(r,z,t)y = 0, \tag{5b}$$

where

$$\Omega^2(r,z,t) = \frac{1}{2k^2 r}\frac{\partial\omega_p^2(r,z,t)}{\partial r}, \tag{5c}$$

and

$$\frac{\partial\omega_p^2}{\partial r} = \omega_{po}^2 \frac{\partial}{\partial r} \left(\frac{n(r,z,t)/n_o}{\sqrt{1+a^2(r,z,t)}}\right). \tag{5d}$$

Equations (5a,b) describe the transverse oscillations of the rays and assume that the rays travel nearly parallel to the z axis, i.e., $|dx/dt|$, $|dy/dt| \ll v_g \approx c$ where v_g is the group velocity, and that the plasma density and laser beam are azimuthally symmetric.

Laser Beam Envelope Equation

In this section we derive an envelope equation which describes the transverse dynamics of the laser beam envelope as it propagates through a plasma. The derivation is similar to that used to obtain a particle beam envelope equation[18] in the sense that the ray equations in (5a,b) have a form which is similar to particle orbits in the paraxial approximation.

We start by taking various moments of the ray equations given by (5). Multiplying (5a) by x and dx/dt, (5b) by y and dy/dt and combining, yields the following virial and energy equations

$$\frac{1}{2}\frac{d^2r^2}{dt^2} - v^2 + \Omega^2(r,z,t)r^2 = 0, \tag{6a}$$

$$\frac{dv^2}{dt^2} + \Omega^2(r,z,t)\frac{dr^2}{dt} = 0, \tag{6b}$$

where $r^2 = x^2 + y^2$ and $v^2 = (dx/dt)^2 + (dy/dt)^2$. Substituting (6a) into (6b) yields

$$\frac{1}{2}\frac{d^3r^2}{dt^3} + \frac{d}{dt}(\Omega^2r^2) + \Omega^2\frac{dr^2}{dt} = 0. \tag{7}$$

To obtain the envelope equation we consider a thin slice of the laser beam lying in the transverse. The transverse segment of the laser beam is assumed to consist of N individual rays traveling through it. We can define the mean square radius of the laser beam envelope as

$$R_L^2(z,t) = \langle r^2 \rangle = \frac{1}{N} \sum_{i=1}^{N} r_i^2 \qquad (8)$$

where r_i is the radial position of the ith ray. Performing an average over all the rays on (7) and using (8) gives

$$\frac{1}{2} \frac{d^3 R_L^2}{dt^3} + \frac{d}{dt} \langle \Omega^2 r^2 \rangle + \langle \Omega^2 \frac{dr^2}{dt} \rangle = 0. \qquad (9)$$

In order to express the averages in (9) in a more manageable form we write the individual radial ray velocity as a sum of a mean velocity term and a residual velocity term. That is the individual ray velocity is written as

$$v_r = \frac{r}{R_L} \frac{dR_L}{dt} + \delta v_r, \qquad (10)$$

where $(r/R_L)dR_L/dt$ is defined as the mean radial velocity at the position r and δv_r is the residual radial velocity. Substituting (10) into the last term in (9) yields

$$\langle \Omega^2 \frac{dr^2}{dt} \rangle = 2\langle \Omega^2 r^2 \rangle \frac{dR_L/dt}{R_L} + 2\langle \Omega^2 r \delta v_r \rangle. \qquad (11)$$

To further simplify (11) we will assume that within any cross-sectional segment of the radiation beam the average radial velocity of the rays at any radial position, say $r = r_0$, is just equal to the mean radial velocity, i.e.,

$$\langle v_r \rangle\Big|_{r=r_o} = \frac{r_o}{R_L}\frac{dR_L}{dt}. \tag{12}$$

This assumption implies that the average deviation in the radial velocity from the mean radial flow (residual velocity) is zero at any position within the beam segment, i.e., $\langle \delta v_r \rangle_{r=r_o} = 0$. Since $\langle \delta v_r \rangle\big|_{r=r_o}$ is assumed to vanish at any point within the radiation beam, it must also be true that $\langle f(r)\delta v_r \rangle$ vanishes where $f(r)$ is an arbitrary function of r and the average is taken over the entire beam segment. Hence, making the assumption that $\langle v_r \rangle\big|_{r=r_o} = (r_o/R_L)dR_L/dt$ we find that $\langle \Omega^2 r \delta v_r \rangle = 0$ and (11) becomes

$$\langle \Omega^2 \frac{dr^2}{dt} \rangle = 2\langle \Omega^2 r^2 \rangle \frac{dR_L/dt}{R_L}. \tag{13}$$

Substituting (13) into (9) yields

$$\frac{1}{2}\frac{d^3R_L^2}{dt^3} + \frac{d}{dt}\langle \Omega^2 r^2 \rangle + 2\langle \Omega^2 r^2 \rangle \frac{dR_L/dt}{R_L} = 0. \tag{14}$$

Since the last two terms in (14) combine to give $R_L^{-2}d(R_L^2\langle \Omega^2 r^2 \rangle)/dt$, (14) can be written as

$$\frac{d}{dt}\left[R_L^2 \frac{d^2R_L^2}{dt^2} - \frac{1}{2}\left(\frac{dR_L^2}{dt}\right)^2 + 2R_L^2 \langle \Omega^2 r^2 \rangle\right] = 0. \tag{15}$$

Integrating (15) and taking the constant of integration to be $2\varepsilon^2$ gives

$$\frac{d^2R_L}{dt^2} + \frac{\langle \Omega^2 r^2 \rangle}{R_L} - \frac{\varepsilon^2}{R_L^3} = 0. \tag{16}$$

Equation (16) describes the self-consistent evolution of the laser beam envelope, R_L. Diffraction effects which result in a spreading of the beam are contained in the constant of integration term ε. The form of the envelope equation for the radiation beam, (16), is similar to the usual particle beam envelope equation.[17-18] The second term in (16) can be either focusing or defocusing depending on the sign of Ω^2. In the present situation, because of the relativistic transverse oscillations of the background plasma electrons, this term results in focusing of the radiation beam. The third term in (16) is always a defocusing term and is due to diffraction effects.

The diffraction constant ε^2 can be evaluated by noting that in the absence of the plasma, (16) has the form $d^2R_L/dt - \varepsilon^2/R_L^3 = 0$ with the solution

$$R_L(z) = R_L(0)(1 + z^2/Z_o^2)^{1/2}, \tag{17}$$

where we have set $z = ct$, $R_L(0)$ is the laser beam radius at $z = 0$ which, for this case, is taken to be the minimum radius and $Z_o = cR_L^2(0)/\varepsilon$ is the characteristic diffraction length (Rayleigh length). The laser beam envelope in (17) evolves like a diffraction limited radiation beam.[19] For a Gaussian radiation beam the diffraction length is the Rayleigh length[19] and is given by $Z_R = \pi R_L^2(0)/\lambda$ where λ is the radiation wavelength $\lambda = 2\pi/k$. Setting $Z_o = Z_R$ gives

$$\varepsilon = \frac{c\lambda}{\pi} = cR_L(0)\theta_d, \tag{18}$$

where $\theta_d = \lambda/\left(\pi R_L(0)\right)$ is the well-known diffraction angle.

Application to the Beat Wave Accelerator

As an illustration we will consider the focusing characteristics of a single radiation beam in the laser beat wave accelerator. The radiation beam is assumed to have a Gaussian profile of the form

$$a(r,z,t) = a_o\left(R_L(0)/R_L\right)e^{-(r/R_L)^2} \tag{19}$$

where a_o is the normalized laser amplitude at z = 0, $R_L(0)$ is the laser beam radius at z = 0 and $R_L(z,t)$ is the laser radius as a function of z and t to be determined by (16). The space charge wave excited by the beating of two laser fields modifies the plasma electron density. In this example, we analyze laser focusing properties due to the relativistic effects and the plasma density wave. We do not include the effect of a radial variation in the equilibrium density. This is valid if $\partial(a^2/2)/\partial r \gg \partial(n_o)/\partial r$, where n_o is the equilibrium electron density. The plasma electron density is assumed to be of the form

$$n(r,z,t) = n_o + \delta n(r,z,t) \tag{20}$$

where

$$\delta n(r,z,t) = \alpha(z,t)a^2(r,z,t)n_o \tag{21}$$

is the excited electron density wave traveling with phase velocity ω_{po}/k_{po}. In general the amplitude of the excited electron density wave evolves in both space and time growing to a maximum value of $\sim n_o$.

The coefficient $\alpha(z,t)$ is given by

$$\alpha(z,t) = g(z,t)\sin(k_{po}z-\omega_{po}t)$$

where ω_{po}/k_{po} is the phase velocity of the space charge wave and $g(z,t)a^2(r,z,t)n_o$ is the amplitude of the traveling space charge wave. Saturation processes limit the amplitude of the electron density wave to values such that $|\delta n(r,z,t)| \lesssim n_o$. If the frequency difference between the two beating laser fields is sufficiently close to the ambient plasma frequency the electron density wave amplitude grows approximately linearly in $t-z/v_g$.

Substituting (19) and (20) into (5c), the second term in the radiation beam envelope equation, given by (16), becomes,

$$\frac{\langle\Omega^2 r^2\rangle}{R_L} = \frac{\omega_{po}^2}{2k^2}\langle\frac{r}{R_L}\frac{\partial}{\partial r}\left(\frac{1+\alpha a^2}{(1+a^2)^{1/2}}\right)\rangle$$

$$= \frac{-\omega_{po}^2}{4k^2}\langle\frac{\left(1-2\alpha(1+a^2/2)\right)}{(1+a^2)^{3/2}}\frac{r}{R_L}\frac{\partial a^2}{\partial r}\rangle. \qquad (22)$$

To make further progress in solving the envelope equation we will assume that the transverse electron velocity is mildly relativistic, i.e., $a^2 \ll 1$. With this assumption (22) reduces to

$$\frac{\langle\Omega^2 r^2\rangle}{R_L} = \frac{\omega_{po}^2/k^2}{R_L(0)}\left(1-2\alpha(z,t)\right)a_o^2\left(\frac{R_L(0)}{R_L}\right)^3\beta, \qquad (23)$$

where

$$\beta = \langle\frac{r^2}{R_L^2}e^{-2(r/R_L)^2}\rangle \lesssim 1.$$

Substituting (23) into (16) yields the final form for the laser beam envelope equation

$$X^3\frac{d^2X}{dt^2} = \Omega_d^2 - \Omega_f^2 + \Omega_{s.c.}^2(z,t), \qquad (24)$$

where $X = R_L(z,t)/R_L(0)$ is the laser beam radius normalized to its initial radius at $z = 0$,

$$\Omega_d^2 = \left(\varepsilon/R_L^2(0)\right)^2, \tag{25a}$$

is a defocusing term responsible for diffraction spreading of the beam,

$$\Omega_f^2 = \left(\frac{\omega_{po} a_o}{R_L(0)k}\right)^2 \beta, \tag{25b}$$

is a focusing term due to relativistic transverse oscillations of the electrons and corresponding mass change and

$$\Omega_{s.c.}^2 = 2\alpha(z,t)\Omega_f^2, \tag{25c}$$

is due to the excitation of the electron space charge wave and results in rapid periodic focusing and defocusing. The magnitude of the space charge term, $\Omega_{s.c.}^2$, can be much greater than the relativistic focusing term Ω_f^2. This can be seen by noting that the space charge wave saturates when $|\delta n| \approx n_o$ which implies that $|\alpha| \approx 1/a_o^2$ at saturation, see Eq. (21).

Envelope Evolution in the Absence of Space Charge Wave

In the absence of the space charge term the solution of (24) is of the same general form as the solution given in (16). Setting $\Omega_{s.c.} = 0$, the solution of (24) is denoted by X_o and is given by

$$X_o(z) = \left[1 + (\Omega_d^2 - \Omega_f^2)z^2/v_g^2\right]^{1/2}. \tag{26}$$

A matched (constant radius) laser beam can be achieved if

$$\Omega_d = \Omega_f, \tag{27}$$

otherwise the beam will either focus ($\Omega_f > \Omega_d$) or defocus ($\Omega_f < \Omega_d$) continuously. If $\Omega_f > \Omega_d$ the laser beam focuses until the higher order relativistic terms which were neglected in (24) dominate. This can be seen by noting the right-hand side of (22), as the radius decreases the peak laser field amplitude increases until a^2 is much greater than unity, see (19). When this occurs the relativistic focusing term in (22) decreases causing the laser beam radius to approach a constant value.

The matched beam condition in (27) can be stated in terms of a laser power requirement. The critical laser power necessary for a matched beam is

$$P_{crit} = \frac{9}{2\beta}\left(\frac{\omega}{\omega_p}\right)^2 10^9 \text{ watts}. \tag{28}$$

The beam will either focus ($P > P_{crit}$) or defocus ($P < P_{crit}$) depending on its power. The critical laser power given in (28) is in agreement with the conclusion reached in Ref. (15) where a completely different approach was taken.

Envelope Evolution with Space Charge Wave

Here we will examine the high frequency modulation on the radiation envelope when a growing space charge wave is excited by two lasers[20] with frequencies $\omega_1 - \omega_2 \simeq \omega_p$. The electron density wave given by (21) is assumed to grow linearly in t for $t > z/v_g$ and have the form

$$\delta n = n_o(\xi/z_g)(a/a_o)^2 \sin(k_{po}\xi), \tag{29}$$

where $\xi = z - v_g t$, $z_g > |\xi|$ and z_g/v_g is the characteristic linear growth time. From the assumed form of δn in (29) the function $\alpha(z,t)$ (see Eq. (21)) becomes

$$\alpha(z,t) = \frac{\xi}{z_g} a_o^{-2} \sin(k_{po}\xi). \tag{30}$$

Note that saturation of the electron density wave occurs around times equal to z_g/v_g. The solution of the laser envelope equation in (24), in the presence of the density wave, can be written in the form

$$X(z,t) = X_o(z) + X_1(z,t), \tag{31}$$

where X_o is an expression similar to in (26) and X_1 represents the effects of the density wave and results in a rapid modulation of the laser envelope. Substituting (31) into (24) and assuming that the high frequency modulations on the envelope are small compared to the average laser radius, i.e. $|X_1| \ll X_o$, the equation for X_1 becomes

$$\ddot{X}_1 = \frac{2\alpha(z,t)\Omega_f^2}{X_o^3(z)}, \tag{32}$$

where $\alpha(z,t)$ and Ω_f^2 are given by (30) and (25b) respectively.

For a laser beam initially matched to the plasma, i.e., $\Omega_f = \Omega_d$, the solution of (31) for $|\xi|$ much greater than a plasma wavelength, $|\xi k_{po}| \gg 1$ is

$$X_1 = -\frac{2\Omega_f^2/c^2}{k_{po}^2 a_o^2} \frac{\xi}{z_g} \sin(k_{po}\xi). \tag{33}$$

The radiation envelope for $\Omega_f = \Omega_d$ is, using (25b) and (33), given by

$$R_L(z,t) = R_L(0)[1 - \frac{2\beta}{R_L^2(0)k^2} \frac{\xi}{z_g} \sin(k_{po}\xi)]. \tag{34}$$

Even though the density fluctuation caused by the electron density wave can be large, the high frequency modulation on the radiation envelope is small. The ratio of the radiation envelope modulation to the equilibrium envelope radius is less than $\beta(\lambda/R_L(0))^2/2\pi^2$ which in turn is much less than unity.

Conclusions and Discussions

An envelope equation has been derived that describes the radial evolution of a radiation beam propagating through a plasma, which supports a large amplitude space charge wave. We find that relativistic effects from the electron's transverse motion can lead to radiation focusing. For a single laser, diffraction spreading dominates the focusing forces when the laser power is smaller than the critical value given by Eq. (28). In this case, the laser beam expands indefinitely. When the laser power is greater than the critical value, the laser will focus until the higher order relativistic effects dominate, see Eq. (22). We examine the effect of the large amplitude density fluctuations on the radiation envelope and find that the radiation envelope develops a small amplitude ripple.

The analysis outlined here can be generalized to multiple laser beams. The envelope equations will be coupled through the effective plasma frequency. The resulting power criteria for matched beams is expected to differ from Eq. (28).

Acknowledgments

This work is sponsored by U. S. Department of Energy, Office of Energy Research, under Interagency Agreement No. DE-AI05-83ER40117.

References

1. Laser Acceleration of Particles, AIP Conf. Proc. No. 91, Ed. by Paul J. Channell, American Institute of Physics, New York, 1982.
2. Challenge of Ultra-High Energies: Ultimate Limits, Possible Directions of Technology, an Approach to Collective Acceleration. ECFA Report 83/68 published by Rutherford Appleton Laboratory (1983).
3. T. Tajima and J. M. Dawson, Phys. Rev. Lett. 43, 267 (1979).
4. T. Tajima and J. M. Dawson, IEEE Trans. Nucl. Sci. NS-26, 4188 (1979).
5. C. Joshi, T. Tajima, J. M. Dawson, H. A. Baldis and N. A. Ebrahim, Phys. Rev. Lett. 47, 1285 (1981).
6. M. Ashour-Abdalla, J. N. Leboeuf, T. Tajima, J. M. Dawson and C. F. Kennel, Phys. Rev. A 23, 1906 (1981).
7. D. J. Sullivan and B. B. Godfrey, IEEE Trans. Nucl. Sci. NS-28, 3395 (1981).
8. J. D. Lawson, Rutherford Appleton Laboratory, Report RL-83-057, 1983. PB83-256297
9. R. Bingham, private communications (1983).
10. C. M. Tang, P. Sprangle and R. N. Sudan, Appl. Phys. Lett. 45, 375 (1984).
11. C. M. Tang, P. Sprangle and R. N. Sudan, NRL Memo Report 5324 (1985), also accepted for publication in Physics of Fluids.
12. C. Joshi, W. B. Mori, T. Katsouleas, J. M. Dawson, J. M. Kindel and D. W. Forslund, submitted for publication.

13. W. Horton and T. Tajima, submitted to Phys. Rev. A.

14. T. Katsouleas and J. M. Dawson, Phys. Rev. Lett. 51, 392 (1983).

15. G. Schmidt and W. Horton, to appear in Comments of Plasma Physics, (1985).

16. C. Joshi, C. E. Clayton and F. F. Chen, Phys. Rev. Lett. 48, 874 (1982).

17. J. D. Lawson, The Physics of Charged-particle Beams, (Clarendon Press, Oxford, 1978) Chap. 4.

18. E. P. Lee and R. K. Cooper, Part. Accel. 7, 83 (1976).

19. A. Yariv, Intro. to Optical Electronics, (Holt, Rinehart and Winston, New York, 1976) 2nd Edition, Chap. 3.

20. M. N. Rosenbluth and C. S. Liu, Phys. Rev. Lett. 29, 701 (1972).

Study of Beat-Wave Growth and Saturation

T. Tajima
Physics Department and Institute for Fusion Studies
University of Texas
Austin, Texas 78712

R.N. Sudan
Laboratory for Plasma Studies
Cornell University
Ithaca, New York 14853

ABSTRACT

We present a particle simulation study of the growth and saturation process of the plasma wave driven by the beating of laser beams and compare the results with theoretical calculations. The observed saturation time and amplitude are in fair agreement with these calculations.. The spectrum of plasma waves at the fundamental wavelength k_p and harmonics nk_p is also studied. A possible interpretation of the spectral shape is discussed in terms of the wave steepening.

INTRODUCTION

The laser beat-wave acceleration mechanism[1] and its recent advancements[2,3] are hotly discussed because of their attractive features for a possible future accelerator.[4] One of the important tasks is to determine the nature of the saturation mechanics to aid experimental diagnosis of the physical processes involved. In the present study we compare our particle simulation runs with earlier calculations.[6,7]

The code we employ is a self-consistent relativistic electromagnetic particle code[5] with $1\frac{2}{2}$ dimensions. Typical parameters are: system length $L_x = 1024\Delta$, number of electrons per cell 10, speed of light $c = 9.99\omega_p\Delta$, electron-to-ion mass ratio $m/M = 10^{-6}$, thermal velocity $v_{th} = 0.3\omega_p\Delta$, particle size $a = 1\Delta$, time step $\Delta t = 0.1\omega_p^{-1}$, laser frequencies $\omega_o = 4.29\omega_p$ and $\omega_1 = 3.29\omega_p$ with frequency separation $\Delta\omega = 1\omega_p$, the laser wavenumbers $k_o = 2\pi \times 68/1024\Delta$ and $k_1 = 2\pi \times 51/1024\Delta$, and the amplitudes of two laser beams are taken to be equal. We typically run 1000 to 1600 timesteps. Here ω_p is the plasma frequency and Δ is the grid spacing. The particles and fields obey periodic boundary conditions.

RESULTS

Figure 1 shows the observed saturated plasma wave amplitude as well as the observed saturation time for the beat-wave as a function of the laser wave amplitude. Theoretical predictions[6,7,8] at $\Delta\omega = \omega_p$ are shown as solid lines for comparison, where $\Delta\omega$ is the frequency separation of two laser beams. These calculations are carried out by utilizing the Lagrangian displacement[9] ξ of a particle postition and incorporating the electron relativistic mass effect[6,7] but neglecting the trapping and heating effects. Here the laser amplitude is written in terms of $a_o = a_1 = eE_o/m\omega_o c = eE_1/m\omega_1 c$ and the plasma wave amplitude by $a_L = eE_L/m\omega_p c$. The agreement between the simulation values of the saturation time and the level of saturated accelerating field is qualitatively

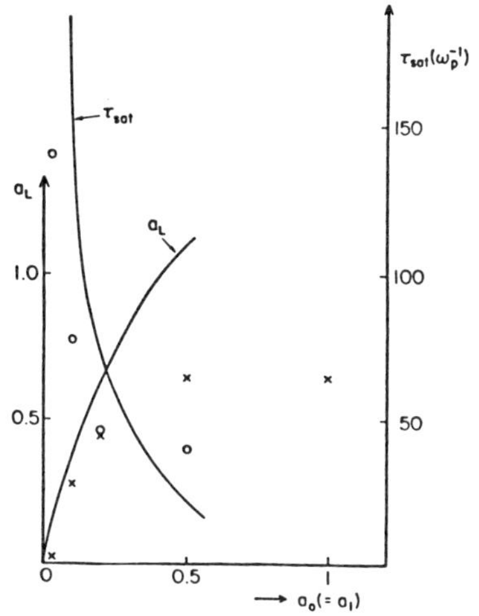

1. The saturation time and amplitude of the plasma wave created by the beat of two laser beams of equal amplitude: τ_{sat} *vs.* $a_o = v_{os}/c$ and $a_L = eE_L/m\omega_p c$ *vs.* a_o. Solid lines are from theory and circles (for τ_{sat}) and crosses (for a_L) are from simulation.

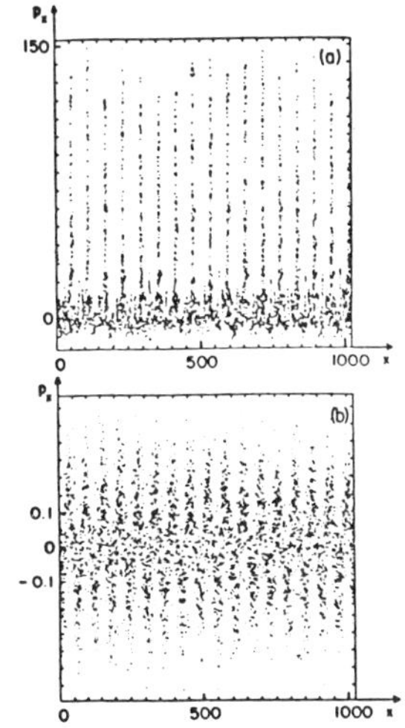

2. Phase space $(p_x - x)$ for electrons at $t = 40\omega_p^{-1}$. (a) $a_o = 0.5$ (b) $a_o = 0.03$.

reasonable, although there remains a quantitative discrepancy. The theoretical saturation time tends to exceed the observed value and the discrepancy grows as the laser fields become weaker. The theoretical saturated accelerating field is always larger than the observed value. The discrepancy reduces as the laser fields become less itense until we reach $a_o = 0.03$, which corresponds to $v_{os}(= eE_o/m\omega_o = ca_o) = v_{th}$ in our simulation. This latter observation may be interpreted as follows: At $a_o = 0.5$ the nonlinearities of laser waves, plasma waves, and particle dynamics are so large that there are some processes which compete with the beat-wave growth, such as the particle trapping in a relativistic fashion (see Fig. 2). Long tongues of electrons stretched in phase space signifies the strong nonlinearities of particle dynamics. On the other hand, if the quivering velocity v_{os} becomes as low as the electron thermal velocity v_{th} at $a_o = 0.03$, the beat-wave growth is hindered by thermal effects.

Figure 3 presents the plasma wave spectrum as a function of wavenumber for the $a_o = 0.5$ case. The fundamental wavenumber $k_p = \omega_p/c = 2\pi \times 17/1024\Delta$ is sharply peaked, while there appear multiple harmonics at $k = nk_p$ with decreasing amplitude. The electrostatic field energy $(|E_k|^2)$ for a given wavenumber k is replotted in Fig. 3. From Fig. 3 we find that $\ell n|E_k|^2$ vs. nk_p lines up linearly, suggesting that $|E_k|^2 \propto exp(-\beta|nk_p|)$, where β is a constant. On the other hand, the spectral function $S(k,\omega)$ for the electrostatic fields behaves differently. See Fig. 4. $|E_k|^2$ and the spectral function $S(k,\omega)$ are related through $|E_k|^2 = \int d\omega S(k,\omega)$. Since $S(k,\omega)$ is integrated over ω to produce $|E_k|^2$, the scaling of $S(k,\omega)$ is not necessarily equivalent to that of $|E_k|^2$. The overall behavior of $S(k,\omega)$ is sketched in Fig. 5. The dominant modes are mainly at harmonics of (k_p,ω_p), i.e., at $k = nk_p$ and $\omega = n\omega_p$. Other modes are overwhelmed by these (see Fig. 5), although some other modes at $k = nk_p$ and $\omega = n'\omega_p(n' < n)$ are also sometimes prominent along with the eigenmodes of plasma $\omega \cong \pm\omega_p$ for all k. It is therefore of interest to plot these harmonics at $k = nk_p$ and $\omega = n\omega_p$ alone (Fig. 4). The closest approximation we find in Fig. 4 for $S(nk_p, n\omega_p)$ is that $S(nk_p, n\omega_p)$ goes proportional to $(nk_p)^{-\alpha}$ where α is a constant.

Finite-amplitude nonrelativistic plasma oscillations may be described by[9]

$$\ddot{\xi} + \omega_p^2 \xi = 0, \tag{1}$$

where $x = x_o + \xi$ with x_o being the original particle position. The solution of Eq. (1) is

$$\begin{aligned} \xi =&\xi_o \sin\left[\Delta k(x-\xi) - \omega_p t\right] \\ &\cong \xi_o\left[\sin(\Delta k(x-\omega_p t) - \Delta k\xi\cos(\Delta kx - \omega_p t)\right], \end{aligned} \tag{2}$$

for $\Delta k\xi \ll 1$, where $t = 0$, $\xi = \xi \sin kx_o$. Solving for ξ, we obtain

$$\xi = \frac{\xi_o \sin(\Delta kx - \omega_p t)}{1 + \Delta k\xi_o \cos(\Delta kx - \omega_p t)}. \tag{3}$$

Furthermore, with $\phi = \Delta kx - \omega_p t$ and $A = \Delta k\xi_o$, we obtain

$$\xi(t) = \xi_o \sin(\phi)[1 - A\cos(\phi) + A^2\cos^2(\phi) - A^3\cos^3(\phi) + \ldots], \tag{4}$$

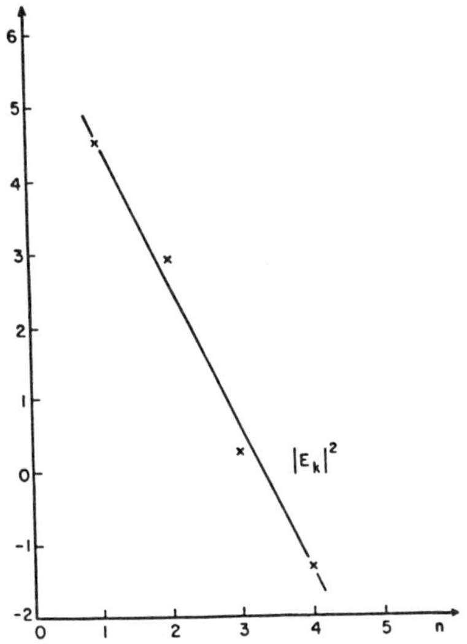

3. Logarithm of $|E_k|^2$ vs. the harmonic number n where the wavenumber k is expressed as $k = nk_p$ for the $a_o = 0.5$ case.

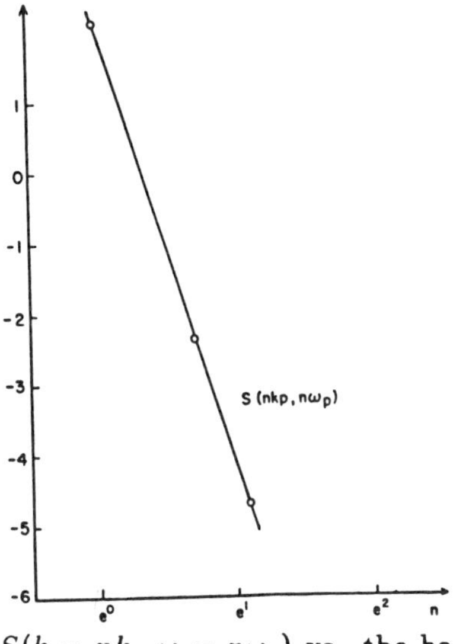

4. Logarith of $S(k = nk_p, \omega = n\omega_p)$ vs. the harmonic number n (or e^n) for $a_o = 0.5$ case.

By Fourier analyzing Eq. (3), we find

$$
\begin{aligned}
|\xi_n| &= \left| \int_0^{2\pi} \frac{d\phi}{2\pi} \exp(in\phi)\xi(\phi) \right| \\
&= \frac{1}{2\sqrt{1-A^2}} \left\{ \frac{1}{A^{n+1}} \left(\sqrt{1-A^2}-1\right)^{n-1} \left[(2-A^2) - 2\sqrt{1-A^2} \right] \right\} \\
&\approx \left(\frac{A}{2}\right)^{n-1} \qquad (\text{for} \quad A \ll 1 \quad \text{and} \quad n > 1)
\end{aligned}
\tag{5}
$$

or

$$|\xi_n|^2 \propto A^{2n-2}.$$

Equation (5) implies that the logarithm of $|\xi_k|^2$ should be linearly proportional to $2n-2$. This statement can be consistent with our simulation observation Fig. 3 if we identify the wave energy for Eq. (5) with $|E_k|^2$ in our simulation. It is well known that a finite-amplitude plasma wave which is described as perfectly harmonic in Lagrangian coordinates, in fact *steepens* when observed from an Eulerian frame, thus generating higher harmonics.

In Fig. 5 we show the observed dispersion relation of the electrostatic modes along with the relative magnitude of the harmonics. As can be seen in Fig. 5, the fundamental wavenumber dominates all the others. We also note that the plasma oscillations with k not very large have frequencies $|\omega|$ less than ω_p except for $k = k_p$. This may be related to the relativistically reduced mass of electrons. We further note that the frequency reduction of these plasma oscillations for $k \neq k_p$ and k not very large is more significant for modes which are backward propagating. This may be attributable to the electron flow due to the laser beams.

Figure 6 shows the observed frequency of the fundamental plasma wave, which shows a frequency up-shift. These data were collected using the auto-correlation technique and its Fourier transform after Blackman and Tukey.[11] Calculations[6,7,8,10] assume or argue that the frequency of the system is picked by the wave frequency difference $\Delta\omega = \omega_o - \omega_1$ and in particular $\omega_L = \Delta\omega = \omega_p$ if we pick $\Delta\omega = \omega_p$. In light of this Fig. 6 might be of interest and surprise, although we do not understand the precise reason of the observed tendency.

Figure 7 displays the electromagnetic energy spectrum as a function of wavenumber. At the time of this snapshot the spectrum cascade has about saturated. It is observed that the degree of cascade is larger when the pump amplitude is larger. Although we have not completely understood the quantitative description of the relative amplitudes, one possible explanation may be that discussed by Cohen et al.[12] We will study the cascade process more quantitatively in the future.

In conclusion we note that: (a) in the region of the validity of the Lagrangian model[6,7] i.e. of $v_{th}/c < a_{0,1} < 1$ the theoretical predictions agree qualitatively with simulation. Although the Lagrangean calculation[7] indicates the plasma wave frequency less than unity for the $\Delta\omega = \omega_p$ case, our observation $\omega_L > \Delta\omega$ may be explained by a current generation by the beating laser beams, as manifested in a slight positive frequency shift in plasma waves of $\omega \simeq \omega_p$ and $\omega \simeq -\omega_p$ in Fig. 5. (b) The spectrum of plasma waves created by the beat of laser beams is studied and it is likely that one of important elements determining the spectrum of harmonics is wave-steepening. *This work was supported by the National Science Foundation and the U.S. Department of Energy.*

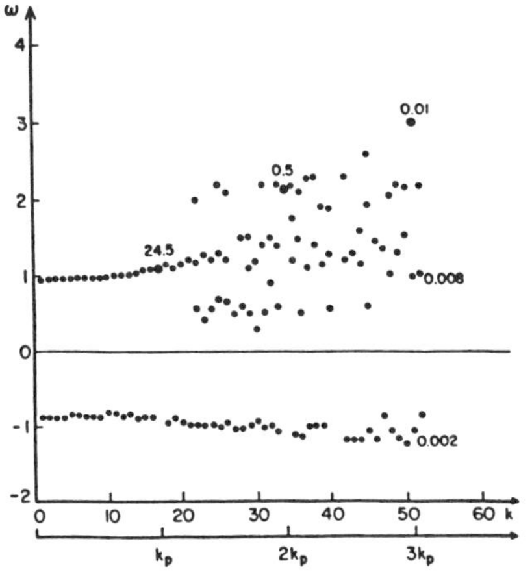

5. Observed dispersion relation ω vs. k for the electrostatic modes in a plasma with two intense laser beams. $a_o = 0.2$. Written in numbers are relative intensities of the harmonics of various plasma wave $\omega \simeq (n-j)\omega_p, k = nk_p$.

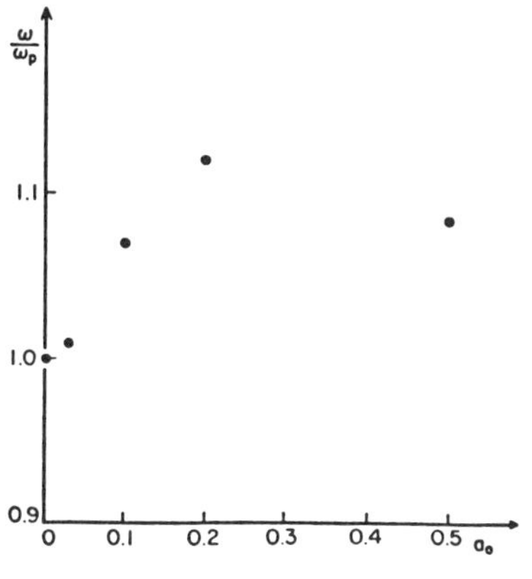

6. Observed frequencies at the fundamental ($\omega \sim \omega_p$) and $k = k_p$ as a function of the pump amplitude a_o.

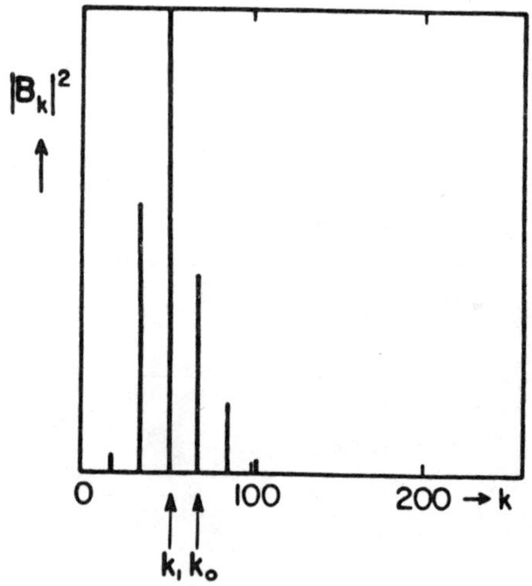

7. Electromagnetic spectrum $|B_k|^2$ vs. k in the $a_o = 0.5$ case at $t = 60\omega_p^{-1}$.

REFERENCES

1. T. Tajima and J.M. Dawson, Phys. Rev. Lett. **43**, 267 (1979).
2. T. Tajima, in *Proceedings of the 12th International Conference on High Energy Accelerators*, eds. F.T. Cole and R. Donaldson (Fermi National Accelerator Laboratory, Batavia, Illinois, 1983) p. 470.
3. T. Katsonleas and J.M. Dawson, Phys. Rev. Lett. **51**, 392 (1983).
4. T. Tajima, to appear in Laser and Particle Beams (August 1985).
5. A.T. Lin, J.M. Dawson, H. Okuda, Phys. Fluids **17**, 1995 (1974).
6. M.N. Rosenbluth and C.S. Liu, Phys. Rev. Lett. **29**, 702 (1972).
7. C.M. Tang, P. Sprangle, and R.N. Sudan, Appl. Phys. Lett. **45**, 375 (1984).
8. W. Horton and T. Tajima, to appear in Phys. Rev. A (May 1985).
9. J.M. Dawson, Phys. Rev. **113**, 383 (1959).
10. E.A. Jackson, Phys. Fluids **5**, 831 (1960).
11. R.B. Blackman and J.W. Tukey, *The Measurement of Power Spectra*, (Dover, New York, 1958).
12. B.I. Cohen, A.N. Kauffman, and K.M. Watson, Phys. Rev. Lett. **29**, 581 (1972).

Effect of Noise and Pump Depletion on the Plasma Beat Wave Accelerator

W. Horton and T. Tajima
University of Texas at Austin
Austin, Texas 78712

ABSTRACT

The limits on the effective growth of the plasma beat wave due to noise in the plasma density and the depletion of the laser driving fields are investigated theoretically.

GROWTH OF THE PLASMA WAVE AND PUMP DEPLETION

We consider the self-consistent growth of the plasma wave ω_p driven up from the beating of two high frequency, ω_1, ω_2, linearly polarized laser beams in the presence of a noisy background plasma density. The laser fields are $E_j \sin(k_j x - \omega_j t + \phi_j)\hat{e}_y$ and $B_j \sin(k_j x - \omega_j t + \phi_j)\hat{e}_z$ with $B_j = (ck_j/\omega_j)E_j$. The plasma wave is $E_p \cos(k_p x - \omega_p t + \phi_p)\hat{e}_x$ with ideal, resonant $\omega_1 - \omega_2 = \omega_p$ growth rate given by $\mathcal{E} = eE_p/m\omega_p c = \lambda t \omega_p$ where $\lambda \cong a_1 a_2/4$ is the ideal growth rate given in terms of the dimensionless quiver velocities defined in Eq. (1). We now investigate how this ideal growth can be limited due to variations in the laser-plasma wave phase relation $\psi = \phi_1 - \phi_2 - \phi_p$ due to variations in the plasma density and pump depletion.

From the transverse electromagnetic wave dispersion relation $\omega_j = (c^2 k_j^2 + \omega_p^2)^{1/2}$ it follows that the phase velocity of the beat wave $v_p = (\omega_1 - \omega_2)/(k_1 - k_2) \simeq d\omega_1/dk_1 = c(1 - \omega_p^2/\omega_1^2)^{1/2}$ equals the group velocity of the laser pulse for $\omega \gg \omega_p$. The Lorentz transformation to the beat wave frame is given by $\beta_p = \Delta\omega/\Delta kc = \omega_p/k_p c$ and $\gamma_p = (1 - \beta_p^2)^{1/2} = \omega_1/\omega_p$. The energy gain in the simple beat wave accelerator[1] occurs from reflection off the electrostatic potential $\varphi_{-wave} = \gamma_p \varphi_p$ in the wave frame and is given by $\Delta E = 2mc^2\gamma_p^2 \mathcal{E}_{max}$ where $\mathcal{E}_{max} = max(eE_p/m\omega_p c)$. We now consider two effects which can determine $\mathcal{E}_{max}$.

The energy density in the transverse laser waves is $W_j = \left\langle (E_j^2 + B_j^2)/8\pi + \frac{1}{2}mnv_{y_j}^2 \right\rangle = (E_j^2/16\pi)\left[1 + (c^2k_j^2 + \omega_p^2)/\omega_j^2\right] = E_j^2/8\pi$ and in the longitudinal plasma wave is $W_p = \left\langle E_p^2/8\pi + \frac{1}{2}mnv_x^2 \right\rangle = (E_p^2/16\pi)(1 + \omega_p^2/\Delta\omega^2) = E_p^2/8\pi$. We define the dimensionless quiver or oscillation velocities by

$$a_j = \frac{eE_j}{m\omega_j c} \quad \text{and} \quad \mathcal{E} = \frac{eE_p}{m\omega_p c} = \frac{\omega_p \xi}{c} \tag{1}$$

where in the last equation we also introduce the amplitude of the Lagrangian displacement ξ of the thermal electrons in the plasma wave. The Poisson equation gives the exact linear relation between E_p and ξ through $E_p = 4\pi en\xi$ wehre $-en\xi$ is the surface charge density introduced by the displacement of ξ of the electrons at x_o from the ions by $x = x_o + \xi(x_o, t)$, unless the wave breaking takes place. The Lagrangian equation of motion for the plasma wave is linear

$\partial_t^2\xi + \omega_p^2\xi = 0$ but describes steepening of the plasma wave through the inversion of the Lagrangian equation for $x_o = x_o(x,t)$. For a sinusoidal oscillation $\xi\sin(k_p x)$ the inversion becomes multivalued when $dx/dx_o = 1 + k_p\xi\cos(k_p x) = 0$ which describes wave breaking at the limit $k_p\xi = 1$. For plasma waves with $v_p = \omega_p/k_p \simeq c$ the breaking limit gives $\mathcal{E} = E_p/m\omega_p c = \omega_p\xi/c = 1$.

The relativistic equation of motion[2] for $\xi(x_o,t)$ follows from calculating the rate of increase of $p_x = m\gamma(\dot\xi)\dot\xi$ due to the resonant frequency part of the Lorentz force $-e\left[E_x + (v_y^{(1)}B_z^{(2)} + v_y^{(2)}B_z^{(1)})/c\right]$ to obtain

$$\gamma_{\dot\xi}^3\frac{\partial^2\xi}{\partial t^2} + \omega_p^2\xi = 2\Delta\omega c\lambda\sin\left[\Delta k(x_o+\xi) - \Delta\omega t + \Delta\varphi\right] \tag{2}$$

where

$$\lambda = \frac{a_1 a_2}{4\beta_p} \qquad \text{and} \qquad \Delta\varphi = \varphi_1 - \varphi_2.$$

For small λ the effective plasma fequency becomes $\omega_p\left[1 - \frac{3}{8}(\omega_p^2\xi^2/c^2)\right]^{1/2}$ and the linear resonant growth $\mathcal{E} = \omega_p\xi/c = \lambda t\omega_p$ stops when $\mathcal{E}_{RL} = 4(\lambda/3)^{1/3}$ at $\omega_p t_s = 3.89\lambda^{-2/3}$.

To close the system we calculate the effect of the plasma wave on the transverse wave amplitude and phase $\hat{E}_j(x,t) = E_j\exp(i\phi_j)/2i$ where $E_j(x,t) = \mathrm{Re}\hat{E}_j\exp(ik_jx - i\omega_jt)$. Reducing the $(\partial_t^2 - c^2\partial_x^2)E_y = -4\pi\partial_t j_y$, we obtain

$$(\partial_t + v_j\partial_x)\hat{E}_j = -2\pi\left\langle j_y^{n\ell}e^{-ik_jx+i\omega_jt}\right\rangle \tag{3}$$

where $j_y^{n\ell} = -e\delta n_p v_y = (4\pi)^{-1}v_y\partial E_p/\partial x$. Introducing the envelope approximation in Eq. (2) with $E_p(x,t) = \mathrm{Re}\hat{E}_p\exp(ik_px - i\omega_pt)$ we obtain for the closed system of waves

$$(\partial_t + v_1\partial_x)\hat{E}_1(x,t) = -\frac{ek_p}{2m\omega_2}\hat{E}_p\hat{E}_2e^{-i\Delta\omega t} \tag{4}$$

$$(\partial_t + v_2\partial_x)\hat{E}_2(x,t) = \frac{ek_p}{2m\omega_1}\hat{E}_1\hat{E}_p^*e^{i\Delta\omega t} \tag{5}$$

$$(\partial_t - i\frac{3}{16}\omega_p|\epsilon|^2)\hat{E}_p(x,t) = \frac{e\omega_p\Delta k}{2m\omega_1\omega_2}\hat{E}_1\hat{E}_2^*e^{+i\Delta\omega t} \tag{6}$$

where $\Delta\omega = \omega_p + \omega_2 - \omega_1$, where only two original electromagnetic waves are retained. The rate of transfer of energy from the lasers to the plasma is determined by

$$T = \hat{E}_1^* \hat{E}_2 \hat{E}_p e^{-i\Delta\omega t} + \hat{E}_1 \hat{E}_2^* \hat{E}_p^* e^{i\Delta\omega t}$$
$$= \frac{1}{4} E_1 E_2 E_p \sin(\psi - \Delta\omega t) \tag{7}$$

where $\psi = \phi_2 + \phi_p - \phi_1$ with rates given by

$$(\partial_t + v_1 \partial_x) W_1 = -\left(\frac{ek_p}{2m\omega_2}\right) T \tag{8}$$

$$(\partial_t + v_2 \partial_x) W_2 = \left(\frac{ek_p}{2m\omega_1}\right) T \tag{9}$$

$$\partial_t W_p = \left(\frac{e\omega_p \Delta k}{2m\omega_1\omega_2}\right) T \tag{10}$$

From Eqs. (8)-(10) it follows that $\Delta(W_1/\omega_1) = -\Delta(W_2/\omega_2) = -\Delta(W_p/\omega_p)$ when $k_p = \Delta k$, a manifestation of the Manley-Rowe relation. For $v_1 \cong v_2 = v_p$ the transformation of laser energy to plasma wave energy is given by

$$(\partial_t + v_p \partial_x)(W_1 + W_2) = -\frac{ek_p \Delta\omega}{2m\omega_1\omega_2} T \tag{11}$$

$$\partial_t W_p = \frac{e\Delta k \omega_p}{2m\omega_1\omega_2} T. \tag{12}$$

We use the conservation of the total energy to calculate the depletion of the laser energy $W_\ell \equiv W_1 + W_2$ by the growth of the plasma wave

$$(\partial_t + v_p \partial_x) W_\ell(x,t) = -\frac{\partial W_p(x,t)}{\partial t}. \tag{13}$$

Since $W_p/W_\ell = (\omega_p^2/\omega^2)(\mathcal{E}^2)/(a_1^2 + a_2^2) \simeq (\omega_p^2/\omega^2)(\mathcal{E}^2/8\lambda) \ll 1$, in the first approximation the solution of Eq. (13) is $W_\ell = W_\ell(x - v_p t) f(t/\tau_D)$ with depletion time τ_D given by

$$\tau_D = W_\ell/(\partial W_p/\partial t) = \tau_\ell(\omega^2/\omega_p^2)(4\lambda/\mathcal{E}_m^2)$$

where $c\tau_\ell$ is the laser pulse length. For $\mathcal{E}_m = \lambda\tau_\ell\omega_p$ the depletion length $L_D = c\tau_D = 4(c/\omega_p)(\omega^2/\omega_p^2)/\mathcal{E}_m$ and for $\tau_\ell\omega_p = 4\lambda^{-2/3}$ (if we set τ_ℓ for the relativistic detuning time) and $\mathcal{E}_m = 4(\lambda/3)^{1/3}$ the depletion length is

$$L_D = (c/\omega_p)(\omega^2/\omega_p^2)(3/\lambda)^{1/3}.$$

A similar estimate is given by T. Katsouleas et al.[3] from energy conservation. The pump depletion length is only slightly larger than the dephasing length. Thus the pump depletion effect may limit the main attractiveness of the surfatron.[4]

Since the pump depletion time τ_D is long compared with growth period $1/\omega_p\lambda$ of the plasma wave, it is convenient to calculate the decay of the laser pulse in the wave frame where

$$\frac{\partial W_\ell}{\partial t'}(x',t') \cong v_p \frac{\partial W_p}{\partial x'}(x', \lambda(x',t')), \tag{14}$$

and the loss of total laser energy $E'_\ell = \int W_\ell d^3x'$ occurs from the flux of plasma waves $v_p W_p(x' = -L_\ell)$ out the rear of the laser pulse

$$\frac{d}{dt'}E'_\ell = \int\int_{-L_\ell}^{0} \frac{d}{dt'} W_\ell da dx' = -v_p A_p W_p(x' = -L_p). \tag{15}$$

Equation (14) shows that the laser pulse decays most rapidly when $\partial W_p/\partial x'$ takes maximum.

We now calculate the growth of the plasma wave in terms of its amplitude $\mathcal{E}$ and phase ψ. The entire closed system has the three degrees of freedom Hamiltonian

$$\begin{aligned} H^{6D}(I_j,\phi_j) = & \frac{\omega_1}{\omega_p} I_1 + \frac{\omega_2}{\omega_p} I_2 + \left(1 - \frac{3}{32}\frac{e^2 I_p}{m^2c^2\omega_p}\right) I_p \\ & - \frac{ek_p(I_1I_2I_3)^{1/2}}{2m(\omega_1\omega_2\omega_p)^{1/2}} \cos(\phi_1 - \phi_2 - \phi_p) \end{aligned} \tag{16}$$

where $d\phi_j/dt\omega_p = \partial H^{6D}/\partial I_j$ and $dI_j/dt\omega_p = -\partial H^{6D}/\partial\phi_j$ reproduces the equations of motion for $(\hat{E}_j) = (\omega_j I_j)^{1/2}$ and $\phi_j = arg(\hat{E}_j)$ with $j = 1, 2, p$. Taking into account the cyclic coordinates $\phi_1 + \phi_2$, and $\phi_1 + \phi_p$ at constant $\psi = \phi_2 + \phi_p - \phi_1$ the system reduces to a four dimensional phase space I_1, I_2, I_p, ψ. The rate of change of ψ is given by

$$\begin{aligned} \dot{\psi} = & \omega_2 + \omega_p\left(1 - \frac{3}{16}\mathcal{E}^2\right) - \omega_1 \\ & - \frac{ek_p\omega_p^{1/2}\cos\psi}{4m(\omega_1\omega_2)^{1/2}} \left[I_p^{1/2}\left(\left(\frac{I_2}{I_1}\right)^{1/2} - \left(\frac{I_1}{I_2}\right)^{1/2} \right) - \frac{(I_1I_2)^{1/2}}{I_p^{1/2}} \right] \end{aligned} \tag{17}$$

For large I_1, I_2 compared with I_p the phase rotation due to the change in the strength of the laser fields is small. The condition for negligible phase rotation due to pump depletion is

$$I_p\left[\left(\frac{I_1}{I_2}\right)^{1/2}-\left(\frac{I_2}{I_1}\right)^{1/2}\right] \ll (I_1 I_2)^{1/2}$$

or approximately

$$\left(\frac{\omega_p}{\omega}\right)\mathcal{E}^2\left[\left(\frac{E_1}{E_2}\right)^{1/2}-\left(\frac{E_2}{E_1}\right)^{1/2}\right] \ll 4\lambda. \tag{18}$$

When condition (18) is satisfied the hamiltonian reduces to the usual fixed laser field hamiltonian $h(I_p, \psi)$ with one degree of freedom

$$h(I_p,\psi) = \left[\omega_1 - \omega_2 - \omega_p\left(1 - \frac{3}{32}\frac{e^2 I_p}{m^2 c^2 \omega_p}\right)\right] I_p - \frac{e k_p \omega_p^{1/2} E_1 E_2}{4 m \omega_1 \omega_2} I_p^{1/2} \cos\psi. \tag{19}$$

The mean plasma density $\langle n \rangle = n$ is taken resonant $\omega_p = \omega_1 - \omega_2$, and we now consider the effect of fluctuations $\delta\omega_p = \omega_p(\delta n/2n)$ on the evolution of the beat wave given by Eq. (19)

$$\frac{d\mathcal{E}}{dt\omega_p} = \lambda \sin\psi \tag{20}$$

$$\frac{d\psi}{dt\omega_p} = -\frac{\delta n}{2n} + \frac{3}{16}\mathcal{E}^2 - \frac{\lambda}{\mathcal{E}}\cos\psi. \tag{21}$$

PLASMA NOISE

We define the dimensionless variance of the density fluctuations $\langle \delta n^2 \rangle = n^2\sigma$ and the correlation time τ_c through $\langle \delta n(t_1)\delta n(t_2 = t_1 + \tau)\rangle = (2\pi)^{-1}\int_{-\infty}^{+\infty} d\omega N(\omega) exp(-i\omega\tau)$ where $\int_{-\infty}^{+\infty} d\tau \langle \delta n_t \delta n_{t+\tau}\rangle = \tau_c \langle \delta n^2 \rangle = N(\omega = 0)$. A typical noise spectrum is the Lorentzian $N(\omega) = 2\delta\omega(\omega^2 + \delta\omega^2)^{-1}$, for which $\tau_c = 1/\delta\omega$.

Taking the ensemble average of Eq. (20) for noise near gaussian statistics gives

$$\frac{d\langle \mathcal{E} \rangle}{dt\omega_p} = \lambda \sin\langle \psi \rangle e^{-\nu t} \tag{22}$$

where

$$\nu = \frac{1}{2t}\int_{o}^{t} dt_1 \int_{o}^{t} dt_2 \langle \dot{\psi}(t_1)\dot{\psi}(t_2)\rangle . \tag{23}$$

When the noise dominates the nonlinear frequency shift in Eq. (21), the lowest order calculation of (23) gives

$$\nu = \frac{\omega_p^2}{4}\int_{-\infty}^{+\infty} \frac{\langle \delta n_t \delta n_{t+\tau}\rangle}{n^2} d\tau = \frac{\omega_p^2 \tau_c}{4}\sigma \tag{24}$$

valid for $\nu\tau_c = \omega_p^2\tau_c^2\sigma/4 < 1$. In Ref. 5 we extend the calculation to $\sigma > \sigma_c = 4/\omega_p^2\tau_c^2$ by using renormalized turbulence theory. In the noisy plasma the plasma wave growth is given by a renormalized propagator given in Eqs (16)-(18) in Ref. 5. For the Lorentzian noise spectrum the frequency integral in the renormalized propagator is performed exactly to give

$$\nu^2 + \nu\delta\omega - \omega_p^2\sigma/4 = 0$$

with

$$\nu = \frac{1}{2}\left[(\omega_p^2\sigma + \delta\omega^2)^{1/2} - \delta\omega\right] = \begin{cases} \omega_p^2\sigma/4\delta\omega & \text{for} \quad \sigma < \sigma_c \\ \omega_p\sigma^{1/2}/2 & \text{for} \quad \sigma > \sigma_c \end{cases} \tag{25}$$

The effective growth time is now $1/\nu$ and the maximum amplitude from Eq. (22) is

$$\mathcal{E}_m = \omega_p\frac{\lambda}{\nu} \qquad \text{for} \qquad \frac{\nu}{\omega_p} > \frac{1}{4}\lambda^{2/3}$$

Numerous examples of the effect of noise on the beat wave growth rate are given in Ref. 5.

We conclude that limits on the amplitude of the accelerating plasma electric field may be set by either the pump depletion length $L_D = c\tau_D$ or the noise limited growth before reaching the ideal accelerator strength[2] under the condition derived in the present work.

ACKNOWLEDGEMENTS

*This work supported by the National Science Foundation and the United States Department of Energy.

REFERENCES

1. T. Tajima and J.M. Dawson, Phys. Rev. Lett. **43**, 267 (1979).
2. M.N. Rosenbluth and C.S. Liu, Phys. Rev. Lett. **29**, 701 (1972).
3. T. Katsouleas et al., *Proceedings of the Laser Acceleration Workshop*, Malibu, CA 1985.
4. T. Katsouleas and J.M. Dawson, Phys. Rev. Lett. **51**, 392 (1983).
5. W. Horton and T. Tajima, Phys. Rev. **A**, 1985 (to appear).

DOUBLE BEAT-WAVE MECHANISM TO KEEP PARTICLE IN PHASE WITH ACCELERATING PLASMA WAVE

Paul L. Csonka*
Brookhaven National Laboratory, Upton, NY 11973

ABSTRACT

Two mechanisms are described by which the phase matching between accelerated particle and accelerating plasma wave in a plasma beat-wave accelerator can be maintained indefinitely. 1) Interference between two beat-waves to cancel a) the fields in the interfering electromagnetic waves, b) electron oscillations which would otherwise be set up by those waves. 2) Alternating plasma density used in conjunction with one or two beat-waves.

INTRODUCTION

The potential well which accelerates particles in a plasma beat-wave accelerator,[1] moves with a speed

$$v_g = c\left[1 - (\omega_p/\omega)^2\right]^{1/2} , \tag{1}$$

where ω_p is the plasma frequency, and ω is the frequency of the electromagnetic wave which sets up the plasma oscillations, which in turn generate the accelerating potential. Particles with speeds $v > v_g$ will eventually overtake the potential well, and that puts on upper limit on the energy they can acquire. This limitation can be overcome by the proposed surfatron mechanism[2,3] in which the particle trajectory makes an angle $\theta \neq 0$ with the line along which the potential well travels. Two difficulties arise in connection with the surfatron mechanism. First, for $\theta \neq 0$ the potential well will tend to curve the particle trajectory. To compensate for that, one has to impose an external magnetic field. Second, since the particle velocity has a component parallel to the plasma wave front, the transverse size of the accelerator has to be larger than if $\theta = 0$ were to hold, and in addition, some of the electromagnetic wave energy will be wasted. The second difficulty can be significantly alleviated through the ingenious device of "transverse optical mixing."[4]

Two alternatives to the surfatron mechanism will be described in the present note. Both insure continuing phase matching between the accelerated particle and the accelerating plasma wave. In both mechanisms phase matching is maintained even if the particle and plasma wave travel parallel to each other ($\theta = 0$), and in neither mechanism is an external magnetic field required.

DOUBLE BEAT-WAVE MECHANISM

To insure continued phase matching through this mechanism, one uses not one but two beat-waves. These beat-waves are arranged so that they interfere constructively whenever the particle is in phase with the accelerating potential well created by both of them. On the other hand, when the particle gets out of phase with the potential well created by, say, the first beat-wave, the second beat-wave will tend to cancel the effect of the first one, so that the effect of deceleration will be correspondingly reduced. The particle will then be allowed to essentially drift through these regions until it again gets into phase with the accelerating potential, where it will be accelerated further, etc.

Choose four electromagnetic plane waves of equal amplitudes, A, and equal polarization state, all traveling parallel to the z axis, which is also the direction of motion of the accelerated particle. Let all four waves be in phase, with the electric field maximum at t = 0 and z = 0. Denote the circular frequency of the ith wave by ω_i (i = 1, 2, 3, 4), the electron density in the plasma by ρ_e, and the plasma frequency by ω_p. Choose

$$\omega_1 - \omega_2 = \omega_p \tag{2a}$$

$$\omega_3 - \omega_4 = \omega_p \tag{2b}$$

$$\omega_3 - \omega_1 = \Delta\omega \ , \tag{2c}$$

The $\Delta\omega$ is to be specified later. With this choice the first two waves will set up a beat-wave with beat frequency ω_p, and so will the third and fourth wave.

Assume, as usual for plasma beat wave accelerators, that $\omega_p/\omega_i \ll 1$, for i = 1, 2, 3, 4. Then the phase velocity of the plasma wave set up by an electromagnetic wave pulse will equal the group velocity of that pulse.[1] Denote the group velocity in the plasma of the first and second beat-wave by v_{g1} and v_{g2} respectively. The wave vector of the plasma wave set up by the first and second beat-waves, respectively, will be

$$\begin{aligned} k_{p1} &= \frac{\omega_p}{v_{g1}} \cong \frac{\omega_p}{c}\left(1 + \frac{1}{2}\left(\frac{\omega_p}{\omega_1}\right)^2\right) \\ k_{p2} &= \frac{\omega_p}{v_{g2}} \cong \frac{\omega_p}{c}\left(1 + \frac{1}{2}\left(\frac{\omega_p}{\omega_1 + \Delta\omega}\right)^2\right) . \end{aligned} \tag{3}$$

We will refer to the beat-wave between the two beat-waves as "super beat-wave." The amplitude (envelope) of the super beat-wave varies as

$$\begin{aligned} |A_s| &= 4\left|A\cos\frac{1}{2}\left((\omega_1 - \omega_3)t - (k_1 - k_3)z\right)\right| \\ &\cong 4\left|A\cos\frac{1}{2}\left\{(\Delta\omega)t - \frac{\omega_p}{c}\frac{1}{2}\left(\left(\frac{\omega_p}{\omega_1}\right)^2 - \left(\frac{\omega_p}{\omega_1 + \Delta\omega}\right)^2\right)z\right\}\right| \\ &\xrightarrow{\Delta\omega/\omega_1 \to 0} 4\left|A\cos\frac{1}{2}\left\{\Delta\omega\left(t - \left(\frac{\omega_p}{\omega_1}\right)^3 \frac{z}{c}\right)\right\}\right| . \end{aligned} \tag{4}$$

If the accelerating potential well is L_+ long and travels with a speed v_g (all quantities in this paper are measured in the laboratory frame), then the time it takes for a particle with speed v to move across the accelerating potential well, from one end to the other, is $t_L = L_+/(v - v_g)$. The distance the particle travels in the plasma during this time is

$$L = L \frac{v}{v - v_g} , \tag{5}$$

assuming v is essentially unchanged during the process.

We describe two distinct methods to realize the double beat-wave mechanism.

a) In the first method $\Delta\omega$ is chosen so that while the particle travels a distance L, one half of a complete super beat-wave period passes over its

$$\pi = \left((\omega_1 - \omega_3)t_1 - (k_1 - k_3)L\right) . \tag{6}$$

Let λ_{b1} be the wavelength of the plasma wave generated by the first electromagnetic beat-wave. Assume $L >> \lambda_{b1}$. Then to insure condition (6), one needs

$$\Delta\omega = \pi \frac{v - v_g}{2L_+} \left(1 - (\frac{\omega_p}{\omega_1})^3 \frac{v}{c}\right)^{-1} . \tag{7}$$

To illustrate, consider the case when $\omega_1 = 1.88 \cdot 10^{15}$ sec^{-1}, $\omega_p = 5.64 \cdot 10^{13}$ sec^{-1}, and $v \not\approx c$. Then $t_L = 2.5 \cdot 10^{-10}$ sec, L = 7.8 cm, and one needs: $\Delta\omega = 1.27 \cdot 10^{10}$ sec^{-1}, so that $\Delta\omega/\omega_1 = 6.8 \cdot 10^{-6}$.

b) In the second method one chooses $\Delta\omega$ to be such that while the particle travels a distance 2L, the second beat-wave moves ahead of the first one by a distance λ_{b1}:

$$\pi = (k_{p1} - k_{p3})L . \tag{8}$$

Then if the two beat-waves are in place when the particle is inside an accelerating potential well, then they will again be in phase when the particle is reaches the next potential well. Halfway between those two events, however, the effect of the two beat-waves cancel, which tends to eliminate any deceleration which would otherwise occur.

As an illustration again choose ω_1, ω_p and v as in the example given under a._ Then from Eq. (8) one finds $\Delta\omega = 0.779 \cdot 10^{15}$ sec^{-1}, i.e. $\Delta\omega/\omega_1 = \sqrt{2}\ \omega_1$.

This method generally requires higher $\Delta\omega/\omega_1$ values than the method described under a), that feature is sometimes desirable.

When method b) is used, the common polarization of the waves with circular frequency ω_1 and ω_2 may differ from the common polarization of the other two waves.

ALTERNATING DENSITY (DOUBLE BEAT WAVE) MECHANISM

In this mechanism the plasma is subdivided into regions with different electron densities. In regions of the first type the electron density is ρ_{eA}, while in regions of the second type, the density of electrons is ρ_{eB}. We will refer to regions of the first and second type as A-regions and B-regions respectively. The plasma frequency in the A-regions and B-regions will be denoted by ω_{pA} and ω_{pB} respectively. Choose two electromagnetic plane waves with equal amplitudes, and equal polarization, travelling along the z axis. Let their frequencies, ω_i and ω_2 satisfy

$$\omega_1 - \omega_2 = \omega_{pA} . \tag{9a}$$

These two waves will set up a beat-wave which will generate resonant electron oscillations in any region-A. On the other hand, if ρ_{eA} and ρ_{eA} are sufficiently different, then this same beat-wave cannot set up resonant electron oscillations in any B-region. Choose the length of any A-region to be L, corresponding to the value of v at the time when the particle passes through that region. Then if the beat-wave is correctly phased relative to the particle, it will continuously accelerate it while the particle moves across the A-region. As the particle leaves the accelerating potential well, and would enter a decelerating well, the particle enters a B-region where it experiences no significant acceleration, drifts through this region, enters the next A-region where it is acclerated further, etc.

One can improve the above scheme by utilizing two more electromagnetic plane waves with equal amplitudes and polarization, both travelling along the z axis. Choose their frequencies according to

$$\omega_3 - \omega_4 = \omega_{pB} ,$$

and Eq. (2c). These two waves will set up a second beat-wave which will generate resonant electron oscillations in any B-region, but not in the A-regions. If the second beat-wave is correctly phased relative to the particle, it will accelerate it across B-regions. Thus the two beat-waves together with the alternating plasma density, will insure continuous particle acceleration while the particle moves along the entire length of the plasma beat-wave accelerator.

The common polarization of the first two electromagnetic waves need not be the same as the common polarization of the other two.

In all the schemes discussed in this note, care must be taken to choose $\Delta\omega$ so that except for the wanted frequencies, no other beat frequency between the various waves should overlap with any resonant frequency regions of the plasma.

ACKNOWLEDGMENTS

I am grateful to John M. Dawson, Chan Joshi and Roger G. Evans for discussion of this problem.

REFERENCES

*Permanent address: Institute of Theoretical Science, University of Oregon, Eugene, Oregon 97403.

(1) T. Tajima and J.M. Dawson, Phys. Rev. Lett. 43, 267 (1979).

(2) T. Katsouleas and J.M. Dawson, IEEE Transactions on Nuclear Science, Vol. NS-30, 3241 (1983).

(3) W. Mori, C. Joshi and J.M. Dawson, IEEE Transactions on Nuclear Science, Vol. NS-30, 3244 (1983).

(4) C. Joshi, W.B. Mori, T. Katsouleas, J.M. Dawson, J.M. Kindel, and D.W. Forslund, Nature 311, 525 (1984).

RELATIVISTIC ELECTRON ACCELERATION BY NET INVERSE BREMSSTRAHLUNG IN A LASER-IRRADIATED PLASMA

S. H. Kim and K. W. Chen
Center for Accelerator Sciences and Technology
The University Texas at Arlington
Arlington, Texas 76019

ABSTRACT

Using the quantum-kinetic method, the net acceleration of relativistic electrons in a laser-irradiated plasma is studied as a function of the relevant paprameters of the incident laser wave and the plasma wave. It is suggested that, in general, the net acceleration in laser-produced turbulent plasmas is primarily due to inverse bremsstrahlung proceses, and the acceleration gradient exceeds several hundreds gigavolt per meter when the electron energy is large (TeV) and the momentum spread of the beam is properly controlled.

During the past few years there has been considerable interest in the possible exploitation of laser waves as a means to accelerate particles to high energies ($\gtrsim$ TeV). The accessibility of its extraordinarily large laser electric field intensities offers a possible means to accelerate electrons to ultra-high energies required[1] in future exploration of subatomic phenomena.

Since the electromagnetic wave is transverse and its phase velocity is c, recent efforts[2,3] have concentrated on ideas in which the laser wave induces a longitudinal wave in a medium with phase velocity close to c. However, since the laser-induced longitudinal wave usually has extremely short wavelengths (~ 10 μm), it is difficult in practice to inject electrons into the proper region in which positive phase matching need to be maintained. Furthermore, it is also difficult to eliminate by-product noise and the co-propagating laser wave, which both could disturb phase matching, from the region in which the longitudinal wave propagates. A different approach without such potential problems is herein proposed.[4] This phase-matching independent approach utilizes a dc force acting on the electrons through actions by inverse and stimulated bremsstrahlung processes in a turbulent plasma field. The idea for the generation of the dc force is presented below.

If an electron beam is injected into a laser-irradiated plasma, the individual electron interacts with the laser light and the plasma field. The potential amplitude of plasma field as seen by the beam electron is enhanced by γ while that of the laser light is not enhanced (since the laser wave is transverse). As long as the potential of the laser field as seen by the beam electron is greater than that of the plasma field as seen by the same electron, the beam electron absorbs or emits a larger number of laser photons than the plasmons. Since the momenta of the laser photons are all in the laser beam direction, the electron is accelerated only in the direction of the laser beam by the emission or absorption of laser photons. Further, if the number of the absorbed photons is greater than the number of the emitted photons, the electron is surely accelerated in the positive direction of the laser beam. The absorption (emission) of the laser photons is possible only by inverse (stimulated) bremsstrahlung. Therefore, the net inverse bremsstrahlung (the absorption by inverse bremsstrahlung minus the emission by stimulated bremsstrahlung) should give rise to a net acceleration (the ponderomotive force).[5] The net inverse bremsstrahlung has been credited for some anomalous phenomena, which have much larger magnitudes than the expected ones in conventional concepts.[6,7] Manheimer[8,9] showed first that the net inverse bremsstrahlung is the dominant energy and momentum transfer mechanism from the laser wave to the plasma electrons. To be specific, consider the simple photon concept in which the force can be expressed as

$$\vec{F} = \left(\frac{dN}{dt}\right)_{\text{net ib}} \hbar\vec{k}_o = \frac{1}{c}\frac{d\varepsilon}{dt}\hat{k}_o \quad , \tag{1}$$

where $(dN/dt)_{\text{net ib}}$ is the number of the net absorbed photons per unit time through the net inverse bremsstrahlung (net ib) in the plasma wave, $\vec{k}_o$ is the wave vector of the laser wave, and $d\varepsilon/dt$ is the energy transferred to the electron per unit time from the laser wave through the net inverse bremsstrahlung. It is stressed here that in laser-plasma acceleration, only a statistical treatment of physical quantities (quantities averaged over many particles) is meaningful, since the plasma is generally turbulent, and inverse and stimulated bremsstrahlung are quantum-mechanical processes. Therefore, $d\varepsilon/dt$ should be properly calculated by the quantum-kinetic method.[10-11] A similar technique was used in the study of anomalous absorption in laser fusion,[12] electron cyclotron resonance heating,[13] and electron cyclotron masing.[13]

Following the standard quantum-kinetic method, consider an electron beam travelling in the z direction in a laser-irradiated plasma. The laser light is assumed to be a circularly polarized electromagnetic wave with the wave vector $\vec{k}_o = k_o\hat{z}$ and the frequency ω_o. Consider an inertial frame K′ moving in the z direction with a mean velocity $\vec{u}$ of the beam electrons relative to the laboratory frame K. The vector potential in the frame K′ is written as

$$\vec{A}(\vec{r}',t') = A_o[\hat{x}' \cos(k_o' z' - \omega_o' t') + \hat{y}' \sin(k_o' z' - \omega_o' t')] \quad , \qquad (2)$$

where $k_o' = (1 - u/c)^{\frac{1}{2}} k_o/(1 + u/c)^{\frac{1}{2}}$ and $\omega_o' = (1 - u/c)^{\frac{1}{2}} \omega_o/(1 + u/c)^{\frac{1}{2}}$.

When the beam electrons have sufficient energies, they do not encounter binary collisions with plasma particles so that the laser-driven plasma field is not significantly perturbed by the electron beam. In the laboratory frame K, the plasma field can be considered longitudinal since the speed of the plasma electron is much less than the speed of light c. The scalar potential of the plasma wave in the frame K is given by,

$$\phi_p(\vec{r},t) = \phi(\vec{k},\Omega)\exp[i(\vec{k}\cdot\vec{r} - \Omega t)] + c.c \qquad (3)$$

where c.c. means the complex conjugate of the preceding term. The scalar and vector potentials of the plasma field as seen by the beam electron in the frame K' are

$$\phi_p(\vec{r}',t') = \phi(\vec{k}',\Omega')\exp[i(\vec{k}'\cdot\vec{r}' - \Omega' t')] + c.c \qquad (4)$$

$$\vec{A}_p' (\vec{r}',t') = -\frac{u}{c}\, \phi_p'(\vec{r}',t')\hat{z} \qquad (5)$$

where

$$\phi'(\vec{k}',\Omega') = \gamma\phi(\vec{k},\Omega),\ k_z' = \gamma(k_z - \Omega u/c\ ),\ k_\perp' = k_\perp, \text{ and } \Omega' = \gamma(\Omega - k_z u). \qquad (6)$$

Here, $\gamma = (1 - u^2/c^2)^{-\frac{1}{2}}$ and $\perp$ is the direction perpendicular to the z′.

In the frame K the wavelength of the laser wave is generally longer than those of the plasma waves and the phase velocities of the plasma waves are much smaller than c, we find from Eqs. (2) and (6) that the wavelength of the laser wave as seen by the beam electrons is far greater than those of the plasma waves. The spatial dependence of the laser waves can thus be neglected in inverse and stimulated bremsstrahlung processes of beam electrons in the plasma field. The vector potential of the laser wave can then be written as

$$\vec{A}'(t') = A_o(\hat{x}' \cos \omega_o' t' + \hat{y}' \sin \omega_o' t') \tag{7}$$

Note that this approximation takes into account the momentum transfer from the laser wave to the electron since $k_o = \omega_o/c$ has been used in the derivation of Eq. (1). We use the Schrödinger equation for the description of the dynamics of the individual beam electrons in the frame K′ since the beam electrons are most likely non-relativistic in the frame K′. Therefore the result from the use of this equation should not be extrapolated into the region where the momentum spread of the beam electrons in the frame K′ is far greater than mc. To continue, the time-dependent Schrödinger equation describing the dynamics of the individual beam electrons interacting with both the incident laser wave and the plasma field in the frame K′ is[15]

$$i\hbar \frac{\partial\psi}{\partial t'} = [\frac{1}{2m} (\frac{\hbar}{i} \vec{\nabla}' + \frac{e}{c} \vec{A}' + \frac{e}{c} \vec{A}_p')^2 - e\phi_p']\psi \tag{8}$$

The solution of Eq. (8) in the absence of the plasma wave is (normalized in a box of unit volume)

$$\psi^o(\vec{r}',t') = \exp(\frac{i}{\hbar} \vec{p}' \cdot \vec{r}' - \frac{i}{2m\hbar} \int_0^{t'} |\vec{p}' + \frac{e}{c} \vec{A}'(t_1)|^2 dt_1) \tag{9}$$

As long as $\gamma|\phi(\vec{k},\Omega)| << |A_o|$, we can consider the plasma wave as the first-order perturber to the motion of the beam electron moving in the laser wave. Using first-order perturbation theory,[12] Eqs. (6)-(9), and the assumption that

$mc > |p_z'|$, the transition probability per unit time for the transition from a state with momentum $\vec{p}_1'$ to a state with momentum $\vec{p}_2'$ is

$$T(\vec{p}_1' \rightarrow \vec{p}_2') = \frac{1}{2T\hbar^2} \left| \iint_{-T}^{T} \psi_2^{o*}(\vec{r}',t')\{\frac{e\hbar}{2mci}[(\nabla'\cdot\vec{A}_p') + \vec{A}_p'\cdot\nabla'] - e\phi_p'\}\psi_1^o(\vec{r}',t')d^3r'dt' \right|^2$$

$$= \frac{2\pi e^2}{\hbar} |\phi'(\vec{k}',\Omega')|^2 \sum_{\substack{n=-\infty \\ n\neq 0}}^{n=\infty} J_n^2(\frac{\mu}{\omega_0})\delta(\varepsilon' - \hbar\Omega' + n\hbar\omega_0')\delta_{\vec{p}_2'-\vec{p}_1',\hbar\vec{k}'} + S(-\vec{k},-\Omega), \quad (10)$$

where $\varepsilon' = (p_2'^2 - p_1'^2)/2m$, $\mu = eA_0 k_\perp/mc$, J_n is the Bessel function of order n, and $S(-\vec{k},-\Omega)$ means the same term as the preceding term except that $\vec{k}$ and Ω are replaced by $-\vec{k}$ and $-\Omega$, respectively.

From Eq. (10) the average energy transfer to the electron having a momentum $\vec{p}'$ from the laser wave per unit time is

$$P(\vec{p}') = \lim_{\hbar\to 0} \sum_{\vec{p}_f'} T(\vec{p}' \rightarrow \vec{p}_f')(\vec{p}'^2 - \vec{p}_f'^2)/2m$$

$$= \pi m e^2 |\vec{\phi}'(k',\Omega')|^2 \sum_{n=-\infty}^{n=\infty} J^2(\frac{\mu}{\omega})(\Omega' - n\omega_0') \frac{\partial\delta[p_{\vec{k}'}' - m(\Omega' - n\omega_0')/k']}{\partial p_{\vec{k}'}'} \quad (11)$$

where $p_{\vec{k}'}' = \vec{p}'\cdot\vec{k}'/k'$. In the derivation of Eq. (11), the terms which change their signs as $(\vec{k},\Omega,n)$ changes to $(-\vec{k},-\Omega,-n)$ have been canceled out.

The average energy per unit time transferred to the electron from the laser wave through the net inverse bremsstrahlung is given by

$$\frac{\partial\varepsilon'}{\partial t'} = \int f(\vec{p}')P(\vec{p}')d^3p' \quad , \quad (12)$$

where $f(\vec{p}')$ is the momentum dist, ibution of the beam electrons as seen in the frame K′, which is assumed to be Gaussian and is given by,

$$f(\vec{p}\,') = [2\pi(\Delta p)^2]^{-3/2}\exp[-p'^2/2(\Delta p)^2] \quad , \tag{13}$$

where Δp is the width of momentum distribution in the frame K'.

The effective acceleration electric field is defined as $E_{eff} = F_z/e$, where F_z is the z-component of the force given by Eq. (1). By combining Eq. (1), (11)-(13), and using the appropriate Lorentz transformation of electric and magnetic fields, we obtain

$$E_{eff} = E'_{eff} = \left(\frac{\pi}{2}\right)^{\frac{1}{2}} \frac{m^2 e}{c(\Delta p)^3} \frac{|\vec{\phi}'(k',\Omega')|^2}{k'} \sum_{\substack{n=-\infty \\ n\neq 0}}^{n=\infty} J_n\left(\frac{\mu}{\omega'_o}\right)\exp\left[-\frac{m^2(\Omega' - n\omega'_o)^2}{2(\Delta p)^2 k'^2}\right](\Omega' - n\omega'_o)^2 \tag{14}$$

Equation (14) can be readily evaluated for two representable cases given below.

For the case where $\vec{k}$ is parallel to $\vec{k}_o$, $\mu/\omega'_o = 0$ and $J_n(\mu/\omega'_o) = 0$ for $n \neq 0$. Accordingly, we have $E_{eff} = 0$ for this case. This implies that the plasma wave whose propagation direction is parallel to the laser beam does not behave as a catalyzer field for either inverse bremsstrahlung or stimulated bremsstrahlung. It is noted that the case in which $\vec{k}$ is parallel to $\vec{k}_o$ typically occurs in the so-called beat-wave concept.[16] In that case, electrons are accelerated by a phase matching with the Lorentz force of the longitudinal plasma wave, as evaluated classically.

For the case in which plasma turbulence is produced by currently available high-power CO_2 lasers, we model the turbulent plasma waves as a single wave having a wave number equal to the Debye wave number k_D, a frequency Ω, a propagation angle of 135° with respect to the direction of the laser beam, and a phase velocity considerably less than the velocity of light. In the model $\mu/\omega'_o \cong 2\gamma e E_o k_D/m\omega_o^2 \gg 1$ and $\gamma\Omega \gg eE_o k_D/m\omega_o$ generally hold for $\gamma \gtrsim 2$, where E_o is the electric field amplitude of the laser wave.[8-9,12] In this case, we can write the approximation

$$\sum_{\substack{n=-\infty \\ n\neq 0}}^{n=\infty} J_n^2\left(\frac{\mu}{\omega_o}\right)(\Omega' - n\omega_o') \frac{\partial\delta[p'_{\vec{k}'} - m(\Omega' - n\omega')/k']}{\partial p'_{\vec{k}'}} = \Omega' \frac{\partial\delta[p'_{\vec{k}'} - m\Omega'/k']}{\partial p'_{\vec{k}'}} \tag{15}$$

Substituting Eq. (15) into Eq. (11) and repeating the same procedure to derive Eq. (15), we have

$$E_{eff} = \left(\frac{\pi}{2}\right)^{\frac{1}{2}} \frac{\gamma^3 k_D (KT_p)^2 G^2}{emc^2 x^3} \exp(-1/2x^2) \text{ for } 2 \lesssim \gamma << |A_o|/|\phi(\vec{k},\Omega| \quad , \tag{16}$$

where $x = \Delta p/mc$, KT_p is the thermal energy of an average plasma electron, and $G = |e\phi(\vec{k},\Omega)|/KT_p$ is the plasma turbulence parameter, which is comparable to (but smaller than) 1 for strongly turbulent plasmas. Note that Eq. (6) has been derived by using the same model and methodology for the derivation of the electron temperature equation in laser induced plasma,[5,12] which had been shown to agree well with experiments[17-20] and other theory.[8-9]

To continue from Eq. (17), we find that the maximum effective electric field occurs at $\Delta p = mc/\sqrt{3} = 0.58$ mc, which is consistent with the assumption for the use of the Schrödinger equation, and that E_{eff} is at the plateau level (saturation) in the E_{eff} versus E_o curve for presently available laser intensities. This saturation is a general characteristic of multiphoton processes. For an order of magnitude estimate of E_{eff}, $\Delta p = 0.58$ mc, $N = 1.6 \times 10^{16}$ cm^{-3}, $KT_p = 10$ KeV, $G = 0.6$, which are representative parameters in laser-fusion and accelerator experiences, we obtain $E_{eff} = 109$ MeV/m for $\gamma = 2$, and $E_{eff} = 872$ MeV/m for $\gamma = 4$. The ratio of the effective acceleration electric field acting on an average plasma electron to the turbulent electric field amplitude KT_pGk_D, is

$$R = \left(\frac{\pi}{2}\right)^{\frac{1}{2}} \frac{KT_p G}{emc^2 x_p^3} \exp(-1/2x_p^2) \quad , \tag{17}$$

where $x_p = (KT_p/m)^{\frac{1}{2}}/c$. This R value for the above case is $R << \left(\frac{\pi}{2}\right)^{\frac{1}{2}} \frac{0.6 \text{ KeV}}{0.5 \text{ MeV}} << 1$, so that the dc force (the ponderomotive force) acting on the plasma electron cannot eject the plasma electrons from the plasma.

Equation (16) also shows that the effective acceleration electric field is proportional to $[\gamma^3/(\Delta p)^3]\exp[-m^2c^2/2(\Delta p)^2]$. As the electron beam travels in the laser-irradiated plasma, not only γ but also Δp increases. The variation of Δp as the beam electron travels in the plasma, calculated by means of the quantum-mechanically extended Boltzmann equation and the Dirac equation, was found to change very slowly. A practical control of Δp may then be possible by the installation of momentum controllers (energy chopper) just before the injection of the electron beam into the plasma. Details of this discussion will be presented separately.

If γ is so great that $\gamma|\phi_p|>|A_o|$, then the electcron absorbs more likely the plasmons of the turbulent plasma field than the photons. Since the directions of the plasmons are random (isotropic for fully developed turbulence), the electron cannot get the net acceleration. Therefore, we should keep the level of the turbulent field $|\phi_p|$ much lower than $|A_o|/\gamma$ for the acceleration of high-energy electrons. Hence, for a CO_2 laser of intensity $I = 10^{16}$ W/cm^2 and electrons with $\gamma = 2 \times 10^6$ (1 TeV), the turbulent electric field should be less than 1.37 KV/cm. If the condition $\gamma|\phi_p| << |A_o|$ is met, the effective acceleration electric field for high-energy electron ($\gtrsim$ TeV) is

$$E_{eff} << \left(\frac{\pi}{2}\right)^{\frac{3}{2}} \frac{\gamma k_D e|A_o|^2}{mc^2 x^3} \exp(-1/2x^2) \quad , \tag{18}$$

which can exceed several hundreds gigavolt per meter for most practically available quiet plasmas and high-intensity lasers.

In conclusion we have derived the effective acceleration electric field acting on beam electrons due to the net inverse bremsstrahlung in laser irradiated plasma using the quantum-kinetic method. The calculation shows that in the laser-plasma acceleration concept, beam electrons can indeed be accelerated to more than several hundred gigavolt per meter level provided if the momentum spread of the beam is properly controlled. This high acceleration electric field is entirely due to the net multiphoton inverse bremsstrahlung.

This work was supported in part by the U.S. Air Force Office of Scientific Research under the grant No. AFOSR-83-0368 and by the Texas Engineering Experiment Station.

REFERENCES

1. A. Salam, The challenge of ultra-high energies, Proceedings of the ECFA-RAL meeting held at New College, Oxford, Sept. 1982, P.1.
2. Proceedings of Laser Acceleration Workshop, Los Alamos National Laboratory, edited by P. Channell, 1982.
3. Proceedings of this conference.
4. S. H. Kim and K.W. Chen, CAST-85-03, Center for Accelerator Sciences and Technology, University of Texas at Arlington.
5 S. H. Kim, Phys. Fluids 27, 675 (1984).
6. J. Dawson and C. Oberman, Phys. Fluids 5, 517 (1962).
7. J. Dawson and C. Oberman, Phys. Fluids 6, 394 (1963).
8. W. M. Manheimer, Phys. Fluids 20, 265 (1977)
9. W. M. Manheimer and D. G. Colombant, Phys. Fluids 21, 1818 (1978).
10. E. G. Harris, in Advances in Plasma Physics, Vol. III, edited by A. Simon and W. B. Tomson (Wiley, New York, 1969), p. 86.
11. J. F. Seely and E. G. Harris, Phys. Rev. A 7, 1064 (1973).
12. S. H. Kim and H. E. Wilhelm, Phys. Fluids 25, 668 (1982).
13. S. H. Kim, Phys. Rev. A 26, 567 (1982).
14. S. H. Kim, Proceedings of the XVI International Conference on Phenomena in Ionized Gases, University of Duesseldorf, Duesseldorf, Germany, 1983, p. 8.
15. H. E. Wilhelm, J. Phys. A 16, 2149 (1983).
16. T. Tajima and J. M. Dawson, Phys. Rev. Lett. 43, 267 (1974); C. Joshi, W. B. Mori, T. Katsouleas, J. M. Dawson, J. M. Kindel, and D. W. Forslund, Nature 311, 525 (1984).
17. B. H. Ripin, Appl. Phys. Lett. 30, 136 (1977).
18. N. G. Basov, A. A. Kologrivov, O. N. Krokhin, A. A. Rupasov, G. V. Skilzkov, and A. S. Shikauov, Zh. Eksp. Teor. Fiz. Pis´ma Red. 23, 474 (1976) [JETP Lett. 23, 428 (1976).
19. T. P. Donaldson, M. Hubbard, and I. J. Spalding, Phys. Rev. Lett. 37, 1348 (1976).
20. R. A. Haas, W. C. Mead, W. L. Kruer, D. W. Phillton, H. N. Kornblum, J. D. Lindl, D. MacQuigg, V. C.Rupert, and K. G. Tirsell, Phys. Fluids 20, 322 (1977).

THE PLASMA WAKE FIELD ACCELERATOR*

PISIN CHEN†
Stanford Linear Accelerator Center
Stanford University, Stanford, California 94305

and

J. M. DAWSON
Department of Physics
University of California, Los Angeles, California 90024

ABSTRACT

A new scheme of electron acceleration, employing relativistic electron bunches in a cold plasma, is analyzed. The wake field of a leading bunch is derived in a single-particle model. We then extend the model to include finite bunch length effect. In particular, we discuss the relation between the charge distributions of the driving bunch and the energies transformable to the trailing electrons. It is shown that for symmetric charge distribution of the driving bunches, the maximum energy gain for a driven electron is $2\gamma_0 mc^2$. This limitation can be overcome by introducing asymmetric charge distributions, in which case energy gains up to $\sqrt{1+(1-\frac{\pi}{2}+k_p|\varsigma_0|)^2}\,\gamma_0 mc^2$ are possible.

I. INTRODUCTION

The main theme of this Workshop is concerned with new acceleration mechanisms that employ lasers in certain ways. Since the first Workshop, there has been tremendous progress towards further understanding of the plasma beat-wave accelerators[1,2] both theoretically and experimentally, as was revealed during this Workshop.[3] However, it is also clear that in order to realize the plasma beat-wave accelerator at a scale beyond laboratory test-of-principle experiments, significant advances in laser technology are needed. For example, the beat-wave acceleration scheme requires fine tuning[4] between the plasma frequency ω_p and the beat-wave frequency of the laser in order for the wake plasma wave excited by the laser beat-wave to grow linearly. This in turn either puts constraints on

⋆ Work supported in part by the Department of Energy, contract DE-AC03-76SF00515 and by the National Science Foundation, Grant PHY 83-04641.

† Permanent address: Department of Physics, University of California, Los Angeles, California 90024.

the uniformity of the plasma density and the linearity of the plasma oscillation, or relies on very high power lasers to shorten the time of growth. In addition, it may be necessary to deliver the laser energy in a pulse shorter than 10 picoseconds in order to avoid competing instabilities.[5] Questions of laser efficiencies are also of considerable concern.

It turns out that if one replaces the lasers by high energy electron bunches traversing the plasma, large energy gradients can still be attained. The idea is to inject a sequence of bunched high energy electrons into a cold plasma. As in the two stream instability, the streaming electrons lose energy to the background plasma by exciting a wake plasma wave. If a late coming electron bunch rides on the wave at a proper phase, it will be boosted to a higher energy due to the longitudinal electric field in the wave.

Chen, Huff and Dawson[6] first studied this scheme using a single particle model under the electrostatic approximation. Later this model was improved by taking full electromagnetic effects into account.[7] Ruth *et al.*,[8] on the other hand, made an important contribution by recognizing the similarity between this scheme and the wake field acceleration scheme using EM cavities. Once this is seen, the "fundamental theorem of beam loading"[9] known in accelerator physics can be readily applied to the "plasma wake field accelerator." Indeed, computer simulations[6,7] have indicated that the maximum energy gain for driven electrons cannot exceed $2\gamma_0 mc^2$, in full agreement with the theorem. But is this really the upper limit of energy gain using wake field acceleration?

In this paper we will review the single particle model with care taken in its detailed derivation, which is a generalization of the nonrelativistic electrostatic method given by Kruer.[10] We then discuss the wake field generated by bunches with finite length. Attention is paid to the relation between the charge distribution of the driving bunch and the maximum energy transformable from the driving bunch to a test charge. We show that the energy gain limitation described above can be surpassed and large energy transforms are possible.

Since the plasma wake field accelerator makes use of already existing accelerators as the source for providing electron bunches, the technical barrier may be lower than that of the plasma beat-wave accelerators. In addition, the accessible free energy of electron beams is comparable to that of the most powerful laser beams. It is thus reasonable to hope that this scheme can meet the more immediate needs of the particle physics community.

II. SINGLE PARTICLE MODEL

Consider a system in which a relativistic electron bunch with initial $\beta_0 = v_b/c \lesssim 1$ streams through a cold, uniform plasma along the z-axis. Assuming that the size of the bunch is much smaller than λ_p^3, where λ_p is the plasma wavelength $(2\pi v_b/\omega_p)$, we can then treat the whole bunch containing q particles as a single particle with charge $Q = eq$.

To the linear approximation the equation of motion and the equation of continuity for the cold, nonrelativistic background plasma are

$$\partial_t \vec{v}_{p1} = -\frac{e}{m}\,\vec{E}_1 \ , \tag{1}$$

and

$$\partial_t n_{p1} + n_{p0}\nabla\cdot\vec{v}_{p1} = 0 \ , \tag{2}$$

where $\vec{E}_1$ is the total electric field contributed from the plasma and the beam: $\vec{E}_1 = \vec{E}_{p1} + \vec{E}_{b1}$, and where the plasma velocity $\vec{v}_p = \vec{v}_{p0} + \vec{v}_{p1}$, $\vec{v}_{p0} = 0$ and the plasma density $n_p = n_{p0} + n_{p1}$, $n_{p0} \gg n_{p1}$ are assumed. The charge and current densities of our beam-plasma system are

$$\rho_1(\vec{x}) = -en_{p1}(\vec{x}) - Q\delta(\vec{x} - \vec{x}_0) \ , \tag{3}$$

and

$$\vec{J}_1(\vec{x}) = -en_{p0}\,\vec{v}_{p1}(\vec{x}) - Q\vec{v}_b\,\delta(\vec{x} - \vec{x}_0) \ , \tag{4}$$

respectively, where $\vec{x}_0$ is the instantaneous position of the beam: $\vec{x}_0 \equiv v_b t\,\mathbf{e}_3$, and $\vec{x} = \rho\mathbf{e}_1 + z\mathbf{e}_3$, in cylindrical coordinates.

We are interested in the wake field $\vec{E}_1$ excited by the beam in the plasma. Our approach is to solve for the scalar potential ϕ_1 and the vector potential $\vec{A}_1$ first. In what follows it is more convenient to introduce a new variable $\varsigma \equiv z - v_b t$ which measures the distance behind the bunch. For the case of an ultra-relativistic electron beam where $\beta_0 \approx 1$, it is a good approximation to take v_b constant over many plasma wavelengths, even though a substantial amount of energy can be transferred to the plasma wave. Under this assumption we put $\partial_t = -v_b\partial_\varsigma$ and $\partial_z = \partial_\varsigma$.

In the Coulomb gauge we have a Poisson equation for the scalar potential,

$$\nabla^2\,\phi_1 = -4\pi\,\rho_1 \ , \tag{5}$$

and an inhomogeneous wave equation for the vector potential,

$$\nabla^2\,\vec{A}_1 - \frac{1}{c^2}\,\partial_t^2\vec{A}_1 = -\frac{4\pi}{c}\,\vec{J}_1 + \frac{1}{c}\,\nabla\partial_t\phi_1 \ . \tag{6}$$

In terms of the new variable ς, and neglecting the term involving the factor $(1-\beta_0^2)$, Eq. (6) can be reduced to

$$\nabla_\perp^2\,\vec{A}_1 = -\frac{4\pi}{c}\,\vec{J}_1 - \beta_0\,\nabla\,\partial_\varsigma\,\phi_1 \ , \tag{7}$$

where $\nabla_\perp^2$ is the two dimensional Laplacian in the transverse direction, and the symbol β_0 in the last term is saved for the purpose of clarity even though we had assumed $\beta_0 \approx 1$.

First we solve for ϕ_1 in Eq. (5). Taking ς-derivative twice and combining with Eqs. (1), (2) and (3) we get, with the gauge condition $\nabla \cdot \vec{A}_1 = 0$,

$$\nabla^2 \left(\partial_\varsigma^2 + k_p^2\right) \phi_1 = 4\pi Q \, \partial_\varsigma^2 \, \delta\left(\vec{x} - \vec{x}_0\right) \ , \tag{8}$$

where $k_p \equiv \omega_p / v_b = (4\pi n_{p0} e^2 / m v_b^2)^{1/2}$. Since $4\pi\delta(\vec{x} - \vec{x}_0) = -\nabla^2(1/|\vec{x} - \vec{x}_0|)$, the solution of this equation requires that we solve

$$\left(\partial_\varsigma^2 + k_p^2\right) \phi_1 = -Q \, \partial_\varsigma^2 \, \frac{1}{|\vec{x} - \vec{x}_0|} \ . \tag{9}$$

One may wonder whether by dropping the Laplacians from both sides of Eq. (5) we risk omitting the homogeneous solutions that satisfy either $\nabla^2 \Lambda(\vec{x}) = 0$ or $(\partial_\varsigma^2 + k_p^2)\Lambda(\vec{x}) = 0$. Actually if we assume that the plasma is quiescent before the bunch entered in the infinite past, then $\Lambda(\vec{x}) = 0$ identically, so no problem arises.

The solution of Eq. (9) is (see Ref. 10)

$$\phi_1(\rho, \varsigma) = -Q \int_\varsigma^\infty d\varsigma' \, k_p^{-1} \sin k_p(\varsigma' - \varsigma) \cdot \partial_{\varsigma'}^2 \, \frac{1}{\sqrt{\rho^2 + \varsigma'^2}} \ , \tag{10}$$

where $|\vec{x} - \vec{x}_0| = \sqrt{\rho^2 + \varsigma^2}$ has been used. Integrating by parts twice we get

$$\phi_1(\rho, \varsigma) = Q \left\{ -\frac{1}{\sqrt{\rho^2 + \varsigma^2}} + k_p \int_\varsigma^\infty d\varsigma' \, \frac{\sin k_p(\varsigma' - \varsigma)}{\sqrt{\rho^2 + \varsigma'^2}} \right\} \ . \tag{11}$$

Next we turn to the vector potential $\vec{A}_1$ in Eq. (7). Taking the ς-derivative on both sides of the equation and invoking the equation of motion for the current term, we obtain

$$\partial_\varsigma \left(\nabla_\perp^2 - \beta_0^2 k_p^2\right) \vec{A}_1 = -\beta_0 \nabla \left(\partial_\varsigma^2 + k_p^2\right) \phi_1 + 4\pi Q \, \vec{\beta}_0 \, \partial_\varsigma \, \delta\left(\vec{x} - \vec{x}_0\right) \ . \tag{12}$$

Combining with Eq. (8), the above equation decouples entirely from the scalar potential. Removing the ς-derivative common to each term, the equation further reduces to a inhomogeneous modified Helmholtz equation in two dimensions for each component of $\vec{A}_1$:

$$\left(\nabla_\perp^2 - \beta_0^2\, k_p^2\right) \vec{A}_1 = Q \left\{ \beta_0\, \nabla \partial_\varsigma\, \frac{1}{|\vec{x} - \vec{x}_0|} + 4\pi Q\, \vec{\beta}_0\, \delta\left(\vec{x} - \vec{x}_0\right) \right\} . \tag{13}$$

When concentrating on the longitudinal component of $\vec{A}_1$, we get

$$\left(\nabla_\perp^2 - \beta_0^2\, k_p^2\right) A_{1z} = -Q\beta_0\, \nabla_\perp^2\, \frac{1}{|\vec{x} - \vec{x}_0|} . \tag{14}$$

We are actually interested in the wake field trailing behind the bunch on the z-axis, i.e., at position $\vec{x} = z e_3$. In that case

$$\phi_1(\varsigma) = -\frac{2\pi Q}{\lambda_p} \left\{ \frac{1}{k_p|\varsigma|} + k_p \int_{\varsigma}^{\infty} d\varsigma'\, \frac{\sin k_p(\varsigma' - \varsigma)}{k_p|\varsigma'|} \right\} , \tag{15}$$

where $\lambda_p = 2\pi k_p^{-1}$, and the corresponding potential in Eq. (14) reads

$$A_{1z}(\varsigma) = -\frac{2\pi Q}{\lambda_p}\, \beta_0^2 \int_0^{\infty} d\rho'\; K_1(\beta_0 k_p \rho') \cdot \frac{\rho'^2}{[\rho'^2 + \varsigma^2]^{3/2}} , \tag{16}$$

where K_1 is the modified Bessel function of order one.

Plots of ϕ_1 and A_{1z} as functions of $|\varsigma|$ are shown in Fig. 1. Notice that A_{1z} diminishes monotonically whereas ϕ_1 remains oscillatory. The longitudinal electric field is computed by taking the ς-derivative since $E_{1z} = \partial_\varsigma(\beta_0 A_{1z} - \phi_1)$.

We first show that the expressions we get in Eqs. (15) and (16) give the correct physical limit when the background plasma is "turned off." To see this we examine a point right behind the bunch, i.e., $k_p|\varsigma| \ll 1$. In that case

$$A_{1z}(\varsigma) \simeq -\frac{Q\beta_0}{|\varsigma|} + Q\beta_0^3\, k_p^2 |\varsigma| \left[\gamma + \ell n\left(\beta_0 k_p |\varsigma|\right) - 1 + \frac{\pi}{2}\, \frac{1}{\beta_0 k_p |\varsigma|} \right] , \tag{17}$$

where γ is the Euler's constant. When turning off the plasma by taking the limit k_p (or ω_p) $\to 0$, only the first terms in Eqs. (15) and (17) survive, i.e.

$$\lim_{k_p \to 0}\, E_{1z}(\varsigma) = \left(1 - \beta_0^2\right)\, \frac{Q}{|\varsigma|^2} \simeq 0 . \tag{18}$$

Thus we recover the well-known expression for the longitudinal electric field of a relativistic charge moving in vacuum with speed $\beta_0 c$. The remaining terms thus correspond to the plasma response to the presence of the relativistic beam.

Figure 2 shows a plot of E_{1z} without making the $k_p|\varsigma| \ll 1$ approximation. It can be seen that E_{1z} is maximum at $|\varsigma| \simeq (n+\frac{1}{2})\lambda_p$, where n is any non-negative integer, and the contribution to the maximum comes predominantly from the scalar potential. If the separation between the driven bunch and the driving bunch is such that $|\varsigma| \simeq (n+\frac{1}{2})\lambda_p$, the energy gradient attainable for each electron in the driven bunch is

$$G = -eE_{1z} \simeq \frac{8\pi^2 eQ}{5\lambda_p^2} \; . \tag{19}$$

As an example, consider a plasma of density $n_{p0} = 10^{16}$ cm^{-3} (which corresponds to $\lambda_p \simeq 0.33$ mm). If the driving bunch consists of $q = 5 \times 10^{10}$ particles, Eq. (19) shows that $G \simeq 4.8$ GeV/m. Note that this treatment ignores nonlinear plasma effects and self-consistent effects that act to slow the driving bunch. It is only valid if the electric field does not approach the cold plasma wave-breaking amplitude, and if the electric energy is small compared to the free energy of the driving bunch. The first condition provides an upper limit on the maximum allowed energy gradient: $G_{\max} \simeq \sqrt{n_{p0}}$ eV/cm $= 10$ GeV/m. Comparing with $G \simeq 4.8$ GeV/m, our linear theory is probably still reasonable. The second condition requires that $(E_{1z}^2/8\pi) \cdot L < q\gamma_0 mc^2/\text{Area}$, where L is the allowable length of the beam-plasma acceleration. Taking the area to be $\pi c^2/\omega_p^2$ and solving for L for the above case gives $L \simeq 0.125\gamma_0$ cm. For $\gamma_0 = 10^5$ (50 GeV) L equals 125 m, so that our constant velocity assumption is extremely well satisfied.

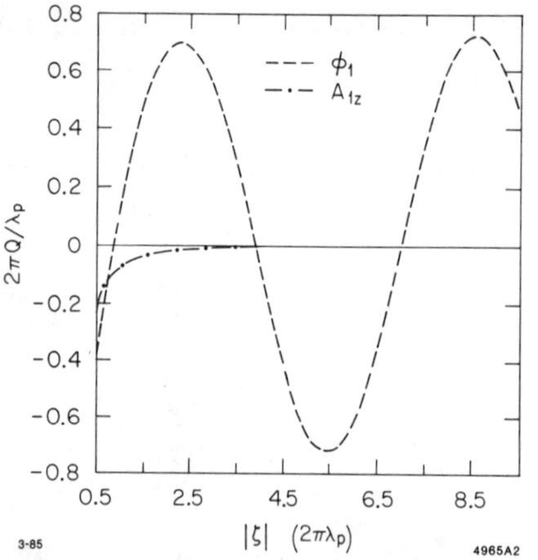

Fig. 1. Potentials as functions of distance behind the driving bunch.

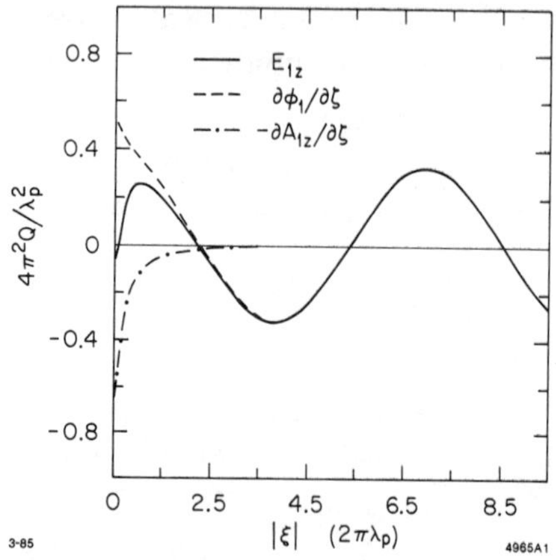

Fig. 2. Longitudinal electric field as a function of distance behind the driving bunch.

III. THE TRANSFORMER RATIO

A. Wake Function and Transformer Ratio

In the previous section, we have calculated the longitudinal electric field in the wake plasma wave excited by a point charge. It is obvious that the electric field per unit charge is a characteristic of the beam-plasma system. Following the analogous situation in the wake field acceleration in metallic cavities, we shall call it the (longitudinal) "plasma wake function":

$$W_z(\varsigma) \equiv -Q^{-1} E_{1z}(\varsigma) , \qquad \varsigma \leq 0 . \tag{20}$$

For a bunch with finite length, the current density associated with the bunch is in general a function of ς, and each charge in the bunch leaves behind it a wake field characterized by $W_z(\varsigma)$. Physically, a trailing charge will either gain or lose energy according to its phase relationship, $\varphi = k_p\varsigma$, with the leading charge that generates the wake. On the other hand, the trailing charge in addition will lose energy by exciting its own wake. The net electric field generated by a longitudinally finite size bunch is a convolution integral of W_z and the current density $I(\varsigma)$:

$$\mathcal{E}(\varsigma) = \int_{\varsigma}^{\infty} W_z(\varsigma - \varsigma')\, I(\varsigma')\, d\varsigma' , \qquad \varsigma \leq 0 , \tag{21}$$

where $W_z(\varsigma - \varsigma')$ acts as a Green's function.

Let the bunch extend from $\varsigma = 0$ to $\varsigma = \varsigma_0$. If $|\varsigma_0|$ is sufficiently large compared to the plasma wavelength, λ_p, then in general the electric field inside the bunch acts to retard some particles and accelerate others but on the average is retarding in the bunch, and $\mathcal{E}(\varsigma)$ behind the bunch is oscillatory. Let the maximum retarding electric field inside the bunch be $\mathcal{E}_m^-$ and the maximum accelerating electric field induced behind the bunch be $\mathcal{E}_m^+$. The ratio $R \equiv \mathcal{E}_m^+/\mathcal{E}_m^-$ is called the transformer ratio. The physical implication of the transformer ratio is as follows: if a monoenergetic driving bunch with particles of initial energy $\gamma_0 mc^2$ excites a plasma wake field, and if within the length L where the particles in the bunch that experience the maximum retarding field $\mathcal{E}_m^-$ come to a stop, i.e., $\gamma_0 mc^2 = eL\mathcal{E}_m^-$, then the maximum possible energy gain for a test charge behind the bunch will be $R\gamma_0 mc^2$ in the distance L.

It is well known in accelerator physics[9] that the transformer ratio for a point charge is equal to two. This fact has been called the fundamental theorem of beam loading. One can also prove that,[11] assuming only one mode in the cavity, $R \leq 2$ for all finite length bunches with symmetric current distribution. Thus it has been the worry that there is a fundamental limitation on the driven electron energy gain in collinear wake field acceleration.

This limitation is also observed in the computer simulation[6,7] of plasma wake field accelerations. Using a one-and-two-halves dimensional (x, v_x, v_y, v_z) relativistic particle code (physically this corresponds to a one dimensional beam-plasma system where both the beam and the plasma extend infinitely in the transverse directions), it was found that, for the driving beam with current density profile $I(x) \sim 1 + \sin kx$, the driven beam gains energy only up to $\Delta U \lesssim 2\gamma_0 mc^2$ (see Fig. 3).

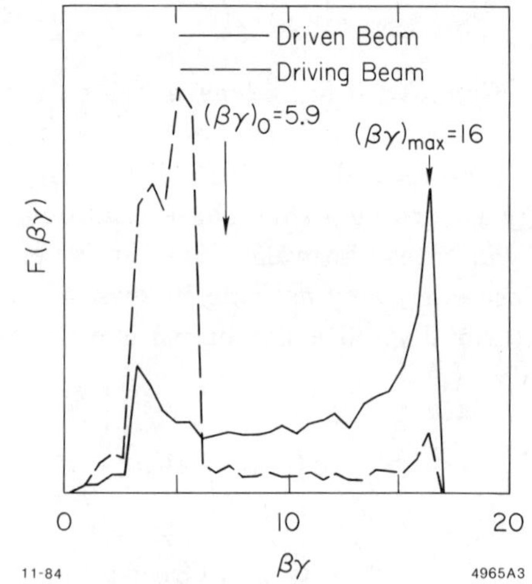

Fig. 3. Momentum distribution of the driving and driven electron beams when the latter has attained its maximum upper limit. The density profiles for both driving and driven electron beams are 1+sin kx, 180° out of phase.

B. The Optimal Transformer Ratio[12]

Can this limitation of energy gain be overcome? It turns out that this can be done. If the current distribution is asymmetric with respect to the mid-point of the bunch length, then the generalized fundamental theorem of beam loading (i.e. $R \leq 2$) for finite length symmetric bunches can be evaded. A simple physical way to look at how this can be accomplished is as follows. If the plasma electrons can move out of the way as the bunch charge builds up, the fields within the bunch can be kept very small. If the bunch charge is suddenly terminated the plasma finds itself very non-neutral just behind the bunch, and a large plasma oscillation exists.

To illustrate this issue further let us consider a one dimensional beam-plasma system. For an infinite thin disk moving with speed $\beta_0 c$ in the normal direction the vector potential $\vec{A}_1$ in Eq. (16) vanishes and[8]

$$E_{1z}(\varsigma) = -\partial_\varsigma\, \phi_1(\varsigma) = \begin{cases} 0 & \varsigma > 0 \\ 2\pi e\sigma & \varsigma = 0 \\ 4\pi e\sigma \cos k_p\varsigma & \varsigma < 0 \end{cases} \tag{22}$$

The corresponding plasma wake function is thus $W_z(\varsigma) = -(e\sigma)^{-1}E_{1z}(\varsigma) = -4\pi\cos k_p\varsigma$. Consider a triangular bunch with current distribution $I(\varsigma)$ rising linearly at the head of the bunch and cut off at the tail, i.e.

$$I(\varsigma) = \begin{cases} 0 & \varsigma > 0 \\ I_0 k_p|\varsigma| & 0 \geq \varsigma \geq \varsigma_0 \;, \quad I_0 > 0 \\ 0 & \varsigma_0 > \varsigma \;. \end{cases} \tag{23}$$

Let the bunch length be $|\varsigma_0| = 2\pi N/k_p$, then it is straight forward to show, from Eq. (21), that inside the bunch,

$$\mathcal{E}^-(\varsigma) = 4\pi k_p I_0 \int_{\varsigma}^{0} \varsigma' \cos k_p(\varsigma - \varsigma')\, d\varsigma' = \frac{4\pi I_0}{k_p}(\cos k_p\varsigma - 1) \;, \quad 0 \geq \varsigma \geq \varsigma_0 \tag{24}$$

whereas behind the bunch,

$$\mathcal{E}^+(\varsigma) = 4\pi k_p I_0 \int_{\varsigma_0}^{0} \varsigma' \cos k_p(\varsigma - \varsigma')\, d\varsigma' = -\frac{8\pi^2 I_0 N}{k_p}\sin k_p\varsigma \;, \quad \varsigma_0 > \varsigma \;. \tag{25}$$

Thus

$$R = \frac{\mathcal{E}_m^+}{\mathcal{E}_m^-} = \pi N \;. \tag{26}$$

This simple calculation was checked by a computer simulation[13] (Fig. 4) which agrees very well with the prediction. Notice that R is proportional to the number of ripples, N, of $\mathcal{E}^-$. This can be easily understood because the smoother the $\mathcal{E}^-$, the more particles experience $\mathcal{E}_m^-$ which bring them to a stop. Thus it allows more energy to be transformed to the driven electrons. The question naturally arises as to whether the triangular bunches give the best transformer ratios.

To look for a better current distribution, it is more convenient to turn the convolution integral of Eq. (21) inside out. If we specify the desired $\mathcal{E}^-(\varsigma)$ and $\mathcal{E}^+(\varsigma)$ in the entire domain, the corresponding current density $I(\varsigma)$ can be obtained. This can be done by first making a Laplace transform of Eq. (21). The convolution (Faltung) theorem says that the Laplace transform of a convolution integral is equal to the product of Laplace transforms of the functionals in the integrand, i.e.

$$\mathcal{L}\{\mathcal{E}(\varsigma)\} = \mathcal{L}\left\{\int_{\varsigma}^{0} W_z(\varsigma - \varsigma')\, I(\varsigma')\, d\varsigma'\right\} = \mathcal{L}\{W_z(\varsigma)\} \cdot \mathcal{L}\{I(\varsigma)\} \;. \tag{27}$$

Furthermore, it can be proved that the inverse Laplace transform of $\mathcal{L}\{I(\varsigma)\}$,

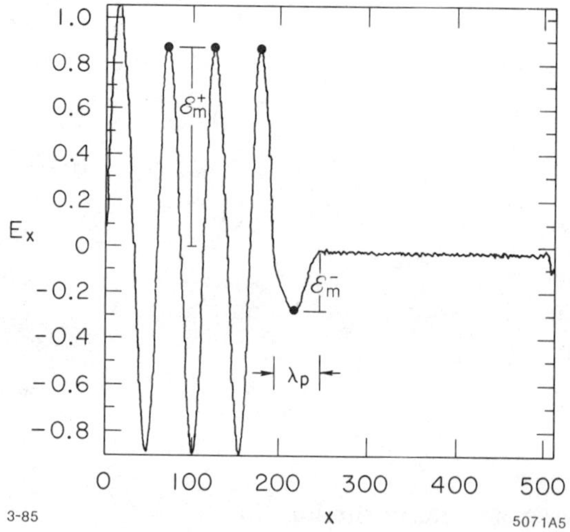

Fig. 4. The electric field generated by a bunch with triangular density profile moving to the right of the picture. The bunch is one wavelength long, and the transformer ratio is $R = \mathcal{E}_m^+/\mathcal{E}_m^- \simeq \pi$.

i.e. the $I(\varsigma)$ itself, can be expressed in the complex plane as

$$I(\varsigma) = \frac{1}{2\pi i} \int\limits_{\gamma - i\infty}^{\gamma + i\infty} \frac{\mathcal{L}\{\mathcal{E}(\varsigma)\}}{\mathcal{L}\{W_z(\varsigma)\}} e^{s\varsigma}\, ds \ , \tag{28}$$

where γ is some negative quantity which avoids the integration to be carried along the imaginary axis. To find I, our ansatz is that there exists some smooth function $\mathcal{E}^-$ inside the bunch and a sinusoidal function $\mathcal{E}^+$ behind the bunch.

Ideally, one would like to have a constant $\mathcal{E}^-$ such that all particles experience the same retarding field and stop at the same distance. But it can be shown by carrying out the calculations in Eq. (28) that the solution of $I(\varsigma)$ for such a situation does not exist. This is actually not too surprising because we had insisted that $\mathcal{E}^-$ be the same nonzero value at $\varsigma = 0$, and this is impossible to prepair. Notice that even a delta-function current at the head of the bunch can only provide $\mathcal{E}^-(0) = \frac{1}{2}\mathcal{E}^-(0^+)$ [cf. Eq. (22)].

We should thus relax our ansatz by allowing for some smoother fall of $\mathcal{E}^-$ at the head of the bunch. For instance, if we let

$$\mathcal{E}^-(\varsigma) = \begin{cases} \dfrac{4\pi I_0}{k_p} \sin k_p\varsigma \ , & 0 \geq \varsigma \geq -\dfrac{\pi}{2k_p} \ , \quad I_0 > 0 \\[2ex] -\dfrac{4\pi I_0}{k_p} \ , & -\dfrac{\pi}{2k_p} \geq \varsigma \geq \varsigma_0 \ , \end{cases} \tag{29}$$

it can be shown that the corresponding current distribution is

$$I(\varsigma) = \begin{cases} I_0 \, , & 0 \geq \varsigma \geq -\dfrac{\pi}{2k_p} \, , \\ \dfrac{2}{\pi} I_0 \, k_p |\varsigma| \, , & -\dfrac{\pi}{2k_p} \geq \varsigma \geq \varsigma_0 \, . \end{cases} \tag{30}$$

The transformer ratio in this case becomes

$$R = \sqrt{1 + \left(1 - \frac{\pi}{2} + k_p |\varsigma_0|\right)^2} \, . \tag{31}$$

For $|\varsigma_0| = 2\pi N/k_p$, $R = \sqrt{1 + (1 - \frac{\pi}{2} + 2\pi N)^2}$. We see that the R in this case is larger than the R for the corresponding triangular current distribution.

At first thought it looks like the energy transformation can be indefinitely improved by increasing the bunch length. However, the absolute value of $\mathcal{E}_m^+$ will not increase proportionally unless the charge density in the bunch is kept constant. Thus one faces the technical limitation of a maximum possible peak current which one can provide near the tail of the bunch. The ultimate limit, however, comes from the cold plasma wave-breaking limit which $e\mathcal{E}_m^+$ cannot exceed. But before reaching this limit, nonlinear plasma effects have already set in, and the previous calculations have to be modified.

IV. DISCUSSION

The single particle model described in Section II is useful for finite size bunches since it gives Green's functions for them. The scalar and vector potentials can be obtained by integrating the Green's functions over the finite size of a bunch. This is exemplified by the discussion of finite length bunches and the corresponding transformer ratios in Section III, although for the sake of simplicity we studied the one-dimensional case. A more realistic situation to consider is a bunch with a finite cross-section. In that case the contribution of A_{1z} to W_z at small distance becomes important when we look for $\mathcal{E}^-(\varsigma)$ inside the bunch. We have not discussed the transverse plasma wake function in this paper, it is important for the study of beam dynamics and should be pursued further.

In summary, large energy gradients over long distances of acceleration are attainable in the plasma wake field accelerator. The study of beam-plasma interaction deserves more attention in the plasma beat-wave scheme as well — because it also affects the accelerated electron bunch and will play an essential part in its beam dynamics.

ACKNOWLEDGEMENTS

We thank Karl Bane and Perry Wilson for allowing us to use some of the yet to be published results in Ref. 12.

REFERENCES

1. T. Tajima and J. M. Dawson, Phys. Rev. Lett. 43, 267 (1979).
2. T. Katsouleas and J. M. Dawson, Phys. Rev. Lett. 51, 392 (1983).
3. See the articles related to the plasma beat-wave accelerator in this proceedings.
4. C. M. Tang, P. Sprangle and R. Sudan, Appl. Phys. Lett. 45, 375 (1984).
5. C. Joshi, W. B. Morri, T. Katsouleas, J. M. Dawson, J. M. Kindel and D. W. Forslund, Nature 311, 535 (1984).
6. P. Chen, R. W. Huff and J. M. Dawson, UCLA Report No. PPG-802, 1984, and Bull. Am. Phys. Soc. 29, 1355 (1984).
7. P. Chen, J. M. Dawson, R. W. Huff and T. Katsouleas, Phys. Rev. Lett. 54, 693 (1985).
8. R. D. Ruth, A. W. Chao, P. L. Morton and P. B. Wilson, SLAC-PUB-3374, 1984 (to be published in "Particle Accelerators").
9. A. W. Chao in *"Physics of High Energy Particle Accelerators"*, AIP Conf. Proc. No. 105 (Am. Inst. Phys., New York, 1983).
10. W. Kruer, Ph.D. Thesis, Princeton University, 1969.
11. K. L. F. Bane, P. B. Wilson and T. Weiland, SLAC-PUB-3528, 1984.
12. K. L. F. Bane, P. Chen and P. B. Wilson, *"Collinear Wake Field Acceleration"*, to be presented at the 1985 Particle Accelerator Conference, Vancouver, B.C., Canada, May 13-16, 1985.
13. P. Chen, J. J. Su, J. M. Dawson, P. B. Wilson and K. L. F. Bane, UCLA Report No. PPG-851, 1985.

A COMPARISON OF THE PLASMA BEAT WAVE ACCELERATOR AND THE PLASMA WAKE FIELD ACCELERATOR*

PISIN CHEN† and RONALD D. RUTH
Stanford Linear Accelerator Center
Stanford University, Stanford, California 94305

ABSTRACT

In this paper we compare the Plasma Beat Wave Accelerator and the Plasma Wake Field Accelerator. We show that the electric fields in the plasma for both schemes are very similar, and thus the dynamics of the driven beams are very similar. The differences appear in the parameters associated with the driving beams. In particular to obtain a given accelerating gradient, the Plasma Wake Field Accelerator has a higher efficiency and a lower total energy for the driving beam.

INTRODUCTION

Recently there have been two similar types of plasma acclerator schemes proposed. The Plasma Beat Wave Accelerator (PBWA)[1,2] employs two laser beams beating at the plasma frequency to drive the plasma while the Plasma Wake Field Accelerator (PWFA)[3,4] replaces the laser beams by a bunched relativistic electron beam. Since the two schemes make use of different sources, the corresponding mechanisms that drive the plasma waves are different. In the PBWA, it is the ponderomotive force which comes from the beating lasers that drives the plasma, whereas in the PWFA the driving bunch is decelerated by the plasma and thus transfers energy to the plasma wave. Other than this difference, however, the two schemes are very similar. In both cases large longitudinal electric fields are generated in the plasma which oscillates at the fundamental plasma frequency ω_p. These fields are then used to accelerate an electron beam. It is interesting to ask how these two schemes compare to each other in detail. To make a fair comparison, in this paper we emphasize self consistency among the various accelerator parameters common to the schemes. We will follow Refs. 2 and 4 in most of the calculations; however, we will include transverse effects in the PBWA to calculate and compare focusing effects.

⋆ Work supported by the Department of Energy, contract DE-AC03-76SF00515.

† Permanent address: Department of Physics, University of California, Los Angeles, California 90024.

FIELDS IN A PLASMA WAVE OF FINITE EXTENT

To find the electric fields in the plasma waves for both schemes, we start with the linearized, nonrelativistic fluid equations,

$$\begin{aligned} \frac{\partial n_1}{\partial t} + n_0(\nabla \cdot \vec{v}_1) = 0 \\ \frac{\partial \vec{v}_1}{\partial t} = \frac{e\vec{\mathcal{E}}_1}{m} + \frac{\vec{F}_{ext}}{m}, \end{aligned} \tag{1}$$

and solve for the perturbed plasma density n_1. $\vec{\mathcal{E}}_1$ is the electric field due to n_1 and $\vec{F}_{ext}$ is the external force due to either a driving beam or a beating laser. In the case of the PBWA the force is most easily calculated from a Hamiltonian which has been averaged over the fast oscillation of the laser frequency. This leaves only the beating effect at a frequency ω_p. The averaged Hamiltonian is given by

$$H = \frac{\vec{p}^{\,2}}{2m} + e\phi_1 + \frac{e^2}{4m\omega^2} E_0^2(r)\cos(k_p z - \omega_p t) \tag{2}$$

where ω and ω_p are the laser and the plasma frequency respectively and k_p is the plasma wave number. The last term is simply the ponderomotive potential due to a beating laser with a finite cross section. For the sake of a comparison with the PWFA later in this paper, we will assume a radial dependence of the ponderomotive potential given by

$$E_0^2(r) = 2E_0^2 \begin{cases} K_2(k_p a)\, I_0(k_p r) + \frac{1}{2} - \frac{2}{(k_p a)^2} - \frac{r^2}{2a^2} & r < a \\ I_2(k_p a)\, K_0(k_p r) & r > a \end{cases} . \tag{3}$$

where K_n and I_n are modified Bessel functions. This radial profile is parabolic near the origin but falls off exponentially for $r > a$. It was chosen to yield a simple parabolic dependence in Eq. (4) below.

To use the above results we need the divergence of the force due to the Hamiltonian in Eq. (2). This is given by

$$\nabla \cdot \vec{F} = 4\pi e^2 n_1 + \frac{e^2 E_0^2 k_p^2}{4m\omega^2}(1 - r^2/a^2)\cos(k_p z - \omega_p t) \quad r < a, \tag{4}$$

where Poisson's equation has been used to substitute for $\nabla^2\phi_1$. Substituting into Eq. (1) yields

$$\frac{\partial^2 n_1}{\partial t^2} + \omega_p^2\, n_1 = \begin{cases} -\left(\frac{\omega_p}{\omega}\right)^2 \frac{E_0^2 k_p^2}{16\pi m}(1 - r^2/a^2)\cos(k_p z - \omega_p t) & r < a \\ 0 & r > a \end{cases}, \tag{5}$$

which has a solution of the form

$$n_1(r,z,t) = f(r,z,t)\sin(k_p z - \omega_p t), \tag{6}$$

where

$$f(r,z,t) = \begin{cases} -\dfrac{E_0^2 k_p^2}{32\pi m\omega^2}(1 - r^2/a^2)(k_p z - \omega_p t) & r < a \\ 0 & r > a \end{cases} . \tag{7}$$

With $n_1(r,z,t)$ in hand, we now must find the electric field $\vec{\mathcal{E}}_1$ due to the plasma oscillation. Since the magnetic field due to a linear plasma wave vanishes, we can simply use Poisson's equation,

$$\frac{1}{r}\frac{\partial}{\partial r}(r\frac{\partial}{\partial r}\phi_1) + \frac{\partial^2\phi_1}{\partial z^2} = -4\pi e n_1 \ . \tag{8}$$

If we have a laser pulse of length τ, at the end of the pulse the amplitude of the plasma density wave will reach its peak value. From Eq. (7) this is given by

$$f_{max}(r) = \frac{\omega_p\tau E_0^2 k_p^2}{32\pi m\omega^2}(1 - r^2/a^2) \qquad r < a, \tag{9}$$

and the potential can be shown to be

$$\phi_1 = R(r)\sin(k_p z - \omega_p t) \tag{10}$$

with

$$R(r) = \frac{\omega_p\tau e E_0^2}{4\omega^2 m}\begin{cases} K_2(k_p a)\ I_0(k_p r) + \frac{1}{2}\left(1 - \frac{r^2}{a^2}\right) - \frac{2}{(k_p a)^2} & , \ r < a \\ I_2(k_p a)\ K_0(k_p r) & , \ r > a \end{cases} \tag{11}$$

The longitudinal and transverse electric fields for $r < a$ for the PBWA are thus given by

$$\mathcal{E}_z = -\frac{\omega_p\tau k_p e E_0^2}{4\omega^2 m}\left\{K_2(k_p a)\ I_0(k_p r) + \frac{1}{2}\left(1 - \frac{r^2}{a^2}\right) - \frac{2}{(k_p a)^2}\right\}\cos(k_p z - \omega_p t) \ ,$$

$$\mathcal{E}_r = -\frac{\omega_p\tau k_p e E_0^2}{4\omega^2 m}\left\{K_2(k_p a)\ I_1(k_p r) - \frac{r}{k_p a^2}\right\}\ \sin(k_p z - \omega_p t) \ . \tag{12}$$

For the case of the PWFA the situation is very similar. We only need to change the laser source term in Eq. (5). For the case of a driving beam of density n_b, the divergence of the force is given by

$$\nabla \cdot \vec{F} = 4\pi e^2(n_1 + n_b) \ . \tag{13}$$

Following Ref. 4, consider a driving beam with density profile

$$n_b = \sigma(r)\delta(z - v_b t) \; . \tag{14}$$

Then the solution for the perturbed density is given by

$$n_1(r) = \begin{cases} k_p\sigma(r)\sin(k_p z - \omega_p t) & k_p z - \omega_p t < 0 \\ 0 & k_p z - \omega_p t > 0 \; . \end{cases} \tag{15}$$

To compare with the PBWA we use a parabolic distribution given by

$$\sigma(r) = \begin{cases} \frac{2N}{\pi a^2}(1 - r^2/a^2) & r < a \\ 0 & r > a \end{cases} , \tag{16}$$

where N is the total number of particles in the driving bunch. Once again it is possible to calculate the longitudinal and transverse electric fields due to the plasma wave.[4] These are given by

$$\begin{aligned} \mathcal{E}_z &= \frac{-16eN}{a^2}\left\{K_2(k_p a)\; I_0(k_p r) + \frac{1}{2} - \frac{2}{(k_p a)^2} - \frac{r^2}{2a^2}\right\}\cos(k_p z - \omega_p t) \; , \quad r < a \\ \mathcal{E}_r &= \frac{-16eN}{a^2}\left\{K_2(k_p a)\; I_1(k_p r) - \frac{r}{k_p a^2}\right\}\; \sin(k_p z - \omega_p t) \; , \quad r < a. \end{aligned} \tag{17}$$

Thus the electric fields for the two schemes turn out to be remarkably similar.

For reasons which we will discuss later the transverse size of the driven beam must be somewhat smaller than the transverse size of the laser beams or the driving electron beam. In addition if $k_p a \gg 1$, then the electric fields for both schemes are of the following form:

$$\begin{aligned} \mathcal{E}_z &\simeq -A(1 - \frac{r^2}{a^2})\cos(k_p z - \omega_p t) \\ \mathcal{E}_r &\simeq \; 2A\frac{r}{k_p a^2}\sin(k_p z - \omega_p t) \end{aligned} \tag{18}$$

where

$$A = \begin{cases} \dfrac{\omega_p \tau k_p e E_0^2}{8\omega^2 m} & PBWA \\ \dfrac{8eN}{a^2} & PWFA \end{cases} \tag{19}$$

Other than different coefficients, the forces that the driven electrons experience share the same physical characteristics in both schemes. To be specific there is

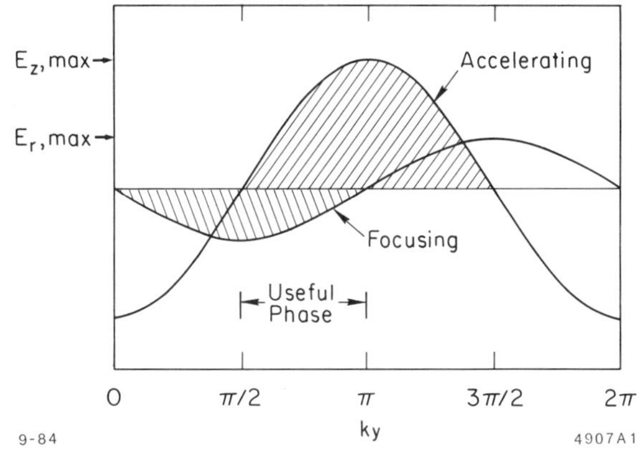

Fig. 1.

a longitudinal force $e\mathcal{E}_z$ that either accelerates or decelerates the driven bunch of electrons, and there is a transverse force $e\mathcal{E}_r$ shifted in phase which either will focus or defocus the driven bunch (see Fig. 1). From Fig. 1 it is clear that we have both acceleration and focusing over 1/4 of the plasma wavelength.

ACCELERATOR PHYSICS ISSUES

In this section we discuss some accelerator physics issues which are relevant to both schemes of plasma accelerators. To begin we concentrate on the quality and intensity of a driven electron bunch with finite transverse extent. In particular we treat the transverse oscillations and the energy spread due to the transverse variation of the accelerating field. We then discuss other issues such as phase slippage, spot size and driving beam energy for the PBWA and PWFA. The details in the discussion of these issues are different for the two schemes since we choose to fix different parameters in the two cases. Finally, in order to address the question of intensity, we discuss the efficiencies of both schemes.

The Beta Function

In this paper the beta function is defined to be the wavelength/2π of the transverse oscillation at some instantaneous phase ϕ along the plasma wave. In the last section we saw that, except for a difference in coefficients, the PBWA and PWFA have the same electric fields. We also pointed out that there is a useful phase between $\pi/2$ and π along the plasma wave. In general there will be some phase slippage between the plasma wave and the driven beam. If this phase slippage is slow, then we can calculate the transverse focusing effects as if the beam were at a fixed phase on the wave. The differential equation governing the transverse oscillations of a highly relativistic particle is

$$\frac{d^2x}{dz^2} = e\frac{\mathcal{E}_x}{\gamma mc^2}, \tag{20}$$

where γmc^2 is the particle's instantaneous relativistic mass. Thus, for small radius from Eq. (18) we have

$$\frac{d^2x}{dz^2} = e\left[\frac{eA\sin\phi}{k_p a^2 \gamma mc^2}\right] x\,. \tag{21}$$

Identifying the coefficient of x above with β^{-2} yields the beta function:

$$\beta = \left[\frac{k_p a^2 \gamma mc^2}{eA\sin\phi}\right]^{1/2}. \tag{22}$$

Energy Spread

From Eq. (18) it is evident that for a driving beam with finite transverse size, the longitudinal field varies transversely. Consider a driven bunch with transverse radius b which moves along the axis of the plasma wave. Since the field varies parabolically in the transverse direction, the average energy gain is reduced slightly and an energy spread is induced. If we assume that the beam is already very relativistic, then the average change in energy for one stage is

$$\Delta E_{ave} = \Delta E\left(1 - \frac{2}{3}\left(\frac{b}{a}\right)^2\right), \tag{23}$$

where ΔE is the energy gain for a particle on the axis of the plasma wave. The corresponding energy spread induced in one stage for the model we have chosen is

$$\left[\frac{\delta(\Delta E)}{\Delta E}\right]_{rms} = \frac{\sqrt{2}}{3}\left(\frac{b}{a}\right)^2. \tag{24}$$

The Trapping Parameter

The trapping parameter is defined to be the ratio of the plasma density perturbation n_1 and the unperturbed density n_0. Physically, this parameter indicates the linearity of the plasma oscillation. Since we work in the linear approximation for the plasma wave in both schemes, α should be kept reasonably small. For the case of the PBWA, we assume that the plasma oscillation saturates at the end of the laser, which corresponds to[5]

$$\alpha \equiv \frac{n_1}{n_0} \simeq \frac{1}{4} \qquad PBWA. \tag{25}$$

For the case of the PWFA we take L and $\mathcal{E}_z$ as chosen parameters. In addition, to scale the transverse effects we fix the ratio between the transverse size of the driving bunch and the plasma wavelength: a/λ_p. This in turn determines the plasma wavelength and the plasma density. In order to check that the plasma wave so generated is indeed a linear wave, we must calculate α, which in this case is given by

$$\alpha = \frac{e\mathcal{E}_z}{mc\omega_p} \qquad PWFA. \tag{26}$$

Phase Slippage

For both accelerator schemes the phase velocity of the plasma wave is not equal to the velocity of the driven bunch. This means that the driven bunch will slip in phase along the plasma wave as it is accelerated. For the PBWA we maximize $\mathcal{E}_z$ for a given L by optimizing the phase shift δ. If we choose a laser frequency ω, an acceleration length L, and a phase slippage δ for speed of light particles; then the plasma frequency is given by[2]

$$\omega_p = \left(\frac{2\delta c\omega^2}{L}\right)^{1/3} . \tag{27}$$

On the other hand, the acceleration gradient that the driven bunch sees varies along L due to the phase slippage. If the total phase slippage over the entire acceleration length is δ, then the average acceleration gradient is related to the ideal gradient by a phase slip form factor $\sin\delta/\delta$, that is

$$e\mathcal{E}_z^{ave} = \alpha mc\omega_p \frac{\sin\delta}{\delta} . \tag{28}$$

Here the phase has been allowed to slip from the top of the cosine down one side so that the bunch is always in a focusing region. The average acceleration gradient can be maximized for a given L if

$$\delta \simeq \frac{5\pi}{16} \quad \text{and} \quad \frac{\sin\delta}{\delta} \simeq 0.85 \quad PBWA. \tag{29}$$

For the PWFA we consider only relativistic driving and driven bunches. In addition we require that the final energy of the driving bunch after the distance

L is still relativistic. In this case we can calculate the phase slippage along the plasma wave since the plasma wave phase velocity is equal to the velocity of the driving bunch. Following Ref. 4 we integrate the relative velocity along the length L to obtain

$$\delta \simeq \frac{\pi L}{\lambda_p}\left[(\gamma_{1i}\gamma_{1f})^{-1} - (\gamma_{2i}\gamma_{2f})^{-1}\right] \quad PWFA. \tag{30}$$

Since in an actual high energy accelerator the second term would be quite small, we will neglect it when using Eq. (30).

The Transverse Size

We need the transverse size to calculate the transverse dynamics of the driven bunch. For the PBWA to make the optimum use of the laser beam it is necessary to match the Rayleigh length R to the acceleration section. We choose the section to be twice the Rayleigh length. This in turn determines the diffraction limited spot size,

$$a^2 = \frac{R\lambda}{\pi} = \frac{L\lambda}{2\pi} = \frac{2\delta c^2 \omega}{\omega_p^3} \qquad PBWA, \tag{31}$$

where Eq. (27) has been used to eliminate L. For the PWFA since we would like to fix the number of particles in the driving bunch, the transverse size is determined by the desired accelerating field,

$$a = \left[\frac{8 r_e N_1 m c^2}{e\mathcal{E}_z}\right]^{1/2} \qquad PWFA, \tag{32}$$

where r_e is the classical electron radius.

The Energy Requirement

In the PBWA the laser beam power for the beam profile given in Eq. (3) is

$$W = \frac{\pi a^2}{2}\frac{E_0^2 c}{8\pi}\ . \tag{33}$$

If we assume that we have a laser pulse length τ, the energy necessary to drive the plasma wave density to αn_0 is[2]

$$W\tau = \frac{\alpha\delta m^2 c^5}{e^2\omega_p}\left(\frac{\omega}{\omega_p}\right)^3 \quad PBWA\ . \tag{34}$$

where Eq. (31) has been used to eliminate a^2. On the other hand, the energy in the driving bunch for the PWFA is simply given by

$$W\tau = N_1 E_1 \qquad PWFA\ . \tag{35}$$

The Efficiency

The overall efficiency of the accelerators here can be divided into three parts. The first part is the efficiency of conversion of 'wall plug' energy to either laser energy or electron beam energy. These two efficiencies may be quite different, however, we will not discuss them here. The second efficiency is the conversion of either laser or electron beam energy to plasma energy. The third efficiency is that for conversion of the plasma energy to the driven electron beam. The efficiency of the transfer of energy from the laser to the plasma has been calculated for the PBWA model we have chosen.[2] For a general phase shift δ the ratio of the plasma energy to the laser energy is given by

$$\eta_1 = \frac{P.E.}{W_T} = \frac{\alpha\delta}{4}. \tag{36}$$

If laser depletion is included in the analysis, this number will be reduced slightly.

The efficiency of the transfer of energy from an electron beam to the plasma is quite different. In this case one must consider the beam loading effects. If we could treat the bunch as a macro-particle, then for a very relativistic driving bunch we could extract nearly all of its energy before it's velocity changed enough to yield a phase slip. However, due to beam loading this is not possible since the leading edge of the driving bunch looses essentially no energy to the plasma while the trailing edge looses twice as much as that calculated for a point like particle. Thus, for very short bunches, we can only extract about 1/2 of the energy

$$\eta_1 = \frac{1}{2} \quad PWFA\ . \tag{37}$$

For longer bunches of electrons, one can improve this factor and also improve the 'transformer ratio'[6] at the expense of the peak field. Since this technique might be quite difficult to realize in the PWFA, we will not consider it here.

The final efficiency to calculate is that from the plasma to the driving bunch. This efficiency is the same for both cases provided that the characteristics of the plasma wave are the same. The total acceleration gradient experienced by a bunch with N_2 particles in a plasma wave is

$$G \equiv \frac{dE_2}{dz} = e\mathcal{E}_z f - 4e^2\frac{N_2}{b^2}. \tag{38}$$

The second 'beam loading' term is due to the plasma wake induced by the trailing bunch. $e\mathcal{E}_z$ is the peak longitudinal electric field, and f is a factor less than unity which takes into account phase slippage or shifts in phase from the peak accelerating field. The efficiency is given by the total energy gained by the bunch divided by the plasma energy,

$$\eta_2 = N_2 G L \left(\frac{\mathcal{E}_z^2}{8\pi} \frac{\pi a^2}{2} L \right)^{-1} . \tag{39}$$

This efficiency has a maximum when

$$N_2 = \frac{f \mathcal{E}_z b^2}{8e} , \tag{40}$$

and the value is given by

$$\eta_2^{max} = f^2 \frac{b^2}{a^2} . \tag{41}$$

For the PWFA f can be taken to be essentially unity while for the PBWA f is given by Eq. (29). This yields

$$\begin{aligned} \eta_2^{max} &\simeq .72 \frac{b^2}{a^2} \qquad PBWA \\ \eta_2^{max} &\simeq \frac{b^2}{a^2} \qquad PWFA \end{aligned} . \tag{42}$$

COMPARISONS AND DISCUSSION

Now we come to a detailed comparison between the PBWA and the PWFA. As mentioned earlier, our guide will be the self consistency among all relevant accelerator parameters within each scheme. Our approach is to choose a set of parameters in each scheme that we fix from the beginning. The remaining parameters in each scheme can then be calculated in terms of those chosen parameters. The scaling to different sets of chosen parameters is straight forward using the results of the previous section. To make a fair comparison we will study two sets of sample accelerators with the same acceleration gradient and the same length L. In addition to make the comparison meaningful to real experiments, we employ only those laser and electron beams that are presently available. Under these considerations, the parameters that should be fixed in the two schemes are quite different. In particular for the PBWA we need to fix the laser frequency ω by choosing a particular laser source. If we then fix the length L of the acceleration section, the phase slippage determines the plasma frequency ω_p. This means that the longitudinal electric field $\mathcal{E}_z$ is a derivable quantity. On the other hand, the energy gradient in the PWFA is chosen so that the intensity and dimensions are not far from realizable values. As we shall see, in spite of this difference it is possible to match the acceleration gradients.

Numerical comparisons

To keep the dimensions to a laboratory scale, we select the acceleration lengths to be 10 cm and 100 cm. These two lengths are then combined with two different laser frequencies, the *Nd*: Glass laser and the CO_2 laser, to form four sets of sample calculations. For the PBWA the parameter α is chosen to be 0.25, which is approximately the saturation value[5] and the phase slippage is taken to be the optimum value given in the previous section. Finally, we assume that the laser pulse length and the growth time for the plasma wave τ is about 159 cycles ($\omega_p \tau = 1000$).

Since the PWFA is not so restrictive in its design, we can now set the parameters to match some of those for the PBWA. In particular we use the same acceleration gradient and the same a/λ_p. The number of particles in the driving bunch is taken from the present number in the SLC and the bunch length is assumed to be somewhat less than the plasma wavelength. The initial and final energies of the driving bunch are selected so that the final energy of the bunch tail is 90% of its initial energy. As we can see from Tables 1 and 2, the phase slippage for the PWFA is much smaller than that for the PBWA. All parameters except the efficiency and the energy in the driving beam turn out to be quite comparable. In particular note that the focusing for both schemes is quite strong. The energy required for the driving bunch is consistently higher for the PBWA; however, because it is less efficient in these examples, the number of particles which can be driven is comparable to the PWFA.

Discussion

The examples above seem to favor the Plasma Wake Field Accelerator especially for the longer accelerator sections. This is due to the divergence of the laser. For longer Rayleigh lengths it is necessary to have a larger spot and thus more peak power to obtain the same intensity at the spot. On the other hand the particle beam is assumed not to diverge. This is true because the emittance of the beam is typically much smaller than the corresponding wavelength/π for the laser. In addition it is possible to use magnetic focusing elements to define the size of a charged particle beam. The problem of the divergence of the laser beam might be solved by using lasers sufficiently intense to self focus in the plasma; however, this possibility was not considered since it lies outside the scope of the simple models given here. In addition, for the PBWA parameters chosen here, the laser power is somewhat below the critical value for relativistic self focusing.[7]

Table 1. Plasma Beat Wave Accelerator

Chosen Parameters	Values			
ω [sec^{-1}]	Nd: Glass 1.78×10^{15}		CO_2	1.78×10^{14}
L [cm]	10	100	10	100
α	0.25	0.25	0.25	0.25
δ [rad]	$5\pi/16$	$5\pi/16$	$5\pi/16$	$5\pi/16$
$\sin\delta/\delta$	0.85	0.85	0.85	0.85
$\omega_p\tau$	1000	1000	1000	1000
Derived Parameters				
ω_p [10^{13} sec^{-1}]	2.65	1.23	.571	.265
n_0 [10^{16} cm^{-3}]	21.7	4.67	1.00	0.22
$e\mathcal{E}_z$ [GeV/m]	9.38	4.36	2.00	0.94
a [mm]	0.13	0.41	0.41	1.30
a/λ_p	1.82	2.70	1.25	1.82
β [$\sqrt{\gamma/\sin\phi}$ mm]	0.18	0.57	0.57	1.80
N [10^{10}]	$1.95\eta_2$	$9.04\eta_2$	$4.19\eta_2$	$1.95\eta_2$
W_T [J]	23.9	515.4	11.1	239.2

Unfortunately, for both schemes the efficiency η_2 and the energy spread induced are directly related. Thus, if a small energy spread is necessary, then η_2 will necessarily be small for both schemes. The efficiency η_1 of the PWFA was better in all cases because the energy transfer from the laser to the plasma is limited by Eq. (36) to quite a small value. There is a possible solution to this problem. Since the laser is not depleted very much, it might be possible to reuse the beam after a suitable amplification. This would yield a very high repetition rate and looks quite attractive; however, this possibility needs much more study.

There is one final problem for the PBWA. We have assumed that the plasma wave would grow over $1000/2\pi$ cycles. If there are density fluctuations greater than about .2%, then the wave would saturate much sooner. This case would require a much larger laser energy in order to drive the plasma to the desired field in a shorter time.

Table 2. Plasma Wake Field Accelerator

Chosen Parameters	Values			
L [cm]	10	100	10	100
$e\mathcal{E}_z$ [GeV/m]	9.38	4.36	2.00	0.94
N_1	5×10^{10}	5×10^{10}	5×10^{10}	5×10^{10}
E_1 [GeV]	1.04	4.84	0.22	1.04
a/λ_p	1.82	2.70	1.25	1.82
Derived Parameters				
a [mm]	0.25	0.36	0.54	0.78
$\delta[10^{-3} rad]$	5.5	2.5	42	18
ω_p $[10^{13}$ sec$^{-1}]$	1.37	1.41	.439	.438
n_0 $[10^{16}$ cm$^{-3}]$	5.90	6.18	.606	.604
α	0.38	0.17	0.25	0.11
β $[\sqrt{\gamma}/\sin\phi$ mm$]$	0.28	0.59	0.73	1.52
N_2 $[10^{10}]$	$2.25\eta_2$	$2.25\eta_2$	$2.25\eta_2$	$2.25\eta_2$
$W_T = N_1 E_1$ [J]	8.33	38.8	1.76	8.33

REFERENCES

1. T. Tajima and J. M. Dawson, Phys. Rev. Lett. 43, 267 (1979).
2. R. D. Ruth and A. W. Chao, in *Laser Acceleration of Particles*, Ed. P. J. Channell, AIP Conference Proceedings No. 91, American Institute of Physics, New York, 1982.
3. P. Chen, R. W. Huff and J. M. Dawson, UCLA Report No. PPG-802, 1984, and Bull. Am. Phys. 29, 1355 (1984); P. Chen, J. M. Dawson, R. W. Huff and T. Katsouleas, Phys. Rev. Lett. 54, 693 (1985).
4. R. D. Ruth, A. W. Chao, P. L. Morton and P. B. Wilson, SLAC-PUB-3374, 1984 (to be published in Particle Accelerators).
5. M. N. Rosenbluth and C. S. Liu, Phys. Rev. Lett. 29, 701 (1972); D. J. Sullivan and B. B. Godfrey, ibid., Ref. 2.
6. K. L. F. Bane, P. Chen and P. B. Wilson, *Co-Linear Wake Field Acceleration for Linear Colliders*, to be presented at the 1985 Particle Accelerator Conference, Vancouver, B.C., Canada, May 13-16, 1985; P. Chen and J. M. Dawson, this Proceedings.
7. H. Hora, *Physics of Laser Driven Plasmas*, John Wiley and Sons, New York (1981) and P. Sprangle and C. M. Tang, this conference.

PLASMA WAKE FIELD ACCELERATION: A PROPOSED EXPERIMENTAL TEST

J. B. ROSENZWEIG, D. B. CLINE, R. N. DEXTER, D. J. LARSON, A. W. LEONARD, K. R. MENGELT, J. C. SPROTT

Department of Physics
University of Wisconsin, Madison, Wisconsin 53706

F. E. MILLS
Fermi National Laboratory, Batavia, Illinois 60510

F. T. COLE
Argonne National Laboratory, Argonne, Illinois 60439*

ABSTRACT

The prospect of achieving very high accelerating fields has led to proposals for using electrostatic plasma waves to accelerate charged particles for high energy physics. It has been predicted theoretically that these plasma waves can be driven by the wake fields of short bunches, or trains of bunches, of charged particles, to accelerate a subsequent bunch; the longitudinal electric fields possible could be of the order of a few GV/m. This note presents an outline of a proposed experimental test of this principle.

I. EXPERIMENTAL MOTIVATION

The use of a plasma to support very high electric fields for accelerating particles has generated considerable interest in recent years, most notably in the the form of the Plasma Beat Wave Accelerator (BWA) concept first formulated by Tajima and Dawson[1]. This scheme in particular, as well as the Surfatron[2] modification proposed later by Katsouleas and Dawson, relies on a sharp resonance condition between the laser frequencies driving the interaction and the density of the ambient plasma.[3] In addition, simulation and experiment indicate that the lasers needed to pump the plasma wave may be required to deliver an exceedingly high power pulse in a few picoseconds in order to grow the wave to the large amplitudes associated with high accelerating gradients.[4,5]

A promising alternative to laser drivers has been proposed by P. Chen, et al.[6], that of utilizing extremely short bunches ($\sigma_z \ll \lambda_p$) of electrons to excite the plasma oscillation. As has been pointed out in the discussion of Ruth, et al.[7], this idea is nearly identical

*On leave from Fermilab.

conceptually to the various wake field accelerators and transformers currently being proposed.[8] This being the case, use of a plasma to couple the wake fields of the leading bunches to the accelerating particles holds the possibility of a much larger amplitude "wake function," or higher accelerating fields than a cavity structure. It is with these considerations in mind that we propose to test this concept experimentally.

II. EXPERIMENTAL ISSUES

In order to project what is needed to execute a successful test of the Plasma Wake Field Accelerator, a short discussion of the physics of scale is necessary. In the idealized limit where the "Wake Field Theorem" holds (colinear beams; short, 'rigid' bunches), a trailing bunch can gain at most twice the energy per particle that the leading bunch particles lose as they are stopped in a wake field device ($dE_2/dz \leq -2dE_1/dz$). This gradient may be improved upon by using several driving bunches at proper phases, or by shaping a single pulse's longitudinal profile.[9] The predicted maximum field is constrained by plasma density to be the linear wave-breaking amplitude of the plasma wave, $e\tilde{E}_{wb}(eV/cm) \simeq \sqrt{n_e}(cm^{-3})$. Using Ruth's model this can be related to the surface number density σ_b of the driving bunch. If one requires that the disk of charge have a radius of one plasma wavelength, $\lambda_p = 2\pi v_b/\omega_p$ (v_b is the velocity of the driving beam and ω_p is the plasma frequency) to curb transverse effects, then the wave-breaking accelerating field ($E_{zmax} = 4\pi e\sigma_b$) is attained if $N_b = (1/2)n_e\lambda_p^3$ where N_b is the number of charges per bunch and n_e is the plasma electron density. This requirement may be relaxed somewhat to drive higher longitudinal fields ($E_{zmax} \sim a^{-2}$, where a is the driving beam radius), but only at the expense of loss in transverse beam quality.

If the driving bunch length is not is not small compared with the plasma wavelength, then the one-dimensional analysis of Ruth, et al., dictates that a convolution integral be performed over the driving bunch charge volume density $\rho(z')$ to yield the beam-induced electric field strength at a point trailing the driving bunch:

$$E_z(z) = 4\pi \int_z^{\infty} dz'\rho(z')\cos\left[k_p(z'-z)\right]$$

For symmetric bunches, this convolution will not exceed the maximum value obtained by treating the bunch as an infinitesimally thin disk. In general, the beam loading due to finite bunch length degrades the maximum field achievable.

Given what is predicted by theory, it seems reasonable to require several attributes of the plasma. First, it should be of the proper density and length to match the characteristics of the driving beam. Furthermore, these quantities should be well diagnosed and tunable. It should also be quiescent and without major density fluctuations. Similarly, the beams used in the experiment should be of high quality: extremely short pulse length, high peak current, and low emittance. Specific proposals and numerical examples are given in the next section.

III. PROPOSED EXPERIMENT AT ANL

(a) ARGONNE WAKE FIELD TEST FACILITY

There is currently considerable interest in performing experiments on wake field acceleration at Argonne National Laboratory.[10,11] A major source of motivation for these ideas has been the existence of an excellent short pulse, L-band subharmonic bunching electron linac at the ANL site, one that appears nearly ideal for wake field studies. The salient features of the linac are listed in Table 1. There exist several distinct ways of generating the smaller witness pulse to follow the more intense driving bunch. The most promising for this experiment is to have the accelerator accept a small number of electrons at a phase slightly behind the main bunch; the final energy of the witness pulse will be slightly off the driving pulse, and they will be separated in time by 0-770 psec. The basic geometry, shown in Fig. 1, includes a trombone leg of adjustable length to allow for control over the delay betweeen the pulses. A more detailed discussion of the facility appears elsewhere in these proceedings (cf. Ref. 11).

Table 1

ARGONNE LINAC CHARATERISTICS

Energy, E	22 MeV
ΔE	$\pm$ 100 KeV
Emittance, ε	$7\pi \times 10^{-6}$ m-rad
Pulse frequency	$\leqslant$ 800 Hz
Pulse length	5 - 30 psec
Bunch length, σ_z	1.5 - 9 mm
Electrons/Pulse, N_b	9 -27 nC

The most exceptional characteristic of this linac is its short pulse length, presently at 30 psec, soon to be modified for operation in the 5-10 psec range. This short pulse corresponds to a spatial length of $\geqslant$ 1.5 mm, which guides the choice of plasma parameters. The emittance is also of consequence, as it gives a lower limit on beam waist size and interaction length.

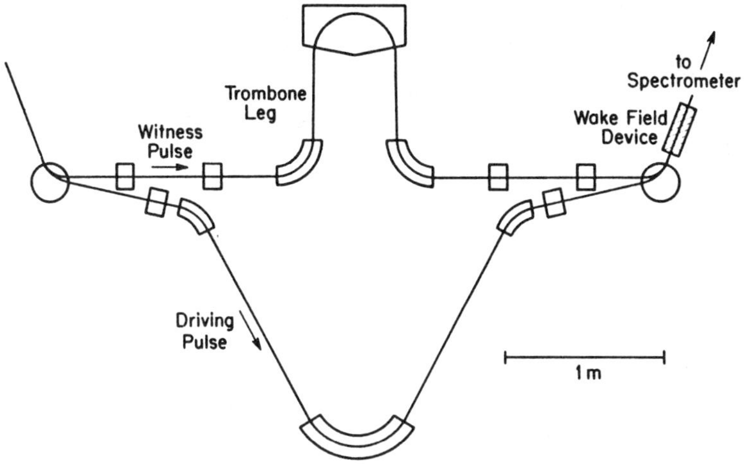

Fig. 1: ANL wake field test facility beam-line.

(b) PLASMA SOURCE

For a given bunch length, it may be desirable to have a plasma wavelength much longer or at least about the same length. The shorter the plasma wavelength, the higher electric field possible; however, if the plasma wavelength is not much greater than the driving bunch length the plasma wave may not be driven coherently, as can be seen by examining the behavior of the convolution integral in Section II. The plasma density associated with $\lambda_p \cong 1.5$ mm is $n_e \cong 5\times10^{14}$ cm^{-3}, which may be taken as an approximate upper bound for our experiment.

A plasma source appropriate for the present round of BWA experiments has been developed at UCLA by Joshi and Clayton, a rail gap discharge augmented by laser ionization. Experimental work has also been performed at UCLA on a θ-pinch[12], a device which is capable of producing of producing long, hot plasmas in the density range $n_e = 10^{15}$-10^{17} cm^{-3}, for possible future use in the next phase of BWA research. F. Chen has also proposed that a rail gap discharge be considered as a source suitable for use in a Surfaton experiment, as it would allow an imposed B-field transverse to the longitudinal axis[13].

Since our experiment requires no laser, transverse magnetic field and considerably less than 10^{17} cm^{-3} density, we propose to utilize a plasma source different from the above. The hollow cathode arc (HCA) produces a DC cylindrical plasma of density $\lesssim 10^{15}$ cm^{-3} which is known to be remarkably quiet and free of density fluctuations[14] (see Fig. 2). This plasma can be well confined, with a long axial density scale length, by an imposed solenoidal B-field of the order of 100-400G. The device and its geometry as shown have some distinct advantages for this experiment: (i) the plasma is cold, quiescent, and nearly fully ionized, (ii) the density is variable, (iii) the interaction length may be changed by repositioning the anode, and (iv) as a DC plasma source it also offers the advantage that the linac may be pulsed rapidly with small shot-to-shot variation in plasma conditions.

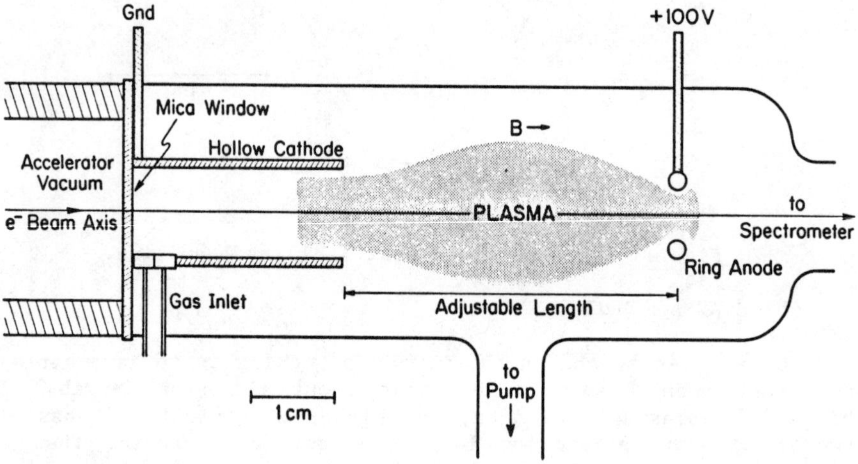

Fig. 2: Hollow cathode arc; experimental setup for plasma wake field accelerator test.

Table 2

PLASMA PARAMETERS

Electron density, n_e	$10^{13} - 10^{15}$ cm^{-3}
Plasma wavelength, λ_p	1 - 10 mm
Interaction length	~ 2 - 10 cm
Wave-breaking limit	0.3 - 3 GeV/m

(c) EXPERIMENTAL GOALS

The experimental setup as outlined above should allow the variation of the defining parameters of the problem. The beam line design admits the possibility of varying:

(i) The pulse delay, for probing the plasma wave many wavelengths behind the driving pulse.

(ii) The beam size at the interaction, to test for larger acceleration and possible induced transverse effects. (For beam radius $a = \sqrt{\beta\varepsilon} = 1.5$ mm, the beam beta function $\beta = 10$ cm.)

(iii) The pulse length and pulse shape, which are likely to be of crucial importance in driving the plasma wave to larger amplitudes.

An HCA plasma should be tunable over a range of densities, allowing longer plasma wavelengths to be examined. The range of the interaction region itself can be adjusted from a minimum of approximately 2 cm to a maximum given by the constraints of the driving beam stopping distance and the dephasing length of the accelerated bunch. The minimum length in which the driving beam stops can be estimated from the maximum decelerating field, that is, one half the maximum accelerating field possible.

$$\frac{dE_1}{dz} = \frac{1}{2}\, eE_{zmax} \simeq 2r_e N_b mc^2/a^2 = 200 \text{ MeV/m}$$

where r_e = classical radius of the electron and $a = \lambda_p$. Thus the driving beam (22 MeV) is expected to stop (i. e., become sufficiently non-relativistic, ~ 2 MeV) in $L_s \simeq 10$ cm. The dephasing length, in which the maximally accelerated electrons outrun the wave, is given approximately by[7] $L_{deph} \simeq (1/2)\lambda_p \gamma_i \gamma_f \simeq 15$ cm, with $\gamma_{i,f}$ the initial and final Lorentz factors (E/mc^2) of the driving beam and requiring less than a $\pi/2$ phase slip. The stopping distance of the driving beam thus seems to be a more stringent constraint.

The success of the test rests ultimately on achieving the correct interaction conditions, but these must be diagnosed as completely as possible. The properties of both bunches must be examined to clarify the physics of wake field acceleration and deceleration. The momentum distribution will be measured with a broad range, high resolution magnetic spectrometer (see Fig. 3). The vertical distribution at detection can be utilized to study the emittance blow up due to strong off-axis radial electric fields. The plasma parameters must also be observed with care. The density profile of the plasma discharge can be measured by interferometric methods. The plasma wave amplitude and phase velocity be diagnosed by Thomson scattering, as is done on the UCLA BWA experiment.[14]

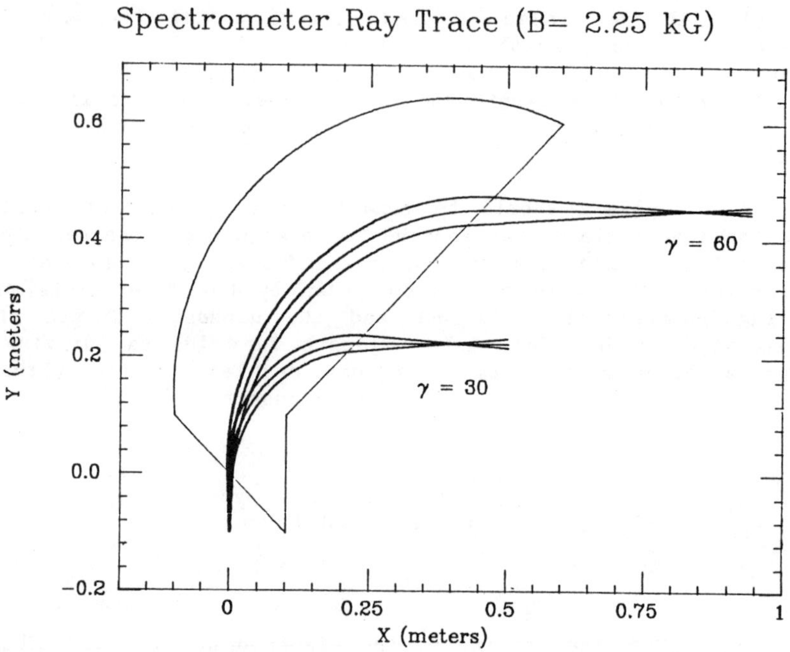

Fig. 3: Ray trace and approximate pole face geometry for proposed high resolution, double focusing, flat-field magnetic spectrometer.

IV. PROJECTED SCHEDULE

The Argonne linac will be modified for 5 psec operation by approximately July, 1985. The facility will then be used for proposed tests on wake field cavity structures (i. e., the Wakeatron concept of Ruggiero, cf. Ref. 10). These experiments should run through the fall. In the intervening time, Wisconsin proposes to build and test the plasma sources, and to develop beam and plasma diagnostics. The tentative schedule then has measurements beginning as early as spring of 1986.

V. CONCLUSION

The experiment proposed here offers a unique opportunity to explore the physics of a new acceleration concept. The properties of the beam at ANL can be matched to a plasma that, from simple theory, could provide an accelerating gradient of 400 MeV/m. This prospect, along with the host of questions surrounding this type of collective acceleration scheme, should make for a very intriguing experiment.

ACKNOWLEDGEMENTS

One of the authors (JBR) would like to thank Pisin Chen and Ron Ruth for helpful discussions during the Workshop.

REFERENCES

1. T. Tajima and J. M. Dawson, Phys. Rev. Lett., 43, 267 (1979).
2. T. Katsouleas and J. M. Dawson, Phys. Rev. Lett., 51, 392 (1983).
3. C. M. Tang, P. Sprangle and R. Sudan, Appl. Phys. Lett. 45, 375 (1984).
4. P. D. Goldstone, presented at Los Alamos Beat Wave Accelerator Peer Review, March 11 - 12, 1984.
5. C. Joshi, W. B. Mori, T. Katsouleas, J. M. Dawson, J. M. Kindel and D. W. Forslund, Nature 311, 525 (1984).
6. P. Chen, J. M. Dawson, R. W. Huff and T Katsouleas, SLAC-PUB-3487, November 1984.
7. R. Ruth, A. W. Chao, P. L. Morton, P. B. Wilson, SLAC-PUB-3374, July 1984.
8. G. A. Voss and T. Weiland, DESY publications 82-015 and 82-074 (1982).
9. R. Ruth, private communication.
10. A. Ruggiero, these Proceedings.
11. J. Simpson, private communication.
12. B. Amini and F. Chen, UCLA PPG-779, May 1984.
13. F. Chen, UCLA PPG-833, November 1984
14. L. M. Lidsky, S. D. Rothleder, D. J. Rose, S. Yoshigawa, C. Michelson and R. J. Mackin, Jr., Journal of Appl. Phys., 33, 8 (1961).

REPORT OF NEAR FIELD GROUP*

R.B. Palmer†(Group Leader), N. Baggett, J. Claus,
R. Fernow, I. Stumer
Brookhaven National Laboratory
Upton, New York 11973

H. Figueroa
University of California at Los Angeles
Los Angeles, CA 90024

N. Kroll
University of California at San Diego
LaJolla, CA 92093

W. Funk and G. Lee-Whiting
Chalk River Nuclear Laboratories
Chalk River, KOj 1jO Ontario

M. Pickup
Cornell University
Ithaca, NY 14853

P. Goldstone and K. Lee
Los Alamos National Laboratory
Los Alamos, NM 87545

P. Corkum
National Research Council of Canada
Ottawa Kla Or6 Ontario

T. Himel
Stanford Linear Accelerator Center
Stanford, CA 94305

1. ABSTRACT AND INTRODUCTION

It is a pleasure to be able to report substantial progress since the Los Alamos Workshop two years ago. A radio-frequency model of a grating accelerator has been tested at Cornell, and extensive calculations compared with observations. Alternative structures consisting of either hemispherical bumps on a plane, or conducting spheres in space, have also been rf modeled. The use of liquid droplets to form such structures has been proposed and a conceptual design studied. Calculations and experiments have examined the effects of surface plasmas, and shown that in this case the reflectivity is low. However, calculations and observa-

*Work done under the auspices of the U.S. Department of Energy.
†On sabbatical leave at SLAC.

tions suggest that gradients in excess of 1 GeV/meter should be obtainable without forming such plasma. An examination of wake fields shows that, with Landau damping, these are indepedent of wavelength. The use of near field structures to act as high gradient focusing elements has been studied and shows promise, independent of the acceleration mechanism. Beamstrahlung in the quantum mechanical limit ahs been shown to scale as $(DN)^{1/3}$. Finally a proposal has been made to establish a facility that would enable "proof of principle experiments" to be performed on these and other laser driven accelerator mechanisms.

2. ACCELERATING STRUCTURES

a) Iris Loaded Linac (Fig. 1a)

Although not practical as a linac for laser wavelengths, the iris loaded linac can serve as a standard for comparison. The SLAC structure, for instance, has a Q of 13,000, and this would scale as the inverse root of the wavelength (10.5 cm). The loss parameter k_l is 19 volts/picocoulomb/meter, scaling as the inverse wavelength squared. The shunt impedance ($r=4k_l\ Q/\omega$) is 56 M ohms/m and this scales as the inverse root of lambda.

b) Grating (Fig. 1b)

This structure was the first proposed for laser acceleration[1], and despite earlier difficulties[2] has been shown[3] to support non-radiating (i.e. resonant) accelerating modes. These, however, are only present when the exciting radiation falls on the grating from the side, and the resulting fields are always periodic across the grating (i.e. perpendicular to the acceleration). Contrary to the hope expressed in Ref. 3, it has been shown[4] that different transverse periodicities cannot be added to restrict the transverse extent of the fields. The periodicity is fixed by the grating profile. Thus although the grating is suitable for accelerating a large number of beams, it would be very inefficient for only one.

c) Grating With Side Walls (Fig. 1c)

This structure has been studied by M. Pickup at Cornell and is discussed in a separate contribution to these Proceedings.[5] The walls, which need be only of the order of a wavelength high, can be placed at any multiple of half the transverse field periodicity. Pickup studied the case where they are one half period apart. Leaving aside the question of how such walls could be constructed, Mike has shown that the Q, scaled to 10.5 cm, would be 16,000 (even higher than in the iris loaded case); however the loss parameter k, again scaled, is 1.7 volts/picocoulomb/meter (much lower). One must remember, however, that as the wavelength gets smaller the loss parameter rises as the square and a high initial value is not necessarily desirable.

d) Inside-Out Iris-Loaded Cavity (Fig. 1d)

Kroll[6] has considered the fields that can be formed on the outside of a structure which is geometrically like a conventional linac. This case can also be thought of as that of a grating in which the two sides have been curled under and joined together. As in the grating case, non-radiating modes exist and, also as in the grating case, these fields must be periodic transverse to the acceleration, i.e., periodic in the azimuthal angle, in this case. The number of periods around the azimuth may be described by the index m. For m = 0 there are no solutions, in analogy with the Lawson theorem for the grating case. He also showed that the m = 1 case (dipole) has a field that extends to infinity. For m = 2 (quadrupole) the fields do fall off but the total energy has a logarithmic divergence. Only for m = 3 and above are the fields truly local, with the structure behaving as a true "open cavity". Kroll also considered structures formed of more than one parallel inside-out cavity, each operating in the m = 1 mode. All these cases give insight into the droplet structures described below.

e) Double Row of Droplets (Fig. 1e)

An rf model consisting of two copper spheres placed between two parallel metal plates demonstrated[7] a mode that would accelerate along the axis between two rows. The spacing between the spheres, both along the rows and between them, was $\lambda/2$, and their diameter was approximately $\lambda/3$. The measured fields were well represented by the assumption that the spheres act as oscillating dipole radiators with their polarization directed in towards the axis. The "measured" loss parameter k, scaled to a wavelength of 10.5 cm, was approximately 2 volts/pC/m, i.e., similar to that for the grating case. However, this case is essentially that of two m = 1 inside out cavities, and the long range fields must have the m = 2 character that, as was pointed out by Kroll, has a divergent energy and thus a zero k parameter. In the measurement however, and in any practical case, a cut-off is in fact imposed either by the surroundings or by the pulse length. Thus despite the divergence this may be a useful case.

f) Four Rows of Spheres (Fig. 1f)

With four rows of spheres the long range fields are octupole (m = 4) and no divergence occurs. Such a mode was also observed with the rf model, but the k has not yet been measured.

g) Rows of Bumps (Fig. 1g)

A second mode observed with two rows of spheres had a symmetry plane such that it would also be present over a double row of hemispheres on an infinite plane. This then represents a "grating" in which no side walls are required. Maximum acceleration in this case occurs along a line over the top of either row of bumps; in fact one row could accelerate electrons while the other accelerated positrons. The logarithmic divergence would still be present in this case, but could, if required, be removed by the use of three or more rows of bumps.

h) Super Bumps (Fig. 1h)

Kroll has proposed a case derived from a double row of inside-out iris cavities (see d above). Each of the inside-out cavities is excited in a mode m = 2 with left-right symmetry and up-down antisymmetry. Half of this arrangement is then placed over a plane conductor to produce the structure illustrated. The long range fields are m = 3 so there is no divergence, and there is even a neutral axis above the surface with quadrupole focusing fields about it.

Although other now radiating modes are possible, I have been discussing π modes. They do not radiate energy out, but also cannot be excited by any incoming radiation. In order to couple to external fields, some perturbation is needed to the symmetry. In the grating case, alternate lines can be made slightly higher. In the case of two rows of droplets, alternate droplets can be displaced out of the plane (Fig. 2). In this case the angular distribution of incoming radiation that would be perfectly absorbed, is shown in Fig. 3.

3. THE PRODUCTION OF DROPLET ARRAYS

Grating structures of almost arbitrary shape can now be made by a number of micro-machining techniques. The structures formed from spheres seem at first a lot less practical. However, liquid jets developed for ink jet printers and other more exotic purposes can place droplets with remarkable precision and at low cost. Structures formed from such droplets, besides having some nice electromagnetic properties, would have the advantage of being "disposable". Damage caused by the radiation or the beam, provided it does not spoil the structure during its few picoseconds of use, need not be considered. It is therefore interesting, as an aside, to give a conceptual design of a section of a droplet structure.

Figure 4a shows a vacuum container with entrance windows, presumably of salt, both above and below the beam. On either sideof the beam are the liquid jet assemblies mounted on micro-manipulators. Pumps are provided to remove vapor given off from the heated droplets. Figure 4b shows a jet assembly with filter and piezo-electric pump. On the right and in Fig. 4c is the jet array itself. The techniques proposed, and being developed at BNL, are extensions of those used in both some ink jet printers and masks. A silicon chip is doped on one side and then anisotropically etched to form the long channel with a thin (circa 2 micron) remaining wall. Through this wall the actual holes are ion etched.

We do not yet know how accurately droplets of the required size (3 microns) can be placed, but it is worth noting that an array used in an ink jet printer[8] was able to make 13-micron jets with an angular accuracy of 1 milliradian. If such angular accuracy could be maintained, droplets could be placed to one tenth of a micron, which would probably be sufficient.

4. LOADING AND EFFICIENCY

The maximum number of particles that can be accelerated in any structure is set by longitudinal and transverse wakes. As the structure gets smaller the wake fields get stronger, but at the same time the stored energy in the fields decreases. For longitudinal wakes it has been shown by Wilson[9] that for a given wake field effect the same fraction of the stored energy can be extracted, independent of wavelength. The situation for transverse wakes is more complicated. It was shown by Wilson[9] that A, the wake amplitude divided by the initial misalignment, is given by

$$A = \frac{N \beta z w e}{4V}$$

where N is the number of particles per bunch
β is the focusing parameter
z is the distance along the accelerator
w is the wake potential
e is the electric charge
V is the beam energy in electron volts.

Since w scales as $w = w_0/\lambda^3$, one might expect the effect to be much worse for small λ, but as we said the situation is more complicated.

Without Landau damping A grows without limit as z increases, but with a finite momentum spread between head and tail the driving frequency gets out of phase with the tail's transverse betatron oscillation and the amplitude reaches a maximum value given by substituting

$$z(\text{Landau}) = \frac{2\beta}{dp/p}$$

For N we can substitute that value that would extract a given fraction η of the stored rf energy,

$$N = \frac{\eta E_a}{4 k_1 e}$$

where E_a is the accelerating gradient
k_1 is the loss factor for the cavity $= k_0/\lambda^2$
k_0 is a dimensionless constant of the cavity geometry.

For the β we will assume RFQ focusing as discussed below in Section 7. Then

$$\beta = \beta_0 \, (\lambda V/E_a)^{\frac{1}{2}}$$

where β_0 is a scale invariant constant of the cavity and focusing geometries. The focus is stronger for a shorter wavelength because the poles are closer to the axis.

Substituting, we obtain

$$A = \frac{1}{8} \eta \left(\frac{w_0}{k_0} \beta_0^2\right) \frac{1}{dp/p}$$

which is independent of λ and, incidentally, also of E_a and V.

Thus we find that both longitudinal and transverse wake considerations set a scale independent limit on the fraction η of rf energy that can be extracted. In practice this limit is about 5%. If only one bunch is accelerated this sets a bound on the accelerator efficiency. With many bunches removing energy in equilibrium with incoming power, however, we know that far higher efficiencies can be achieved; even as high as 80%.

The relevance of these remarks arises because a collider probably requires pulses of the order of 1 mm in length, and it is certainly simpler if they are single. If a conventional wavelength is used this pulse can only consist of a single rf bunch; in the 10 micron case, however the pulse will contain 100 micro bunches, each with only a small charge, and in these circumstances much higher efficiency (say 50% instead of 5%) may be expected. This may then offset the lower power source efficiency of a laser compared with a klystron or lasertron (5-10% vs 40-80%).

5. BEAMSTRAHLUNG SCALING

Himel and Siegrist[10] have studied the scaling of the quantum mechanical corrections to beamstrahlung. Using the approximation that the spectrum remains as in the classical calculation up to $E = E_c$ and then is cut off one obtains

$$\delta(\text{quantum mechanical}) \approx \delta(\text{classical}) \times \left(\frac{E}{E_c}\right)^{\frac{4}{3}} \quad .$$

In fact this is always a conservative estimate. Here δ is the average fraction of beam energy lost to beamstrahlung. E is the beam energy and

$$E_c = \frac{3}{2} \hbar c \frac{\gamma^3}{\rho} = \frac{3 \hbar c \gamma^2 N r_e}{2 \quad r d}$$

where ρ = radius of curvature in field
d = bunch length (assumed uniformly filled)
N = number of particles
r_e = classical electron radius
r = radius of bunch (assumed uniformly filled).

Substituting one obtains:

$$\delta_{QM} = \frac{8}{\sqrt{3}} \left(\frac{r_e m c^2}{2 \sqrt{3} \hbar c}\right)^{\frac{4}{3}} (D N)^{\frac{1}{3}}$$

where $D = \dfrac{N r_e d}{\sqrt{3} \gamma r^2}$ = disruption parameter.

The simplicity of this relation is almost startling. If D = 1 is chosen to assure significant self-focusing, $\delta_{QM} \leq 0.3$ to keep the energy spread less than 10%,, and if we are in the quantum mechanical regime ($\frac{E}{E_c} < 1$), then

$$N \leq 1.2 \times 10^7$$

This is a very small number and though well suited to a laser accelerator will not match the large stored energy in cavities using larger wavelengths.

6. ACCELERATING FIELD LIMITS

At the last laser acceleration workshop limits were shown for electrical breakdown and surface heating (Fig. 5). In this case the surface heating was calculated for pulse lengths equal to the filling time of a copper cavity, and the assumption was made (correctly in this case) that for such relatively long pulses the temperature is limited by thermal conduction away from the surface. But it is not necessary to use such long pulses if adequate power sources are available. For instance a wake field accelerator uses only a single half wave. In such cases the temperature is found to be limited by the specific heat of the materials and depends only on this and the number of cycles. Kroll[6] has calculated the maximum electric fields over a plane mirror for which the temperatures do not exceed the melting point of various materials. The results are plotted on Fig. 6, and the limits for a half cycle on tungsten also indicated on Fig. 5. One sees that for tungsten this field is 28.5 GeV/m; an astonishingly high value. For currently available laser pulses of 100 cycles, the limit is 1.8 GeV/m. The shortest pulses that may be possible, using isotopic gas mixtures, were given as about 10 cycles, which would give fields of 5.6 GeV/m. Note that the accelerating field in a real structure will be less than these numbers by at least 2, nevertheless the conclusion is that very high gradients may be possible with a grating accelerator without destroying the surfaces.

A check on the above calculation is provided by an experimental observation by Corkum[11] that a gold mirror was not visibly damaged by a 3 picosecond pulse with the order of .5 terawatts per cm^2, corresponding to fields of about 4 GeV/m. This field is even higher than the tungsten calculation (1.8) and may indicate that melting for these very short times does not damage; it may be the boiling point that is relevant. It must also be pointed out that the measurement was very preliminary.

It had been hoped by this reporter that efficient acceleration would occur with these structures even when the fields were such as to form a surface plasma and subsequently destroy the surface altogether. Another observation of Corkum's is discouraing to this hope. At least at a field level of 100 terawatts/cm^2, corresponding to 60 GeV/m, he found that only 30% of the incoming light was reflected by a plane mirror. Ken Lee also presented a theoretical calculation that also predicted relatively large losses when a surface plasma is present. Such high absorption implies that the resonant structures we have been discussing would not work. There may of course be an intermediate field region above the melting point limit but below the field needed to produce a plasma, which would destroy the structure but still be suitable for a resonant structure. It should also be noted that acceleration can, in principle, still be obtained in non-resonant structures even when the losses are high. More experiments are required.

The conclusion at this point is that while the very high fields that produce plasmas will not be suitable for resonant structures, yet the fields that do not produce plasmas, and do not even visibly damage the surfaces, are very high (providing well above 1 GeV/m acceleration). For a high energy physics accelerator it may well not be necessary to go above such limits. For focusing however, higher fields may be desirable, but then an efficient resonant structure is not important.

7. FOCUSING STRUCTURES

Several people at the workshop started independently thinking of the use of these laser mechanisms as focusing elements. Already at the Frascati Conference the importance of focusing was emerging. The beam, and thus the wall plug, power needed to run a high luminosity, high energy collider can be very high, even prohibitively high. This power can be lowered if the beams can be brought to a finer focus at the collision point. In order to do this one needs higher gradient focusing elements.

We have been discussing structures that might, given short enough laser pulses or allowing plasma production, achieve average acceleration of the order of 5 GeV/m. Many of these structures would also provide quadrupole focusing average fields of the same order of magnitude at the "pole tips" only a few microns from the axes. The deflecting magnetic field corresponding to 5 GeV/m is 15 Tesla or 150 kG. This is a very high field, and when combined with the small aperture would provide quadrupole gradients equivalent to 5 million Tesla/meter. This is about 3 orders of magnitude higher than about the smallest conventional quadrupole magnet one can think of. The beta that can be produced at a focus goes as the inverse root of the gradient. The beam power goes linearly as the beta. So these high field structures offer the possibility of reducing the beam power by more than one order of magnitude.

We now review some possible focusing structures:

a) <u>Simple grating</u>

If the phase of the particles with respect to the fields is set for zero acceleration, then the particles see a deflection field combined with a quadrupole focusing field. In the last workshop it was proposed that the deflecting field could be corrected by a fixed magnet but at this workshop, Pickup[5] has shown that if the grating azimuthal positions with respect to the beam are rotated, then strong focusing is obtained without excessive undulation of the beam (Fig. 7a).

b) <u>Double row of droplets (Fig. 7b)</u>

As in the grating case the field along the axis for off phase particles is a quadrupole, only this time there is no dipole to cause deflection. In this case there is of course also acceleration for the other phase.

c) <u>Four rows of droplets (Fig. 7c)</u>

The accelerating mode discussed in Section 2 (Fig. 1f) contains no quadrupole fields, but another mode that can be excited in the same structure (Fig. 7c) does have such fields. This mode does not have acceleration. If a mixture of acceleration and focusing is required, one would alterate the modes between the two.

d) <u>Super bumps (Fig. 7d)</u>

The simple bump structures of Fig. 1g will, as in the grating case, have a combination of focusing and deflection. The super bump case (Fig. 1h) however, quadrupole focusing, no deflection, and thus a true stable orbit. It may not be quite as good as the droplets since the essential one sidedness of the structure will always introduce sextupoles or some other up-down symmetry.

At the workshop various calculations of the effect of these fields were made by M. Pickup and R. Fernow. It was shown, for instance, that with any of the above structures giving gradients of 5 million Tesla per meter, and with a phase 10° from maximum acceleration, then at 5 TeV the beta in the structure would be only .36 m. This is very small compared with that (100 m) in the SLC.

A very simple conceptual design of a final focus was worked out by J. Claus (Fig. 8). In this example the square root of the products of the initial beta and the beta at the intersection were 6.5 and 21 cm in the horizontal and vertical directions. It is reasonable to suppose that a more complicated design, symmetrical in the two directions, would have a value for this product of about 15 cm. This in turn implies that a final beta of 1 mm would involve maximum betas of the order of 22.5 m. If the invariant emittance $\leq 10^{-6}$, then the maximum beam size would be ≤ 1.5 μm. which would fit in the structure. We conclude therefore that this super high gradient RFQ focusing could give final betas of the order of 1 mm: at least 10 times smaller than by conventional means.

Pickup and Fernow checked that the synchrotron radiation with these high gradients was not a problem.

$$dE/d\ell = 1.5\ 10^{-15} \frac{A^2\gamma^2}{L^4}$$

$$L=2\ \pi\beta$$

$$A = (\eta\ \beta/\pi)^{\frac{1}{2}}\ .$$

With an invariant emittance of 10^{-6} the loss per meter in the accelerator (β=.36 m) at 5 TeV is only .04 MeV/m.

At the final focus just described the total loss would be only 200 MeV.

8. A POSSIBLE EXPERIMENTAL FACILITY

There was discussion within the group of the desirability and design of a facility where proof of principle experiments could be carried out. It was agreed that there was much work that could and should be done without a real test beam but that the difficulties and long time needed to design and build such a facility justified going ahead now with such a proposal.

The facility should provide a test beam with of about 10^4 particles focused to a spot of the order of one micron diameter, with a beta of at least 1 cm, a momentum spread of less than 1 MeV, and a pulse length of less than 2 mm. Such a specification demands a very high brightness beam. In principle this could be provided at any cooling electron storage ring of the order of 1 GeV, but in practice it seems likely that the SLAC source and cooling ring may well be the only such source that would be accessible for this work. A possible location of the experiment would be at the one third point where an extraction port and tunnel do now exist.

The second requirement for the experiment would be a laser capable of amplifying 3-6 picosecond pulses and delivering about 100 mJ. This specification is essentially that already demonstrated by the high pressure CO_2 laser at NRC (Ottawa, Canada).

Finally, and non-trivially, one requires a mechanism to synchronize the beam pulse with the laser. Two schemes were studied at the workshop. In the first (Fig. 9a, Pellegrini, Slater) the electron beam is used in a FEL to amplify a short section of a much longer pulse from a low power atmospheric pressure laser. This short and synchronized section is then further amplified in a high pressure amplifier and finally brought down to accelerate particles. In the second (Fig. 9b, Fernow, Himel, Corkum) the initial high intensity beam is passed through a gas Cerenkov and the light from this is focused onto a semiconductor "switch" used to cut a short section out of a larger CO_2 laser pulse. Both schemes seemed possible but require further study.

Due to the delays involved in these processes the light will be used not to accelerate the same pulse of electrons, but a second pulse will be extracted from the same cooling ring.

Some preliminary considerations were given to the design of both the collimator (Himel) and the spectrometer (Baggett). The collimator will require very small gaps and it was asked if scattering in the jaws of these gaps would give a "fuzzy" edge. An EGS calculation was quoted indicating that such "fuzziness" should be only a few microns and thus present no problem. The spectrometer, it was suggested would observe both vertical deflection and energy change. A conceptual design with a two dimensional (possibly solid state) array readout was suggested.

With such a spectrometer different phases between beam and laser would generate a combination of deflections and acceleration such as to form a hollow ellipse on the array (Fig. 10).

9. OTHER WORK TO BE DONE (Goldstone)

It is clear that the despite the great progress since the last workshop, much work remains to be done. In general qualitative understandings need developing to quantitative knowledge.

a) Polaritons (Corkum)

More study is needed on polaritons. These are surface fields that can exist over any plane dielectric or conducting surface in the presence of a complex dielectric constant. These fields can contain accelerating components and, since they are slow waves, can couple to relativistic particles when at an angle to them. They can be excited by slight surface ripples at 1 λ periodicity.

b) Structures

More study is needed of bump grating structures. We know the general concept but have not begun to identify the optimum structures. Study is needed on dielectric structures for a water droplet or solid grating accelerator.

c) Coupling

More work is needed on the coupling mechanisms including efficiency and tolerance studies.

d) Plasmas

Not only is the onset of plasmas not yet known but the mechanism and effect of the observed losses is not yet understood. Much work is needed if gradients in the 10 GeV/m are to be attempted.

e) Experiments

In connection with the proof-of-principle experiment described above, we will need many new diagnostics:

1) Beam position and shape monitors operating on the 1 micron scale: silicon strip detectors, scintillating thin layers, film, etc.

2) Structure position and shape monitors: flash TV, electron microscopy.

3) Energy flow observation: measurement of reflected and scattered light from the structures.

4) Wave front monitoring: measurements by interferometry.

5) Smith-Purcell check: measurement of light emitted by beam passing over structures.

f) Accelerator Optimization

Studies are needed on the overall optimization not only of the 10 TeV, 10^{34} machine discussed here but also of more modest stepping stone designs such as a 200 + 200 GeV, 10^{31} machine.

REFERENCES

1. Y. Takeda and I. Matsui, "Laser Linac with Grating", Nucl. Instrum. Methods 62, 306 (1968).
2. J.D. Lawson, Rutherford Lab. Rept. RL-75-043 (1975); IEEE Trans. Nucl. Sci., NS-26, 4217 (1979); P.M. Woodard, J. IEE 93, Part III A, 1554 (1947).
3. R.B. Palmer, "A Laser-Driven Grating Linac", Part. Accel. II, 81 (1980).
4. M. Tigner and J.D. Lawson, private communication (1984).
5. M. Pickup, these proceedings.
6. N. Kroll, reported at this workshop; to be published.
7. R.B. Palmer and S. Giordano, "Preliminary Results on Open Accelerating Structures", BNL 35981 and these proceedings.
8. E. Bassous, IEEE Trans. Electron Devices ED-25, 1178 (1978).
9. P. Wilson, Proc. Laser Acceleration of Particles, Ed. P.J. Channell, AIP Conf. Series No. 91 (1982).
10. T. Himel and J. Siegrist, SLAC Pub. 3572 and these proceedings.
11. P. Corkum, preliminary and private communications (1984).

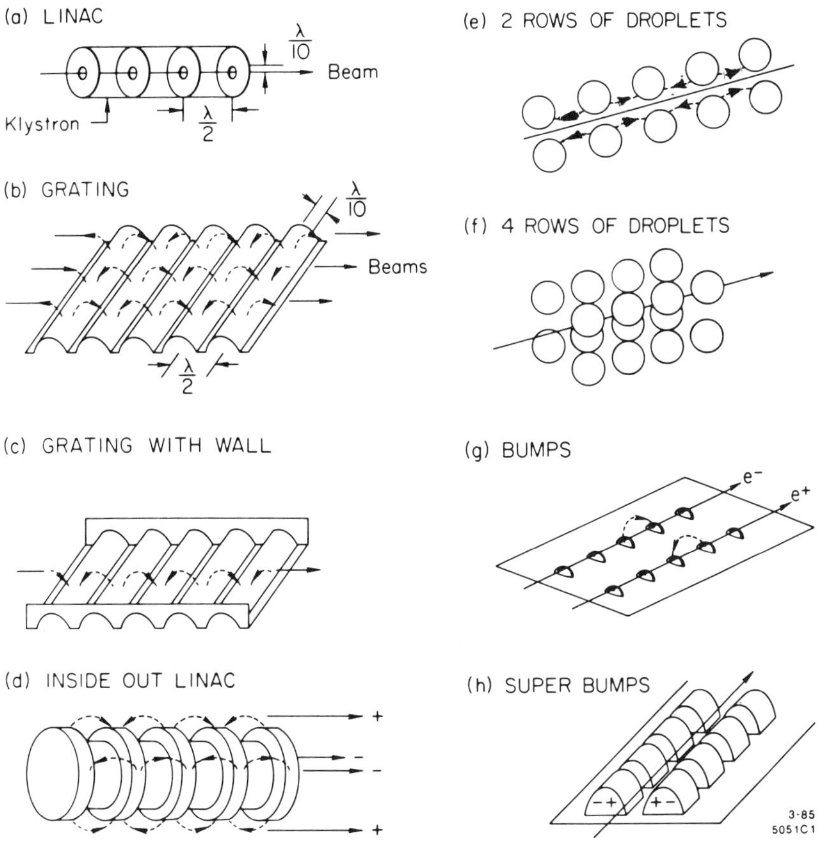

Fig. 1. Near field accelerating structures.

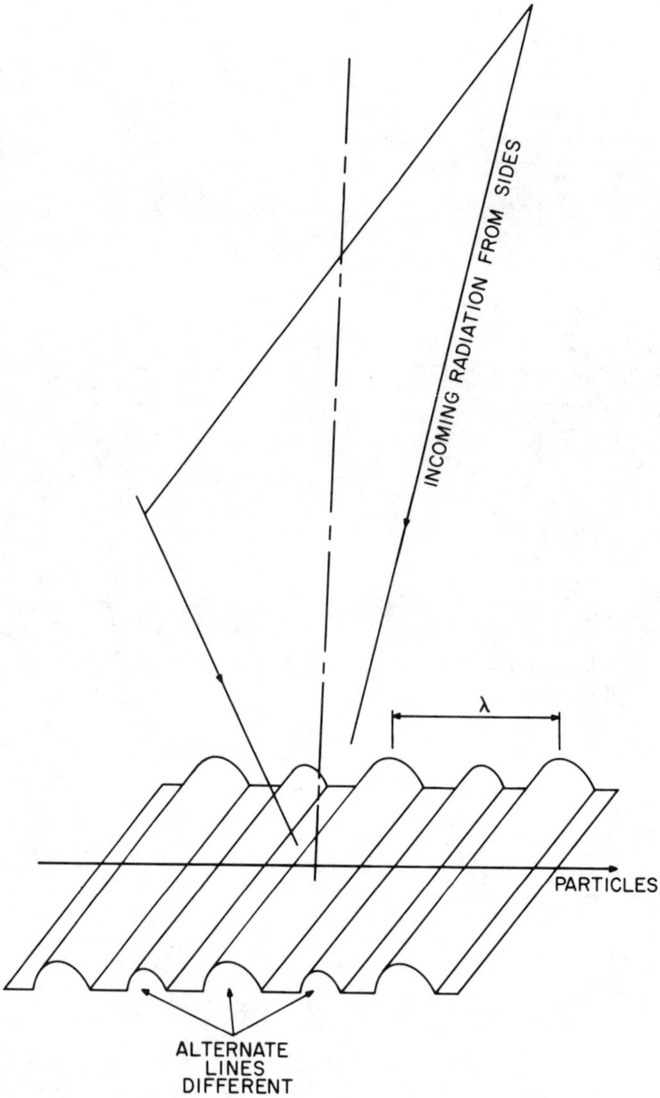

Fig. 2. Coupling to structures. (a) Grating.

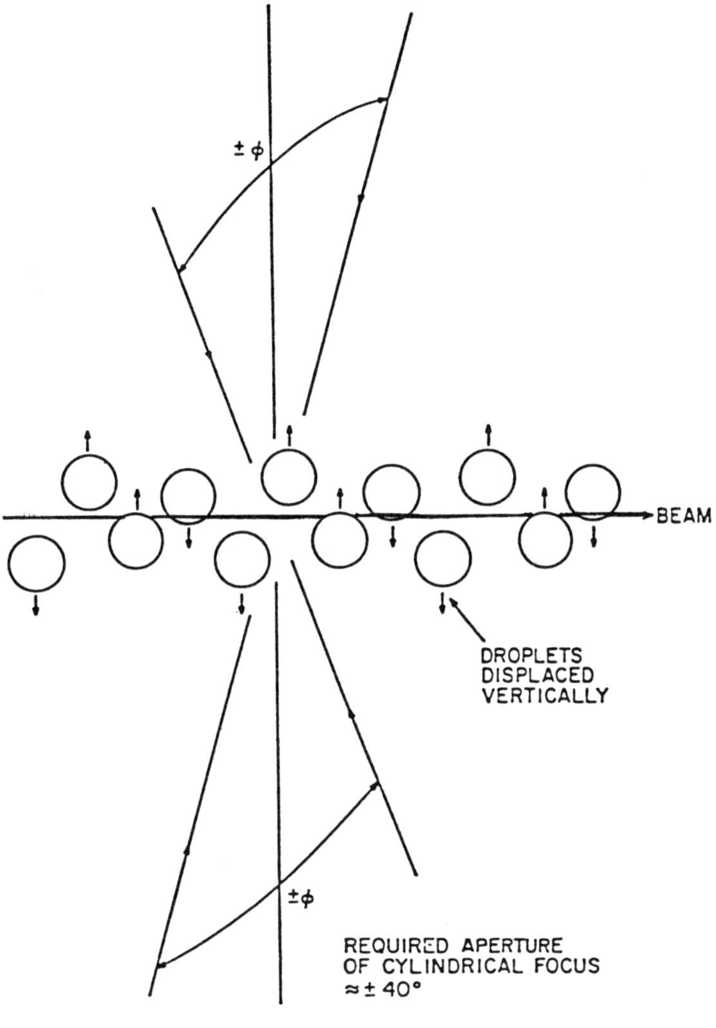

Fig. 2. Coupling to structures. (b) Double droplet rows.

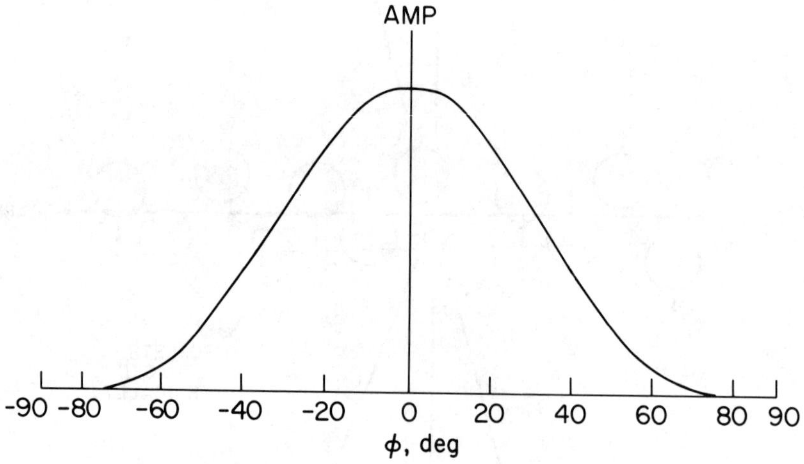

Fig. 3. Angular distribution of radiation to or from droplet structure.

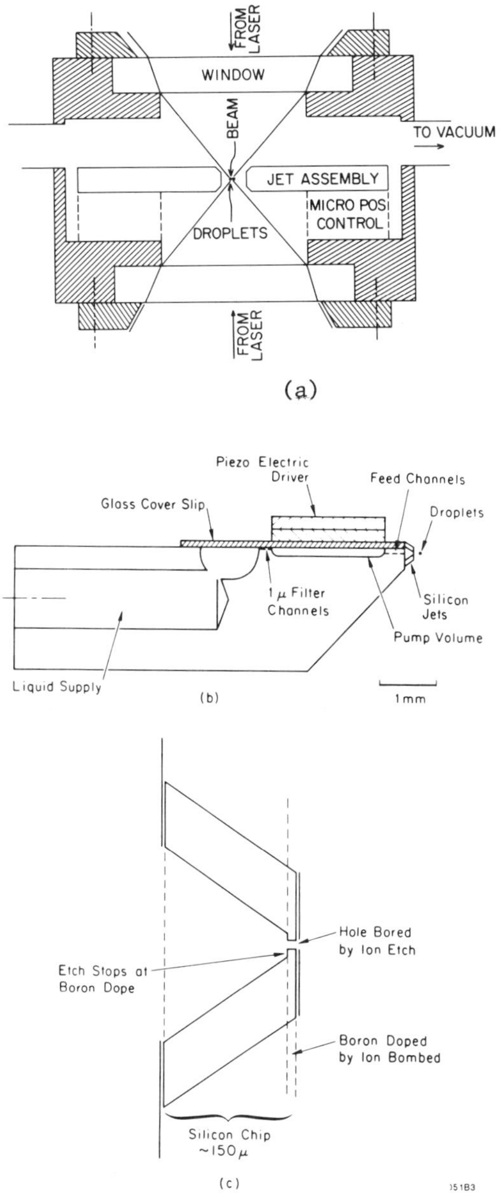

Fig. 4. Conceptual design of droplet accelerator: (a) overall section, (b) jet assembly, (c) nozzel assembly.

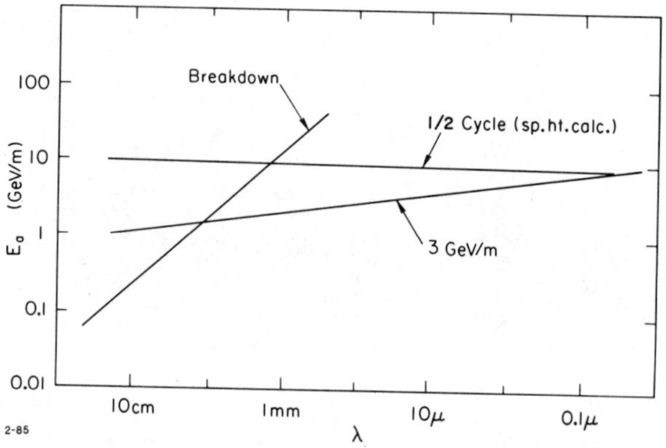

Fig. 5. Limits on acceleration gradients vs wavelength.

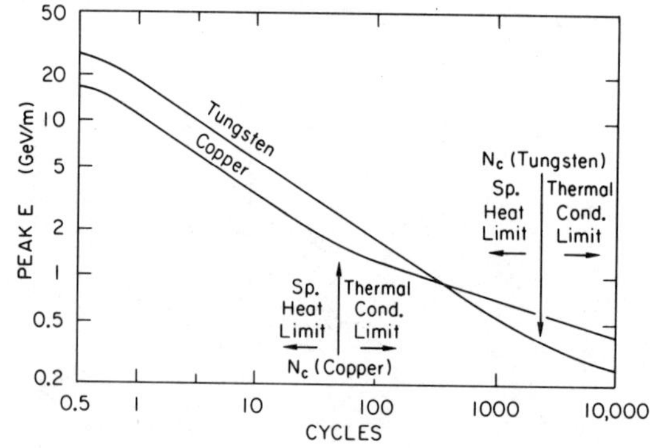

Fig. 6. Melting point limit vs number of cycles.

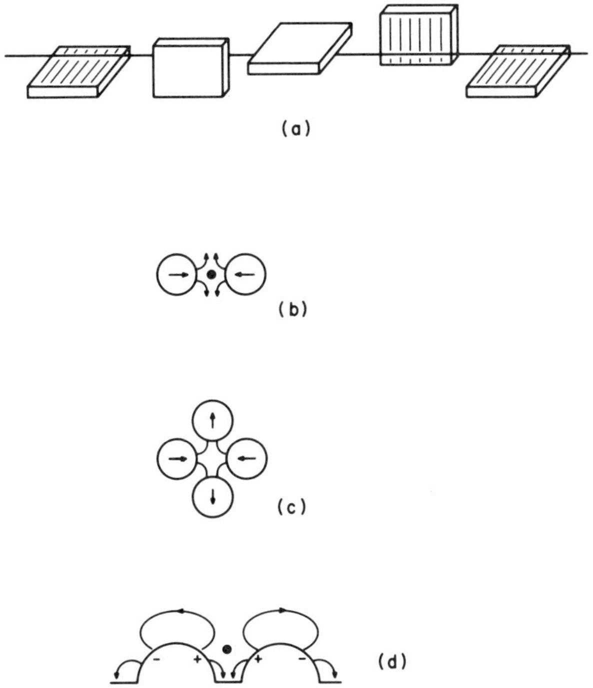

Fig. 7. Focusing structures: (a) Rotating grating arrangement, (b) fields between double rows of droplets, (c) four rows of droplets, (d) Kroll bumps.

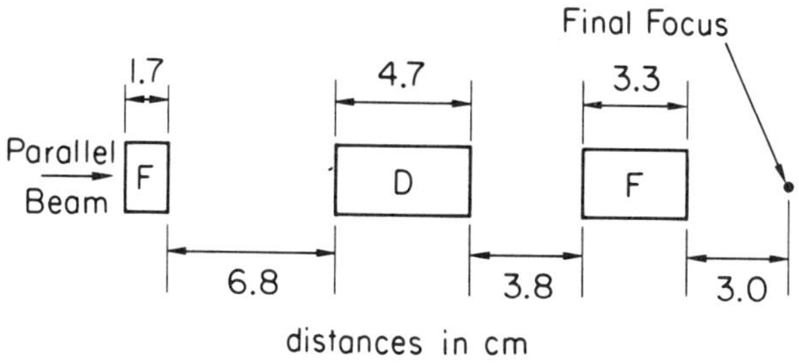

Fig. 8. Final focus.

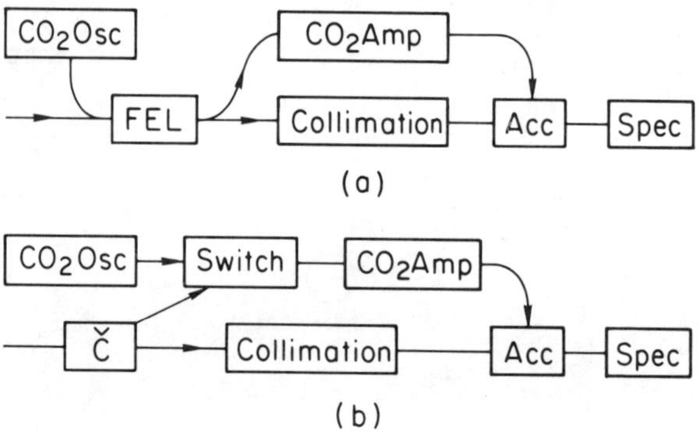

Fig. 9. Experimental facility: (a) with synchronization by FEL, (b) with synchronization by Cerenkov detector.

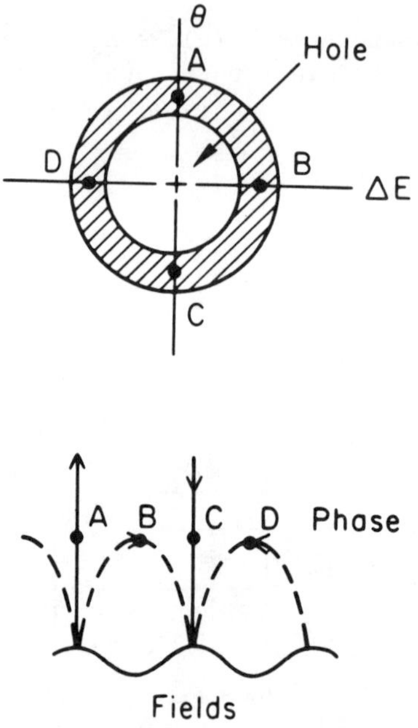

Fig. 10. Expected distribution in spectrometer.

General Features of the Accelerating Modes of Open Structures

Norman M. Kroll

University of California, San Diego
La Jolla, California 92093

ABSTRACT

Acceleration modes are defined, and their general characteristics in open structures are discussed. Some simple examples are given. Circumstances under which the number of such modes is small are elucidated. Various approaches to the transverse stability problem are discussed.

1. Introduction

As one extends linear accelerator technology to shorter and shorter wavelengths, it seems clear that the conventional closed cavity structures used at ten centimeter wavelengths will eventually become impractical. In recognition of this fact, a number of open structures which support bound modes suitable for accelerating extremely relativistic particles have been proposed. Notable examples are the gratings proposed by R. Palmer[1] and discussed at this conference by M. Pickup,[2] the droplet structure discussed by Palmer,[3] and the plane dielectric structure described by T. Weiland.[4] The purpose of the following is to present a number of related alternate structures and to emphasize the common features which all such structures exhibit.

An acceleration mode of an open structure has, by definition, the following properties.

1. It is bound to the structure. That is to say the stored energy, W, per unit length is finite, and there is no transverse radiation.

2. The fields have a fourier component, exp $i(kz - \omega t)$ with $k = \omega/c$, which we call the synchronous wave.

3. The z component of the electric field is non zero. Thus the loss factor $k_1 \equiv 4E_a^2/W$ is non zero for an acceleration mode. Here E_a, the accelerating field, is the value of E_z at the location of the particle beam.

The above definition may be somewhat more restrictive than necessary as will be discussed later.

Generally speaking, we shall be dealing with structures which are periodic in the z direction, which includes z independent structures as a special case.

2. Properties of the Synchronous Wave[5]

Every cartesian field component of the synchronous wave is of the form $f_i(x,y)\exp i(kz-\omega t)$ with $k = \omega/c$ and $\nabla^2 f_i = 0$. In addition the following relations between components are satisfied. (We take $\vec{\nabla} \equiv \hat{x}\frac{\partial}{\partial x} + \hat{y}\frac{\partial}{\partial y}$, *i.e.* the transverse gradient.)

$$\vec{\nabla} B_z = -\hat{z} \times \vec{\nabla} E_z \tag{2.1}$$

so that the modes are hybrid modes rather than TM.

$$\vec{\nabla} \cdot \vec{E}_T = -ik\ E_z \tag{2.2}$$

$$\vec{\nabla} \cdot \vec{B}_T = -ik\ B_z$$

where the subscript T refers to the transverse part of the field.

$$\vec{\nabla} \times \vec{E}_T = ik\ B_z\ \hat{z} \tag{2.3}$$

$$\vec{\nabla} \times \vec{B}_T = -ik\ E_z\ \hat{z}$$

$$\vec{B}_T = \hat{z} \times (\vec{E}_T + \frac{i}{k}\vec{\nabla} E_z) \tag{2.4}$$

$$\vec{E}_T = -\hat{z} \times (\vec{B}_T + \frac{i}{k}\vec{\nabla} B_z)$$

It is instructive to exhibit the solutions of the above equations which separate in polar coordinates. The complete set of solutions of laplace's equations in two dimensions, separated in polar coordinates, are of the well known form $r^{\pm m}\cos m\theta$, $r^{\pm m}\sin m\theta$ and $ln r$, where we take $m \geq 0$. ($\sin m\theta$ vanishes of course for $m = 0$.) An acceleration mode is obtained by setting E_z equal to one of these solutions and then satisfying equations (2.1) to (2.4). We also associate a TEM mode with each of the above solutions by setting $\vec{E}$ equal to the transverse gradient. Thus we may write

Acceleration modes: ($m \neq 1$ for the "r^{-m}" mode)

$$E_z = r^{\pm m} \cos m\theta$$

$$B_z = \pm r^{\pm m} \sin m\theta$$

$$E_r = \frac{\pm ikr}{2(m \pm 1)} r^{\pm m} \cos m\theta$$

$$B_r = i\left(\frac{kr}{2(m \pm 1)} + \frac{m}{kr}\right) r^{\pm m} \sin m\theta \tag{2.5}$$

$$E_\theta = \frac{-ikr}{2(m \pm 1)} r^{\pm m} \sin m\theta$$

$$B_\theta = \pm i\left(\frac{kr}{2(m \pm 1)} - \frac{m}{kr}\right) r^{\pm m} \cos m\theta$$

For the $m = 1$ case excluded above

$$E_z = \frac{1}{r} \cos\theta$$

$$B_z = \frac{1}{r} \sin\theta$$

$$E_r = -ikr \; lnr \; \cos\theta \tag{2.6}$$

$$B_r = -ik\left(lnr - \frac{1}{k^2 r^2}\right) \sin\theta$$

$$E_\theta = ik \; lnr \; \sin\theta$$

$$B_\theta = -ik\left(lnr + \frac{1}{k^2 r^2}\right) \cos\theta$$

The associated TEM modes are

$$\vec{E} = \vec{\nabla} r^{\pm m} \cos m\theta \tag{2.7}$$

$$\vec{B} = \hat{z} \times E \qquad (m > 0)$$

and for the $m = 0$ case

$$\vec{E} = \frac{\hat{r}}{r} \tag{2.8}$$

$$\vec{B} = \frac{\hat{\theta}}{r}$$

The substitution ($\cos m\theta \rightarrow \sin m\theta$, $\sin m\theta \rightarrow -\cos m\theta$) produces the remaining solutions (obtainable from the above by appropriate rotation of the coordinates). Note that an arbitrary linear combination of an acceleration mode with its associated TEM mode yields a similar structure. The form arrived at above is obtained by (arbitrarily) requiring the radial functions for the electric field components to contain a single power of r. Fields for which the magnetic rather than the electric field satisfies this requirement are obtained by taking an appropriate linear combination of the above form with its associated TEM mode. Note also that there is no single valued acceleration mode with $E_z = lnr$ as equation (2.1) then requires $B_z = -\theta$.

In the region exterior to the accelerating structure, the synchronous wave can be decomposed into a linear combination of the above solutions. The condition that the stored energy per unit length be finite implies that only the "r^{-m}" solutions appear. Since for these components the transverse fields behave as $r^{-(m-1)}$ (lnr for $m = 1$), it also implies $m \geq 3$. For $m = 2$ the stored energy integral diverges only logarithmically at large r, and, as will be discussed later, this circumstance leads us to consider this case to be quasi bound. For any synchronous wave the component of smallest m dominates at large distances. The "r^{+m}" ($m \geq 0$) solutions are of interest for discussing the fields in the vicinity of the acceleration path.[6]

3. Cylindrical Structures

Structures which are uniform in the z direction are called cylindrical structures. These are particularly simple because the synchronous wave comprises the entire field. That is to say the t and z dependence of all of the field components can be taken to be given by the common factor $\exp i\,(kz - \omega t)$. As a consequence the synchronous wave formulas of the preceding section now apply to the field as a whole. One immediate consequence is the fact that a cylindrical structure formed entirely of perfect conductors can not support an acceleration mode. This follows from the fact that E_z must be a solution of the two dimensional laplace's equation which vanishes on all boundaries and at infinity and hence must vanish everywhere.

The simplest example of a cylindrical structure which supports acceleration modes is a dielectric rod of circular cross-section with radius a. Taking $\epsilon > 1$, of course, and $\mu = 1$ for simplicity,[7] the solutions for $r < a$ are of the well known bessel's function form of argument $\sqrt{(\epsilon - 1)}\, kr$. Boundary conditions at $r = a$ determine the proper linear combination of an exterior acceleration mode and a TEM mode and a mode (or eigen) frequency. We get eigenfrequencies rather than a dispersion relation because we have assumed a phase velocity c from the start. The eigenfrequency equation for the acceleration modes is

$$y\, J_{m-2}(y) = -(m-1)(\epsilon-1)\, J_{m-1}(y) \quad ; \quad m \geq 2 \tag{3.1}$$

where $y = \sqrt{(\epsilon-1)}\, ka$.

There are an infinite set of solutions, y_{mn} , with each y_{mn} lying between the n'th zero of J_{m-2} and the n'th zero of J_{m-1}. For example $3.832 < y_{31} < 5.136$ and for $\epsilon = 2.315$ (for example) $y_{31} = 4.3$ which requires a radius of 0.60 λ. There is an additional set of modes, determined by $J_m(y) = 0$, which, being TEM for $r > a$, are not acceleration modes.

The set of modes described above are, of course, related to the bound modes of the dielectric cylinder which have $k > \omega/c$ and have a phase velocity less than c . These modes are, in general, characterized as TE_0 and TM_0 for $m = 0$ and HE_m and EH_m for $m > 0$, corresponding to the fact that both E and H have z components for $m > 0$. The frequency for which $k = \omega/c$ is called the cutoff frequency of the mode because for $k < \omega/c$ the mode is no longer bound. Note that since the binding condition is $k > \omega/c$ which corresponds to phase velocity $v_\phi = \omega/k < c$, the situation is contrary to that in hollow metallic wave guides where v_ϕ always exceeds c for the modes which propagate. The propagating non radiating modes in a dielectric waveguide typically satisfy $c/n < v_\phi \leq c$.

Each of the modes has a non vanishing loss factor k_1 for $v_\phi < c$. The acceleration modes differ from the others in the respect that as $v_\phi \rightarrow c$ the loss factor remains finite rather than vanishing. In the case of the $m = 2$ mode given by equation (3.1), k_1 vanishes only as $|1/\ln(1-v_\phi/c)|$, and since v_ϕ need never precisely equal c , it is possible that $m = 2$ modes (and more generally non-circular modes which are dominated by an $m = 2$ component at large distances) may be suitable for acceleration. It is for this reason we have referred to these modes as quasi bound. In addition acceleration making use of these open structures may make use of very short pulses. The transient problem requires much more thorough analysis than will be provided here, especially since one is proposing to operate at a center frequency which is equal to a cutoff frequency, but it seems safe to say that the penalty in effective k_1 for using an $m = 2$ mode is unlikely to exceed a factor $\ln\ c\tau/a$, where τ is the pulse width.

We have discussed the case of a dielectric rod largely for illustrative purposes. We have not investigated such questions as breakdown strength, dielectric losses, and fabrication problems. One point is worth making, however. Because any accelerating structure would have to be tens to thousands of meters long, and rod diameters are likely to be of the order of one wavelength of the accelerating wave, a question of mechanical rigidity naturally occurs. The solution, of course, here as in other examples to be discussed later, is to replace a dielectric rod of circular cross-section with a dielectric rod of semicircular cross-section mounted on a broad conducting plane. The circular rod solutions for which $E_z \sim \sin m\theta$ are also solutions for rods mounted on conducting planes. (Here, as in what follows, "mounting on a conducting plane" will always be understood to imply halving the structure along a suitable symmetry plane before mounting it on the conducting plane.)

In the limiting case of a plane interface, $m\,\theta$ becomes replaced by qx, where q is a transverse wave number, and, for $k = \omega/c$, fields fall off on the free space side of the dielectric as $\exp -qy$. On the other hand for $k \neq \omega/c$ the y dependence is $\exp\,(q^2+k^2-\omega^2/c^2)^{1/2}\,y$. Radiation occurs in the y direction only when $q^2 + k^2 < \omega^2/c^2$, so that the condition $k = \omega/c$ does not correspond to cutoff in the sense that the word is used in dielectric wave guides. This is in contrast to the situation described for the dielectric rod, and indeed for any cylindrical structure of finite cross-section. In practice the transverse wave number q is fixed by means of a pair of conducting planes separated a distance d, parallel to the (y,z) plane, and extending, in principle, from $y = -\infty$ to $+\infty$. The boundary conditions on the planes then yield $q = m\,\pi/d$, where m is the transverse mode number. Realistically, of course, the planes do not extend to $y = \pm\,\infty$, and as a result radiation in fact ensues as soon as ω/c exceeds k. The amount of radiation can, however, be kept as small as desired by making the planes wide enough. The weak onset of radiation in structures of this type at $k = \omega/c$ may well be preferable to the stronger onset that occurs in the dielectric rods, because the transient behavior is likely to be smoother. This is an issue which requires further investigation.

For completeness we note that for a practical realization of the plane dielectric one might coat the bottom of a rectangular groove in a metal plate with a strip of dielectric. Such a structure would have two sets of modes characterized by $H_y = 0$ (E modes) and $E_y = 0$ (H modes). Setting the dielectric thickness equal to a, one easily finds for the dispersion relation relating k and ω:

$$\tan q_2 a = \frac{\epsilon q_1}{q_2} \qquad \text{(E modes)}$$

$$\tan q_2 a = \frac{-q_2}{q_1} \qquad \text{(H modes)}$$

where

$$q_1 = \sqrt{q^2 + k^2 - \frac{\omega^2}{c^2}}$$

$$q_2 = \sqrt{\frac{\epsilon\omega^2}{c^2} - k^2 - q^2}$$

Outside the dielectric we have

$$E_y = E_{y0} \sin qx \; e^{-q_1 y} \; e^{ikz} \tag{3.2}$$

$$\frac{E_z}{E_y} = \frac{-ikq_1}{q^2 + k^2} \tag{3.3}$$

$$\frac{E_x}{E_y} = \frac{qq_1}{q^2 + k^2} \cot qx$$

for the E modes, and

$$E_z = E_{z0} \sin qx \; e^{-q_1 y} \; e^{ik_1 z} \tag{3.4}$$

$$\frac{E_z}{E_x} = -i \, \frac{q}{k} \tan qx \tag{3.5}$$

for the H modes. Here the dielectric interface is at $y = 0$, and the sidewalls are at $x = 0$, and with $qd = m\pi$, $m = 1, 2, \cdots$. There is also a $q = 0$, TEM mode for each case, not given by the above formulas, and of no interest for acceleration. These formulas illustrate the disappearance of the accelerating field as $q \to 0$ at $k = \omega/c$, and the fact that the dispersion relation is smooth in the neighborhood of $k = \omega/c$. If the finite depth of the groove were taken into account this would no longer be the case because some radiation would then occur, but this effect would also usually be smoothed out by conduction losses and dielectric losses.

4. Periodic Structures

Structures which are periodic (with period p) in the z direction rather than uniform (as in the previous section) differ from the latter in that the z dependence of the fields is given by a sum of space harmonics rather than a single component. The propagation properties can be fully characterized by a phase factor $\exp i\gamma$, $-\pi \le \gamma \le \pi$, which describes the relative phase between the (otherwise identical) fields in successive periods (the m+1'st to the m'th). This phase factor gives rise to a field consisting of a sum of space harmonics of the form $\exp i\, k_\nu z$ with

$$k_\nu = k_0 + \frac{2\pi}{p}\nu \qquad \nu = 0, \pm 1, \pm 2, \cdots$$

$$k_0 = \frac{\gamma}{p}$$

In general, γ is related to frequency by a dispersion relation. For closed structures such as those typically used in linacs (we assume perfectly conducting walls), there are an infinite number of modes, each with its own dispersion relation, with the frequency for each mode an even periodic (period 2π) function of γ. While it is sufficient to restrict γ to the range $-\pi < \gamma < \pi$, and hence k_0 to $-\pi/p < k_0 < \pi/p$, it is convenient here to drop the subscript on k and think of ω as a periodic function of k with unrestricted range, it being understood that each value of k carries with it the full sequence of $k + (2\pi\nu/p)$ space harmonics. It follows from this that there is for every propagating mode a frequency satisfying $\omega(k) = kc$, and, as in the preceding section, there are an infinite number of acceleration modes.

The situation is quite different for periodic open structures of finite cross-section. As discussed in the preceding section, the condition $\omega(k) = kc$ always occurs at the boundary between a bound region and a radiation region and in this sense corresponds to a cutoff condition. It is easy to see that if such a boundary exists at all in a periodic structure then it must occur in the region $0 < \omega < c\,\pi/p$. To see why this is the case let us suppose that $\omega_c > c\,\pi/p$ is the value of ω at the boundary. Then at least one of the space harmonics $k_\nu = (\omega_c/c) + (2\pi\nu/p)$ satisfies $|k_\nu| < \omega_c/c$ and hence couples to the radiation field. This contradicts the hypothesis that ω_c is on the boundary, since no radiation is supposed to occur at the boundary. The fact that the available frequency range is finite implies that the number of acceleration modes for the structure is finite. It is typically small and can easily be zero. Bound propagating modes are fully characterized by their behavior in the interval $\omega/c < k < \pi/p$, the rest of the Brillouin plot (*i.e.* ω vs. k plot) is then determined by the periodicity and symmetry conditions. Note that if the acceleration mode is a π mode ($\gamma = \pi$), there is no bound branch for the $\omega(k)$ curve. The bound π mode is simply an isolated resonance at $\omega = c\,\pi/p$ (or equivalently, $\lambda = 2p$). The various features of the Brillouin plots for open and closed periodic structures are illustrated in Fig. 1.

A simple example of the above discussion is provided by the conducting slotted rod of circular cross-section. The structure, together with its more practical "mounted on a plane" configuration are shown in Fig. 2. Sufficiently narrow slots behave like open circuit radial transmission lines, which have characteristic resonances. Considered individually they would radiate and have low Q, but if the phase from slot to slot varies rapidly enough, the radiation is suppressed. Since the slots are coupled together by the exterior fields each such resonance will be shifted and broadened so as to cover a band along which γ varies. Because of the circular symmetry the modes can be characterized by $\cos m\theta$, $\sin m\theta$ behavior of the various field components. In order to have a strictly bound accelerating mode we require at least an $m = 3$ mode. A sequence of discs is a special case of the slotted rod in which the radius of the inner conductor vanishes. For this case a

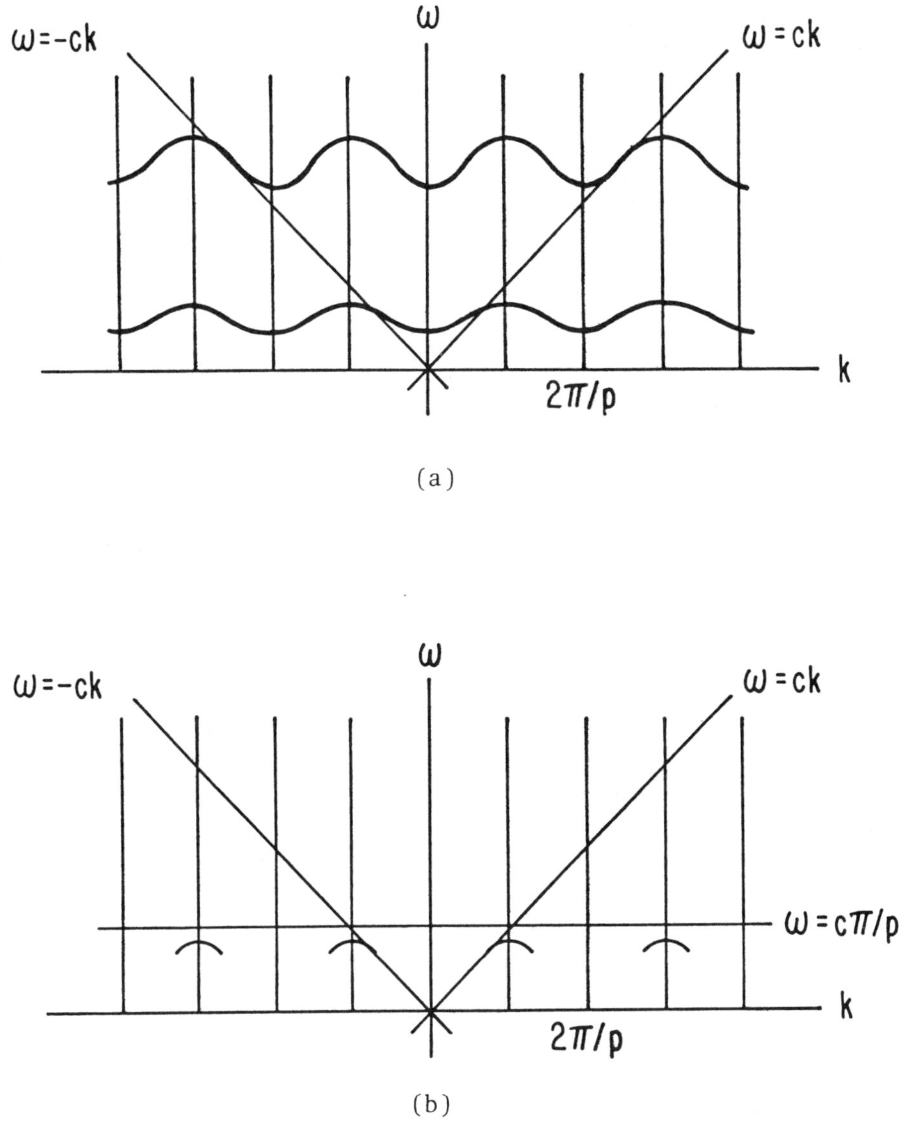

Fig. 1 Brillouin plots for periodic structures. (a) Closed cavities. Every mode intersects the $\omega = kc$ line. (b) Open cavities. Every non-radiating mode must have $\omega \leq c\,\pi/p$.

crude estimate indicates that the disc radius $a \sim (2/3)\lambda$. Since radiation is avoided only if $\lambda \geq 2p$ we see that $a > (4/3)p$ is necessary. If this condition is satisfied, there will also be $m = 2, 1,$ and 0 propagating branches all of which have vanishing loss factors k_1 at their $\omega = kc$ points. If one reduces a, keeping p fixed, the same crude estimate indicates that the modes become radiative successively, in the order 3, 0, 2, 1, so that for sufficiently low a/p no non radiative modes are expected. Palmer has suggested that the fact that the number of synchronous non radiating modes is small in these structures may alter the wake field problem in a favorable way, especially in the context of bunch to bunch instabilities.

Another example of a periodic structure with circular symmetry is a row of conducting spheres. This configuration has been studied by Palmer[3] in connection with the development of his droplet accelerator proposal. Experiments performed with a single sphere between two conducting planes indicated the presence of only one resonant frequency, which was found to have $m = 1$ symmetry. The results obtained imply that a row of spheres has a single (doubly degenerate) non radiating mode for $a > \sim p/4$, where a is the radius of the sphere, and none otherwise. Since the spherical geometry implies $a < p/2$, it is, qualitatively speaking, not surprising that no other non radiative modes exist.

Another class of periodic structures, which have been discussed extensively elsewhere, are plane conducting gratings. The synchronous waves for plane gratings are similar in form to those given by eq. (3.2), (3.3), (3.4) and (3.5) for the plane dielectric sheet except that for the grating E modes and H modes are characterized by $H_x = 0$ and $E_x = 0$ respectively. It is the H modes which have received the most attention. Here, in addition to the $\exp i(\omega/c)(z - ct)$ component, one has the usual set of space harmonics $k_\nu = (\omega/c) + (2\pi\nu/p)$, and hence to avoid radiation we require $k_\nu^2 > (\omega^2/c^2) - q^2$ for all ν. As in the plane dielectric case, a transverse wave number q has been assumed to be fixed by a pair of conducting planes (separated by a distance d) and a transverse wave number m ($q = m\pi/d$). Because q^2 can be arbitrarily large, there are an infinite number of non radiating modes. As discussed in the case of the plane dielectric, components for which $|k_\nu| < \omega/c$ are in fact weakly radiating due to the fact that the conducting planes are of finite width in practise, but the radiation loss can be made small. The use of parallel conducting planes as a radiation shield is, of course, a device which is adaptable to any of the circular structures that we have discussed heretofore and may be helpful in providing smoother propagation properties.

A general method of coupling into periodic structures, which has been exploited by Palmer,[1,3] is based upon a weak doubling of the period length. If the structure is only approximately periodic in p, but precisely periodic in $2p$, then the set of space harmonics is given by $k_\nu = (\omega/c) + \pi\nu/p$, but the components of odd ν are much smaller in amplitude than those of even ν. For example one might alternately widen or shorten the slots in Fig. 1. It is easy to arrange matters so that only a single component (say $\nu = -1$) violates the "no radiation" condition. At large distance from the structure such waves will radiate at an angle ϕ with respect to the forward direction given by $\cos\phi = 1 - \pi c / p\,\omega$, and by sending a wave towards the structure in the reverse direction, the desired mode in the structure can be excited. In Palmer's examples the synchronous mode was a pi mode so that the direction of irradiation must be broadside in that case. Since pi modes do not propagate, irradiation along the entire length would be required. For a non-pi mode (SLAC uses a $2\pi/3$ mode) the inputs can be localized by restricting the period perturbation to local regions. By adding additional "matching structures" it should be possible to design localized inputs and outputs that are analogues of standard microwave coupling ports.

5. Transverse Forces and Focusing

As shown by Panofsky and Wentzel,[8] the transverse forces acting on a particle in an accelerator can be related to the longitudinal accelerating force by rewriting the transverse part of faraday's law as

$$\vec{\nabla} E_z = \frac{\partial}{\partial z} \vec{E}_T - \frac{1}{c} \frac{\partial}{\partial t} \hat{z} \times \vec{B} \ . \tag{5.1}$$

Now assuming that the z, t dependence of the transverse components is given by $\sin(kz - \omega t) \equiv \sin\phi$ we may rewrite (5.1) as

$$\vec{\nabla} E_z = k\,(\vec{E}_T + \frac{v_\phi}{c} \hat{z} \times \underset{\sim}{B})\ cot\ \phi \tag{5.2}$$

or in terms of the forces

$$\vec{\nabla} F_z = k\ cot\ \phi\ \vec{F}_T \left[1 + 0\,(\frac{v_\phi - v}{c}) \right] \tag{5.3}$$

For the synchronous wave $v_\phi = v$, or if we take $v_\phi = c$ as we have in earlier sections, the correction is of order $1/\gamma^2$ and completely negligible. We shall consider some cases later where $v_\phi \neq c$, but the correction will still be negligible.

Applying the above to the polar coordinate separated synchronous waves of Section 2, we have

$$F_z = F_0 \left(\frac{a}{r}\right)^m \sin m\,\theta \cos\phi \tag{5.4}$$

$$\vec{F}_T = F_0 \frac{m}{kr} \left(\frac{a}{r}\right)^m (-\hat{r} \sin m\,\theta + \hat{\theta} \cos m\,\theta) \sin\phi \tag{5.5}$$

where a may be chosen for convenience (*e.g.* say corresponding to the nominal radial coordinate of the beam). It follows that transverse forces vanish for structures with circular symmetry only where $\sin \phi = 0$. In order to obtain a significant phase acceptance width $\delta\phi$ and to establish a mean distance from the structure about which stable oscillations might take place, some sort of supplementary focusing fields would have to be supplied. Palmer[1] has suggested applying a uniform transverse magnetic field to produce a ϕ dependent transverse equilibrium point. Letting F_B equal the radial force due to the magnetic field we have, at $\theta = 0$

$$-F_0 \frac{m}{kr} \left(\frac{a}{r}\right)^m \sin \phi + F_B = 0$$

for transverse equilibrium. The equilibrium value $r(\phi)$ is then given by

$$r(\phi) = a \left(\frac{\sin \phi}{\sin \phi_m} \right)^{1/m+1} \qquad \sin \phi_m \equiv \frac{F_B}{F_0} \frac{ka}{m}$$

$$\phi_m < \phi < \pi - \phi_m \quad \text{for } \sin \phi_m > 0$$

$$-\pi - \phi_m < \phi < \phi_m \quad \text{for } \sin \phi_m < 0$$

In this scheme the equilibrium points occur at ϕ values different from zero and hence away from the point where the accelerating field is a maximum. For $F_0 = 1$ *Gev/m*, a B_θ of 2.4 KG yields $\sin \phi_m = .1$ for $m = 3$, $ka/m = 1.4$.

Examination of the stability of the above equilibrium points shows that for $F_B/F_0 < 0$ the points are stable for radial motion but unstable for θ motion while for $F_B/F_0 > 0$ the reverse is the case. Alternating gradient focusing can be achieved by shifting the phase of the synchronous wave back and forth[9] by an amount $2\phi_0$ in successive accelerating sections while simultaneously alternating the sign of F_B/F_0. Particles at phase $\pm\phi_0$ are stable at fixed $r(\phi_0)$, the alternating gradient focusing principle having stabilized motion in both the r and θ directions. Particles in a small band $\delta\phi$ around $\pm\phi_0$ move in nominal stable periodic radial orbits. An example worked out in reference 9 yielded the following:

$F_0 = 1$ *Gev/m* $\qquad \lambda = 10\ \mu \qquad q = 0.2\ k$

$\phi_0 = 45^\circ$

stable range: $\quad 41^\circ < \phi < 56^\circ$

The phase is shifted 90° every $L = 1.2\,[E\,(Gev)]^{1/2}$ cm., alternately back and forth. All particles within the acceptance band have approximately equal mean values for $<\cos \phi>$ so that a nearly monoenergetic acceleration of .7 Gev/m takes place.

The requirement that one balance the transverse electromagnetic force with a transverse magnetostatic force clearly presents practical difficulties. More symmetric structures which have an axis along which the transverse electromagnetic force vanishes avoid this problem. Such structures can be constructed from symmetric combinations of two or more structures of the sort already considered. Thus if two identical structures with circular symmetry are aligned parallel to one another it is reasonable to suppose that their modes can be thought of as primarily symmetric and antisymmetric combinations of modes of the individual structures having the same dispersion relation, the dispersion relation of the combination being somewhat altered from that of the modes out of which they are formed and depending upon the symmetry. Figure 3a illustrates two $m = 1$ modes coupled together antisymmetrically. The resultant mode will have $m = 2$ behavior at large distances and be quasi bound at $v_{\phi} = c$. Midway between the two the transverse force vanishes, providing an appropriate axis for acceleration of the particle beam. Focusing and defocusing quadrupole forces appear in the neighborhood of the beam axis. Figure 3b achieves a similar result with a pair of $m = 2$ modes mounted on a conducting plane. A realisation of Fig. 3a, using rows of conducting spheres for the individual structures has been proposed by Palmer[3] as a potential configuration for a droplet accelerator, and the existence of the designated mode confirmed by RF modeling.[10] Figure 3b could be realised by a pair of structures of the sort illustrated in Fig. 2, and provides a candidate for a survivable accelerator structure.

Although the above structures solve the transverse equilibrium problem without the imposition of an external static magnetic field, the equilibrium is still an unstable one. One can think of at least three methods for dealing with this problem. The first is the scheme of ref. 9, discussed earlier, in which the phase in successive accelerator sections is alternately shifted back and forth $2\phi_0$, where ϕ_0 is the mean acceleration phase. A second method is an adaptation of RFQ. In this method one drives the desired mode at two frequencies. The main wave, at the lower frequency, is synchronous with the particle beam and provides the acceleration. In the vicinity of $\phi = 0$ the accelerating force is at a maximum and the focusing and defocusing forces from the main wave are weak. The secondary wave, at the somewhat higher frequency, moves at a somewhat slower phase velocity so that its phase drifts steadily with respect to the particles as they advance along the accelerator. As a result the accelerating forces of the secondary wave average to zero while the quadrupole forces regularly alternate, thus providing alternating gradient focusing.[11] Because the effectively static quadrupole forces of the synchronous wave interfere with the RFQ focusing, stability is confined to a narrow region around $\phi = 0$ where these forces are weak. The phase acceptance width depends of course upon the amplitude of the secondary wave relative to the main wave. A crude estimate, in which the

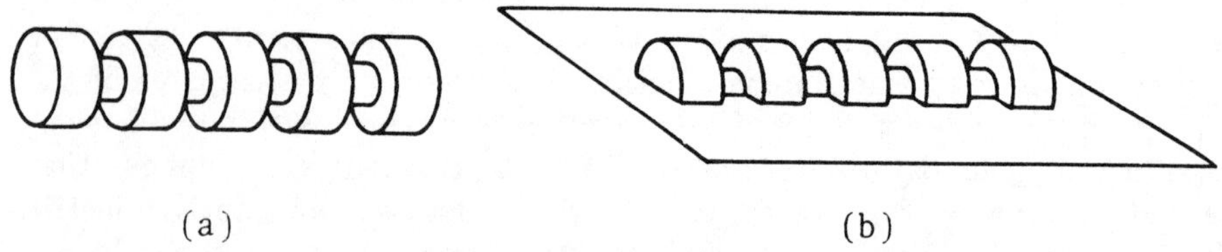

Fig. 2 The slotted rod open cavity structure. (a) With full circular symmetry. (b) Halved and mounted on a conducting plane.

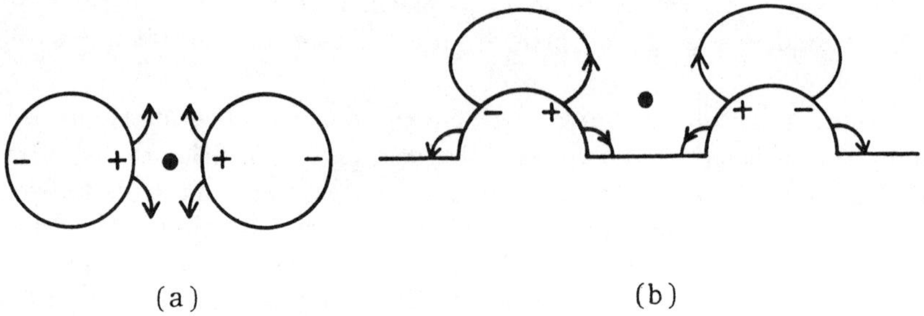

Fig.3 Coupled structures with circular symmetry which have an acceleration axis free of transverse forces. (a) Two $m=1$ modes coupled asymmetrically. (b) Two $m=2$ modes coupled asymmetrically, and mounted on a conducting plane.

sinusoidal phase alternation which would actually take place has been replaced by an alternating step function yields an acceptance width of about 20° for an amplitude ratio of 3/7 for the secondary to the primary wave.

Both of the methods described so far have the defect that the heating limit on the maximum acceleration gradient is less than it would be if the focusing procedure were not applied. In addition the power required for a given effective acceleration gradient is greater than it would otherwise be. In the phase alternation method both effects arise from the fact that the operating point phase ϕ_0 is sufficiently large to cause a significant deviation of $\cos \phi_0$ from unity. While the operating phase is at $\cos \phi_0 = 1$ in the RFQ method, power is required to drive the secondary wave. In addition, because the phase drift of the secondary wave is slow, points of constructive interference between the two waves at the structure surfaces provides an excess of heating over that which would ensue in the absence of the secondary wave. On the basis of the crude calculations performed so far the penalty is comparable in the two methods but probably more severe for the phase alternation method.

A third method is to arrange matters so that the orientation of the structure is z dependent, say by twisting it with a very small pitch about the beam axis. This method has the advantage that the transverse motion is stable for all values ϕ although beam emittance would make it necessary to avoid a band around ϕ equal to zero, because the focusing forces vanish at $\phi = 0$. In addition, of course, there is no obvious power or acceleration gradient penalty imposed by this method.[12] Technical problems which may arise in implementing any of these schemes may prove to be more decisive than the considerations we have mentioned.

A phase independent RFQ solution of the transverse stability problem can also be achieved by combining four circularly symmetric structures as illustrated in Fig. 4. The mode indicated in Fig. 4a, which provides the acceleration, has no transverse or quadrupole forces at the beam location. The mode indicated in Fig. 4b is driven at a frequency for which $v_\phi < c$, and provides the RFQ focusing.[13] It has a quadrupole field but no acceleration field at the beam location. It provides uniform focusing at all particle phases.[14] Because the synchronous wave has no quadrupole fields to interfere with the focusing, the amplitude of the RFQ wave relative to the accelerating wave can be reduced, thus ameliorating both the efficiency and heating penalties associated with the RFQ method.

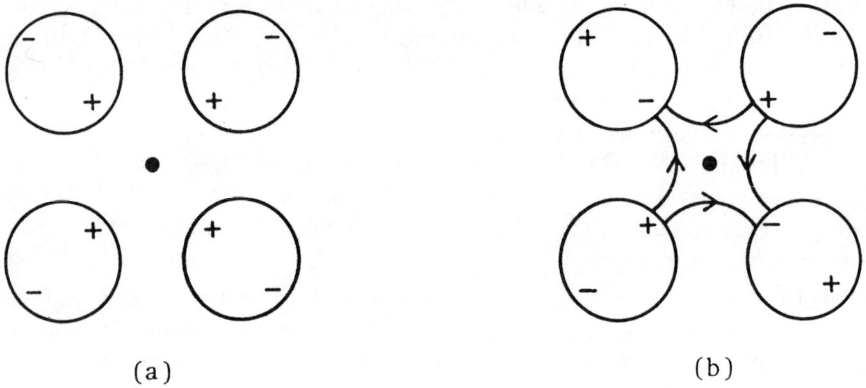

Fig. 4 Four coupled $m = 1$ modes. (a) Coupled symmetrically to yield an acceleration mode with no transverse or quadrupole focus at the symmetry axis. (b) Coupled push-pull to yield a mode which can provide RFQ quadrupole focusing along the symmetry axis, without participating in the acceleration process.

6. Concluding Remarks

The discussion given above has been confined to qualitative features of open structure acceleration modes. It is intended to provide physical insight which may prove useful in the design and invention of desirable configurations. Real progress in the field will, however, ensue only as the result of detailed study. The time has come for detailed investigation, both theoretical and experimental, of some generic open cavity structures. While ultimate utility is important, substantial weight should be given to calculability and experimental convenience in the selection. While mode structure and field configurations are important issues, such questions as coupling, transient induced radiation loss, wake fields, and especially the propagation properties of the very short pulses which will be needed to permit structures with proposed high acceleration gradients to survive are equally important and should begin to receive comparable attention.

Acknowledgments

The author wishes to acknowledge having received much stimulation from R. Palmer and especially from his preconference communication to members of the near field working group.

This research was supported by the United States Department of Energy under DOE contract DEAT03-81ER40029.

Footnotes and References

1. R. B. Palmer, "A Laser-Driven Linac", Particle Accelerators **11** , 81 (1980).
2. M. Pickup, these proceedings.
3. R. B. Palmer, *et al.,* "Report of Near Field Group", these proceedings.
4. T. Weiler, "A Thin Layer Dielectric Near Field Laser Accelerator", Proc. Laser Acceleration of Particles, Ed. P. J. Channell, AIP Conf. Series No. 91 (1982).
5. N. Kroll, "A Note on Cylindrical Waves which Propagate at the Velocity of Light", Proc. Laser Acceleration of Particles, Ed. D. J. Channell, AIP Conf. Series No. 91 (1982).
6. As, for example, in A. Chao, "Coherent Instabilities of a Relativistic Bunched Beam", Physics of High Energy Accelerators, Ed. M. Month, AIP Conf. Series No. 105 (1983).
7. See, for example, J. D. Jackson, Classical Electrodynamics (Wiley, New York, 2nd Edition 1975), Ch. 8.

8. W. K. H. Panofsky and W. A. Wenzel, Rev. Of. Sci. Inst. **27** , 967 (1956).

9. K. Kim and N. Kroll, "Some Effects of the Transverse Stability Requirement on the Design of a Grating Linac", Proc. Laser Acceleration of Particles, Ed. P. J. Channell, AIP Conf. Series No. 91 (1982).

10. R. B. Palmer and S. Giordano, "Preliminary Results on Open Accelerator Structures", BNL 35981 and these proceedings.

11. The deviation of v_ϕ from c that is required is very small so that the $(v_\phi - c)/c$ correction in eq. 5.3 can be neglected.

12. M. Pickup in a report to this conference has analyzed a discretised version of the twist scheme applied to the grating structure. He has made the important observation that it is not necessary to perform the rotation about a line of zero transverse force. The rotation of the structure causes the transverse force to rotate as well, leading to a stable periodic nominal orbit.

13. Both modes have been shown to exist (Ref. 10) for an RF modeled spherical droplet realization.

14. That is for a given particle energy. The particle energy is of course phase dependent.

PRELIMINARY RESULTS ON OPEN ACCELERATING STRUCTURES*

R.B. Palmer† and S. Giordano
Brookhaven National Laboratory
Upton, New York 11973

I. INTRODUCTION

Conventional accelerating structures consist of cavities or enclosed transmission lines with the particles to be accelerated passing along the axis of the enclosure. Electromagnetic modes can be excited within such structures with electric components in the direction of the particle motion. Their wave velocity is inevitably less than that of light unless the cavity is loaded with a dielectric or periodic structure. Conventionally some periodic structure is used.

In a plasma linac the conducting surfaces of the structure are replaced by plasma surfaces. The dimensions of the structure are drastically smaller (with wavelengths less than 100 microns) and the electric fields can be much higher (of the order of 10 GeV per meter or more). For this application, a closed cavity is probably not suitable. A grating surface has been suggested[1] in which the surface of the grating is turned into a thin layer of plasma. Such a structure may be described as a semi-open structure. The accelerating fields are restricted to within a few wavelengths of the grating surface, but inevitably spread in the two dimensions over that surface. (The hope expressed in Ref. 1 that the fields could be restricted to a narrow band along the grating by the use of cylindrical optics now appears to have been in vain.[2])

In this paper we consider periodic structures consisting of rows of spherical conductors. In a plasma linac, these spherical conductors would be formed from liquid droplets on whose surfaces a plasma would be formed. For this paper, the field configurations have been investigated using copper spheres approximately 11 cm diameter and microwave radiation of approximately 30 cm wavelength. No suitable accelerating mode was found for relativistic particles using a single row of spheres, but with two parallel rows of spheres both accelerating and focusing modes were found.

In Section II we re-examine the accelerating modes over a grating surface, including a grating formed of parallel conducting rods. In Section III we discuss the coupling of these structures to incoming radiation.

*Work done under the auspices of the U.S. Department of Energy.
†On sabbatical leave at SLAC.

II. GRATING STRUCTURES

In Ref. 1 it was shown that a grating surface existed over which a standing wave pattern could be established, with the fields restricted to a limited region above the grating. The fields were periodic both with respect to the grating periodicity and along the direction of the grating lines. Such an electromagnetic field was shown to include an accelerating mode, accelerating particles perpendicular to the grating lines just above the surface, providing the particle trajectory was not at one of the nodes in the periodic field. In fact if a trajectory were chosen in the next periodic maximum and the phase is left the same, deceleration would occur. Perpendicular to the surface the fields fall off exponentially, these fields being evanescent waves. If a rectangular section of grating is surrounded by vertical conducting walls, see Fig. 1a, then this structure behaves as an accelerating cavity with standing waves within the walls and losses only due to resistive wall effects. Since no lid is required on this cavity, it is open in one dimension.

A natural extension of the grating cavity of Fig. 1a is shown in Fig. 1b. Here the grating is replaced by a series of cylindrical bars and the resulting cavity is open in both the upward and downward directions. When the electric field lines are examined, it is seen that the charge and current distributions in the rods approximate a series of dipole antennas with vertical polarization and phase alternating both along the rods and from one rod to the next. It is a natural further extension from Fig. 1b to speculate whether an array of spherical conductors with each one behaving as a dipole oscillator would also provide an accelerating non-radiating solution (see Fig. 1c).

An experiment to test this idea was set up using one copper sphere and four aluminum conducting surfaces (see Fig. 2a). A resonant (i.e. non-radiating) solution was found. The observed Q is relatively low because of an inadequate wall height H. It is clear that if this height were extended to infinity, the Q would be limited only by resistive losses.

Such an array of droplets could easily be produced by an array of liquid jets, but still suffers from the objection that the radiation will leak out transversely unless constrained by a conducting wall on either side of the particle trajectory. We therefore examined a structure consisting of a single row of droplets.

III. ROWS OF CONDUCTING SPHERES

We first examined a single row of spheres and searched for a non-radiating π mode. In such a mode, all the field directions reverse from one sphere to the next and thus all field lines must be perpendicular to a plane half way between any two spheres. We can thus investigate such modes by using the experimental arrangement shown in Fig. 2b, which has one sphere placed between two conducting surfaces. One can think of these surfaces as behaving like mirrors and thus producing an infinite number of multiple

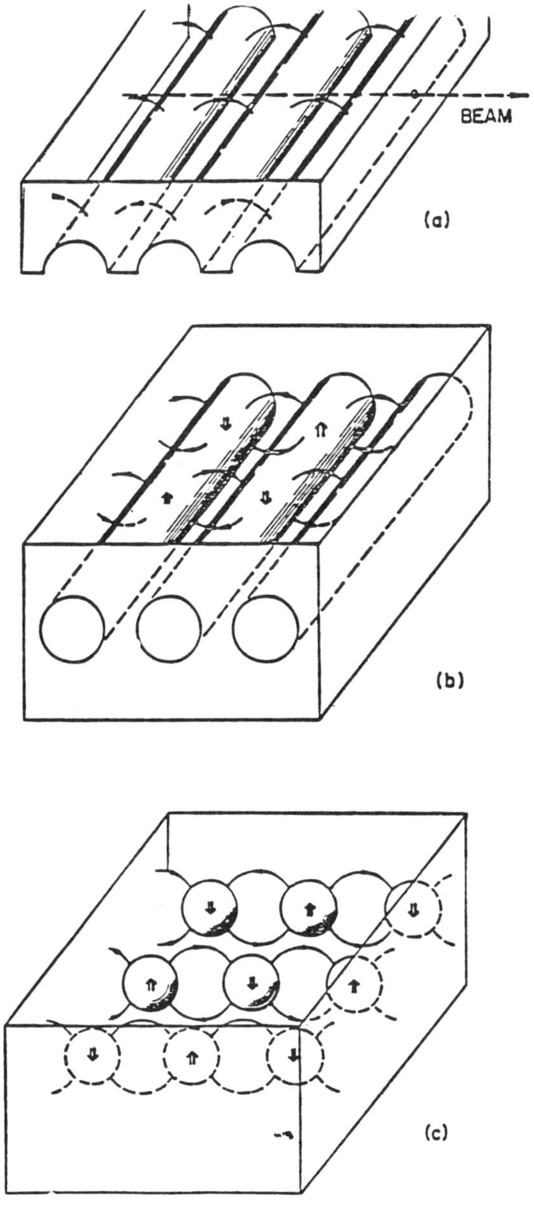

Fig. 1. Grating solutions: (a) original one-sided grating, (b) a two-sided rod grating, and (c) array of spheres grating.

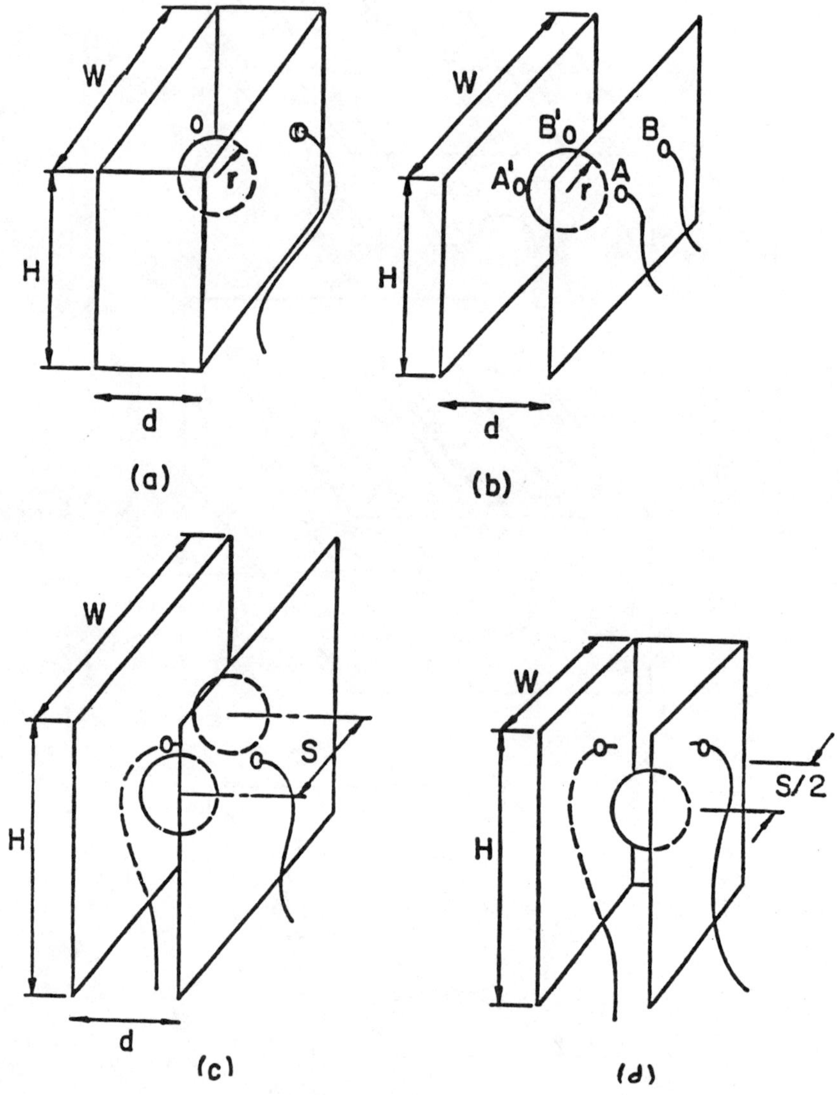

Fig. 2. Experimental arrangements: (a) for array of droplets grating, (b) for single row of droplets, (c) for double row of droplets with acceleration on axis, and (d) for double row of droplets with antisymmetry.

images of the single sphere. First a search was made for axially symmetric modes by placing an exciting probe at A' and a pickup probe at A. No such symmetric modes were seen. An exciting probe was then placed off center at B' and pickup at B. Now non-radiating modes were observed for various spacings d, as shown in Fig. 3. The sphere diameter is plotted in units of the observed resonant wavelength and the spacing between the plates (d) is given in units of half the wavelength. This latter fraction corresponds to the velocity β of a particle passing along the row of spheres that would be in phase with the oscillating electric fields. The observed excited modes are illustrated in Fig. 4a. To first approximation the spheres act as transverse dipole oscillators and the fields provide good acceleration above and below these resonators. Unfortunately, as is seen from Fig. 3 the Q of the observed resonance states falls as the velocity of the in-phase particles (β) approaches one. The mode is only suitable for accelerating particles when they are less than relativistic. The low Q was found to correspond not to resistive losses but to a large forward radiation from the structure. Indeed it can be seen that the radiation from all spheres is coherent in the forward and backward directions when $\beta = 1$. This situation is analogous to that obtained with a grating exposed to radiation from directly above. In that case we know from Lawson's theorem[2] that there is no acceleration and that the electrical excitation of each line of the grating is in phase to radiate in the forward and backward directions. We remember that the solution[1] to this problem was to bring the light not from directly above, but from either side. In this way a transverse periodicity is introduced on the grating surface such that along one line one may have acceleration, then along a nearby parallel line one has deceleration. Correspondingly the forward radiation from one of these lines is exactly out of phase with the neighboring one, and no net forward radiation occurs. The corresponding situation with spheres is of course to have a complete grid as discussed in Section II where we did find satisfactory resonant conditions at $\beta = 1$ without forward radiation. A possible simpler solution, however, would be to employ two rows of spheres with one row out of phase with respect to the other so that again the forward radiation is suppressed.

A double rows of spheres was therefore investigated using the experimental arrangements of Fig. 2c and d. In both cases resonant excitation was observed for $\beta = 1$ for various different sphere spacings (S) (see Fig. 5). The modes excited in the two cases are illustrated in Figs. 4b and c.

In the first case (Fig. 4b and also Fig. 6a) we see that the spheres are acting to first approximation as dipoles oscillating in the plane of the double row, 180° out of phase, and we see that a very favorable condition occurs for acceleration along the axis between the rows.

In the second case, the spheres are acting as dipoles oscillating transverse to the plane, also 180° out of phase. In this case there are no accelerating fields along the axis but there are such fields above and below the poles of the individual rows of droplets

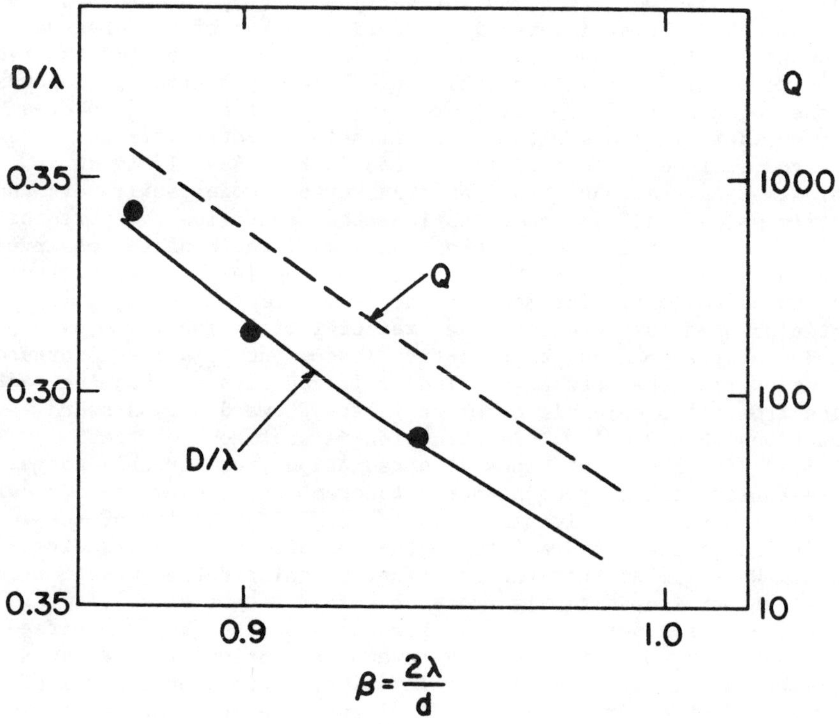

Fig. 3. Results of single row experiment: the Q and droplet diameter D as a function of the synchronous velocity β.

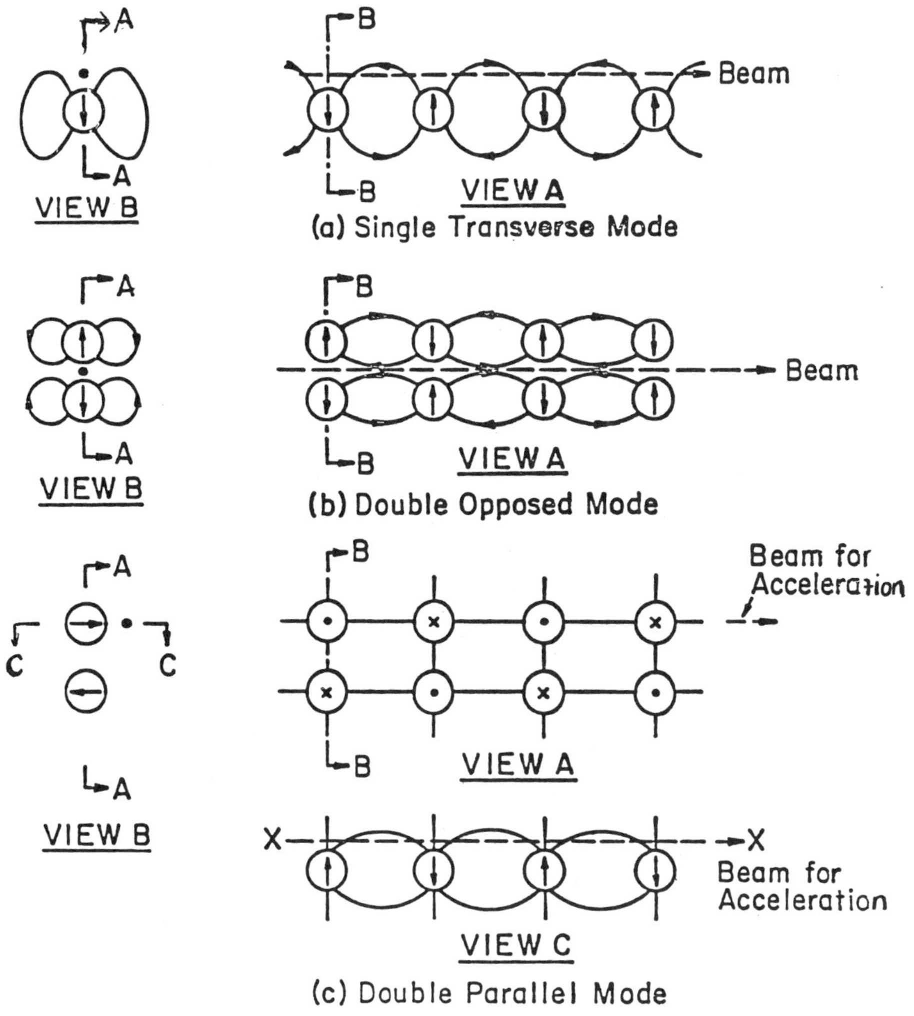

Fig. 4. Fields of resonant solutions: (a) single row, (b) double row with acceleration on axis and (c) double row with acceleration on the side.

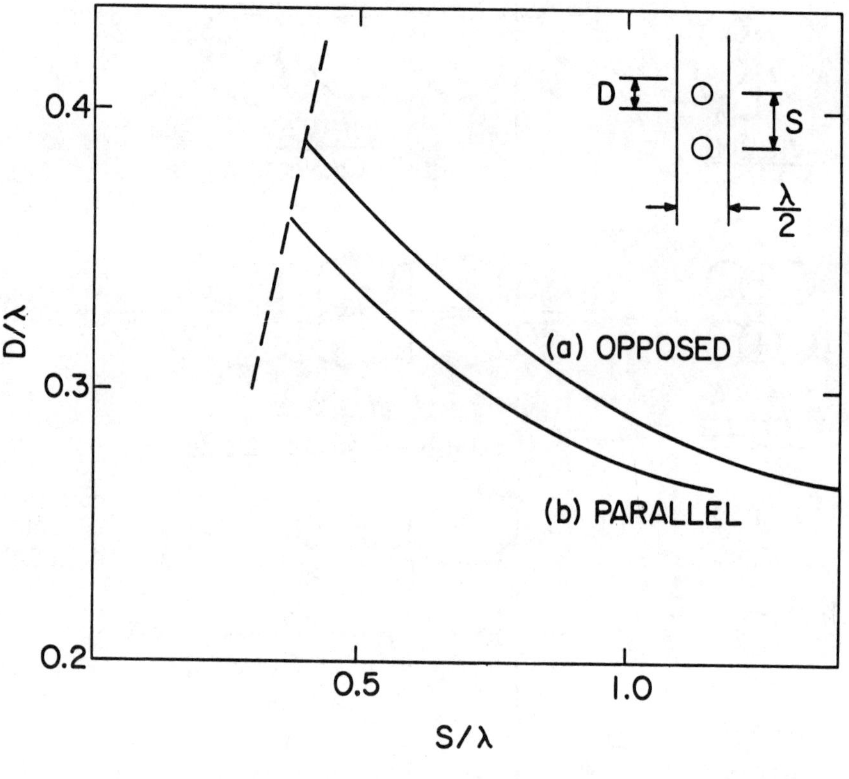

Fig. 5. Resonant conditions for double row. The sphere diameters are given as function of distance between the two rows for a) acceleration on axis, b) acceleration on side.

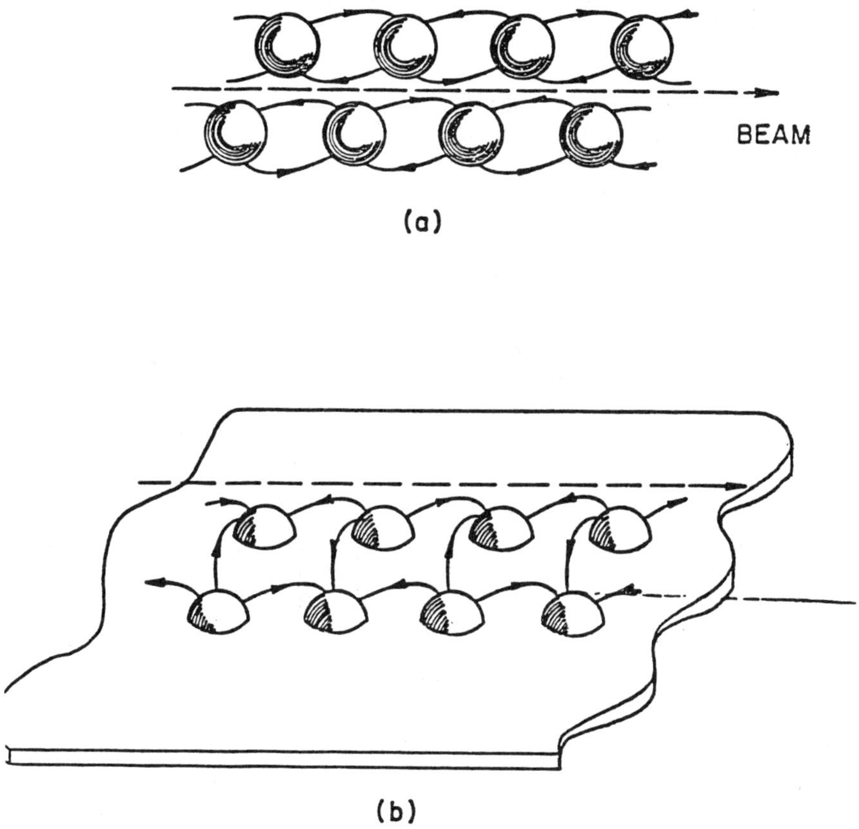

Fig. 6. Double row solutions: (a) case with acceleration on axis-opposed mode and (b) case with acceleration above rows-parallel mode. In this case a conducting plane is inserted without changing the solution.

(Fig. 4c). An interesting feature of this mode is that the fields are all perpendicular to a plane passing through the centers of all spheres. The mode would thus exist over a conducting surface with two rows of hemispherical bumps on it as indicated in Fig. 6b. Such a surface could be generated by ion etching or other techniques and provides a solution to the grating problem of radiation leaking out sideways.

One final arrangement of spheres was also tried consisting of four rows of spheres arranged in a square about the axis. Again, a resonant non-radiating solution was found. This mode involved the four spheres acting as transverse dipole oscillators with all the dipoles facing in towards the axis.

III. COUPLING TO INCOMING RADIATION

The arrangements of spheres that we have described above have resonant non-radiating modes which would provide acceleration if excited. They are, however, like closed cavities in that we have provided no mechanism for introducing radiation into them. What is required is a perturbation on the structures such as to couple the near field modes, that are restricted to small distances from the structure, to outgoing or incoming waves. The addition of these perturbations may be regarded as equivalent to providing slots in a conventional cavity in order to couple a waveguide to it.

Almost any perturbation on the position or size of any of the structures described will cause it to radiate. If for instance, all the spheres in one row were slightly larger than the spheres in the parallel row, then the cancellation of the forward radiation would not be exact and the structure would couple to forward radiation.

Another interesting case is that in which alternate spheres are either slightly lifted or slightly depressed from the plane in which they initially lay. Such perturbations couple the modes to radiation propagating either upward or downward perpendicular to the plane containing the spheres. Calculations of azimuthal angular distributions of radition from such structures are in progress.

REFERENCES

1. R.B. Palmer, "A Laser-Driven Grating Linac", Particle Accelerators, Vol. II, pp. 81-90 (1980).
2. J.D. Lawson, Rutherford Lab. report RL-75-043 (1975); IEEE Transactions on Nuclear Science, NS-26, 4217 (1979).

A GRATING LINAC AT MICROWAVE FREQUENCIES

Michael Pickup*
Cornell University, Ithaca, NY 14853

ABSTRACT

Current work on the numerical simulation of a grating linac is described. Surface waves above a grating are reviewed with emphasis placed on rectangular grooved gratings. A geometry that achieves transverse focusing without external magnetic fields is shown. The conversion of a standard wakefield program for cylindrical cavities to a planar grating configuration is discussed. Areas of future study are listed.

INTRODUCTION

Current schemes for laser driven particle accelerators fall into two categories; far-field and near-field. Far-field methods, such as the inverse free electron laser tend to bend the particles' paths to obtain acceleration perpendicular to the direction of propagation of the laser light. Near-field schemes, such as the grating accelerator, depend on obtaining electric fields parallel to the particles' motion. Conventional accelerating structures are near-field devices. Laser driven near-field structures differ from conventional accelerator cavities in that they depend on guided surface waves rather than waveguide modes.

Many of the features of a laser driven grating accelerator may be studied at microwave wavelengths. A microwave grating can support both slow and fast waves with longitudinal electric field components. It is an open structure that can be coupled to by 'illumination' and higher order modes may radiate away. The propagating waves are evanescent, dying off exponentially away from the surface. This decay results in strong transverse forces due to the transverse field gradients present at all points. This feature is an important difference between gratings and closed cavities. Transverse focusing effects must be balanced to obtain net focusing. Also, the passage of charged particles close to the grating will excite fields that may have adverse effects on both longitudinal and transverse beam stability. These wake effects are present in closed cavities, but they may be more harmful here due to the small distance between beam and structure. These problems may be studied profitably using methods developed for cylindrical cavities.

In the 1970's, the electromagnetic theory of gratings was developed to explain their behavior at optical wavelengths.[1] Wood's anomalies are sharp dependencies in the reflection

*Supported by the National Science Foundation

coefficient of a grating with respect to the angle of incident radiation. The prediction of these detailed reflection curves over a wide range of wavelengths and grating shapes is a major success of the electromagnetic theory. In particular it was found that the behavior of gratings in the infrared could be accounted for by a simple model of the grating as a nearly perfect conductor. Thus results obtained by studying microwave gratings may simply scale to infrared, i.e. CO_2 laser wavelengths.

SURFACE WAVES

At first look, one is tempted to use a simple transverse magnetic surface wave. This proves not to be useful for acceleration but is a good place to begin. Figure 1 shows a rectangular grooved grating in profile and the coordinate system to be used.

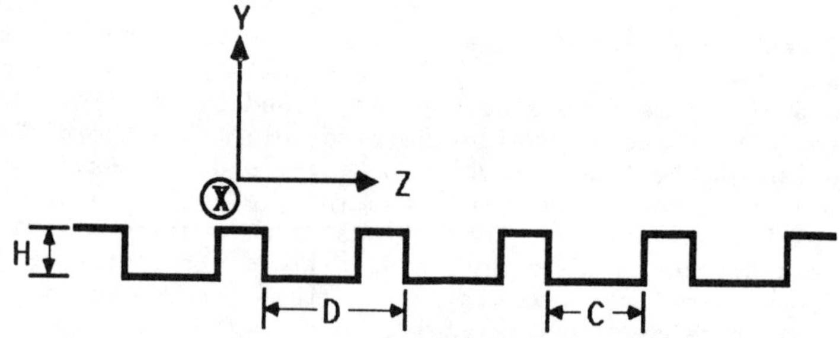

Fig. 1. Grating profile

A TM wave has no variation along the x direction and contains H_x, E_y, and E_z field components only. By Floquet's theorem, fields above any grating may be represented as a summation of space harmonics.[2] The magnetic field is

$$H_x = \sum_n A_n e^{-q_n y} e^{-i\beta_n z}$$

$$\beta_n = \beta_0 + \frac{2\pi n}{D} \qquad\qquad k_0^2 = \beta_n^2 - q_n^2 \tag{1}$$

and the electric field components may be obtained from H_x by partial differentiation. When q_n is imaginary, that space harmonic is a plane wave. If q_n is real and positive, the term is evanescent.

Fields within the grooves may be represented by expansions for a few simple cases. For rectangular grooves,

$$H_x = \sum_m B_m \cosh(h_m y) \cos(\frac{m\pi z}{C})$$
$$k_0^2 = (\frac{m\pi}{C})^2 - h_m^2 \tag{2}$$

Solutions are found by demanding i) continuity across the plane separating space above the grating from that in the groove and ii) that the tangential electric field go to zero on the conducting surfaces. These boundary conditions produce a set of matrix equations that may be solved for A_n and B_m. If all q_n are real and positive, the solution is a bound surface wave that propagates without radiative loss.

Each space harmonic has a different phase velocity,

$$v_p = \frac{k_0}{\beta_n} c \tag{3}$$

Only the space harmonic with phase velocity equal to the speed of light accelerates relativistic particles. Choosing n = 0 as the synchronous harmonic,

$$v_p = c = \frac{k_0}{\beta_0} c \longrightarrow k_0 = \beta_0$$
$$\text{but} \quad k_0^2 = \beta_0^2 - q_0^2 \longrightarrow q_0 = 0 \tag{4}$$

This wave is not bound. It is a plane wave traveling parallel to the grating. Plane waves have only transverse electric fields. As B_0 approaches k_0, E_z/E_y goes to 0. TM solutions will not work.

The next simplest case is the longitudinal electric wave.[2] These have the following general dependencies,

$$F(x,y,z) \;\alpha\; e^{\pm iKx}\, e^{-q_n y}\, e^{-i\beta_n z} \tag{5}$$

If we include both positive and negative iK_x terms we get standing waves in x.

$$E_x = 0$$
$$H_x, E_y, E_z \quad \alpha \quad \cos(Kx)\, e^{-q_n y}\, e^{-i\beta_n z}$$
$$H_y, H_z \quad \alpha \quad \sin(Kx)\, e^{-q_n y}\, e^{-i\beta_n z} \tag{6}$$
$$\beta_n = \beta_0 + \frac{2\pi n}{D} \qquad k_0^2 = \beta_n^2 - q_n^2 + K^2$$

When we ask for a phase velocity of c we get a condition on K.

$$v_p = c \longrightarrow k_0^2 = \beta_0^2 \longrightarrow q_0^2 = K^2 \qquad (7)$$

Now we can have a bound surface wave with a synchronous space harmonic. Applying the same boundary conditions as before, we can find bound surface wave solutions. Shunt impedance, Q and group velocity can then be calculated. Groove period, width and depth may be varied to optimize performance.

The following observations come from solving rectangular groove problems, but are probably general. When choosing a surface wave there are four variables, wavelength(k_0), phase velocity(v_p), decay constant(q_0) and transverse wave constant(K). Demanding a phase velocity of c sets q_0 equal to K. The grating shape determines a functional dependence,

$$v_p = f(k_0^2 - K^2) \qquad (8)$$

After fixing the grating profile, choosing any two of v_p, k_0 or K determines the third. Waves that have the same frequency but differing transverse constants will have differing phase velocities. It is not possible to 'assemble' a monochromatic set of bound surface waves into a non-dispersive wave of gaussian cross-section. A 'line focus' of incident radiation will excite surface waves that disperse both longitudinally and transversely.

For a microwave grating linac, this problem is solved by adding sidewalls to guide the wave. Several copper models have been constructed with both sidewalls and endwalls. The endwalls restrict measurement to standing waves. These models may be excited using a coaxial probe in one endwall. A second probe on the far endwall is used to make transmission measurements. Figure 2 shows a typical model. Table I lists predicted and measured properties for several standing wave modes.

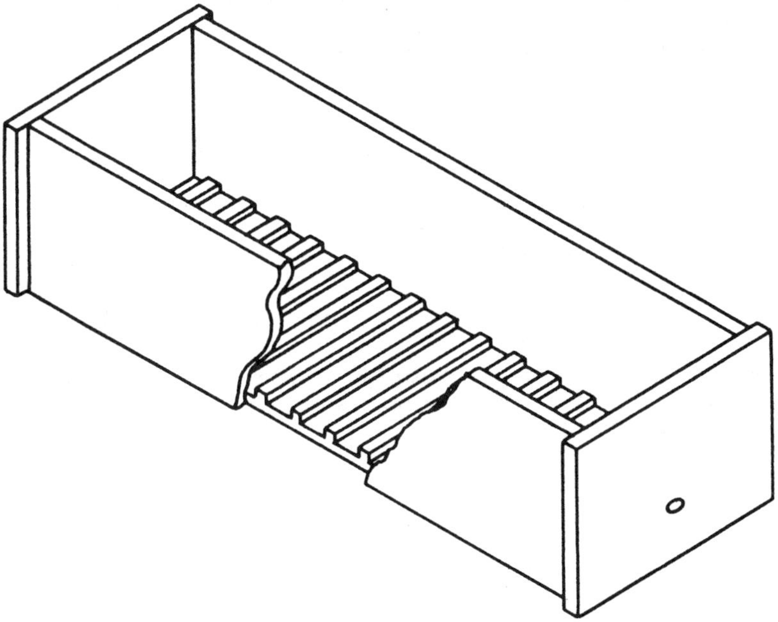

Table I Standing wave modes

$\beta_0 D$ = (mode)	predicted frequency GHz	predicted R/Q Ω/meter	measured frequency GHz	measured R/Q Ω/meter
π	9.89	690	9.87	380
$\frac{15}{16}\pi$	9.85	950	9.84	480
$\frac{7}{8}\pi$	9.74	1200	9.72	620
$\frac{13}{16}\pi$	9.56	1400	9.54	780
$\frac{3}{4}\pi$	9.33	1500	9.31	890
$\frac{11}{16}\pi$	9.06	1400	9.04	970
$\frac{5}{8}\pi$	8.76	1300	8.75	1000

Table II gives an example with dimensions similar to the SLAC cylindrical cavities. We see that the overall efficiency of the structure, R/Q, is roughly half that of the SLAC cavities.

Table II SLAC-like example

$$D = 3.5 \text{ cm} \quad C = 2.915 \text{ cm} \quad H = 2.5 \text{ cm}$$

$$\beta_0 * D = \frac{2\pi}{3} \quad K * D = 1.53 \quad \text{beam height} = 1.625 \text{ cm}$$

$$\text{frequency} = 2.857 \text{ GHz} \quad v_p = c \quad v_g = 0.118\ c$$

$$R = 25\ \frac{\text{meg-ohm}}{\text{meter}} \quad Q = 13{,}500 \quad \frac{R}{Q} = 1860\ \frac{\text{ohm}}{\text{meter}}$$

Table III shows an X-band example. Notice that the decay of the fields away from the grating compels us to operate as close to the grating as possible.

Table III X-band example

$$D = 1.2 \text{ cm} \quad C = 0.8 \text{ cm} \quad H = 0.44 \text{ cm}$$

$$\beta_0 = 2.0 \text{ cm}^{-1} \quad K = 1.0 \text{ cm}^{-1} \quad Q = 9000$$

if beam height = 0.5 cm:

$$R = 22\ \frac{\text{meg-ohm}}{\text{meter}} \qquad \frac{R}{Q} = 2400\ \frac{\text{ohms}}{\text{meter}}$$

if beam height = 0.25 cm:

$$R = 36\ \frac{\text{meg-ohm}}{\text{meter}} \qquad \frac{R}{Q} = 3900\ \frac{\text{ohms}}{\text{meter}}$$

SINGLE PARTICLE MOTION

The transverse motion of accelerated particles above a grating was described by Kim and Kroll at the first workshop.[3] The following is a slight extension of their treatment. The full equations for the fields of the synchronous space harmonic are,

$$\vec{E}(\vec{x},t) = \left\{ -\left(\frac{\omega\mu_0}{k_l^2} \right) \cos(q_0 x) \left[k_0 \hat{y} + i q_0 \hat{z} \right] \right\} e^{-q_0 y}\, e^{i(\omega t - k_0 z)}$$

$$\vec{H}(\vec{x},t) = \left\{ \cos(q_0 x)\hat{x} + \left(\frac{q_0}{k_l^2} \right) \sin(q_0 x) \left[q_0 \hat{y} + i k_0 \hat{z} \right] \right\} e^{-q_0 y}\, e^{i(\omega t - k_0 z)}$$

$$k_l^2 = k_0^2 - q_0^2 \tag{9}$$

It is convenient to replace the z dependence with a phase angle in advance of the electric field peak.

$$\Phi = \omega t - k_0 z + \frac{\pi}{2} \tag{10}$$

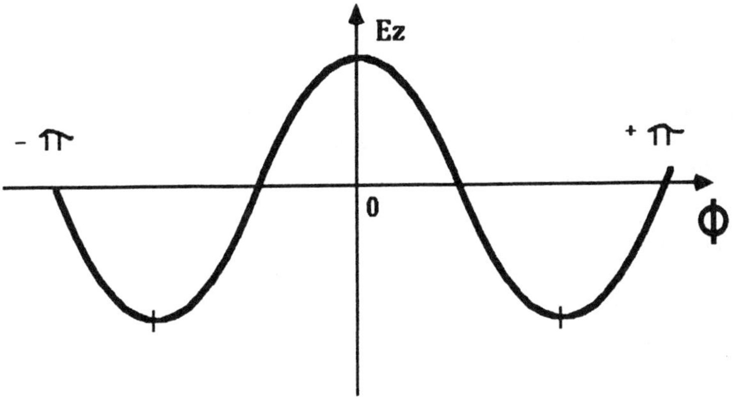

Fig. 3. Definition of Phi

The equations of motion follow from the Lorentz force on a charged particle.

$$\vec{F} = -e\,(\vec{E} + c\hat{z} \times \vec{B})$$

$$\vec{F} = \frac{d}{dt}\,\vec{p} = \frac{d}{dt}\,(\gamma m\vec{v}) \qquad \text{but } \frac{dz}{dt} = c$$

$$\frac{d}{dt}\,(\gamma m\vec{v}) = mc^2\,\frac{d}{dz}\,(\gamma\,\frac{d\vec{x}}{dz})$$

$$mc^2\,\frac{d}{dz}\,(\gamma\,\frac{dx}{dz}) = eE_0\,\frac{q_0}{k_0}\,\sin\Phi\,\sin q_0x\,e^{-q_0y}$$

$$mc^2\,\frac{d}{dz}\,(\gamma\,\frac{dy}{dz}) = eE_0\,\frac{q_0}{k_0}\,\sin\Phi\,\cos q_0x\,e^{-q_0y} \tag{11}$$

$$mc^2\,\frac{d\gamma}{dz} = eE_0\,\cos\Phi\,\cos q_0x\,e^{-q_0y}$$

The first two equations determine the transverse motion. The third governs the longitudinal motion. These equations may be simplified by two assumptions: i) γ varies slowly w.r.t. x and y motions, ii) x and y make small deviations from the z-axis.

Expanding about x, y = 0,

$$\ddot{x} = \frac{q_0^2}{\ell k_0} x \sin \Phi \qquad \ddot{y} = \frac{q_0}{\ell k_0} (1 - q_0 y) \sin \Phi$$

$$\text{where} \quad \ell \equiv \frac{mc^2\gamma}{eE} \qquad \text{and} \quad \ddot{x} = \frac{d^2}{dz^2} x \tag{12}$$

In this approximation the x and y motions are independent and linearized. When $\sin \Phi > 0$, the fields are defocusing in x and focusing in y. When $\sin \Phi < 0$, the fields are focusing in x and defocusing in y. This suggests that net focusing may be obtained by periodically switching $\sin \Phi$. This possibility was studied by Kim and Kroll.

An alternative way to achieve focusing is to rotate the orientation of the grating periodically along the axis, as shown in Figure 4.

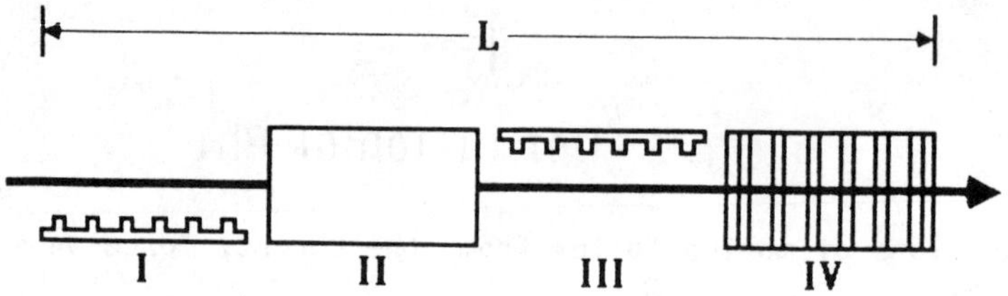

Fig. 4. Rotated grating cell

This trick achieves strong focusing over the cell length L. The particular solution of the above equations determines a reference orbit of period L. The orbit shape depends on particle energy, surface wave field amplitude, transverse constant, cell length and phase. Everything contributes. Figure 5 shows a typical orbit.

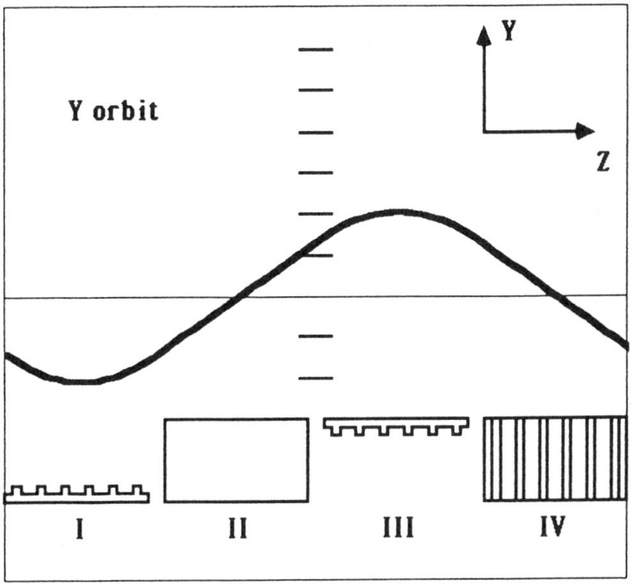

Fig. 5. Reference orbit

Solution of the homogeneous parts of the equations of motion leads to betatron oscillations about these orbits. The linearizing approximations are valid only when x and y are small.

$$\sin q_0 x = q_0 x \quad , \quad \cos q_0 x = 1 \quad , \text{ and } \quad e^{-q_0 x} = 1 - q_0 x$$

are good approximations only if

$$x \text{ and } y < \frac{0.2}{q_0} \tag{13}$$

If we use this restriction to set a maximum believable emittance, the results in Table IV follow.

Table IV Example emittances

X-band 20 GeV electrons

$|x| < 2$ millimeters $L = 8$ meters $5° < \Phi < 15°$

27 meters $< \beta <$ 80 meters

$\gamma\epsilon = 5 \times 10^{-3}$ meter-radians

X-band 120 GeV electrons

$|x| < 2$ millimeters $L = 20$ meters $5° < \Phi < 15°$

60 meters $< \beta <$ 200 meters

$\gamma\epsilon = 1 \times 10^{-2}$ meter-radians

10-micron 20 GeV electrons

$|x| < 1.5$ microns $L = 10$ centimeters $5° < \Phi < 15°$

30 centimeters $< \beta <$ 60 centimeters

$\gamma\epsilon = 2 \times 10^{-7}$ meter-radians

10-micron 1 TeV electrons

$|x| < 1.5$ microns $L = 60$ centimeters $5° < \Phi < 15°$

2.5 meters $< \beta <$ 5 meters

$\gamma\epsilon = 1 \times 10^{-6}$ meter-radians

For comparision, the SLC will have $\gamma\epsilon = 3 \times 10^{-5}$ meter-radians

The combined orbit and betatron envelope looks like this,

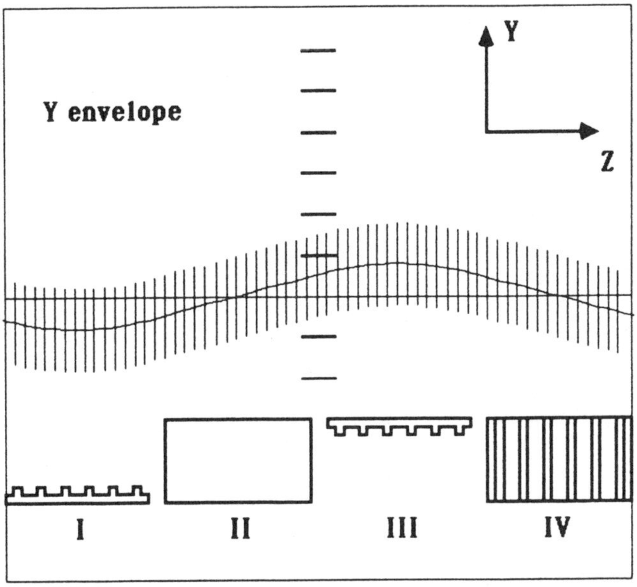

Fig. 6. Beam envelope

Since the equilibrium orbits are curved paths, particles lose energy to synchrotron radiation. The total power radiated by an accelerated charge is[4]

$$P = \frac{2}{3}\frac{e^2\gamma^2}{m^2c^3}\left[\left(\frac{d\vec{p}}{dt}\right)^2 - \frac{1}{c^2}\left(\frac{dE}{dt}\right)^2\right] \quad \text{CGS} \qquad (14)$$

Orbits may be approximated by

$$x(z) = A\sin\left(\frac{2\pi}{L}\right) \qquad \frac{dz}{dt} = c$$

$\frac{d\vec{p}}{dt}$ and $\frac{dE}{dt}$ follow, giving power radiated

$$P = 7.2 \times 10^{-17}\frac{A^2}{L^4}\gamma^4 \text{ watts}$$

and energy lost per meter (15)

$$\frac{\delta E_s}{\delta l} = -1.5 \times 10^{-12}\frac{A^2}{L^4}\gamma^4 \frac{\text{MeV}}{\text{meter}}$$

A, L in meters

Table V gives some typical results. The energy loss to synchrotron radiation is a small fraction of the gain due to acceleration. But the absolute value for the laser-driven case is by no means negligible.

Table V Synchrotron radiation examples

X-band gradient $eE_0 = 100 \frac{\text{Mev}}{\text{meter}}$

Energy	A	L	$\frac{\delta E_s}{\delta l}$
20 GeV	0.67 mm	8 meters	$-4.0 \times 10^{-4} \frac{\text{MeV}}{\text{meter}}$
120 GeV	0.67 mm	20 meters	$-1.3 \times 10^{-2} \frac{\text{MeV}}{\text{meter}}$

10-micron gradient $eE_0 = 1 \frac{\text{GeV}}{\text{meter}}$

Energy	A	L	$\frac{\delta E_s}{\delta l}$
20 GeV	0.33 μ	10 cm	$-4.0 \times 10^{-3} \frac{\text{MeV}}{\text{meter}}$
1 TeV	0.25 μ	60 cm	$-1.1 \times 10^{+1} \frac{\text{MeV}}{\text{meter}}$

WAKEFIELDS

Wakefields may be calculated by numerical integration of Maxwell's equations in the time domain.[5] Programs like DBCI simplify the full three-dimensional problem by treating only modes that have cylindrical symmetry. A similar simplification may be applied in the grating case. The driving currents, and thus the resulting fields are assumed to be periodic along the grooves. DBCI may be converted to a grating version (GBCI?) by the following substitutions,

$$
\begin{array}{lcl}
\text{DBCI} & \text{--->} & \text{GBCI} \\
J_z \;\alpha\; \cos m\Phi & & J_z \;\alpha\; \cos\left(\frac{2\pi x}{\lambda}\right) \\
E_z, E_r, H_\Phi \;\alpha\; \cos m\Phi & & E_z, E_y, H_x \;\alpha\; \cos\left(\frac{2\pi x}{\lambda}\right) \\
H_z, H_r, E_\Phi \;\alpha\; \sin m\Phi & & H_z, H_y, E_x \;\alpha\; \sin\left(\frac{2\pi x}{\lambda}\right)
\end{array}
\qquad (16)
$$

Figure 7 shows that the driving current and its image charges may be considered to be a function periodic in x. This current density function may be decomposed into it's Fourier components w.r.t. x. GBCI is then used to generate wake tables for each λ. These tables may then be summed to obtain a wake table for the general driving current. Figures 8 and 9 show typical wakes using the SLAC-like grating.

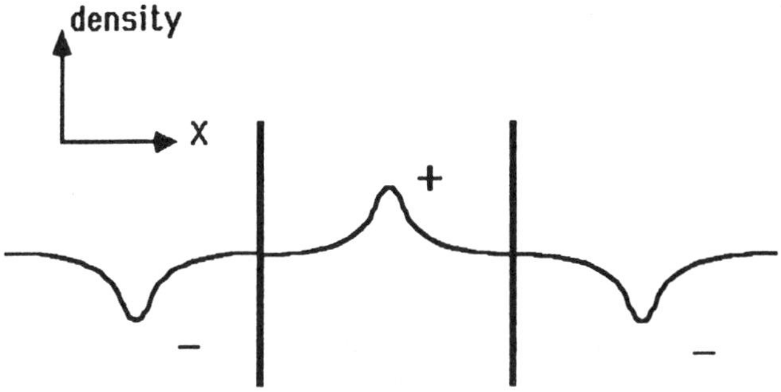

Fig. 7. Driving current

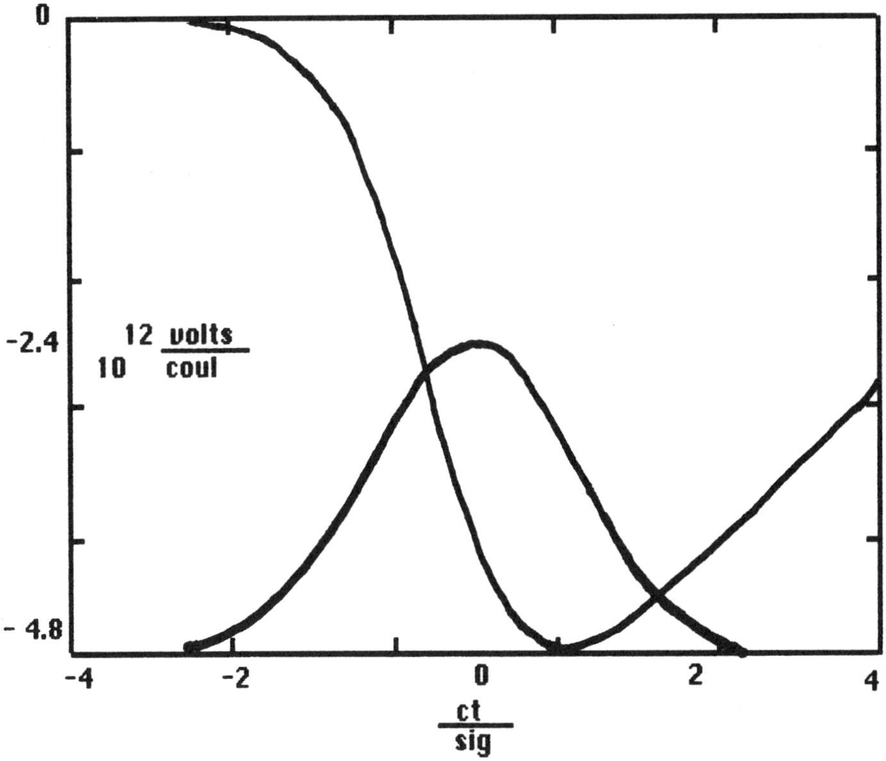

Fig. 8. Longitudinal wake

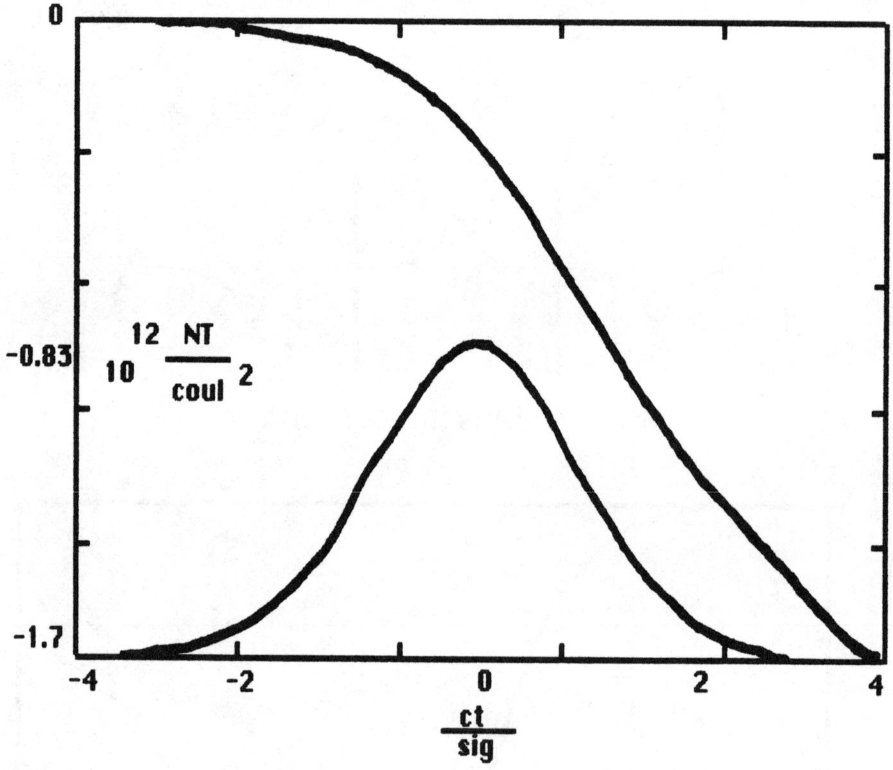

Fig. 9. Transverse wake

Evanescent space harmonics are the principal contributors to the wake. The magnitude of the wakefields depends on both the driving particles and test particle height as follows,

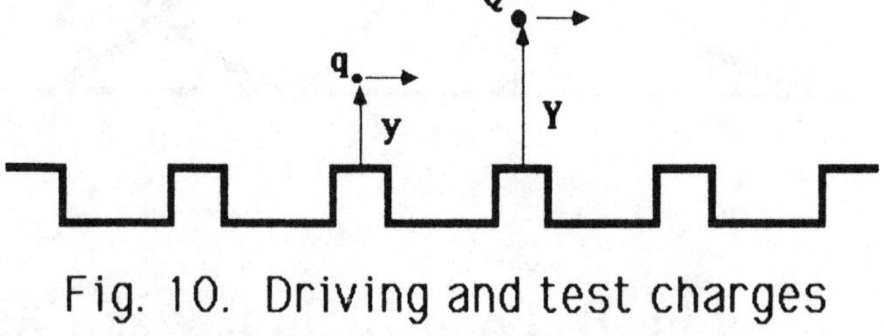

Fig. 10. Driving and test charges

$$\text{wakefields} \quad \alpha \quad e^{-\frac{2\pi}{\lambda} Y} \, e^{\frac{2\pi}{\lambda} y} \tag{17}$$

For the SLAC-like example,

$\lambda = 7.3$ cm $\qquad Y = 0 \qquad y = 1.625$ cm

The peak longitudinal value is

$$\left(-\ 4.8 \times 10^{12}\ \frac{\text{volts}}{\text{coul}}\right) \times \left(0.25\right) = -\ 1.2 \times 10^{12}\ \frac{\text{volts}}{\text{coul}}$$

The transvers value at 4σ is

$$\left(-\ 1.7 \times 10^{12}\ \frac{\text{NT}}{\text{coul}^2}\right) \times \left(0.25\right) = -\ 0.4 \times 10^{12}\ \frac{\text{NT}}{\text{coul}^2}$$

For comparison, the 4σ transverse wake value in the SLAC cavity is approximately $= 1.5 \times 10^{12}\ \frac{\text{NT}}{\text{coul}^2}$. [6]

FUTURE WORK

Work on the simulation of bunch passage through a linac made of rotated grating cells is in progress. The physical bunch is represented by a macro-particle bunch in the simulation. The macro-particles' motion through phase space will be determined by the single particle equations of motion and wake effects. Of particular interest will be the current limits imposed by the transverse wake. Other areas of interest include optimization of the groove profile for best realizable efficiency. The grating shape chosen must strongly couple incident plane-wave radiation to the surface waves. The effects of alignment errors and groove dimensional tolerances must also be studied. The results obtained from this 'paper linac' will hopefully aid the design of possible experiments discussed at this workshop.

ACKNOWLEDGEMENTS

I would like to thank J. Lawson, P. McIsaac, R. Palmer, R. Siemann, and M. Tigner for many helpful discussions.

REFERENCES

1. R. Petit, ed., Electromagnetic Theory of Gratings (Springer-Verlag, N. Y., 1980).
2. R. E. Collin, Field Theory of Guided Waves (McGraw-Hill, N. Y., 1960), chapter 11.
3. K. Kim and N. M. Kroll, Proceedings of the Workshop on the Laser Acceleration of Particles (Los Alamos, 1982), AIP No. 91.
4. J. D. Jackson, Classical Electrodynamics, Second Edition (John Wiley and Sons, N. Y.,1975), page 660.
5. G. Aharonian, R. Meller and R. H. Siemann, NIM 212, 23 (1983).
6. K. Bane and T. Weiland, SLAC-PUB-3173, July 1983.

SURFACE HEATING BY SHORT PULSES OF RADIATION

Norman M. Kroll
University of California, San Diego, La Jolla, CA 92093

Experimental results presented by Paul Corkum[1] indicate that it may be possible to produce intense laser pulses only a few cycles in length. Furthermore, W. Willis[2] has suggested an acceleration scheme based upon short intense electric field pulses, which may be thought of as "half-cycles" of radiation. Such short pulses should lead to relaxation in the limit on accelerating fields imposed by surface melting. It is well-known[3] that heating of conduction cooled surfaces by pulses of electromagnetic radiation leads to a temperature rise proportional to the square root of the pulse width. In the following, we reexamine this problem with special attention to very short pulses for which heat conduction may be ineffective. We find that the square root law is modified and that the dependence of the temperature rise on material properties is also modified. Accelerating fields in excess of 10 GeV/m may under certain circumstances be consistent with the absence of surface melting.

We first consider the case of a plane wave normally incident upon a plane highly conducting half space of conductivity σ, with permeability $\mu = 1$, and dielectric constant $\epsilon = 1$. We assume σ much greater than the frequencies which characterize the pulse. (We are using Gaussian units.)

Let $x < 0$ be free space and $x > 0$ be the conductor. Then an incident wave of the form

$$E_0(x,t) = \int E(\omega)\, e^{i(kx-\omega t)}\, d\omega$$

is almost perfectly reflected to produce the approximate field

$$E(x,t) = 2i \int E(\omega) \sin kx\, e^{-i\omega t} d\omega \;, \quad x < 0$$

$$E(x,t) = 2 \int \frac{\omega}{\alpha c} E(\omega)\, e^{i(\alpha x - \omega t)} d\omega \;, \quad x > 0$$

where

$$\alpha = \sqrt{4\pi i \omega \sigma}\,/\,c \;, \quad \mathrm{Im}\,\alpha > 0$$

Heat is deposited at the rate per unit volume $A(x, t) = \sigma E^2(x, t)$, and is a maximum at $x = 0$.

For a sufficiently short and smooth pulse, thermal conduction can be neglected. In this case we have for the maximum temperature rise ΔT_M,

$$\Delta T_M = \frac{\sigma}{\rho C} \int_{-\infty}^{\infty} E^2(0, t)\, dt = \frac{8\pi\sigma}{\rho C} \int_{-\infty}^{\infty} \frac{\omega^2}{|\alpha|^2 c^2} |E(\omega)|^2 d\omega$$

$$= \frac{2}{\rho C} \int |\omega|\ |E(\omega)|^2 d\omega$$

$$\equiv \frac{2\overline{\omega}}{\rho C} \int |E(\omega)|^2 d\omega = \frac{\overline{\omega}}{\pi \rho C} \int E_i^2(0, t)\, dt$$

$$\equiv \frac{\overline{\omega}\,\overline{\tau}}{2\pi} \frac{1}{\rho C} E_0^2 \qquad (1)$$

where ρ is the density, C the specific heat, and E_0 is the peak incident field. The mean frequency $\overline{\omega}$ and pulse width $\overline{t}$ are defined as indicated in the above equations, and the quantity $\overline{\omega}\,\overline{t}/2\pi$ is a measure of the number of oscillations in the pulse. If the pulse involves one or more oscillations, then E_0 is approximately one half the peak field outside the conductor, which we take to be the accelerating field. We note the striking fact that the result is independent of electric conductivity (so long as we are in the high $4\pi\sigma/\omega$ regime) and depends upon the number of oscillations rather than explicitly on the frequency.

As a useful example, we consider the square pulse

$$\begin{aligned} E_i(0, t) &= E_0 \sin \omega_0 t \quad , \quad 0 < t < \overline{t} \\ &= 0 \quad , \quad \text{otherwise} \; . \end{aligned} \qquad (2)$$

With $\omega_0 \overline{t} = 2\pi N$, $N = 1/2, 1, \ldots$. Note that we have chosen the form of the square pulse so as to guarantee that there is no discontinuity in $E_i(0, t)$. If this is not done $\int |\omega|\ |E(\omega)|^2 d\omega$ is logarithmically divergent. Since this divergence in ΔT occurs only at $x = 0$ it simply means that one can never neglect the effect of thermal conductivity for such a case. Practically speaking, however, the rise time of the pulse is unlikely to be less than the quarter period automatically introduced here, so we will consider our results to hold for any square pulse with such a rise time. The $\bar{t}$ in Eq. (2) is clearly equal to the $\bar{t}$ defined in Eq. (1). One then finds

$$E(\omega) = \frac{\omega_0 \exp(-i\omega\bar{t}/2)}{2\pi(\omega_0^2 - \omega^2)} [\exp i\omega\bar{t}/2 - (-1)^{2N} \exp(-i\omega\bar{t}/2)]$$

$$\Delta T_M = \frac{2}{\pi\rho C} \left[N\,Si(2\pi N) - \frac{1-(-1)^{2N}}{2\pi} \right] \quad (3)$$

where

$$Si(y) \equiv \int_0^y \frac{\sin x}{x}\, dx \ .$$

Note that if one approximates the integrations over ω by setting $\bar{\omega}$ in (1) equal to ω_0 (the narrow bandwidth approximation) one obtains $\Delta T_M = N/\rho C$, which is an upper bound to (3), and approximates it for large N. The error is less than 10% for $N \geq 1$, while for $N = 1/2$ we have

$$\Delta T_M = \frac{0.387}{\rho C} E_0^2$$

To obtain limits on the accelerating field we set E_0 equal to one half the accelerating field and limit ΔT_M to the melting point less 25 °C (the melting limits). It is, however, not clear that the very transient melting of a very thin surface layer would cause any actual surface damage as the time available for flow is so very short. It may well be that surface damage occurs only when the boiling point is reached. We therefore include in the table below a maximum accelerated field limited by boiling. The heat of fusion is taken into account in the values given, but the effect of the increase of specific heat with temperature (which would increase the accelerating field limit somewhat) has not been included.

As a second example, consider the Gaussian pulse,

$$E_i(0,t) = E_0\, e^{-t^2/2\tau^2} \cos \omega_0 t \quad (4)$$

then

Table 1. Maximum accelerating fields in GeV/m

	$N = \frac{1}{2}$		"Large N"	
	Melting Limit	Boiling Limit	Melting Limit	Boiling Limit
Copper	18.4	39.2	$11.4/\sqrt{N}$	$24.3/\sqrt{N}$
Molybdenum	24.8	49.5	$15.4/\sqrt{N}$	$30.8/\sqrt{N}$
Tungsten	28.5	53.6	$17.7/\sqrt{N}$	$33.3/\sqrt{N}$

$$\Delta T_M = \frac{E_0^2}{\pi \rho C} \left(\exp(-\omega_0^2 \tau^2) + \frac{\omega_0 \tau}{2} \sqrt{\pi} \operatorname{erf} \omega_0 \tau \right) \tag{5}$$

$$\approx \frac{E_0^2}{2 \rho C \sqrt{\pi}} \omega_0 \tau$$

for $\omega_0 \tau > 2$. For this example, $\bar{t} = \sqrt{\pi}\,(1 + \exp(-\omega_0^2 \tau^2)\,\tau$, so that, as expected, we again get $\Delta T_M \approx (N/\rho C) E_0^2$ for "large N" ($N \geq 1$ is sufficient). On the other hand, a single half wave pulse corresponds to $\omega_0 = 0$, which yields $\Delta T_M = E_0^2/\pi \rho C$, which is somewhat smaller than the temperature rise found for the half sine wave. Note that while the temperature rise for these single pulses depends mildly on their shape, being smaller for smoother pulses, it does not depend upon the pulse width.

The temperature rises computed above are upper bounds, as they will be reduced by heat conduction. To estimate this effect, we first observe that the narrow band approximation is quite good even for a single cycle pulse. Since thermal conductivity is expected to become significant for longer pulses, we can make use of a narrow band approximation when taking it into account. Thus we take

$$E_i(0,t) = \epsilon(t) \cos(\omega_0 t + \phi) \tag{6}$$

and assume $E(\omega)$ sharply peaked about $\omega = \pm \omega_0$. Then we find, ignoring rapidly oscillating terms,

$$A(x,t) = \frac{2\sigma\omega_0^2}{\alpha_0 \alpha_0^* c^2} \epsilon^2(t) \exp i(\alpha_0 - \alpha_0^*)x$$

$$= \frac{\omega_0}{2\pi} \epsilon^2(t) e^{-\beta x} \tag{7}$$

where the inverse skin depth β is given by

$$\beta = \frac{4\pi}{c} \sqrt{\omega_0 \sigma / 2\pi}$$

The temperature rise is then found by solving the differential equation

$$\frac{dv}{dt} - \kappa \frac{d^2 v}{dx^2} = \frac{1}{\rho C} A$$

where $\kappa = K/\rho C$ is the thermal diffusivity, and v represents the temperature increase as a function of x and t. Using standard formulas[4] one easily sees that the maximum temperature occurs at $x = 0$ and is given by

$$v(0,t) = \frac{\omega_0}{2\pi\rho C} \int_{-\infty}^{t} \epsilon^2(t_1)\, dt_1 \, e^{\beta^2 \kappa (t - t_1)} \operatorname{erfc} \beta \sqrt{\kappa (t - t_1)} \tag{8}$$

If the thermal conductivity K is set equal to zero then the expression reduces to

$$v(0,t) = \frac{\omega_0}{2\pi\rho C} \int_{-\infty}^{t} \epsilon^2(t_1)\, dt_1$$

which has its maximum ΔT_M at $t = \infty$. Noting that for our narrow band pulse

$$\int_{-\infty}^{\infty} \epsilon^2(t)\, dt \cong 2 \int E_i^2(0,t)\, dt = \bar{t} E_0^2$$

we obtain

$$\Delta T_M = \frac{\omega_0 \bar{t}}{2\pi} \frac{1}{\rho C} E_0^2 ,$$

our previous result.

The opposite limit, large κ, is

$$v(0,t) = \frac{\omega_0}{2\pi\rho C\sqrt{\pi}} \int_{-\infty}^{t} \frac{\epsilon^2(t_1)}{\beta\sqrt{\kappa(t-t_1)}} dt_1 \tag{9}$$

Returning now to (8) and using the square pulse $\epsilon = E_0$ for $0 < t < \bar{t}$, we have

$$\Delta T_M = \frac{\omega_0 E_0^2}{2\pi\rho C} \int_0^{\bar{t}} e^{\beta^2 \kappa(\bar{t}-t_1)} \operatorname{erfc} \beta\sqrt{\kappa(\bar{t}-t_1)}\, dt_1$$

$$= \frac{\omega_0 E_0^2}{2\pi\rho C \beta^2 \kappa} \left(\frac{2}{\sqrt{\pi}} \sqrt{y} - 1 + e^y \operatorname{erfc} \sqrt{y} \right)$$

where

$$y = \beta^2 \kappa \bar{t} = \frac{16\pi^2 \sigma\kappa}{c^2} N \equiv N/N_c , \text{ and } N_c \equiv c^2/16\pi^2\sigma\kappa$$

characterizes the role of heat conduction. Hence

$$\Delta T_M = \frac{E_0^2 N_c}{\rho C} \left(\frac{2}{\sqrt{\pi}} \sqrt{\frac{N}{N_c}} - 1 + \exp\frac{N}{N_c} \operatorname{erfc} \sqrt{N/N_c} \right) \tag{10}$$

$$= \frac{E_0^2 N_c}{\rho C} (0.556) \quad \text{at } N = N_c$$

$$\approx \frac{E_0^2 N}{\rho C} \left(1 - \frac{4}{3\sqrt{\pi}} \sqrt{\frac{N}{N_c}} + \frac{1}{2}\frac{N}{N_c} + \cdots \right), \; N/N_c \text{ small} \tag{11}$$

$$\approx \frac{2E_0^2\sqrt{NN_c}}{\sqrt{\pi}\rho C}\left(1 - \sqrt{\frac{\pi N_c}{2N}} + \frac{N_c}{2N} - \frac{1}{4}\left(\frac{N_c}{N}\right)^2 + \ldots\right), \tag{12}$$

N/N_c large.

Table 2. Values for N_c

	N_c (20 °C)	N_c ($\leqslant$ Melting Point)
Copper	9.3	66
Molybdenum	69	1200
Tungsten	52	2200

Since the temperature is x dependent, our theory cannot really take the variation of N_c with temperature into account. The values at the melting point are estimates and are included to indicate the magnitude of the temperature variation.

For completeness we again consider the case of the Gaussian pulse $\varepsilon = E_0 \exp(-t^2/2\tau^2)$. Then from equation (8) we have

$$v(0,t) = \frac{\omega_0 E_0^2}{2\pi\rho C}\int_{-\infty}^{t} e^{-t_1^2/\tau^2 + \beta^2\kappa(t - t_1)} \operatorname{erfc} \beta\sqrt{\kappa(t - t_1)}\, dt .$$

Using $\bar{t} \cong \tau\sqrt{\pi}$, and defining $\gamma = \beta^2\kappa\tau = N/\sqrt{\pi}N_c$ and $y = t/\tau$ we find

$$v(0, y\tau) = \frac{NE_0^2}{\rho C}\frac{1}{\sqrt{\pi}}\int_{-\infty}^{y} e^{-z^2 + \gamma(y-z)} \operatorname{erfc}\sqrt{\gamma(y-z)}\, dz$$

To obtain ΔT_M we must maximize the above expression in y. We note that again the final result depends only upon N and N_c. We proceed only with the large N limit. Here we obtain

$$v = \frac{\sqrt{NN_c}\, E_0^2}{\pi^{3/4}\, \rho C} \int_0^y \frac{e^{-z^2}}{\sqrt{y - z}}\, dz$$

The integral is a parabolic cylinder function with maximum value 1.998 at $y = 0.4207$. Thus we obtain

$$\Delta T_M = 0.846 \frac{\sqrt{NN_c}}{\rho C} E_0^2$$

which is somewhat less than the corresponding result for the square pulse (Eq. 12).

While we have confined our attention in the above to the case of a normally incident plane wave, the results in the large $4\pi\sigma/\omega$ regime which we have been considering are similar for the case of a wave at grazing incidence with the electric field perpendicular to the surface. Here it is appropriate to identify the electric field at the surface with the accelerating field, and with this understanding the results expressed in terms of the accelerating field are identical for the two cases. The $N = 1/2$ results tabulated previously are of particular interest for this case as the incident field doubling effect assumed in constructing the table does not occur for $N = 1/2$ in the normal incidence case.[5]

SUMMARY

To facilitate application of the above, we summarize our principal results here.

(1) We find that limits on the accelerating field can be most conveniently expressed in terms of the total number of oscillation cycles N rather than the frequency.

(2) When N is small the effects of thermal conduction can be neglected, and one finds that the accelerating field limit is independent of electric conductivity so long as the high electric conductivity regime is applicable. Table 1 then provides estimates of maximum accelerating fields for structures made of copper, molybdenum, or tungsten. The $N = 1/2$ column is intended to apply to pulse accelerators of the type described by Willis, while the $N \geq 1$ column is intended to apply to the various open cavity propagating structures that have been discussed at this conference such as gratings, multiple rows of spheres, and the like. Our calculations have been carried out for plane surfaces. Since surface dimensions will typically be much larger than the skin depth, they should, qualitatively speaking, be of more general applicability. To obtain

more exact results, each proposed structure should be examined specifically. Reductions in maximum accelerating fields by perhaps a factor two may well occur.

(3) The effect of thermal conduction is determined by the characteristic cycle numbers N_c (see Table 2). For $N \ll N_c$ its effect can be ignored, while for $N = N_c$, it increases the maximum allowable field by a factor 4/3. Unfortunately, N_c is very temperature dependent so that the results given should be applied only qualitatively. Nevertheless, one obtains upper limits on the temperature rise if one uses the value of N_c associated with the maximum temperature allowed.

Acknowledgment: We have profited greatly from discussions with R. B. Palmer and T. Himel. This research was supported in part by the United States Department of Energy.

REFERENCES

1. Paul Corkum. Presented at this workshop.

2. W. Willis. Presented at this workshop.

3. In the context of laser acceleration of particles, see Y. Takeda and L. Matsui, Nuclear Instruments and Methods 62, 306-310 (1968).

4. H. S. Carslaw and J. C. Jaeger, Conduction of Heat in Solids, Oxford at the Clarendon Press, 2nd Edition, 1959. Especially page 80, Eq. (13) and Duhamel's Theorem, page 30.

5. A similar result for a single pulse at grazing incidence was obtained independently at this conference by T. Himel.

ON ACCELERATION BY THE TRANSFER OF ENERGY BETWEEN TWO BEAMS

J. S. Wurtele [1]
Plasma Fusion Center
Massachusetts Institute of Technology, Cambridge, Ma. 02139

ABSTRACT

Work on the powering of a high gradient structure utilizing a Free Electron Laser (FEL) is reviewed. The concept envisions a dense low energy beam producing power via the FEL mechanism at wavelengths substantially shorter than the $10cm$ used at SLAC. The energy of this beam must be periodically replenished by induction units. The microwave power produced by the FEL is then used to drive a high gradient structure where a second beam is accelerated to very high energies. The theory of FEL operation in waveguides is discussed and possible coupling schemes between the two beams are presented. An overview of the near and far term plans of the LBL/LLNL group is given. The parameters for a $1TeV \times 1TeV$ collider with luminosity $4 \times 10^{32}\, cm^{-2}\, sec^{-1}$ are presented.

INTRODUCTION

The development of efficient high power sources, in the centimeter wavelength region, is of considerable importance to the design of the next generation collider. This is well illustrated by the Stanford Linear Collider (SLC) which will produce $50GeV \times 50GeV$[1] beams utilizing a $10cm$ accelerating structure with accelerating gradients of $17MeV/m$. Many authors[2,3,4] have pointed out the advantages of utilizing a shorter wavelength accelerating structure. Two major benefits of operating at shorter wavelengths are that the stored energy required to achieve a given accelerating gradient scales as the square of the wavelength (λ^2) and the breakdown field, in the $1 - 10cm$ wavelength region, scales as $\lambda^{-7/8}$. The improvement in stored energy requirements corresponds to substantial power savings. The higher breakdown field permits larger accelerating gradients and hence shorter accelerators. It therefore seems likely that the next generation linear colliders, which may attain energies in excess of $1TeV \times 1TeV$[5], will operate at shorter wavelengths than the SLC.

Different designs for the accelerator , depending on the power source, have been considered. One is to keep the present SLAC configuration of a high gradient structure powered by rf-sources. In this case the SLAC Klystrons would have to be replaced by Gyrotrons (or some other power source). While

[1] Work supported by the Office of Naval Research

there is much work to be done on source development, this scheme is quite similar to present machines in that it requires multiple phase-locked power sources. A second scenario, originally suggested by Sessler[3], is to replace the numerous Klystrons with one Free Electron Laser. The electron beam in the FEL would propagate *parallel* to the beam being accelerated, and continuously produce power to drive a conventional high gradient structure. The operation of the FEL with a higher order waveguide mode, and the coupling of power between the beams has been studied[6]. A recent overview of the Two-Beam Accelerator (TBA) has also been made[7].

In this paper we examine the Two-Beam Accelerator. The TBA would consist of a Low Energy Beam (LEB) which undergoes FEL action to produce microwave power. To replenish its energy, this beam would be periodically reaccelerated by induction units. The microwave power would then be coupled into a high gradient structure where a High Energy Beam (HEB) is accelerated. (By comparison, one might say SLAC is a 241-beam accelerator.) The TBA therefore has a novel aspect, namely the operation of very long "steady state" FELs with periodic reacceleration of the electron beam and extraction of microwave power. The aspects of the TBA which involve induction modules and slow wave structures are more conventional. Figure I shows the TBA configuration.

The LEB must pass through hundreds of sections of the TBA without undergoing beam degradation (emittance growth) and also remain trapped by the ponderomotive beat wave of the wiggler and radiation fields. Otherwise, if there is substantial detrapping, the LEB will not be able to produce the needed power (since only the trapped electrons contribute to the FEL interaction).

This paper is organized as follows. We first review the operation of the FEL. The analysis includes the FEL coupling to the waveguide modes, allowing for the longitudinal electric fields of TM modes and the transverse variation of higher order modes. The discussion of FELs ends with the description of a possible driver for the TBA.

We next present three possible coupling schemes between the overmoded waveguide and the slow wave structure. In the following section we report on the LBL/LLNL experimental program, and present some problems for future research. We conclude by giving parameters for a $1TeV \times 1TeV$ collider with luminosity of $4 \times 10^{32} cm^{-2} sec^{-1}$.

FREE ELECTRON LASERS

The power source for the TBA is a Free Electron Laser. The requirements on the FEL ($2GW/m$ in the example considered in Table 2) require that the Low Energy Beam be reaccelerated periodically (since a $1kA$ beam at $20MeV$

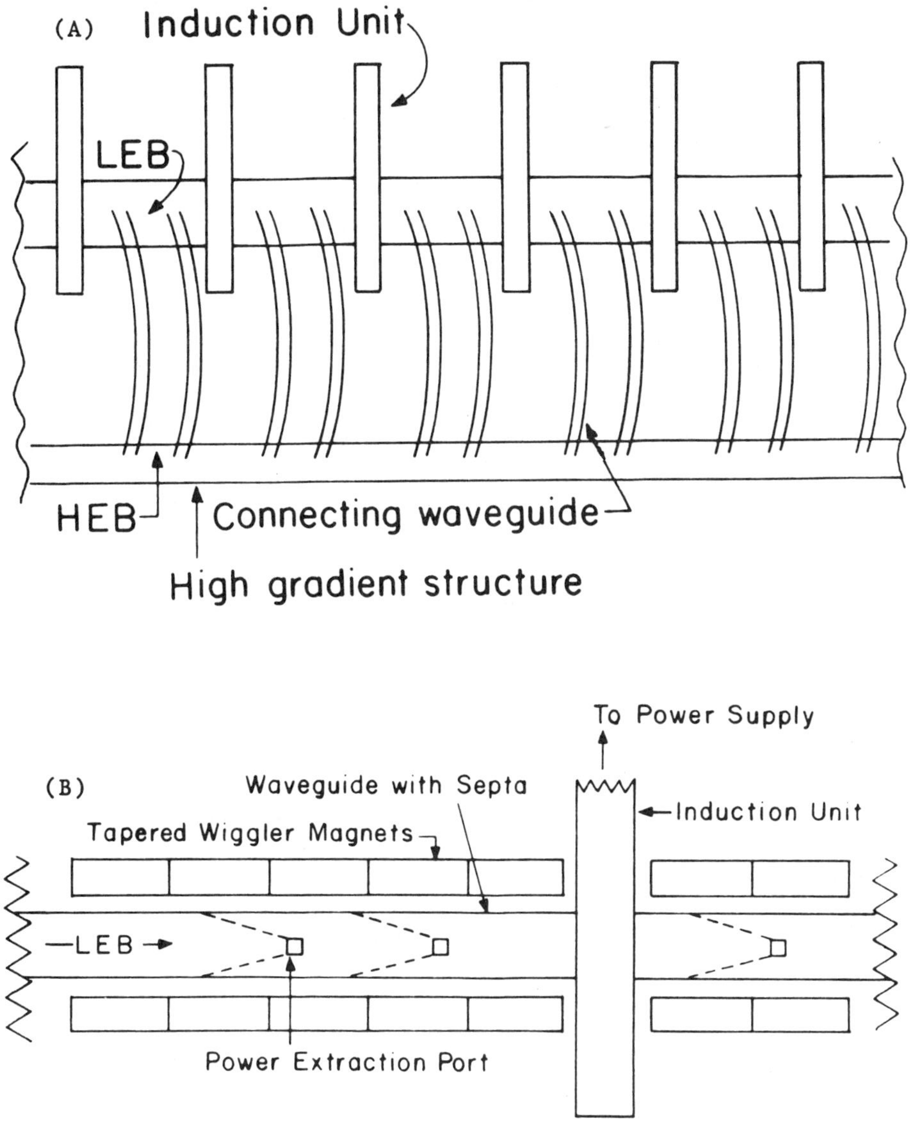

Figure I: The Two-Beam Accelerator. (a) Schematic of the TBA. (b) The Low Energy Beam and steady-state FEL.

has only $20GW$ total power and hence would exhaust itself after $10m$). An FEL designed for a TBA would begin with a start-up segment in which the field power is built-up and the electron beam bunched in the ponderomotive potential well. During the start-up phase the FEL would not power the high gradient structure. The length of this section may be tens of meters. After the beam is bunched, the "steady state" FEL begins. It would be composed of numerous sections, each of which consists of a tapered FEL followed by an induction unit. The induction unit should replenish the beam energy to its value at the beginning of the section.

Before discussing the FEL equations in detail, we briefly examine the transverse motion of the electron beam including external quadrupole focusing. This treatment neglects the effects of the FEL and induction units on the transverse motion. Coupling between the transverse and longitudinal motion, which might lead to detrapping or reduced FEL performance, is a topic for future research.

Planar wigglers, which are mechanically much easier to fabricate than helical wigglers, seem a likely choice for the TBA. The planar wiggler, in order to have vanishing divergence and curl, must vary in at least one transverse direction. For example the magnetic field arising from the vector potential

$$\vec{A}_w = \tilde{A}_w \cosh(k_w y) \cos(k_w z)\hat{x} \tag{1}$$

would be

$$\vec{B}_w = -\tilde{B}_w \cosh(k_w y) \sin(k_w z)\hat{y} - \tilde{B}_w \sinh(k_w y) \cos(k_w z)\hat{z} \ . \tag{2}$$

In Eq. 2, $B_w = k_w A_w$. This field, when evaluated in the $y = 0$ plane, is

$$\vec{B}_w = -\tilde{B}_w \sin(k_w z)\hat{y} \ . \tag{3}$$

The magnetic field of Eq. 3 is called an *ideal* wiggler field and can only be realized in a two dimensional plane. The use of this field is valid only when the electron beam satisfies $k_w r_b \ll 1$.

The motion of the electrons is governed by the Lorentz equations

$$\frac{d\vec{x}}{dt} = \frac{\vec{p}}{\gamma} , \tag{4}$$

and

$$\frac{d\vec{p}}{dt} = -e\vec{E} - e\frac{\vec{p} \times \vec{B}}{\gamma} \ . \tag{5}$$

Here $e = |e|$ is the absolute value of the electron charge.

Examination of Eq. 2 shows that, since $\vec{B}_w$ is independent of x, the x component of canonical momentum is conserved. The planar wiggler therefore provides no focusing in the wiggle (x, z) plane. By retaining only the linear terms in $k_w y$, we see that B_{wz} increases linearly with y, thus providing focusing in the non-wiggle plane. The electron beam must be confined in both planes. There are various ways to achieve this. External quadrupoles can be used to focus in the wiggle plane; the defocusing of the quadrupoles in the non-wiggle plane being overwhelmed by the wiggler focusing. Alternatively, the planar wiggler can be built to provide focusing (by utilizing curved pole faces, for example) in both transverse planes. This has been examined and been shown to be advantageous for FEL operation[8].

An electron with vanishing perpendicular canonical momentum (corresponding to motion in $y = 0$ plane with vanishing average v_x) has

$$x(t) = x_w(t) = \frac{a_w\sqrt{2}\sin(k_w v_0 t)}{k_w\gamma} \,. \tag{6}$$

Here $v_0 = (1 - (1 + a_w^2)/\gamma^2)^{1/2}$ is the average longitudinal velocity. Here, and in the rest of this paper, we use the dimensionless variables:

$$a_s = \frac{e\tilde{E}_s}{\sqrt{2}mc\omega} \quad \text{and}$$

$$a_w = \frac{e\tilde{B}_w}{\sqrt{2}mc^2k_w} \,. \tag{7}$$

The motion of particles with small perpendicular momenta can be found by inserting $x = x_w + x_1$, $y = y_1$ in the equations of motion, linearizing, and solving for x_1 and y_1 by averaging over a wiggle period. The result is:

$$x = x_w + x_0\cos(\beta_q z) + \frac{x_0'}{\beta_q}\sin(\beta_q z) \,, \tag{8}$$

$$y = y_0\cos(\beta_y z) + \frac{y_0'}{\beta_y}\sin(\beta_y z) \,, \tag{9}$$

$$z = v_0 t + \frac{a_w^2\sin(2k_w v_0 t)}{4\gamma^2 k_w} \,. \tag{10}$$

Here $\beta_q = \sqrt{eB_q/\gamma mc^2 L_q}$ is the quadrupole focusing, Bq/Lq the quadrupole gradient, $\beta_y^2 = a_w^2 k_w^2/\gamma^2 - \beta_q^2$, $'$ denotes d/dz, and we have neglected terms of higher than second order in a_w/γ.

WAVEGUIDE FEL EQUATIONS OF MOTION

In this section we examine FEL amplifier operation in a waveguide and derive one-dimensional nonlinear equations of motion for the particles and field. Since TBA should only couple into one waveguide mode, only single mode operation is assumed. This is a good approximation if the modes are well separated. The amplifier model assumes that the frequency ω of the mode is fixed by an input signal. The amplitude and phase of the mode are assumed to be slowly varying and are driven by the electron beam. The waveguide modes have different transverse profiles and longitudinal wave number k_z. The effect of transverse mode variation during a wiggle oscillation and beam-mode overlap are included in this one-dimensional model of the FEL. More sophisticated models include the transverse motion of the particles. Much of the basic physics can be understood within a one dimensional formulation.

The wiggler field is assumed to be given by Eq. 3. The radiation field is assumed to be of the form

$$\vec{E} = (\vec{q}_{\perp s}(x,y)\sin\alpha_s + q_{zs}(x,y)\hat{z}\cos\alpha_s)\tilde{E}_s(z) \tag{11}$$

where $\alpha_s = k_s z - \omega t + \varphi_s$, φ_s and $\tilde{E}_s$ are slowly varying amplitudes and phases, and $\vec{q}_s(x,y)$ is the waveguide mode profile (q_z is nonzero only for TM modes). The slowly varying amplitude and phase approximation implies $d\ln(E_s)/dz, d\varphi_s/dz << k_s$ and $E_s''/E_s, \varphi_s'' << k_s^2$. Also,

$$k_s = (\frac{\omega^2}{c^2} - k_{\perp,s}^2)^{\frac{1}{2}}, \tag{12}$$

where $k_{\perp,s}$ satisfies $\nabla_\perp^2\vec{q}_s = k_{\perp,s}^2\vec{q}_s$. Electrons are labeled by their phase with respect to the beat wave between the wiggler and the waveguide mode.

An FEL works by wiggling the electrons (in the (x,z) plane here) so that $\vec{v}\cdot\vec{E}$ is always of one sign. As we will soon see, this imposes a resonance condition which equates the longitudinal electron velocity with the phase velocity of a beat wave between the wiggler and radiation fields. We allow the amplitude and wavelength of the wiggler field to be slowly varying in order to maintain resonance between electrons (which lose energy between induction units) and the electromagnetic wave.

The particle energy, which evolves according to

$$\gamma' = -t'\vec{v}\cdot\vec{E}(\frac{e}{mc^3}), \tag{13}$$

can now have two resonant contributions – one from $v_x E_x$ and one (from TM modes) from $v_z E_z$. For some modes $v_z E_z$ has a slowly varying component.

The resonant phase is defined by

$$\psi = (k_s + k_w)z - \omega t(z) + \varphi_s \,. \tag{14}$$

Note that the product $\vec{v} \cdot \vec{E}$ also yields a rapidly varying beat with phase $\psi - 2k_w z$. The phase ψ defined by Eq. 14 is the electron phase relative to the slow beat wave of the wiggler and radiation fields.

An appropriate averaging of the particle equations over a wiggler period eliminates the rapidly varying terms. We find , for the longitudinal motion,

$$\gamma_i' = -\frac{\omega}{c} C_s a_s a_w \frac{\sin \psi_i}{\gamma_i} \tag{15}$$

and

$$\psi_i' = (k_w - \delta k_s) - \frac{w}{2c\gamma_i^2}(1 + a_w^2 - 2C_s a_w a_s \cos \psi_i) + \frac{d\varphi_s}{dz} \,, \tag{16}$$

where $\delta k_s = \frac{\omega}{c} - k_s$ is the shift of the longitudinal wavenumber from its value in vacuum. The coefficient C_s depends on the waveguide and mode geometry, along with the amplitude of wiggle oscillations. Equations 15-16 are an extension of the standard FEL equations from vacuum to the waveguide[9]. The notation is such that the vacuum equations result by setting $C_s = J_0(R) - J_1(R)$, where $R = a_w^2/2(1 + a_w^2)$ and J_n is the Bessel function of the first kind of order n. As an example, we now evaluate the coupling to the TM_{21} mode.

The electric field of the TM_{21} mode is given by

$$\begin{aligned} \vec{E} &= E_0 \cos(\frac{2\pi x}{a}) \cos(\frac{\pi y}{b}) \sin(kz - \omega t + \varphi)\hat{x} \\ &- R E_0 \sin(\frac{2\pi x}{a}) \cos(\frac{\pi y}{b}) \cos(kz - \omega t + \varphi)\hat{z} \,, \end{aligned} \tag{17}$$

where a(b) is the waveguide width (height) and $R \approx 2\pi a/4k_\perp b^2$. The $v_z E_z$ term gives a contribution, with y = 0 and $v_z \approx v_0 =$ constant, of

$$v_z E_z = -R E_0 v_0 \sin(\frac{2\pi x_w(z)}{a}) \cos(kz - \omega t + \varphi) \tag{18}$$

Expanding $\sin(2\pi x_w/a) \approx 2\pi x_w/a$ and using $x_w \approx (a_w/\gamma k_w)\sin(k_w z)$ we have

$$v_z E_z = -\frac{R E_0 v_0 \lambda_w a_w \sqrt{2}}{a\gamma} \cos(kz - \omega t + \varphi)\sin(k_w z) \,. \tag{19}$$

The resonant part of Eq. 13 is then

$$\gamma' = \frac{-\omega}{c}\frac{a_s a_w}{\gamma}(1 - \frac{R\lambda_w}{a}) \sin \psi \,. \tag{20}$$

MODE	TE_{01}	TE_{21}	TM_{21}
C_s	$K(R)$	$\frac{k_y}{k_\perp}J_0(\rho)K(R)$	$\frac{k_x}{k_\perp}J_0(\rho)K(R) - \frac{k_\perp}{k_z}J_1(\rho)(J_0(R)+J_1(R))$

Table 1: Mode Coupling Coefficients. Here $\rho = k_x x_w$, $R = a_w^2/2(1+a_w^2)$, $k_\perp^2 = k_x^2 + k_y^2 = \omega^2/c^2 - k_z^2$, $k_x = m\pi/a$, $k_y = n\pi/b$, $a(b)$ is the waveguide width (height), J_0 ,J_1 are the usual Bessel functions, and $K(R) = J_0(R) - J_1(R)$. The term $K(R)$ is present in an analysis in vacuo.

The modification due to $v_z E_z$, namely $R\lambda_w/a$, can be substantial. Table 1 contains the results of a more detailed calculation of the coupling factors for a few low order modes.

The FEL system is self-consistent; the current drives the electromagnetic mode. To find the evolution equations for the mode, we insert Eq. 11 into Maxwell's equation

$$\left(\frac{\partial^2}{\partial t^2} - \nabla^2\right)\vec{E} = -\frac{4\pi}{c}\dot{\vec{J}} . \tag{21}$$

In order to get a closed set of equations, we multiply by the form factor $\vec{q}_s(x,y)$ and integrate over perpendicular dimensions using the orthogonality relation

$$\int \vec{q}_s \cdot \vec{q}_{s'} dx dy = \frac{ab(1+c_0)\omega}{4k_s Z_s}\delta_{s,s'} \tag{22}$$

where $Z_s = k_s/\omega\ (\omega/k_s)$ for $TM\ (TE)$ modes, and $c_0 = 0$ unless $n = 0$ or $m = 0$, in which case $c_0 = 1$. We recall that there are no TM_{mn} modes where an index vanishes. This gives, for the TE_{01} mode,

$$\tilde{E}_s' k_s \cos\alpha_s - \tilde{E}_s k_s \varphi_s' \sin\alpha_s = \frac{4\pi}{cab}\int dx dy \vec{q}_{01} \cdot \dot{\vec{J}} . \tag{23}$$

The current in Eq. 23 is determined from the particles. Higher order modes are slightly more complicated. In general, one finds an equation like Eq. 23 and averages the current density to find

$$a_s' = \frac{\omega_{p\,s,eff}^2 a_w C_s Z_s}{2\omega c}\left\langle\frac{\sin\psi}{\gamma}\right\rangle \tag{24}$$

$$\varphi_s' = \frac{\omega_{p\,s,eff}^2 a_w C_s Z_s}{2\omega c a_s}\left\langle\frac{\cos\psi}{\gamma}\right\rangle \tag{25}$$

In the case of a beam much smaller than the guide, for the TE_{01} mode, we have

$$\omega_{p\,s,eff}^2 = \frac{8\pi e I}{mcab} . \tag{26}$$

In Eq. 24and 25 $\langle()\rangle = \frac{1}{N_p}\sum_{i=1}^{N_p}()_i$, where N_p is the number of electrons per wavelength of the ponderomotive wave. More generally the effective plasma frequency includes an overlap integral between the beam and the mode.

While the coupling strengths to the beam are mode dependent, the full system of equations for the FEL (Eqs. 15- 16, 24- 25) conserve the total (beam+mode) power. (Note that in free space the power per unit area is conserved). The "fill factor" which enters into the effective plasma frequency (or effective current density) is seen, from Eq. 26, to vary as the inverse of the waveguide (not electron beam) area when the electron beam is small compared to the guide.

Equations 15- 16 ($2N_p$ of them) and 24- 25 are one-dimensional single mode FEL equations in a waveguide. Until now we have not considered the effects of the power loss from coupling to the high gradient structure or the reacceleration by the induction units. At least two modifications must be made. First, since the waveguide dimensions are likely to be changing, the effective plasma frequency will be a (periodic) function of position. Secondly, the power loss corresponds to a resistive term on the RHS of the field amplitude equation (Eq. 24).

For example, we consider the waveguide with septa pictured in Fig. V, and assume the width, a, varies as $a(z) = a_{min} + z\, da/dz$ where da/dz is constant and z is the distance from the last jump. Then $\omega^2_{p\,s,eff}$ is given by Eq. 26 with $a = a(z)$, and an additional term $-(da/dz)a_s/(2a(z))$ must be added to the RHS of the field amplitude equation (Eq. 24). There is no change in δk_s since the phase velocity of the TE_{01} is determined only by the waveguide height b. The power is lost at each jump in $a(z)$, the guide width, because we ignore the delta function nature of da/dz at the jump. The induction units are represented, to the approximation that they give particles an instantaneous kick in energy, by a periodic delta function term in the γ' equation.

The synchronous particle model[10] is useful in designing the FEL. The electron beam is replaced with a synchronous particle interacting with the mode. The equations of motion are:

$$\gamma_r' = -\frac{\omega}{c}C_s a_s a_w \frac{\sin\psi_r}{\gamma_r} \tag{27}$$

$$\psi_r' = (k_w - \delta k_s) - \frac{\omega}{2c\gamma_r^2}(1 + a_w^2 - 2C_s a_w a_s \cos\psi_r) + \frac{d\varphi_s}{dz} \tag{28}$$

$$a_s' = \frac{\omega^2_{p,eff} a_w C_s Z_s}{2\omega c}\frac{\sin\psi_r}{\gamma_r} \tag{29}$$

$$\varphi_s' = \frac{\omega^2_{p,eff} a_w C_s Z_s}{2\omega c a_s}\frac{\cos\psi_r}{\gamma_r} \tag{30}$$

We define the resonant energy of a mode by setting $\psi' = 0$ in Eq. 28 :

$$\frac{k_w - \delta k_s}{\omega/c} = \frac{1 + a_w^2}{2\gamma_r^2}, \tag{31}$$

where we have ignored the (small) $d\varphi_s/dz$ term. From Eq. 31, we see that a mode has a resonant energy so long as $k_w > \delta k$. This is an important point, since it implies that many modes can be below cutoff (i.e. can propagate), and still not be resonant with an electron beam. That $k_w > \delta k_s$ is required for resonance can be seen from the phase velocity of the ponderomotive wave

$$v_{ph} = \frac{\omega}{(k_w + k_s)}, \tag{32}$$

or

$$v_{ph} = \frac{\omega}{(\frac{\omega}{c} + (k_w - \delta k_s))}. \tag{33}$$

From Eq. 33 we have $v_{ph} = c$ when $\delta k_s = k_w$.

From Eq. 27 we see that ψ must be positive for the synchronous particle to loose energy. Of course, as the energy (and perhaps δk_s through the waveguide) changes the value of ψ will change. The tapered FEL uses slowly varying a_w and k_w to maintain resonance. Equations 27 and 28 can be linearized in $(\gamma - \gamma_r)/\gamma_r$; the result is a pendulum equation for ψ. The synchrotron frequency deep in the potential well and the condition for a particle near in energy to the synchronous particle are thus found to be

$$L_s = \frac{\sqrt{1 + a_w^2}\lambda_w}{2\sqrt{C_s a_s a_w}} \tag{34}$$

and

$$\frac{\gamma - \gamma_r}{\gamma} < 2\frac{\lambda_w}{L_s}. \tag{35}$$

In Eqs. 34- 35 we have not included the corrections for $\psi_r \neq 0$. For the TBA, $\psi_r << 1$, and these corrections are unimportant.

For the TBA design, the electrons undergo many synchrotron oscillations between induction units and the energy extracted per section is small compared to a bucket height.

STEADY STATE FEL DESIGN

We now discuss specific parameters for a steady-state FEL to drive the high gradient structure. The performance figure for the FEL is the power produced per meter, which depends on the desired accelerating gradient. New experimental results at SLAC have found that the breakdown field at short pulse lengths, with $\lambda = 10cm$, is greater than $100MeV/m$. We therefore believe that gradients of $500Mev/m$ are achievable at $1cm$. Production of the higher gradient requires more stored energy ($W \approx E_a^2$). With the gradient $500MeV/m$, the FEL must produce 4 times the power cited previously[6], or $2.2GW/m$. While this is considerably larger than present experimental results, we believe it can be achieved. Finally, since experimental interest seems to be in the $1TeV \times 1TeV$ range we use that for our design goal.

Previous discussions of the TBA[3,6] have considered low energy beams of $3-5MeV$. We believe now that an energy of $\approx 20MeV$ is advantageous. Our reasons are two-fold. First, transverse betatron oscillation amplitudes will be decreased.This reduces the beam size and allows smaller waveguides. This has the advantage of permitting larger "fill factors" (for the TE_{01} mode) and larger mode separation. Secondly (and more importantly), at fixed power production and current, higher energy corresponds to smaller fractional energy change between induction units. Thus $2GW/m$ at $2kA$ implies energy loss of $\Delta\gamma = 2$. For a $5MeV$ beam this is a 20% loss per meter, while for a $20MeV$ beam it is a 5% loss. With smaller fractional energy extraction between sections, particle detrapping can be minimized.

Having chosen the electron energy ($\gamma = 40$), we can estimate the beam size and find minimal waveguide dimensions. The beam will be about $1cm$ diameter, so we pick a $5cm \times 2cm$ (width $\times$ height) waveguide. The additional width is to allow for wiggling and power extraction. The wiggler parameters which need to be chosen are the wavelength and field strength. The resonance condition Eq. 31 provides one relation between them. The second relation relates the slippage of the LEB relative to the HEB and the high gradient structure fill time (T_f). The fractional slippage is $(c - v_z)/c = (1 + a_w^2)/2\gamma^2$, where γ is the energy of the LEB. For efficient accelerator design, we need the LEB pulse length equal to cT_f. Therefore, in the length of the TBA, the slippage cannot exceed the LEB length cT_f. With $cT_f = 6m$, and given a length of the TBA, we can fix λ_w and a_w. It is reasonable to assume that the $4km$ TBA will consist of 4 LEB injectors, and thus allow a slippage of $6m$ in $1km$. This, with $\gamma = 40$, yields $a_w = 4.3$; Eq. 31, with $\delta k \approx \pi\lambda/4b^2$ and $b = 2cm$, yields $\lambda_w = 27cm$. For these parameters the peak wiggler field is $2.5kG$. In order to extract 5% of the beam power per meter, we design with a resonant phase of $\psi_r = 0.15$ and an average microwave power of $5GW$. From the power requirements of the high gradient structure, $2.2GW/m$, we find the current to be $2.1kA$.

Using a one dimensional numerical simulation Sternbach and Sessler[11] have begun a study of steady state FEL's. For parameter values near to those of interest in this work, the simulation shows particle detrapping at the rate of 1% per $100m$ after an initial loss of 25% of the beam during the start-up of the steady state FEL. In Fig. II (a) the phase space (γ, ψ) is plotted at the start of the next to last section of TBA, and in Fig. II (b) the phase space is shown at the end of that section (note the lower mean energy). The phase space at the start of the last section is shown in Fig. III; the beam energy was jumped by an induction unit between Fig. II (b) and Fig. III. Figure IV shows the percentage of electrons trapped in the ponderomotive potential, as a function of z.

COUPLING BETWEEN THE BEAMS

The coupling of power between the overmoded waveguide and the slow wave structure has been investigated in three schemes[6]. One of the schemes has been realized in sufficient detail and simplicity for testing on the LLNL-LBL free electron laser (ELF) FEL (See the section on future plans below).

The original configuration[3] employed "scoops" which would be periodically inserted into the overmoded guide to extract the power. Keunning[12] has implemented this idea by introducing septa into the waveguide. This is shown in Fig.V. The septa taper down to the size of the fundamental guide, and the power is fed into the high gradient structure. These septa extract power from alternate sides of the LEB waveguide and essentially form a series of slowly varying walls separated by jumps. The septa are almost perpendicular to the waveguide walls, which will minimize mode conversion. This configuration, in which the HEB and LEB are in separate waveguides, is called a Dual Structure TBA.

An alternative idea is to use the longitudinal component of the TM modes to accelerate the HEB while the transverse component interacts with the LEB via the FEL mechanism. Thus one mode would be utilized for both FEL interaction and acceleration, and this scheme is a Single Structure TBA. This places some requirements on the mode:

1. The HEB must be located around a null of the transverse field to minimize deflections.

2. The LEB and HEB must be well separated.

3. The guide must be overmoded to allow transport of the LEB and slow wave to allow acceleration of the HEB.

The TM_{21} mode in a loaded guide satisfies these requirements. The Single Structure configuration suffers a serious drawback. In order to achieve significant accelerating gradients a very large power flow must be established in the guide. For example[6], if the mode has a group velocity near c, $10^{11}W$ are required to produce an accelerating field of $250MeV/m$.

A third possibility is to build the overmoded guide and high gradient structure adjacent to each other. The coupling between them would then be accomplished by a series of holes that act as directional couplers. The differing phase velocities in the guides determine the spacing between the

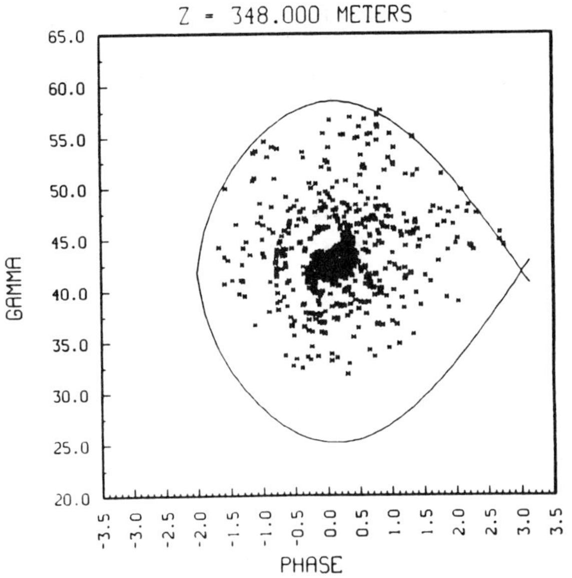

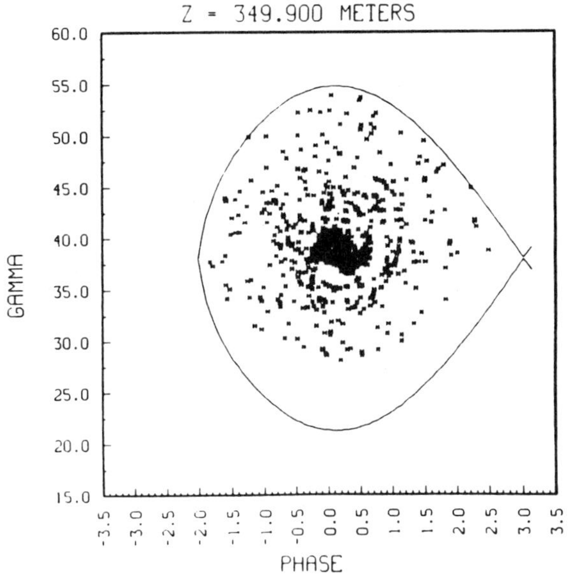

Figure II: (a) The particle phase space (γ, ψ) at the beginning of the next to last section of the $300m$ steady-state FEL, (b) at the end of the same section. This figure and Fig. III-IV are from Sternbach and Sessler[11].

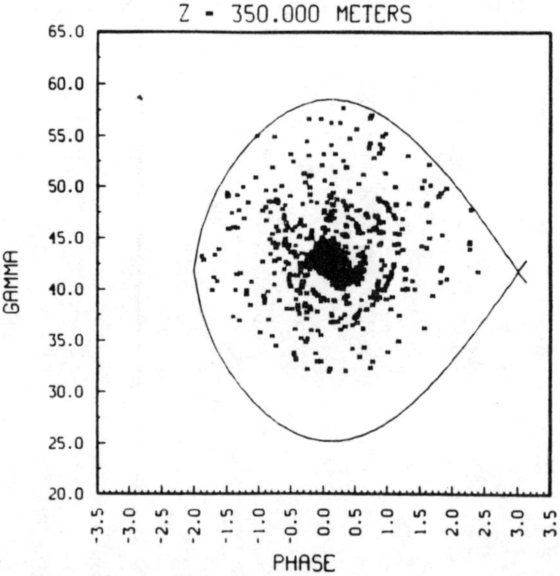

Figure III: The phase space at the start of the last section. Note that the induction unit has raised the energy between Fig. III (b) and Fig. IV.

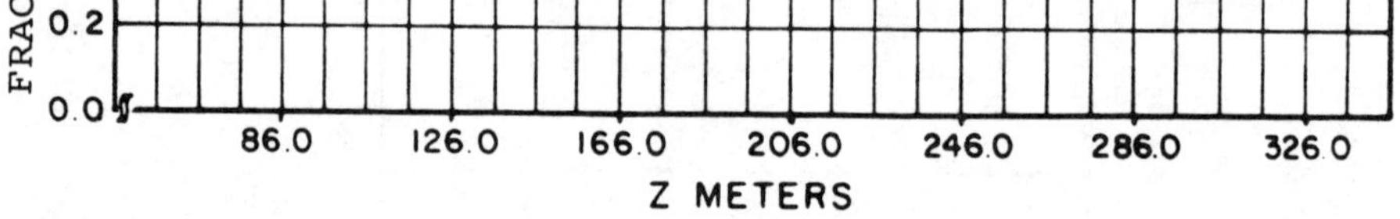

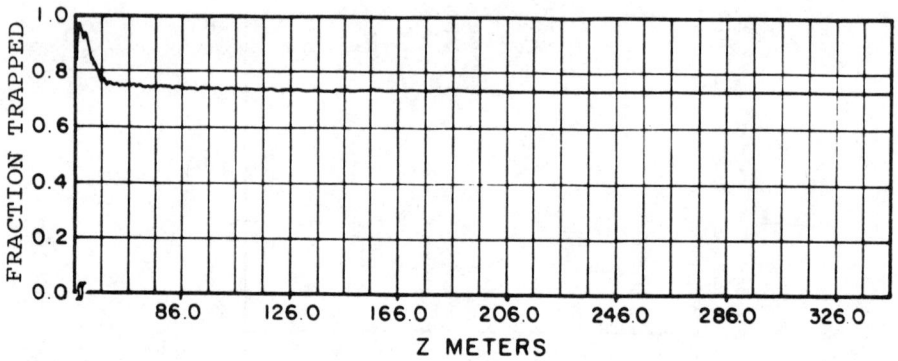

Figure IV: Fraction of trapped particles in the steady state FEL. After capturing 75% of the beam in the start-up section (the first $50m$), the losses were less than 3% in the $300m$ TBA. Inductions units were located every $2m$ and $\Delta\gamma$ per section was nearly 4.

couplers; the percentage of power to be transmitted from the overmoded guide to the high gradient structure determines the size of the holes. For the purpose of efficient coupling, the LEB beam should amplify the higher mode with a longitudinal field component (such as the TM_{21} mode described above). In contrast to the single structure, the requirements on stored energy are not enormous and the high energy beam does not wiggle. More details on this configuration, the Composite Structure, can be found elsewhere[6]. The disadvantages of this scheme are that it utilizes a higher order mode and has no obvious way to tune the phase shift between the FEL and the high gradient structure. The higher order mode must be isolated from other nearby modes, which can be accomplished by employing elliptical waveguides.

FUTURE PLANS FOR THE TBA

The only sites of TBA research at present are Lawrence Berkeley Laboratory and Lawrence Livermore National Laboratory. The project has both short and long term objectives[13], and we begin by describing the former.

The short term objectives utilize ELF, a millimeter wave high gain FEL amplifier[14] (see Figure VI). The ELF FEL uses the Experimental Test Accelerator at Livermore which provides an electron beam of $10kA$ at an energy of $3.2MeV$. The beam is propagated through a long narrow tube immersed in a solenoidal magnetic field (an "emittance selector") which allows passage of only a small emittance. This limits the usable current to $\approx 1 - 1.5kA$. The beam then moves through a quadrupole transport channel and is matched into the planar wiggler (with continuous quadrupole focusing in the wiggle plane). The FEL consists of three meters of a pulsed electromagnetic wiggler with each two periods independently controlled. The beam pipe in this region is an overmoded rectangular waveguide (presently $3cm \times 10cm$). In amplifier operation a $34.6GHz$ signal is injected from a magnetron source and coupled into the guide by a tapered guide and a reflecting foil located at the entrance to the interaction region. Efficiencies of up to 5% have been achieved without tapering of the wiggler.

The near term goals of the TBA group are to investigate disc loaded waveguides at 17 and $34GHz$. A seven cell disc loaded waveguide, corresponding to the SLAC structure reduced by a factor of 11, has already been fabricated. The power produced by the FEL will be coupled into the structure, and breakdown levels observed. Gradients of several hundred MeV/m are expected. Low power bench tests will be performed as well.

Subsequent to these studies, the smooth ELF waveguide will be replaced by a septum channel like that in Figure V. The channel will be made as narrow as possible, consistent with the electron beam dimensions, in order to

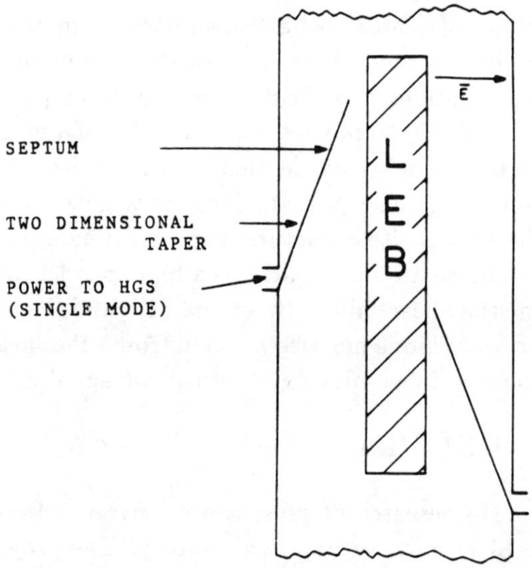

Figure V: A waveguide with power extraction. The LEB couples to the TE_{01} mode and the power is extracted by the septa. In the TBA the single mode waveguide feeds the high gradient structure. In initial studies the power extraction port will be used for diagnostics.

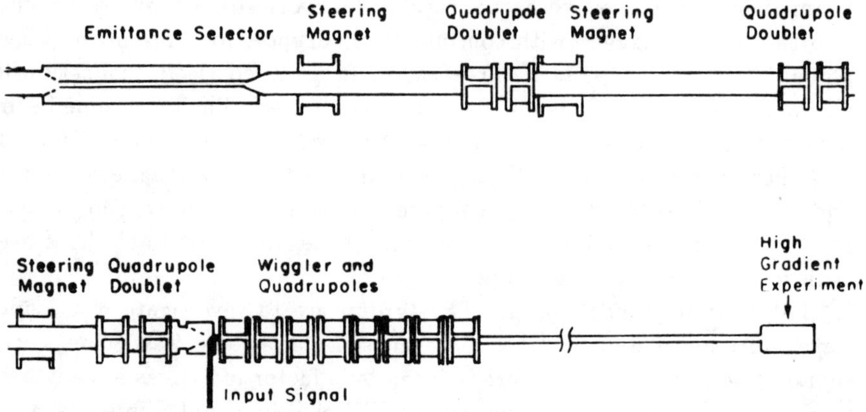

Figure VI: The ELF beamline. The emittance selector reduces the the ETA current to about 1 $1.5kA$, and the quadrupole channel transports the beam into the wiggler. Power produced in the $3m$ of wiggler will be used for breakdown studies of high gradient structures.

maximize the coupling (i.e., increase $\omega^2_{p_s,eff}$). The gain, mode composition, and phase stability of the FEL in the channel will be investigated. The final plan for the TBA work at ELF includes insertion of small accelerating modules between each meter of wiggler (there is room for up to $6m$ of wiggler) and operating short periodic FEL. In parallel, some meters of high gradient structure installed along side the FEL will permit high power testing of the amplitude and phase stability of the FEL with both the septum channel and periodic reacceleration.

A longer term effort consists of building a TBA prototype, of approximately $30m$ length. This involves much numerical and theoretical work, as discussed in the next paragraph. An injector for the FEL must be built, presumably incorporating the recent magnetic switching technology, and optimized for TBA applications. The injector would be used to provide the LEB for the steady state FEL. The facility should be located near a good injector for the high gradient structure. The $1GeV$ low emittance beam leaving the SLAC cooling ring is a possible candidate. With gradients of $500MeV/m$ the $30m$ prototype would be a $15GeV$ machine.

We mention some theoretical and numerical challenges, beginning with those related to the FEL. The steady state FEL requires particles to be trapped through many sections of tapered wiggler and induction unit. In particular, particle motion must not be stochastic and reliable stochasticity criteria must be developed. Criteria are needed not only for stable operation in a single mode model, but also to establish tolerances on the amplitudes of other waves that may be produced. The effect of the "fast" ponderomotive wave ($\psi_{fast} = \psi - 2k_w z$), other waveguide modes, sidebands and space charge needs to be examined. Numerical modelling, in both one and two dimensions, should progress to the point where a specific realization of the TBA, such as that in Fig. V, is analyzed. This requires theoretical work on coupling schemes. For example, the mode structures of waveguides with power extraction needs to be calculated, along with wall losses, mode conversion losses, and reflections. (Reflections are important because the LEB will not be transit time isolated from the coupling to the high gradient structure.)

High gradient structures need to be optimized at short wavelengths and the optimal frequency for the TBA (taking into account both the LEB and HEB) chosen. Additionally, there is much engineering work to be done on establishing tolerances and fabrication of structures.

The acceleration of the high energy beam to a fixed final energy imposes a shot-to-shot tolerance on the variation of the electromagnetic phase,φ_s, and amplitude, a_s. The phase and amplitude are driven by the FEL (see Eqs. 24-25). The sensitivity of the amplitude and phase to errors in (LEB) current, energy, wiggler field amplitude, and electromagnetic field amplitude are examined elsewhere in these proceedings[15]. It is shown that a $5km$ TBA requires a control system to limit the accumulation of amplitude and phase error to acceptable levels. Three systems which can accomplish this are proposed.

PARAMETERS FOR A TBA-COLLIDER

We present, in Table II, the parameters for a $1TeV \times 1TeV$ collider with luminosity $\mathcal{L} = 4 \times 10^{32} cm^{-2} sec^{-1}$. Our analysis follows Wilson[16], and the reader should consult that reference for details. We have increased the gradient and the final beam energy from previous TBA designs. The average power output of the steady state FEL is $f\,40J/m \times 4000m$, where f is the repetition rate. We set $f = 500kHz$ so that the steady-state FEL produces $80MW$. Assuming an induction unit with a 50% efficiency between the mains and the beam, the average power consumption is about $160MW$. With $N = 10^{11}$ particles per bunch, the (single) HEB power is $8MW$ and the overall efficiency $\approx 10\%$.

In order to achieve the design luminosity of $\mathcal{L} = fN^2H_D/4\pi\sigma_x\sigma_y = 4 \times 10^{32} cm^{-2} sec^{-1}$ we need to have, with a luminosity enhancement $H_D = 2$, transverse dimensions with $\sigma_x\sigma_y \approx 0.2\mu^2$. With a circular beam this would lead to an unacceptably high beamstrahlung parameter δ. We therefore consider a flat beam with $R = \sigma_x/\sigma_y = 10$ and $\sigma_y = 0.14\mu$. Setting the longitudinal bunch length $\sigma_z = 1mm$ results in a disruption parameter $D = 1.3$, and $\delta = 0.2$. The luminosity enhancement for a flat beam with these parameters is $H_D \approx 2$. The estimate for the beamstrahlung may be pessimistic; it can be reduced by a factor of 2 by using a beam with the same cross sectional area and an aspect ratio $R = 20$.

Acknowldgments

The author is pleased to thank A. Sessler for encouragement and advice. The author also thanks D. Hopkins, R. Keunning, R. Palmer, F. Selph, E. Sternbach and P. Wilson for useful conversations and other communications.

REFERENCES

1. B. Richter, in Proc. of the 11'th Int'l Conference on High Energy Accelerators, Birkhauser Verlag,Basel, (1980), p168.
2. P. B. Wilson, IEEE Trans. on Nuclear Science, NS-28,2742 (1981).
3. A.M. Sessler, "The Free Electron Laser as a Power Source for a High Gradient Structure", in Laser Acceleration of Particles, P.J. Channell ed., AIP Proceedings No. 91, New York 1982, p163-189.
4. D. Prosnitz, IEEE Trans. on Nuclear Science, NS-30, 2754 (1983).

5. B. Richter in Laser Acceleration of Particles II, AIP Conf. Proc., New York 1985.
6. D.B. Hopkins, A. M. Sessler, and J. S. Wurtele, Nucl. Instr. and Meth. in Phys. Res. 228,p15 (1984)
7. F. B. Selph, "The Two Beam Accelerator", LBL-18403,and to be published in Proceedings of the Third International Summer School on High Energy Particle Accelerators, (1984).
8. E. T. Scharlemann, in preparation.
9. J. S. Wurtele, PhD Thesis, Univ. of Calif., Berkeley (1985)
10. N. M. Kroll, P. L. Morton, M. W. Rosenbluth, IEEE Journal of Quantum Electronics,QE-17,1436 (1981).
11. E. Sternbach and A. M. Sessler, personal communication.
12. R. W. Keunning, unpublished notes.
13. D. B. Hopkins and F. B. Selph, personal communication.
14. T. J. Orzechowski et. al., Microwave Radiation from a High Gain Free Electron Laser Amplifier, Phys. Rev. Lett. 54, pp889-892 (1985).

15. R. W. Keunning, A. M. Sessler, and J. S. Wurtele in Laser Acceleration of Particles II, AIP Conf. Proc., New York 1985
16. P. B. Wilson in Physics of High Energy Particle Accelerators, R. A. Carrigan, F. R. Huson, and M. Month Eds., AIP Conf. Proc. No. 87, New York (1981) p450.

Low Energy Beam	
Average Beam Energy (Units of mc^2)	40
Beam Current	$2.15kA$
Bunch Length	$6m$
Wiggler Wavelength	$27cm$
Average Peak Wiggler Field	$2.4kG$
Beam Power	$43GW$
Beam Energy	$0.8kJ$
Power Production	$2.2GW/m$
Number of FEL Injectors	2×2
Power from mains	$160MW$
High Gradient Structure	
Wavelength	$1cm$
Gradient	$500Mev/m$
Stored Energy	$40J/m$
Fill Time	$18ns$
High Energy Beam	
Injection Energy	$2GeV$
Repetition Rate(f)	$0.5kHz$
Final Energy	$1TeV$
Length	$2 \times 2km$
Luminosity	$4 \times 10^{32} cm^{-2} sec^{-1}$
Beam height (σ_y)	0.14μ
Beam width (σ_x)	1.4μ
Single Beam Power	$8.0MW$
Number of Particles	10^{11}
Disruption (D)	1.3
Beamstrahlung (δ)	0.2
Overall efficiency (from mains to HEB)	10%

Table 2: Parameters for a $1TeV \times 1TeV$ Two-Beam Accelerator Collider

PHASE AND AMPLITUDE CONSIDERATIONS FOR THE TWO-BEAM ACCELERATOR*

R. W. Kuenning and A. M. Sessler
Lawrence Berkeley Laboratory, University of California
Berkeley, CA 94720

J. S. Wurtele+
Plasma Fusion Center, Massachusetts Institute of Technology,
Cambridge, MA 02139

ABSTRACT

Phase and amplitude considerations are made for a Two-Beam Accelerator and analytic formulas are obtained expressing the phase and amplitude errors in terms of magnetic wiggler errors, beam energy errors, beam current errors, and microwave field amplitude errors. The necessity of phase and amplitude control is shown and schemes are proposed which can accomplish this control.

I. THE TWO-BEAM ACCELERATOR

The Two-Beam Accelerator (TBA) was first proposed some years ago.[1] Further descriptions of this device have already been given[2,3] and a rather comprehensive description can be found in this very volume.[4]

We have, for the considerations of this paper, taken the parameters given in Ref. 4. Note, that these are somewhat revised over that given in the earlier papers. The major differences are the following. Firstly, we have gone to a top energy of 1 TeV, rather than 300 GeV, because physics interest has moved to the higher energy and, consistent with this increase in energy, we have increased the luminosity to 4×10^{32} cm^{-2} sec^{-1}. We have, in addition, adopted a gradient of 500 MeV/m, rather than 250 MeV/m, because recent theoretical analysis and experiments suggest that this larger value can be achieved.

As a consequence of these changes, and taking a final focus beam size of 0.1 μm, we have the parameters listed in Table I. Note that we have kept the radiation wavelength at 1 cm. We considered raising this to 2 cm, so as to ease the manufacturing problems associated with making a small structure, and believing that we could obtain the high gradient of 500 MeV/m even at this lower frequency, but the increased power demand on the FEL seemed excessive to us: The required power went from 2.2 GW/m to almost 4(2.2)GW/m.

*This work was supported by the Division of High Energy Physics, Office of Energy Research, U. S. Department of Energy under Contract Contract No. DE-AC03-76SF00098.

+Supported by the Office of Naval Research

Table I Parameters for a Two-Beam Accelerator

Low-Energy Beam	Energy/Rest Energy (γ)	40
	Beam Current (I)	2.2 kA
	Bunch Length (ℓ)	6 m
	Number of FEL Sections	2 x 2
	Power Requirement	2.2 GW/m
Wiggler	Wavelength (λ_w)	27 cm
	Average Wiggler Peak Field (B_w)	2.4 kG
High-Energy Beam	Repetition Rate (f)	500 Hz
	Final Energy (E_f)	1 TeV
	Gradient	500 MeV/m
	Length of Accelerator ($2xL_A$)	2 x 2 km
	Luminosity ($\mathscr{L}$)	$4x10^{32} cm^{-2} sec^{-1}$
	Single Beam Power (P)	8.0 MW

II. PHASE AND AMPLITUDE ERRORS

We start with the FEL equations:

$$\frac{d\gamma_i}{dz} = -\frac{\omega}{c} a_w a_s \frac{\sin\psi_i}{\gamma_i} , \qquad (1)$$

$$\frac{d\psi_i}{dz} = k_w - \frac{\omega}{2c\gamma_i^2}(1 + a_w^2 - 2a_w a_s \cos\psi_i) + \frac{d\phi}{dz} , \qquad (2)$$

$$\frac{da_s}{dz} = \frac{\omega_p^2 a_w}{2\omega c}\left\langle\frac{\sin\psi_i}{\gamma_i}\right\rangle , \qquad (3)$$

$$\frac{d\phi}{dz} = \frac{\omega_p^2 a_w}{2\omega c a_s}\left\langle\frac{\cos\psi_i}{\gamma_i}\right\rangle , \qquad (4)$$

where we have used standard notation.[5] For a TBA, in the simplest model, we model the beam by one macro particle and modify these equations by adding to Eq. (1) the term

$$+ 2\alpha\frac{\omega^2}{\omega_p^2} a_s^2 , \qquad (5)$$

and to Eq. (3) the term

$$- \alpha a_s . \qquad (6)$$

In this model, α represents the continuous energy taken from the low energy beam to the high energy beam, while the induction units are modeled with a continuous source which puts this very same energy back into the low energy beam. The discrete nature of the energy extraction and the induction units are, of course, not included in this model.

From Eqs. (3) and (4) we can compute the error in the amplitude and phase of the signal wave:

$$\left(\frac{\Delta a_s}{a_s}\right) = k_1 L \left(\frac{\Delta a_w}{a_w} + \frac{\Delta \omega_p^2}{\omega_p^2} - \frac{\Delta \gamma}{\gamma}\right)(\sin \psi) \quad , \tag{7}$$

$$\Delta \phi = k_1 L \left(\frac{\Delta a_w}{a_w} + \frac{\Delta \omega_p^2}{\omega_p^2} - \frac{\Delta \gamma}{\gamma} - \frac{\Delta a_s}{a_s}\right)(\cos \psi) \quad , \tag{8}$$

where

$$k_1 = \frac{\omega_p^2 a_w}{2\omega c a_s \gamma} \quad . \tag{9}$$

In these equations [Eqs. (7), (8), (9)], all of the quantities such as a_w, a_s, ψ, γ, ω_p^2, ω are evaluated for the macro particle (equilibrium particle); the quantity L is the length one is considering. The fractional deviations in a_w, ω_p^2, γ, and a_s are explicitly indicated.

Numerical evaluation of the phase and amplitude deviations which one can expect in a TBA can now be done using the parameters of Section I. One has $a_s = 0.19$, $a_w = 4.3$, $\omega = 1.9\text{x}10^{11}\text{sec}^{-1}$, $\omega_p = 1.7\text{x}10^{10}\text{sec}^{-1}$ and hence $k_1 = 1.4$ rad/m. Taking $\psi = 0.15$ and L = 100 meters we see that a 0.1% relative error in any of the quantities leads to $|\Delta\phi| \approx 0.14$ radians and $(|\Delta a_s|/a_s) \approx 2.1\%$. Thus, without some sort of control on the phase and amplitude of the signal wave we cannot have an L of 2 km.

In these estimates of the effect of errors, Eqs. (7), (8), (9), we have not considered the differential coupling between the variables a_s, ϕ, ψ, and γ as described by Eqs. (1) - (4). Of course any deviation will "propagate" through these variables, and a proper treatment of errors must involve solution of the coupled differential equations. We leave such study to the future, believing that our first estimates are adequate for this note.

III. FEEDBACK CONTROL

Proper operation of a TBA will require a master oscillator (a "clock") to which phase and amplitude is compared. This signal wave is sent down the accelerator in a third waveguide.

One possibility for control of phase and amplitude is simply not to control them, but put great effort on reducing the errors Δa_w, $\Delta\omega_p^2$, $\Delta\gamma$, and Δa_s. The Eqs. (7) and (8) can be employed to deduce the length L, once one knows the acceptable values of $\Delta\phi$ and $\Delta a_s/a_s$. The last are set by the acceptable variation in the energy of the high energy beam and, typically, are a few percent. (Since beamstrahlung will introduce an energy spread of this magnitude.) Probably, and this depends on how successful one is in practice in controlling Δa_w etc., L is of the order of 100 meters. Thus the TBA has become a multi-beam accelerator with the low energy beam going through an FEL which then powers (about) 100 meters of the high gradient structure. This is a significant modification of the TBA idea, but may be a quite acceptable concept.

A second possibility (suggested by Donald Prosnitz) is to remove all of the signal wave after a distance L (where the errors in a_s and ϕ have grown to a large value), but not to remove the low energy electron beam. Then one starts the FEL again, with the proper phase as given by the clock. The electromagnetic wave can be removed, while not removing the electron beam, by means of a thin reflecting foil. In this approach one has 2 x 2 low energy beam FELs as contrasted with the first possibility where one has 2 x 20 FEL power sources.

A third possiblity is the use of "feed back" (in this case "feed forward") to control phase and amplitude. The energy of the low energy beam is a quantity that can be readily controlled in order to dynamically correct phase errors. This could be done by small added induction accelerator units, driven by hard tubes. The hard tube driver chain could be similar to a pulser designed for the ASTRON accelerator cathode to give a 20 kV, 1000 A pulse, with a nominal 5 ns rise time.[6]

Closed loop regulation during the pulse would require gain-bandwidths larger than the state-of-the-art permits. Therefore, open-loop correction is required. Since the rf energy travels, according to waveguide propagation theory, at 0.985c and the low energy beam travels at about 0.95c, the correction of LEB energy cannot affect the portion of the rf energy on which the phase was measured. Furthermore, phase error is a cumulative effect, occurring over axial distance. It is not feasible to measure phase at one location, and apply the correction many meters downstream where an electrical signal could catch up with the same portion of the rf on which the measurements were made, since more phase errors have accumulated during the transit. Thus the correction will always be late, by the delay time in the amplifier system plus connecting cables.

We propose a feed-forward system. Obviously, the phase error accumulation during the amplifier and cable delay time must be less than the allowable error, which implies that if we have correction units every 100 meters, the error change in 10 ns must be less than 0.1% for $\Delta\phi = 0.14$ rad and $\Delta a_s/a_s = 2.1\%$. The LEB captured current and the voltage of the induction accelerator modules must not vary at a faster rate than 1.5% over the 150 ns pulse. This is reasonable to achieve but will require some extra effort in flattening the pulses. Phase measurement, within a few nanoseconds, is a subject that requires further study, and which we leave for the future.

REFERENCES

1. A. M. Sessler, "The Free Electron Laser as a Power Source for a High Gradient Structure," in Laser Acceleration of Particles, P. J. Channell, ed., AIP Conf. Proc. No. 91, New York, p. 163 (1982).

2. D. Prosnitz, IEEE Trans. on Nuclear Science NS-30, 2754 (1983).

3. D. B. Hopkins, A. M. Sessler and J. S. Wurtele, "The Two-Beam Accelerator," Nuclear Instr. & Methods in Physics Research, (to be published, 1985).

4. J. S. Wurtele, "Progress on Acceleration by the Transfer of Energy Between Two Beams" in Laser Acceleration of Particles II, AIP Conf. Proc., New York (1985).

5. N. M. Kroll, P. L. Morton and M. W. Rosenbluth, IEEE Journal of Quantum Electronics, QE-17, 1436 (1981).

6. R. W. Kuenning and S. D. Winter, "Pulser for Accelerator Cathode," Lawrence Radiation Laboratory Report UCID-15156, April 25, 1967 (unpublished).

A GAS-LOADED TRANSVERSE-FIELD ACCELERATOR[1]

M. A. Piestrup
Adelphi Technology, 13800 Skyline Blvd., Woodside, California 94062

J. A. Edighoffer
TRW, 1 Space park, Redondo Beach, California 90278

ABSTRACT

The introduction of a gas into the transverse-field accelerator allows the use of realizable helix magnetic fields while still maintaining good acceleration gradients for ultrarelativistic electron beams. Synchrotron radiation losses can then be kept to acceptable values. To account for elastic scattering in the gas, a Monte-Carlo simulation has been done, and it shows that the introduction of the gas does not effect the interaction appreciably.

INTRODUCTION

The transverse-field accelerator, or inverse free electron laser (IFEL), has several outstanding characteristics that make it an excellent candidate for laser-driven particle acceleration.[1] The laser beam can be coaxial with the electron beam and the acceleration gradient can be larger than conventional klystron-driven linear accelerators. Its feasibility has already been demonstrated.

Unfortunately, the IFEL is limited to moderate electron energies ($E < 20$ GeV) by the amount of synchrotron losses generated by the helical trajectories of the electrons.[2] For single particles, the losses from synchrotron emission scale as the square of both the electron's energy and the magnetic field of the undulator. Thus for extremely relativistic electrons, the losses can be large.

To achieve high-energy gradient acceleration, the pitch angle (the angle between the electron's direction and the optical wave) must be large. This in turn requires that the magnetic field of the helix be large in order to maintain phase synchronism. This results in large synchrotron losses.

The introduction of a low-scattering gas into the IFEL causes a change in the wave-particle velocity synchronism condition, resulting in a lowering of the needed magnetic field to achieve synchronism and hence a reduction of the synchrotron losses.

----

[1] This work was supported by the U. S. Department of Energy under the Small Business Innovative Research Program, contract No. ER80201.

The theory of the gas-loaded IFEL is the same as the conventional vacuum IFEL with the primary exceptions being that the phase matching condition has been changed, and multiple scattering has been introduced. We will first solve for the phase-matching conditions for the vacuum and gas-loaded IFEL's and then solve for the acceleration gradient. The effects of multiple scattering will be studied using both a computer simulation and the Rutherford scattering theory.

Some of the work given here has been presented elsewhere.[3]

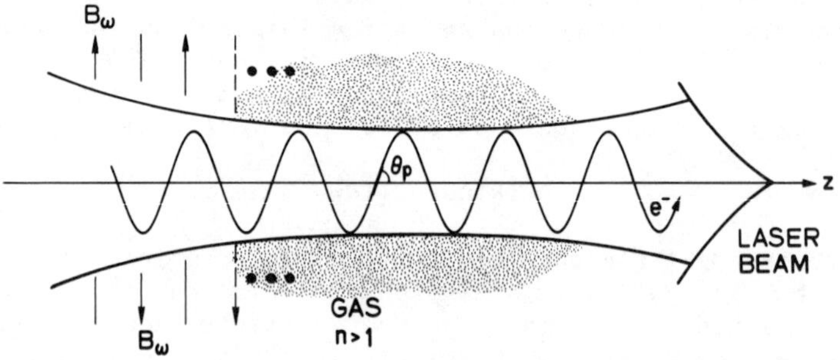

Figure 1. The Gas-loaded Transverse-Field Accelerator. The configuration is identical to the conventional IFEL with the exception that there is a gas in the interaction region.

PHASE MATCHING CONDITIONS

The condition for the greatest amount of work to be done by the optical wave on the electron requires a constant phase relationship between the optical wave and the electron's transverse motion. For this to occur in an IFEL, the distance traveled by the optical wave should differ by an optical wavelength from the distance traveled by the electron as it traverses one undulator period λ_w. If a gas is introduced into the undulator, the optical wave is slowed to c/n. The ratio of path lengths of the optical wave to that of the electron's is given by[3-6]

$$\frac{(c/n)t}{\beta ct} = \frac{\lambda_w \pm \lambda/n}{\lambda_w/\cos\theta_p} \tag{1}$$

where θ_p is the pitch angle of the electron helical motion in the undulator; n, the index of refraction; c, the speed of light; and t, the time. Solving for λ_w:

$$\lambda_w = \frac{\lambda\beta\cos\theta_p}{(1 \pm n\beta\cos\theta_p)} \tag{2}$$

For $\theta_p << 1$, $\beta \simeq 1 - 1/2\gamma^2$, and $n = 1 + \alpha$, then:

$$\lambda_w = \frac{2\gamma^2\lambda}{1 - \gamma^2(2\alpha - \beta^2\theta_p{}^2)} \simeq \frac{2\lambda}{\theta_p{}^2 - \theta_c{}^2} \tag{3}$$

where γ is the ratio of the total particle energy to its rest energy, $\beta = v/c$, $\theta_c = \cos^{-1}(1/\beta n) \simeq \sqrt{2\alpha}$ is the Cherenkov condition and v is the particle velocity. This is the phase-matching or resonance condition for the gas-loaded IFEL.

Setting $n = 1$ (α=0) and assuming $\gamma >> 1$, we then obtain the phase-matching condition for the vacuum IFEL:

$$\lambda_w = \frac{2\gamma^2\lambda}{(1 + \beta^2\gamma^2\theta_p{}^2)} \simeq \frac{2\lambda}{\theta_p{}^2} \tag{4}$$

where the last approximation assumed $a(\omega) = \beta\gamma\theta_p >> 1$. $a(\omega)$ is the normalized vector potential.

ACCELERATION GRADIENT

The amount of power dW/dt transferred to an electron from an optical wave in synchronism with the electron is given by the dot product between the electric field and the electron velocity. The electron gains energy from the tangential electric field:

$$\frac{dW}{dt} = e\bar{E}\cdot\bar{v}_o - \frac{dW_{loss}}{dt} \tag{5}$$

where e is the particle charge, $\bar{E}$ is the electric field vector, and $(dW/dt)_{loss}$ is included to account for losses due primarily to synchrotron emission. The angle between the direction of the optical field and the instantaneous velocity vector of the electron is called the pitch angle, θ_p, and is given by[7]

$$\theta_p = \frac{eB\lambda_w}{2\pi\gamma m v_o} \tag{6}$$

where B is the magnetic field, λ_w is the period of the helix, and m is the mass of the electron.

Setting v = dz/dt, Eqn.(5) becomes:

$$\frac{dW}{dz} = E\theta_p - \frac{dW_{loss}}{dz} \tag{7}$$

where E and dW/dz are now in eV/m.

To first order, the electron loses energy $(dW/dz)_{loss}$ from a variety of processes. These include synchrotron, Cherenkov and ionization losses. The dominant loss term for high energy electrons comes from synchrotron emission. This reradiation loss is calculated from Jackson to be[8]

$$\frac{dW_{loss}}{dz} = \frac{2er_eB^2\gamma^2}{3mc^2} \simeq 3.25\times10^{-6}\gamma^2B^2 \tag{8}$$

where B is in kG, r_e is Bohr electron radius, and $(dW/dz)_{loss}$ is in eV/m.

For large acceleration gradients, the electric field and the pitch angle must be kept as large as possible. The electric field is limited by breakdown, while the pitch angle is limited by the amount of synchrotron loss that can be tolerated. From Eqn.(6), the pitch angle is seen to vary linearly with the magnetic field B of the helix. However, from eqn(8) we see that a large B field results in a large synchrotron loss.

From Eqn.(6) it appears that the needed magnetic field can be reduced if the helix period λ_w is allowed to increase. For the vacuum IFEL, the pitch angle is constrained by its phase matching condition Eqn.(4). On the other hand, the introduction of a gas allows a third variable (namely the index of refraction of the gas) to be introduced into the phase matching condition. For the case where θ_p is close in value to θ_c, the phase matching condition Eqn.(3) allows the helix period to be large. Thus for equal pitch angles, the gas-loaded IFEL can have substantially smaller synchrotron losses.

To obtain the total amount of acceleration, we integrate Eqn.(7) over the interaction length and assume that the electric field at the waist of a gaussian laser beam of power P is given by[9]

$$E = \left[\frac{4PZ_o}{\lambda L}\right]^{1/2} \qquad (9)$$

We also assume that λ_w is kept constant while the magnetic field is varied. Neglecting the loss term, this gives the final electron energy, E_f, to be

$$E_f - E_o = \left[\frac{4PZ_oL}{\lambda}\right]^{1/2} \theta_p \mathrm{Sin}\phi \qquad (10)$$

where we now include the synchronous phase term ϕ, and E_o is the initial energy of the electron beam.

SOME EXAMPLES

The design goal is to maximize the acceleration gradient without violating several constraints on the variables. These constraints are:

1. The peak electric field at the waist is smaller than the breakdown field of the gas.

2. The laser waist is greater than the amplitude of the transverse oscillation of the electron trajectory.

3. All parameters have values that are physically realizable using conventional technology (for example, the magnetic field of the undulator is held to be less than 20 kG for a conventional magnet, and 30 kG for a superconducting magnet).

In table I and II we present a comparison between a gas and a conventional IFEL. The pitch angle for the gas loaded IFEL is a variable, whereas for conventional IFEL's it is determined by the phase matching condition. In the case of the gas-loaded IFEL, the phase matching can be changed by varying the Cherenkov angle, i.e. the index of refraction can varied by changing the gas pressure. This allows the undulator period to be changed along with its magnetic field.

In Table I we compare the vacuum and gas-loaded IFEL'S using a 20 Terawatt laser. The values presented for the vacuum IFEL are approximately equal to the ones derived by Pellegrini in ref(10). The pitch angle of the gas-loaded IFEL is adjusted to be identical to that of the vacuum IFEL's (θ_p = 4.62 mr). The accderation gradients are identical; however, most importantly, the final magnetic field for the undulator is an order of magnitude smaller than that of the vacuum IFEL. The value is a reasonable 10 kG, and the final loss gradient is 680 keV/m. However, Pellegrini's values are for an electron-beam energy of modest value (500 MeV to 4 GeV). Physicists are interested in final acceleration values of 100 GeV or more for the next generation of accelerators.

TABLE I
COMPARISON AT MODERATE ELECTRON ENERGIES

LASER PARAMETERS	GAS-LOADED IFEL	VACUUM IFEL
LASER POWER (Gw)	2.00E+04	2.00E+04
ELECTRIC FIELD (eV/cm)	2.8E+08	2.8E+08
LASER WAIST (cm)	0.25	0.25
LASER WAVELENGTH (MICRONS)	1.06	1.06
LENGTH OF INTERACTION (m)	37.62	37.62
ELECTRON BEAM PARAMETERS		
INITIAL GAMMA	500.00	500.00
ELECTRON BEAM WAIST (cm)	0.20	0.20
UNDULATOR PARAMETERS		
OSCILLATION AMPLITUDE (cm)	0.05	0.01
PITCH ANGLE (mr)	4.62	4.62
MAGNET PERIOD (cm)	68.00	9.93
CHERENKOV ANGLE (mr)	4.27	not applicable
INITIAL MAGNETIC FIELD (kG)	0.36	2.49
SYNCHRONOUS PHASE	1.05	1.05
FINAL PARAMETERS		
ACCEL. GRADIENT (MeV/m)	110.03	110.03
TOTAL ACCELERATION (GeV)	4.14	4.14
FINAL MAGNETIC FIELD (kG)	6.26	42.85
FINAL LOSS GRADIENT (MeV/m)	0.01	0.44

Using the same laser parameters as in Table I, but using an electron beam of energy 20 GeV (γ = 40,000), we calculate a final electron-beam energy of 27 GeV for a single pass through the laser. The final magnetic field is now 30 kG, requiring superconducting magnets for implementation. The fields required for the vacuum IFEL (170 to 203 kG) are too large for any practical use.

In order to operate at higher energies, the oscillation amplitude of the electron beam must be increased in order to reduce the synchrotron losses. A larger radius allows a weaker magnetic field and, hence, lower synchrotron losses. However, in order to keep a large acceleration gradient, the electric field at the waist must remain high, thus the power of the laser must be increased. This is shown in Table III for five extremely high energy electron beams ranging from 50 GeV to 1 TeV.

TABLE II
COMPARISON AT HIGH ELECTRON ENERGIES

LASER PARAMETERS	GAS LOADED IFEL	VACUUM IFEL
LASER POWER (Gw)	2.00E+04	2.00E+04
ELECTRIC FIELD (eV/cm)	2.8E+08	2.8E+08
LASER WAIST (cm)	0.25	0.25
LASER WAVELENGTH (MICRONS)	1.06	1.06
LENGTH OF INTERACTION (m)	37.62	37.62
ELECTRON BEAM PARAMETERS		
INITIAL GAMMA	40000.00	40000.00
ELECTRON BEAM WAIST (cm)	0.20	0.20
UNDULATOR PARAMETERS		
OSCILLATION AMPLITUDE (cm)	0.05	0.01
PITCH ANGLE (mr)	4.36	4.38
MAGNET PERIOD (cm)	72.05	11.05
CHERENKOV ANGLE (mr)	4.01	not applicable
INITIAL MAGNETIC FIELD (kG)	25.92	169.80
SYNCHRONOUS PHASE	1.05	1.05
FINAL PARAMETERS		
ACCEL. GRADIENT (MeV/m)	103.84	104.31
TOTAL ACCELERATION (GeV)	3.91	3.92
FINAL MAGNETIC FIELD (kG)	30.88	202.41
FINAL LOSS GRADIENT (MeV/m)	7.03	302.70

The peak laser powers for these higher energy electron beams (Table III) are extremely large and beyond present day technologies. The pulse length of the laser power must be subnanosecond in duration in order to prevent breakdown and nonlinear processes.

TABLE III
PARAMETERS FOR EXTREMELY HIGH ELECTRON BEAM ENERGIES
50 GeV to 1 TeV

LASER PARAMETERS					
LASER POWER (Gw)	2.00E+04	4.00E+05	4.00E+06	5.00E+06	6.00E+06
ELECTRIC FIELD (eV/cm)	1.5E+09	1.7E+09	1.9E+09	2.0E+09	2.2E+09
LASER WAIST (cm)	0.05	0.18	0.52	0.55	0.55
LASER WAVELENGTH (MICRONS	1.06	1.06	1.06	1.06	1.06
LENGTH OF INTERACTION (m)	1.26	19.69	157.63	177.83	176.36
ELECTRON BEAM PARAMETERS	50 GeV	200 GeV	600 GeV	800 GeV	1 TeV
INITIAL GAMMA	100000.00	400000.00	1200000.00	1600000.00	2000000.00
ELECTRON BEAM WAIST (cm)	0.05	0.18	0.52	0.55	0.55
UNDULATOR PARAMETERS					
OSCILLATION AMPLITUDE(cm)	0.05	0.18	0.52	0.55	0.55
PITCH ANGLE (mr)	3.00	1.60	0.90	0.70	0.60
MAGNET PERIOD (cm)	96.74	715.73	3600.18	4916.40	5712.05
CHERENKOV ANGLE (mr)	2.61	1.50	0.87	0.67	0.57
INITIAL MAGNETIC FIELD(kG	33.21	9.58	3.21	2.44	2.25
SYNCHRONOUS PHASE	1.05	1.05	1.05	1.05	1.05
FINAL PARAMETERS					
ACCEL. GRADIENT (MeV/m)	389.71	235.56	148.09	121.24	114.32
TOTAL ACCELERATION (GeV)*	0.49	4.64	23.34	21.56	20.16
FINAL MAGNETIC FIELD (kG)	33.53	9.79	3.34	2.50	2.29
FINAL LOSS GRADIENT(MeV/m	37.26	52.17	56.10	54.96	71.16

*Note that the total acceleration calculated here does not include the synchrotron loss which, for these energies, is large.

Theoretically, the total peak power can be reduced by using a ring or segmented laser mode. This is shown in Fig. 2. Since the electrons are traversing a helical orbit of radius $r_o = \lambda\!\!\!^{-}_w \theta_p$, the power in the center of the regular single mode (see Fig. 1) is wasted. A segmented or circular mode would allow a reduction in the total power by the ratio of the annulus width to the radius of the helical trajectory:

$$P_c = \frac{2\omega_o P_o}{r_o} \tag{11}$$

where P_c is the power in the circular mode, P_o is the power in the Gaussian mode, and $2\omega_o$ is the thickness of the annulus. Thus an annulus of 1 mm width and a 5 mm radius would have a factor of 5 reduction in total power needed to accelerate the electron at the acceleration gradient given for Table III. However, the power can be additionally reduced by noting that r_o can be increased indefinitely (within the physical constraits of the wiggler and vacuum system). If a segmented laser beam (multiple laser beams that are in phase and have the proper polarization) is used, each beam is separate and can be focused to obtain a maximum peak electric field to maintain a high acceleration gradient. The Rayleigh range and the interaction length would be changed as compared to the case of the single large Gaussian mode of Fig. 1. Dramatically lower peak-power but more tightly focused laser beams could then be used, resulting in high acceleration gradients but shorter interaction lengths as compared to the Fig. 1 configuration.

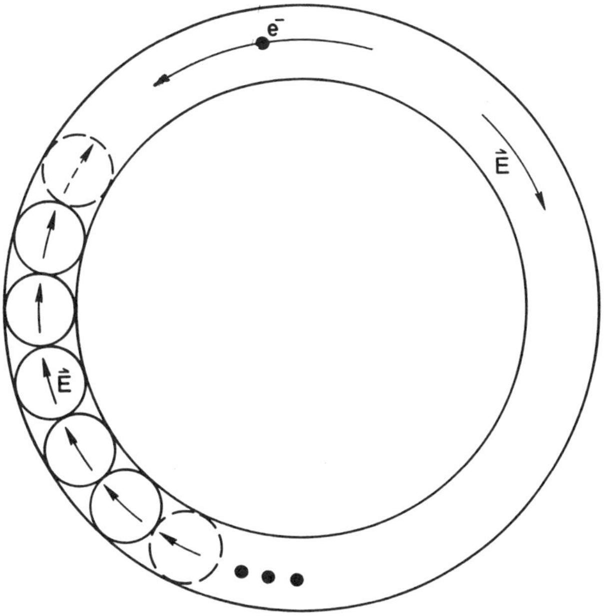

Figure 2. The circular or segmented laser mode for the reduction of laser power.

As a final example, we design a relatively modest experiment for the SLAC beam using an electron-beam energy of 2 GeV. A new Nd:glass slab laser technology has been developed that can produce laser powers in the 10 to 1000 Gw range.[11] These lasers are relatively inexpensive, compared to the large klystrons now used. The experiment would be designed to show that these lasers could produce high gradient acceleration comparable to or larger than conventional klystron systems. Assuming that laser systems of 100 or even 1000 Gw are available, and that we wish to use conventional magnets ($B < 20$kG), Table IV presents realistic values for such an experiment. We see from this Table that gradients of 50 MeV/m and 110 MeV/m can be achieved using the 100 and 1000 Gw laser systems, respectively, without requiring an undulator field exceeding 20 kG. Thus, using realizable experimental values, a modest experiment could be performed.

TABLE IV
PARAMETERS FOR PROOF-OF-PRINCIPLE EXPERIMENTS

LASER PARAMETERS	100 Gw		1 Tw
LASER POWER (Gw)	1.00E+02	1.00E+02	1.00E+03
ELECTRIC FIELD (eV/cm)	5.0E+07	5.0E+07	1.6E+08
LASER WAIST (cm)	0.10	0.10	0.10
LASER WAVELENGTH (MICRONS)	1.06	1.06	1.06
LENGTH OF INTERACTION (m)	5.69	5.69	5.56
ELECTRON BEAM PARAMETERS			
INITIAL GAMMA	4000.00	4000.00	4000.00
ELECTRON BEAM WAIST (cm)	0.08	0.10	0.10
UNDULATOR PARAMETERS			
OSCILLATION AMPLITUDE (cm)	0.08	0.08	0.08
PITCH ANGLE (mr)	8.00	12.00	8.00
MAGNET PERIOD (cm)	61.56	41.04	60.84
CHERENKOV ANGLE (mr)	7.78	11.78	7.78
INITIAL MAGNETIC FIELD (kG)	5.57	12.53	5.63
SYNCHRONOUS PHASE	1.05	1.05	1.05
FINAL PARAMETERS			
ACCEL. GRADIENT (MeV/m)	34.64	51.96	110.85
TOTAL ACCELERATION (GeV)	0.20	0.30	0.62
FINAL MAGNETIC FIELD (kG)	6.11	14.34	7.33
FINAL LOSS GRADIENT (MeV/m)	.00	0.01	.00

EFFECTS OF MULTIPLE SCATTERING

A more realistic estimate of the energy exchange requires the inclusion of coulomb scattering from atomic nuclei and ionization of the atomic electrons of the gas. The coulomb scattering results in two detrimental processes: trajectory changes for the electrons and Bremsstrahlung losses. Ionization contributes to random energy losses, with both a mean energy loss and an incoherent broading of the energy spectrum (straggling).

Angular divergence results both from the coulomb scattering and the variations in trajectories of the incident electron beam (beam emittance). The result of such trajectory changes causes the electrons to slip out of phase with the electromagnetic wave. If this slippage is sufficiently great then the electron will receive no net coherent acceleration.

For the vacuum IFEL, the angular acceptance of the IFEL corresponding to a π phase-slippage over a total length L is[12]

$$\Delta\theta_q = \left(\frac{\lambda}{2L}\right)^{1/2} \tag{12}$$

A reasonable approximation to the amount of scattering in a length L is given by the rms angular spread due to multiple scattering. This is given by[13]

$$\Delta\theta_s = (11/E_b)\sqrt{L/X_o} \tag{13}$$

where E_b is in MeV and X_o is the radiation length of the gas (X_o = 7X10^3 meters for H_2 at S.T.P.). To achieve efficient energy exchange between the electrons and the electromagnetic wave, the condition:

$$\Delta\theta_s < \Delta\theta_q \tag{14}$$

should hold. Using this inequality and eqns. (12) and (13) the maximum length of interaction can be calculated:

$$L < \frac{E_b}{11}\left(\frac{\lambda X_o}{2}\right)^{1/2} \tag{15}$$

For a 100 GeV electron beam interacting with a 1 µm laser field in 1 atmos. of H_2, the length of interaction must be L < 100 meters. More energetic electrons can stay in phase for longer distances.

A MONTE-CARLO COMPUTER SIMULATION AT MODERATE ELECTRON-BEAM ENERGIES

A Monte-Carlo simulation of the gas-loaded IFEL has been done. Single electrons interacting with a Gaussian laser beam inside a gas-loaded undulator are tracked. The simulation includes discrete Rutherford-scattering events; losses due to Bremsstrahlung, synchrotron, Cherenkov, and ionization of the gas atoms; electron-beam energy spread and phase space; laser-beam wavelength, shape and power; gas pressure, temperature, index of refraction; and helix period, taper and magnetic field values.

Two cases showing the effects of changing the taper of the undulator are presented in Figs. 3 and 4. In Fig. 3 the initial electron beam energy is 100 MeV, while in Fig. 4 it is 2 GeV. Both cases show that increasing the taper allows a larger acceleration for the same interaction length, while the number of trapped electrons decreases.

For the simulation shown in Fig. 3, the 100 MeV electron beam is injected into a 100 Mw laser field. Other parameters of this simulation are given in the figure caption. Since elastic scattering will cause thermalization of the electron beam, the interaction length was limited to 107 cm.

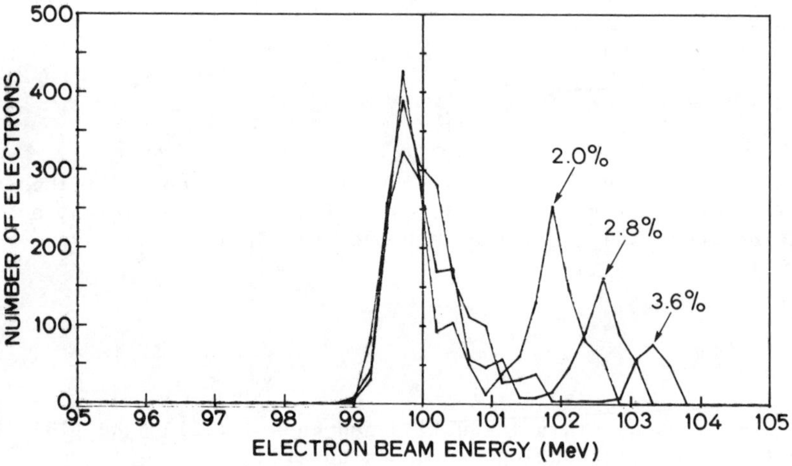

Figure 3. The number of electrons as a function of energy for three taper values. The parameters of this simulation are as follows: electron-beam energy, 100 MeV; θ_p, 8 mrad; laser power, 100 Mw; laser wavelength, 1.06 μm; interaction length, 107 cm; laser-beam waist, 0.4 mm; electron-beam waist, 0.3 mm; wiggler period, 3.8 cm. The total number of electrons used in the simulation is 2000.

Injecting the electrons at a higher energy allows a longer interaction length and a larger (E_f - E_o). The case of a 2 GeV electron beam injected into a 100 Gw laser field is shown for five tapers. The parameters for this simulation are in the second column of the 100 Gw case of Table IV. As before, the number of electrons trapped and the total acceleration energy is dependent upon the value of the taper. A taper of 5.5 % allows 14.4 % of the electrons to be trapped. The maximum acceleration gradient shown in Fig. 4 is 37 MeV/m for the 9.9 % taper. As given in Table IV, the ideal case (an 11 % taper) would give 52 MeV/m.

In Fig. 5 the percentage of particles trapped is plotted as a function of the ideal taper. The ideal taper is defined as the change in B_w needed to maintain perfect phase synchronism between the optical wave and the electron. For the 100 MeV case a 4 % taper is ideal, while for the 2 GeV case an 11 % taper is ideal. As can be seen from this figure, the number of particles trapped decreases linearly with increasing taper. The total number of trapped particles at the ideal taper is practically zero.

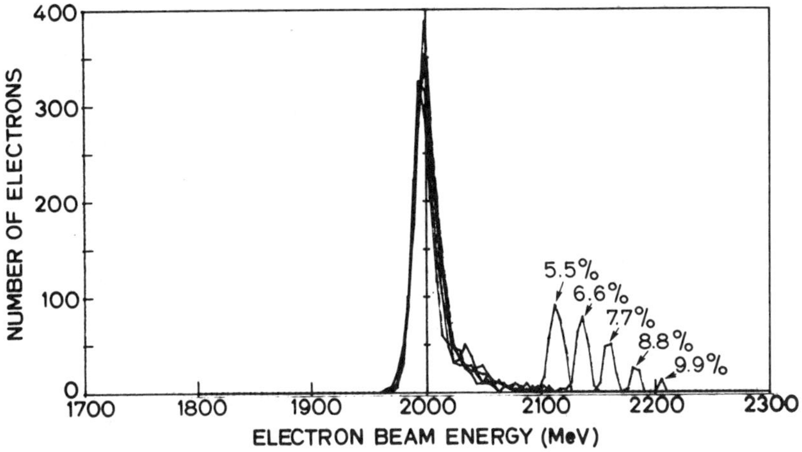

Figure 4. The number of electrons as a function of energy for five taper values. The parameters of this simulation are as follows: electron beam energy, 2 GeV; θ_p, 12 mrad; laser power, 100 GW; laser wavelength, 1.06μm, interaction length, 569 cm; laser-beam waist, 1 mm; electron-beam waist, .8 mm; wiggler period, 41 cm. The total number of electrons used in the simulation is 2000.

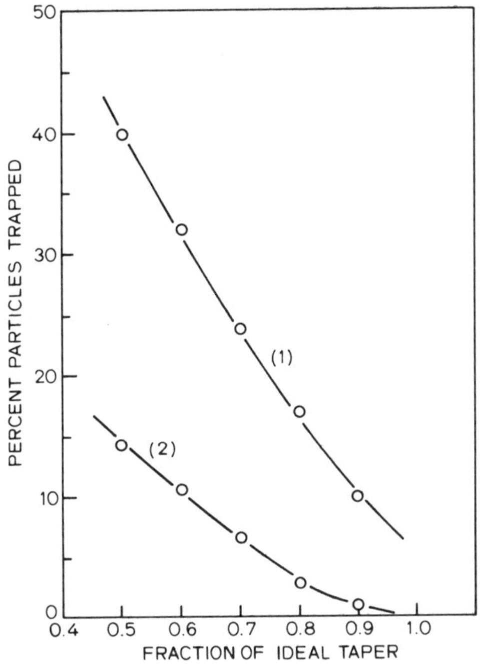

Figure 5. The percentage of particles trapped as a function of the fractional ideal taper. Case (1) is for the 100 MeV beam, while (2) is for the 2 GeV beam.

To see how multiple scattering affects the trapping of the electrons, the laser power was increased while the pitch angle was held constant. This allows the Rayleigh range and the interaction length to increase. The percentage of the trapped electrons as a function of the interaction length was then plotted for the two cases of (1) no scattering and (2) scattering. From (15), the maximum length of interaction before thermalization of the beam occurs is 71 cm at 100 MeV. Fig. 6 shows that deviation from the non-scattered case does not occur until 100 cm. The percentage of trapped particles drops by a factor of two at 250 cm. Thus (15) appears to give a worst case limit for scattering.

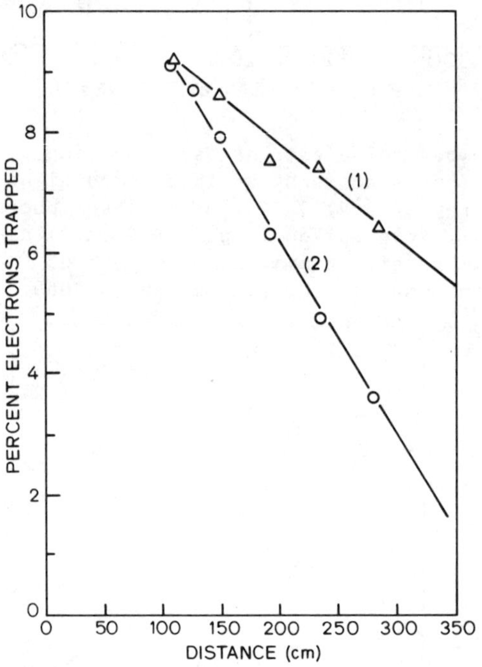

Figure 6. The percentage of particles trapped as a function of interaction length. Two cases are presented (1) no scattering and (2) scattering. The pitch angle is held to 8 mr. The initial electron beam energy is 100 MeV. The laser electric field is also held to 3.7 X 10^6 eV/cm, while the power and waist (and hence the Rayleigh range and interaction length)of the laser beam are allowed to increase. The percentage of particles trapped as a function of interaction length is made above for a constant acceleration gradient.

Both the examples presented in Figs. 3 and 4 can be seen as parameters for proof-of-principle experiments. Ideally, one would like to inject all the electrons in at the proper phase with the optical wave. For all of the cases presented here, the electrons injected at random phase. Few electrons are trapped in the high energy case. An untapered IFEL could be used as a first stage injector section in which a majority of the particles would be trapped. These are then injected in the proper phase into a second stage where high gradient acceleration takes place.

CONCLUSION

The gas-loaded IFEL can achieve large acceleration gradients at relatively high electron-beam energies (E < 50 GeV). This is because the new phase-matching condition allows the use of large pitch angles and long period undulators with lower synchrotron losses than conventional vacuum IFEL's.

At extremely relativistic energies (100 GeV < E <1 TeV) the gas-loaded IFEL allows possible operation using extremely high-peak-power lasers. However, the use of a ring or segmented mode laser beam can lower the needed peak power by one or two orders of magnitude.

The addition of a gas into the IFEL results in coulomb scattering of the electrons, which results in Bremsstrahlung losses and trajectory changes. A Monte Carlo simulation shows the effect of multiple scattering on a moderate energy case. The simple estimate of the maximum interaction length given by (14) gives a good estimate when compared with the results of the Monte Carlo simulation.

REFERENCES

1. R. B. Palmer, J. Appl. Phys., 33, 3014 (1972).
2. R. H. Pantell "Electron (and Positron) Acceleration with Lasers," in Laser Interactions and Related Plasma Phenomena, 6, (New York: Plenum Pub. Co., 1983), pp.1083-1092.
3. M. A. Piestrup, and J. A. Edighoffer, "Increasing the acceleration gradient of the transverse-field accelerator using a dielectric medium," accepted for publication IEEE Quant. Electr., July 1985.
4. J. A. Edighoffer, and Z. G. T. Guiragossian, "Theory and experiment for a new FEL approach (a Cherenkov radiation enhanced free electron laser)" Report #DE-TN-006, TRW, Bldg. R-1/1184, 1 Space Park, Redondo Beach, CA, 90278, Sept. 1981.
5. W. J. Cocke, Opt. Comm., 28, 123 (1979).
6. A.-M. Fauchet, J. Feinstein, A. Gover and R. H. Pantell, SPIE, 453, 423 (1984).
7. W. R. Smythe, Static and Dynamic Electricity, (New York: McGraw-Hill, 1950), p. 277.
8. J. D. Jackson Classical Electrodynamics, (New York: John Wiley and Sons, 1962).
9. A. E. Siegman, An Introduction to Lasers and Masers, (New York: McGraw-Hill, 1971), ch.8.

10. C. Pellegrini, "Report of the working group on far field acclerators," in Laser Acceleration of Particles, (AIP Conference Proceedings #91, Los Alamos, 1982), pp. 138-149.
11. J. M. Eggleston, T. J. Kane, K. Kuhn, J. Unternahrer and R. L. Byer, Optics Letters, 7, 405 (1982).
12. T. I. Smith and J. M. J. Madey, Appl. Phys. B 27, 195 (1982).
13. V. L. Highland, Nuc. Instr. and Meth., 129, 497 (1975).

LASER WIGGLER BEAT WAVE

J.L. BOBIN

L.P.O.C., Université Pierre et Marie Curie, 75230 PARIS, FRANCE.

ABSTRACT

It is shown that electrons can be accelerated to high energies by combining a laser wave and an electron beam propagating through a plasma inside a wiggler. For usual laser frequencies and wiggler wavelengthes, plasma densities are in the range 10^{15} - 10^{16} cm^{-3} . Although the interaction is off-resonance, the plasma density fluctuation in the longitudinal wave suffices to obtain electron energies of several hundred MeV over short distances.

INTRODUCTION

In the beat wave scheme for accelerating electrons[1], two laser waves with frequencies ω_1 ω_2 , separated by about one plasma frequency, propagate in the same direction. High energies require high laser intensities on both lines, typically 10^{18} W/cm^2 at the wavelength of the Nd laser ; 1 μm. It is proposed here to replace one of the waves by an undulator whose field is equivalent to a very intense electromagnetic oscillation when relativistic electrons are passing through it.

WAVEMATCHING

In the laboratory reference frame, the undulator is a zero frequency device. The only way an incoming electromagnetic wave can match an electron plasma wave is in presence of a relativistic electron beam, according to the beam plasma dispersion relation as shown on fig.1.

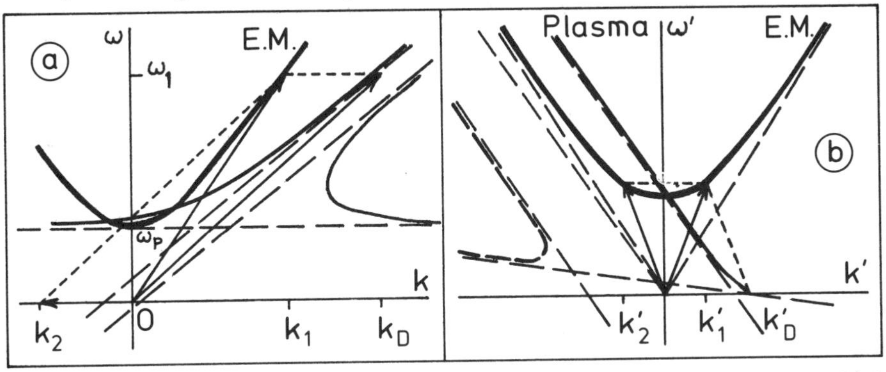

Fig. 1. Matching of E.M. and plasma waves : a) in the laboratory frame, b) in the moving frame.

Let u_R be the phase-velocity of the plasma wave. It is related to the frequency and wave number ω_1, k_1 of the laser wave and to the undulator's wavenumber k_2 by

$$u_R = \omega_1/(k_1+k_2) \ . \qquad (1)$$

It is to be noted that u_R is the velocity of a reference frame (moving frame) in which the longitudinal beat between the laser and the undulator is time independant. The electron beam velocity u_b is a function of u_R deduced from the beam plasma dispersion relation in the laboratory frame

$$u_b = u_R - \omega_{pb}/(k_1+k_2)(1-\omega_p^2/\omega_1^2)^{1/2} \qquad (2)$$

where ω_p is the plasma frequency and ω_{pb} is the plasma frequency of the beam. In the laboratory frame, the electron energy is $m_o \gamma_b c^2$ with

$$\gamma_b^2 = \gamma_R^2/(1+2\omega_{pb}\gamma_R^2/\omega_1) \ . \qquad (3)$$

Now γ_R is given by

$$\gamma_R^2 = (k_1+k_2)^2/((k_1+k_2)^2-\omega_1^2) \qquad (4)$$

which since

$$k_1^2 = (\omega_1^2/c^2)-(\omega_p^2/c^2) \qquad (5)$$

implies a divergence for

$$\omega_p^2 = 2c\omega_1 k_2 - c^2 k_2^2 \sim 2c^2 k_1 k_2 \ , \qquad (6)$$
$$n_0(cm^{-3}) \sim 2.2\ 10^{13}/\lambda_1(cm)\lambda_2(cm) \ .$$

Given ω_1 and λ_2 there is an upper boundary for the density of the plasma inside the undulator. Thus the laser associated with the electron beam are able to drive plasma waves with phase velocities as close as one wishes to the velocity of light in vacuum.

Furthermore, at given γ_R^2 and k_2, ω_p^2 is a function of k_1 which passes through a maximum for

$$k_1 = (\gamma_R^2-1)k_2 \ . \qquad (7)$$

At this point, it turns out that the E.M. wave frequencies

$$\omega_1' = \omega_2' = 2u_R\gamma_R k_2 \qquad (8)$$

are close to the relativistic invariant plasma frequency

$$\omega_1'^2 = \omega_2'^2 \sim \omega_p^2 = c^2 k_1 k_2 \ . \qquad (9)$$

The plasma density is then half of the value corresponding to an infinite γ_R (fig.2).

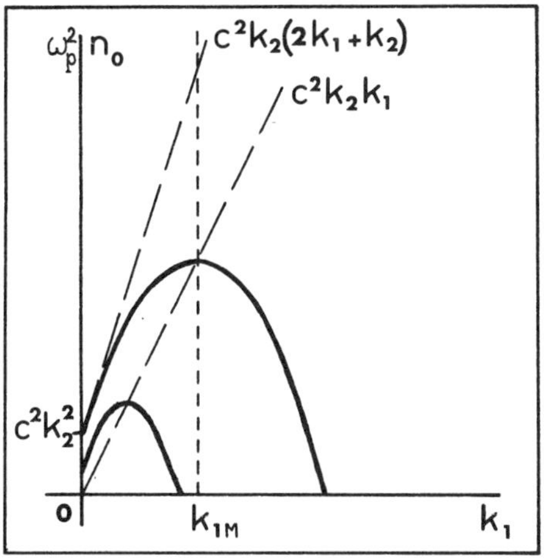

Fig. 2. At given wiggler wavelength, the plasma density depends upon the laser frequency and the velocity of the moving frame.

PLASMA WAVE AMPLITUDE ; ELECTRON ENERGY

In order to evaluate the amplitude of the longitudinal wave which is created by this mechanism, one can use the equations set up by Rosenbluth and Liu[2] for the displacement in the z direction in the laboratory frame :

$$\begin{aligned}\ddot{\xi}+\omega^2\xi &= -(e/m_0)^2(A_1A_2/4)(k_1+k_2)\cos\{(k_1+k_2)z-\omega_1 t\} \\ &= -\Lambda c^2(k_1+k_2)\cos\{(k_1+k_2)z-\omega_1 t\}\end{aligned} \tag{10}$$

where A_1 and A_2 are the vector potentials of the laser wave and the undulator respectively. This is the equation for a forced oscillator. Here we are far from resonance. The amplitude is then simply

$$Z = c^2(k_1+k_2)\Lambda/\omega_1^2 \tag{11}$$

which corresponds to an amplitude of the density oscillation

$$N = n_0(k_1+k_2)Z = c^2(k_1+k_2)^2\Lambda n_0/\omega_1^2 \sim \Lambda n_0 \quad . \tag{12}$$

In the frame moving with the velocity u_R, the accelerating field is

$$E'_D = en'_0\Lambda/(\varepsilon_0 k'_D) \tag{13}$$

which is relativistically invariant provided the electron oscillatory velocity is much smaller than c. $k'_D = 2k'_2 = 4\gamma_R k_2$, is the plasma wave wavenumber.

In order to trap the whole population of electrons in the plasma (wave breaking condition), the longitudinal field should be large enough viz.

$$eE_D'/k_D' = m_0\gamma c^2 \tag{14}$$

which implies

$$\Lambda = (\omega_1'/\omega_p)^2 . \tag{15}$$

The energy an electron is able to acquire in the potential well of the longitudinal wave is a fraction $\Lambda(\omega_p/\omega_1')^2$ of the maximum $m_0\gamma_R c^2$ for which the whole electron population would have been trapped. By applying the usual Lorentz transform, the acceleration energy in the laboratory frame is

$$W_A = 2m_0c^2\gamma_R^2\Lambda(\omega_1'/\omega_p)^2 \sim 2m_0\Lambda(\omega_p/k_2)^2 \tag{16}$$

and the corresponding acceleration length is given by

$$l_A = W_A/eE_D = 2m_0(\omega_p/k_2)(\varepsilon_0 k_D'/n_0'e^2)(\omega_1'/\omega_p)^2$$

$$= (8/k_2)(\omega_1'/\omega_p)^2 = (4\lambda_2/\pi)(\omega_1'/\omega_p)^2 = (4\lambda_2/\pi)(\gamma_R/\gamma_R^m)^2 \tag{17}$$

where γ_R^m refers to the case $\omega_1' = \omega_2' \sim \omega_p$. This length might be reduced by phase slippage considerations.

In real life, the length and the wavelength of the wiggler are given and, once constructed, cannot be altered. The free parameters are the frequency and intensity of the laser beam and the energy of the colinear electron beam. For instance, take an undulator 1m long with wavelength 10 cm. Fix the phase velocity of the plasma wave by $\omega_1' = \omega_2' = \omega_p$. Thus

$$\gamma_R^2 = k_1/k_2 = \lambda_2/\lambda_1 . \tag{18}$$

A magnetic field of 6000 gauss is not completely unreasonable. $\Lambda = 1/20$ is a convenient value for trapping of the driving electron beam. This fixes the intensity of a laser beam with given frequency. For CO_2 radiation (λ_1= 10 µm) one finds

$$\gamma_R = 10^2 , \quad E_1 = (3/320)(m_0c/e)^2(\omega_1 k_2/B_2) \sim 4\times10^8 V/m , \tag{19}$$

i.e. a comparatively modest intensity value : 5.10^{10} W/cm^2. The plasma density in these conditions is

$$n_0 = 1.1\times10^{13}/(\lambda_1\lambda_2) = 1.1\times10^{15} cm^{-3} . \tag{20}$$

The energy of the accelerated electrons is accordingly

$$W_A = 8.1\times10^{11} J \sim 450 \text{ MeV} . \tag{21}$$

The acceleration length in this case is about 1 wiggler wavelength. It might be better to accelerate the electrons over the whole length of the wiggler. Then, at the same laser frequency,

$$\gamma_R = 280 \quad , \; n_0 = 1.97\times10^{15}\text{cm}^{-3} . \tag{22}$$

The acceleration energy does not change. The electron energy is 15 MeV at a density 10^{13} cm^{-3}.

CONCLUSION

A plasma wave can be set up inside an undulator. The basic mechanism is a beating between a laser beam and the alternating transverse magnetic field in presence of a low density moderate energy electron beam. The corresponding plasma wave is non resonant. However, the energies and acceleration lengthes, look attractive. A given choice of parameters yields a limited acceleration energy in a single stage. The final energy could be increased by increasing the laser intensity. An alternative possibility would combine the present scheme with a D.C. magnetic field, similarly to the surfatron concept[3].

REFERENCES

1. T. Tajima, J.M. Dawson, Phys. Rev. Lett. 43 p.267 (1979).
2. M.N. Rosenbluth, C.S. Liu, Phys. Rev. Lett. 29 p. 701 (1972).
3. T. Katsouleas, J.M. Dawson, Phys. Rev. Lett. 51 p.392 (1983).

REPORT OF THE WORKING GROUP ON OTHER ACCELERATION SCHEMES

Andrew M. Sessler
Lawrence Berkeley Laboratory
University of California
Berkeley, CA 94720

The first thing our group did was compile the various acceleration schemes, of which we know, so that we could determine which were being covered by the other groups and which were truly "other schemes". The compilation proved useful to us and would, probably, prove useful to readers of this volume. It is presented in Table I.

Perhaps the most revealing thing that this Table discloses is the large number of particle driven schemes. This is in sharp contrast with the conference three years ago and, also, I might add, in sharp contrast with the name of this workshop. It was decided, however, to consider novel schemes; not just laser acceleration schemes.

Both lasers and particle beams are a source of high peak power, but particle beams can be an inexpensive (especially if they are of low energy and induction-accelerator-produced) source of high average power. Therefore it is most natural that the novel acceleration schemes invoke either lasers or particle beams; the necessary "trick" is to use this power; i.e. to convert the power to a proper accelerating field.

After examining the schemes listed in Table I; the group determined that they only needed to consider the various devices listed in Table II. Fortunately, invited talks (and hence invited papers for these Proceedings), contributed papers to these Proceedings, or published papers cover the schemes listed in Table II. Consequently, this summary can be brief. We shall, simply, take the schemes in the order of Table II.

1. INVERSE CHERENKOV ACCELERATOR

Employing a cylindrically symmetric configuration one finds

$$E_r \Big|_{max} = \frac{0.582\ E_0}{\tan\theta} \quad ,$$

$$E_z\ (r=0) = E_0 \cos\psi \quad ,$$

where ψ is the phase angle. Gas breakdown limits the maximum value of E_r, which for picosecond pulse lengths in hydrogen gives maximum accelerating gradients (E_0) of up to a few GeVs/m.

A numerical example, using a large CO_2 laser ($\lambda = 10\,\mu m$) of $P = 7 \times 10^{13}$ W, an accelerating length of 50 meters, and a Cherenkov angle of 20 mrad (H_2 at 1.5 atmos), yields an accelerating gradient of 500 MeV/m and a net energy increase of 25 GeV. If this were

Table I Novel Accelerator Concepts

1. Plasma Accelerators (Beat-Wave, Surfatron)
   a. Laser excited (L)
   b. Particle beam excited (PB)
2. Inverse Cherenkov Accelerator (L)
3. Inverse Free Electron Lasers (L)
   a. Regular kind
   b. Gas loaded
   c. Two-wave
   d. Three-wave, etc.
4. Droplets, Gratings, Open Structures
   a. Laser excited (L)
   b. Transverse Electron Resonance Accelerator (PB)
5. Plasma Focus (L)
6. Two-Beam Accelerator (PB)
7. Wake-Field Accelerator (PB)
   a. Electron ring excited
   b. Electron beam excited
   c. Photon beam excited
   d. Intense electron beam (plus laser) excited
   e. Radial implosion of electrons
   f. Photo diode initiated pulse
8. Improved Power Sources (PB)
9. Periodic Plasma Waveguides
10. Collective Accelerators (PB)
    a. Ionization - Front Accelerator
    b. Moving Potential Well Accelerator
11. Laser Focusing Schemes (L)

employed as an "after burner" at the Stanford Linear Accelerator Center (SLAC) it would raise the beam energy from 50 GeV to 75 GeV while increasing the emittance (due to gas scattering) by 10^{-5} mrad. (The present SLC emittance is $3x10^{-5}$ mrad. The gas scattering effect, while not negligible, is acceptable.) More details can be found in Ref. 1.

Table II Devices Considered By The Other Schemes Group

1. Inverse Cherenkov Accelerator (Fontana)
2. Three-Wave Accelerator (Abedi)
3. Transverse Electron Resonance Accelerator (Csonka)
4. Plasma Focus (Hora)
5. Radial Implosion Accelerator (Channell)
6. Laser Focusing (Channell)

2. THE THREE-WAVE ACCELERATOR

It has been observed by Abedi that the Two-Wave Accelerator (where the static field wiggler is replaced with an electromagnetic wave and the other wave is the accelerating beam) can be improved by employing three waves.[2] Two of the waves play the role previously played by the wiggler; i.e. they produce a dynamical wiggler.

If we compare a Two-Wave Accelerator and a Three-Wave Accelerator we see that the two waves which produce the wiggle motion in the Three-Wave Accelerator can constructively interfere and, hence, produce twice the gradient. If we look at the energy efficiency; i.e. how much gradient one gets for a given amount of wave power, then one can show that the Three-Wave Accelerator is $\sqrt{2}$ times as efficient as a Two-Wave Accelerator.

3. TRANSVERSE ELECTRON RESONANCE ACCELERATOR

A near-field accelerator has the advantage, in comparison with a plasma accelerator, that the longitudinal, or accelerating, field can be of the order of the transverse field (E_T) in the laser beam (ω_0). Thus very high accelerating fields can be obtained rather than $(\omega_p/\omega_0)\ E_T$ as in a plasma accelerator (which could nevertheless be adequately large).

To accomplish this one needs to have microstructures of the order of a wavelength (λ) in size and to be within λ of them with the accelerated particles. Such small dimensions and high laser power (needed for the large gradient) bring up the questions of the size of beams and the integrity of the microstructures in this environment. Closing our eyes to these practical questions, for we are firstly interested in matters of principle, we see that an accelerator can be envisioned which has small microstructures of solid density spaced longitudinally so that there is an accelerating wave propagating along them.

Such a device has been proposed by P. Csonka,[3] with the added features that he proposes the microstructures be excited by a particle beam and that the excitation be resonant. In this case one can "build up" very large fields, much larger fields than in the laser itself. One needs a resonance between the plasma frequency (ω_p) and the frequency of the exciting microbunches (ω_0).

Csonka has proposed generating the microbunches by a free electron laser (FEL) or a transverse optical klystron (TOK). The advantage of particle beam excitation, besides that of decreased capital cost and increased efficiency of producing the requisite power, is that a focusing (quadrupole) mode can be excited rather than a dipole mode (as would be generated by a laser).

A numerical example[3] with microbunches, of length 10 μm, radius 10 μm, and containing 10^9 electrons, would excite microstructures to 10 GeV/m even if there is no resonant excitation. With resonant excitation the gradient becomes correspondingly larger.

Of course, many questions need to be addressed such as: 1) How small can one make microbunches? 2) How close to the microstructures can one send them? 3) At what value do various non-linear effects saturate the resonant excitation? The concept does appear, however, to merit further study.

4. PLASMA FOCUS

It is well-known that when a powerful laser is shown onto a slab of material the laser light is self-focused down to a waist which is very small. It has been observed, at Los Alamos National Laboratory (LANL), that electrons are produced with energies greater than 50 keV and that ions are produced with energies greater than 100 MeV and that the energy of these ions is proportional to their atomic number.

Theoretical explanation of these facts have been given by Hora and co-workers.[4] In fact, even as early as at the first Workshop, Hora emphasized that plasma foci of laser light could be employed to accelerate particles and that this effect was most interesting for the acceleration of ions.

The theoretical explanation proceeds from a two-fluid hydrodynamic code in which charge neutrality is <u>not</u> assumed. (Clearly, it is necessary to remove this usual assumption if one is interested in studying accelerating electric fields.) The analysis predicts two interesting, and important, features. Firstly there are density depressions, named cavitons, which are extensive in length (100 optical wavelengths) and; secondly there is significant charge separation, called a double layer, which is actually inverted in sign from what one might expect. The combination leads to high fields over large distances; i.e. to significant acceleration.

On the basis of this theory, which agrees with the present experiments, Hora has predicted that a powerful CO_2 laser (2×10^{14}W) with a short rise-time (150 psec) will accelerate an ion with Z=50 to 30 GeV (an energy of 600 MeV/nucleon). In this case the caviton is 100 laser wavelengths, i.e. 0.1 cm in length, and the longitudinal accelerating field is 3×10^7 MeV/m. Furthermore, it appears possible to stage this acceleration many times.

Experiments using the Antares laser, whose pulse length is longer than 150 psec, can be expected in the near future. They will be most important for the Plasma Focus Accelerator.

5. RADIAL IMPLOSION ACCELERATOR

It was observed, by Channell, that a gradient of 3 GeV/m is "equivalent" to a magnetic field of 100 kG; and that such a field can be made available for acceleration if a magnetic field of this magnitude is changed with a velocity approaching that of light.[5] Of course, moving a magnetic field is the basis for all accelerators (except the DC machines), so this concept is readily accepted by accelerator physicists.

The proposal is to use an axial current to make an azimuthal magnetic field which then is imploded by means of an intense radial current. The result is an axial electric field, the accelerating field. Rough estimates,[5] obtained from a snow-plow model, show that a radial current, of electrons of 10 MeV, of 160 kA/cm^2 is needed. This is about the magnitude of current densities obtained in the light ion inertial fusion program at Sandia and so appears attainable with present technology.

Of course the concept, which is quite new, needs further theoretical study. Questions which need to be studied include: 1) How stable is the implosion? 2) How is the magnetic field initiated? 3) What is the wall damage? Perhaps the first question is the most important, for the proposal seems to be subject to 2-D Rayleigh-Taylor instabilities (heavy electrons on top of light magnetic field). The second question also needs to be addressed; perhaps an electron beam (axially directed) is used to set-up the azimuthal magnetic field.

The scheme, if it can be made into a practical accelerator, seems to offer a number of advantages. Perhaps paramount, is the fact that the use of induction accelerated electrons as the main power source suggests a high efficiency for the device. Also, because the scheme is non-resonant it should be good for accelerating low velocity particles and it should be rather easy to stage accelerating sections.

6. LASER FOCUSING

The necessity for high luminosity in linear colliders was emphasized at this workshop. To achieve this requires large beam power, and consequently very efficient accelerators (so as to keep the demand for average power, and hence the operating costs, within bound), or it requires very small beams at the crossing point. To achieve the latter requires a tight focus; i.e. a powerful lens with small aberrations (both chromatic and spherical). It also requires adequate control of the beam position.

It was pointed out, by Channell,[6] that a laser beam can be employed to give very strong focusing. In a vacuum, as is well-known, the electric and magnetic forces of the light beam, upon a particle, just cancel. In a gas this cancellation no longer occurs, but now the changed velocity of the light wave implies that after some distance the light and the particles will be out of phase. Thus, the device must be of finite, and appropriate, length.

Of course, some particles will have a phase relative to the laser light such that they are defocused, while others will have a phase such that they are focused. Channell proposes having two lenses, separated by 180° (plus a large number of 360° phase changes), so as to produce net AG focusing. (Measuring, and correcting, relative phase is quite within present capability.)

Channell has produced a numerical example, employing a laser of only 10^{10}W, a gas pressure of 10 atmos, a waist size of 0.2 mm, and a length of 0.6 cm. The focal length, for a 50 GeV particle, is 30 m and corresponds to an equivalent gradient of 90 kG/cm (for a magnet of the same longitudinal extent).

A different configuration, namely a cylindrical configuration such as he has developed for the Inverse Cherenkov Accelerator, has been proposed by Fontana. The focal length of a lens of length L, for a particle with relativistic factor γ, subject to laser light of wavelength λ and power P is given by

$$f \simeq (2.5 \times 10^3)\ \frac{\gamma\lambda^{3/2}}{\theta^3 L^{1/2}\ P^{1/2}} ,$$

where θ is the Cherenkov angle and all quantities are in MKS units. If, for example, we have hydrogen at 10 atmos (θ=52 mrad), $\lambda=10^{-5}$m (CO_2 laser), $\gamma=10^5$ (50 GeV), L=1 cm, and P=10^{10}W then f=5.3 meters.

The advantages of these laser focusing schemes is the high field gradient which can be achieved, with even a modestly sized laser, and the fast accurate control of the focusing. The speed with which the focusing can be turned on or off should prove most useful to linear collider designers and linear collider users. If the field breakdown of the gas can be increased, or even exceeded, without degrading the lens, then really high focusing gradients can be achieved.

7. CONCLUSIONS

The working group came to the following conclusions:

a. None of the schemes is revolutionary; i.e. changes drastically what we think or what we are doing.

b. Experimental work will be done which will be relevant to the Plasma Focus Accelerator, but no work is being done on any of the other ideas.

c. More theoretical work is needed on the Radial Implosion Accelerator and Laser Focusing concept before experimental work is initiated.

d. The Inverse Cherenkov Accelerator is ready for further experimental study and such study would teach one about high-power laser optics and gas media behavior under high fields. (When does it break down, and is breakdown in the form of a ring of fire bad for acceleration along the axis?)

e. The development of focusing schemes looks like an area in which more effort will pay off. These are not laser, or even novel, accelerators, but could be important for attaining a high-luminosity, high-energy collider.

REFERENCES

1. J. R. Fontana and R. H. Pantell, J. Appl. Phys. 54 (8), 4285 (1983); and these proceedings.
2. M. J. Abedi, these proceedings.
3. P. Csonka, these proceedings.
4. H. Hora, P. Lalousis, and S. Eliezer, Phys. Rev. Lett. 53, 1650 (1984); H. Hora, Lasers and Part. Beams 3, 59 (1985); H. Hora, these proceedings.
5. P. Channell, these proceedings.
6. P. Channell, these proceedings.

INVERSE CHERENKOV ACCELERATION

J. R. Fontana
Department of Electrical & Computer Engineering
University of California, Santa Barbara, CA 93106

ABSTRACT

This paper describes the use of interactions between laser fields and charged relativistic particles inside a passive gas medium for high-energy acceleration. Suitable geometrical configurations are discussed quantitatively, and consideration is given to gas breakdown and collision effects. It is shown how currently available laser powers can yield large electron acceleration gradients and total energy gain while controlling beam spreading by the inherent focusing properties of the fields.

INTRODUCTION

A charged particle moving in a gas at speed β c and a plane electromagnetic wave propagating at angle Θ relative to this motion will interact cumulatively provided the Cherenkov condition is fulfilled:

$$n \beta \cos\Theta = 1 \tag{1}$$

where n is the index of refraction of the gas. If β and Θ do not change, the fields seen by the particle remain constant in time, and an electric field of magnitude E $\sin\Theta \cos\Psi$ continuously increases or decreases the particle energy, according to the sign of the phase angle Ψ of the fields. For $\gamma \equiv (1-\beta^2)^{-\frac{1}{2}} >> 1$ and small angles Θ, the Cherenkov condition may be written

$$2 (n - 1) \cong \Theta^2 + \frac{1}{\gamma^2} \tag{2}$$

This method of interaction is particularly suited to highly relativistic particles because

(a) The condition of Eq. (1) remains valid for any energy as long as

$$\Theta^2 >> \frac{1}{\gamma^2} \text{ ; and}$$

(b) The effect of collisions with gas molecules is less at large γ.

Experimental confirmation of the process described above was obtained at Stanford by R. H. Pantell, M. A. Piestrup and their

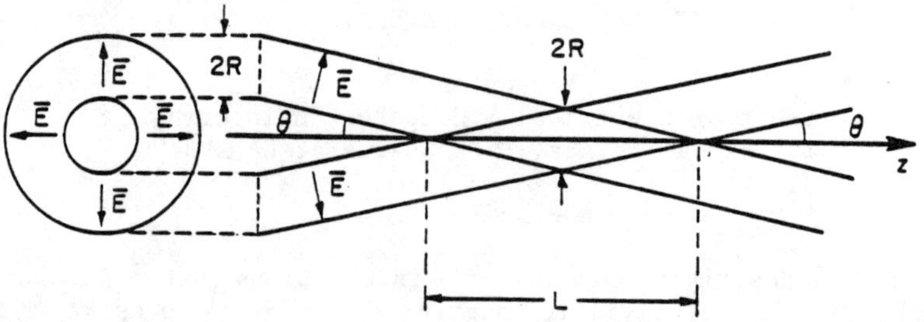

Fig. 1: Cylindrical symmetry configuration for laser interaction.

co-workers[1]. Using a $\lambda = 1.06$ micron laser to energy modulate $\gamma \cong 100$ electrons in hydrogen gas at near atmospheric pressure they observed cumulative interaction over a distance of seven centimeters.

Inside a gas, several combinations of plane waves have the properties of near c phase velocities and accelerating fields in the same direction. Limiting the study to monochromatic radiation, one such configuration is a planar symmetry situation produced by two linearly polarized plane waves converging on the z-axis at angles $\pm\Theta$ with their k-vectors and their E fields contained in the xz plane. The resulting fields are given by:

$$E_x(x,\Psi) = -2E_m \cos\Theta \sin[nkx \sin\Theta] \sin\Psi \quad (3)$$

$$E_z(x,\Psi) = 2E_m \sin\Theta \cos[nkx \sin\Theta] \cos\Psi \quad (4)$$

$$H_y(x,\Psi) = -2E_m n\sqrt{\frac{\varepsilon_o}{\mu_o}} \sin[nkx \sin\Theta] \sin\Psi \quad (5)$$

where $$\Psi = kct - nkz \cos\Theta \quad (6)$$

and E_m is the peak value for each plane wave. There is no y-dependence. The whole pattern moves in the z direction at speed β c where β is given by Eq. (1), so a particle moving at this particular

β will see constant axial and transverse forces. The maximum axial field E_z is only $\tan\Theta$ as large as the maximum transverse E_x.

A cylindrical symmetry configuration consists of a continuous distribution of linearly polarized plane waves, all converging on the z-axis at Θ with their k-vectors forming a cone. The fields are in this case:

$$E_r(r,\Psi) = -E_0 \cot\Theta \, J_1(nkr \sin\Theta) \sin\Psi \tag{7}$$

$$E_z(r,\Psi) = E_0 \, J_0(nkr \sin\Theta) \cos\Psi \tag{8}$$

$$H_\phi(r,\Psi) = -E_0 n \operatorname{cosec}\Theta \sqrt{\frac{\varepsilon_0}{\mu_0}} \, J_1(nkr \sin\Theta] \sin\Psi \tag{9}$$

Again this pattern moves at a β c given by Eq. (1). Now the maximum axial field E_0 is 1.72 $\tan\Theta$ as large as the maximum E_r.

This cylindrical symmetry arrangement is particularly suitable for high-energy electron acceleration. As will be discussed later, the field gradients compatible with gas breakdown limitations are larger than those achievable in microwave linacs, and there is no requirement of a "wiggling" in the trajectory like for IFEL's, so no synchrotron radiation occurs. As the fields have focusing properties, gas collision scatter can be controlled, particulary at large γ's, even over long interaction paths.

A configuration embodying this geometry is sketched in Figure 1. The field pattern, identical to that in a circular wave guide operating in a TM_{0m} mode with large radial variation index m, can be obtained by an axicon-type optics. All dimensions are large compared to λ, so diffraction effects are unimportant. An electron moving with β c on the z-axis at constant phase Ψ of the fields will increase its energy by[2]:

$$\Delta W = -68.8 \sin\Theta \left[\frac{LP}{\lambda}\right]^{\frac{1}{2}} \cos\Psi \tag{10}$$

where ΔW is in electron-volts, the interaction length L and the wavelength λ are in meters, and the total laser power P is in watts.

For instance, a 1 Gev energy gain can be obtained over L = 1 meter by using a $CO_2(\lambda = 10^{-5}$ meter) laser power $P = 2 \times 10^{12}$ watts at Θ = 32 milliradians, which corresponds to hydrogen at about 4 atmospheres.

Simple scaling laws are obvious from Eqs. (2) and (10); as n-1 is proportional to molecular density for each gas, ΔW increases as the square of the pressure. The wavelength comes in because the fields are more closely concentrated around the z-axis at shorter λ's. But the practical scaling range is limited by physical constraints of the medium. In the example above, a radial field $E_{r,max} = 1.76 \times 10^{10}$ volts/meter occurs at $r \cong 90$ microns from the axis; increasing P may cause gas breakdown and a disruption of the whole process. Also, increasing the pressure to increase Θ implies enhanced gas collision effects as well as an unfavorable change in breakdown threshold.

BREAKDOWN EFFECTS

An important consideration in determining the largest accelerating gradient available is the maximum field the gas can stand before breaking down into a plasma.

Gas breakdown under short, intense laser pulses is a complex phenomenon. There are several distinct regimes, dominated by one or the other of several competing physical processes, and current theory doesn't give clear numerical predictions. The following parameters are significant:

(a) The nature and pressure of the gas;
(b) The pulse length of the laser; and
(c) The laser wavelength.

Also of importance but harder to express are such parameters as the dust and ion contents of the gas before the pulse, and the pulse repetition rate. Relevant experimental data available in the literature is sparse and reflects a wide range in both measurement conditions and results[3]. CO_2 lasers present a number of advantages for acceleration, among them high overall power efficiency and a wavelength suitable for optics design within easily achievable tolerances. But until recently no short (picosecond range) pulses had been obtained and there is yet little experimental information on their breakdown-inducing properties. On the other hand, some meaningful data has been published for $\lambda = 1.06$ micron short pulses in nitrogen near atmospheric pressure. It indicates a critical dependence on pulse length. Below a few picoseconds breakdown is almost pressure-independent, indicating that electron cascade multiplication is no longer dominant, and that processes such as multiphoton ionization, Keldish tunnelling[4] and inverse bremsstrahlung become the key factors.

Figure 2 is a breakdown power density versus pulse duration plot showing three data points published by different authors,[5,6,7] all for N_2 or air at $\lambda = 1.06$ microns and 1 atmosphere. The curve that joins them expresses a clear trend. For instance, at 5 picoseconds, typical of S-band linac electron bunches, the gas may be expected to withstand some 2×10^{15} watts/cm^2, i.e. 1.2×10^{11} volts/meter. For hydrogen rather than nitrogen, the breakdown threshold should be even higher due to the higher ionization potential. Regarding wavelength scaling, as less energetic photons are less efficient in producing electron-ion pairs, the breakdown threshold should be larger at longer wavelengths. Reference 5 gives comparative measured values for λ = 1.06, 0.69 and 0.53 microns which confirm this expectation.

More complete experimental work is needed to predict the limitations imposed by breakdown on the inverse Cherenkov acceleration method. As the overall field configuration may be significant, measurements should be made with the cylindrical symmetry laser geometry, and they should involve parameters such as repetition rate and pre-ionization as well as pulse length. A related area of study is the actual effect of plasma formation on the acceleration process. As the largest field, responsible for breakdown, is radial and way from the axis, the plasma may not affect directly the interaction region. Changes in index of refraction associated with the plasma electrons would certainly perturb the Cherenkov condition, but the importance of this effect near the axis, where accelerating electrons are located, must be studied experimentally. The same applies to phenomena such as wavefront instability and scatter above the breakdown threshold.

COLLISION EFFECTS

For high-energy electrons, in the Gev range or above, inelastic collision processes resulting in molecular excitation or ionization are a relatively unimportant cause of energy loss. More significant is the radiation loss (bremsstrahlung) caused by deflections due to atomic nuclei; it gives rise to an exponential decay in average electron energy, characterized by a radiation length X_0, as well as to a spread in energy distribution. Typical values of X_0, however, are much larger than the interaction region lengths considered for inverse Cherenkov acceleration; for instance in hydrogen at one atmosphere, $X_0 \cong 7 \times 10^3$ meters. Consequently radiation losses may also be neglected as a first approximation.

A more troublesome process involves multiple elastic collisions[8]. It produces a scatter in transverse position and in angular direction which results in emittance growth and also deeply affects the acceleration process by displacing the particles to unfavorable positions within the laser field pattern.

Consider again the two-dimensional geometry described in the Introduction, with two plane waves converging on the z-axis and with no y-variations. If at z = 0 there is an incoming electron beam, perfectly collimated and of zero width in the x-direction, multiple collision scatter will cause it to spread so at any z it will have acquired a transverse distribution density which is no longer a delta-function. In the absence of laser fields, this distribution is gaussian, characterized by a 0.368 density point[9]

$$\pm \frac{z^{3/2}}{\sqrt{3}} \Theta_s \tag{11}$$

where

$$\Theta_s \cong \frac{6}{\gamma\sqrt{X_0}} \qquad \text{for } \gamma >> 1 \tag{12}$$

X_0 being the radiation length in meters.

With the laser fields present, this collision transverse distribution is modified by both the change in γ due to acceleration (positive or negative according to Ψ) and the focusing effect (again positive or negative) of the transverse forces. From Eqs. (3) and (5):

$$F_x(x,\Psi) = - e\,[E_x(x,\Psi) - \beta c\mu_0 H_y(x,\Psi)] =$$

$$= - 2eE_m \sin\Theta \tan\Theta \sin[nkx \sin\Theta] \sin\Psi \tag{13}$$

For electrons near the z-axis $\frac{\pi}{2} < \Psi < \frac{3\pi}{2}$ is the range for positive acceleration, and $0 < \Psi < \pi$ the range for positive focusing. For phase angles $\frac{\pi}{2} < \Psi < \pi$ the electrons are accelerated and also subject to focusing forces which counteract the elastic collision spreading effect. This is also the stable acceleration range where the lagging particles in a bunch are more accelerated than the leading particles.

Figures 3, 4 and 5 show the result of some Montecarlo calculations involving these effects. They show electron distributions at $z = 10$ meters for the case of $\lambda = 10$ microns, $\Theta = 20$ milliradians and laser fields producing $E_z = 5 \times 10^8$ volts/meter on the z-axis. In this pattern, $x = \pm 125$ microns is the first zero of the axial field. The initial electron energy is $\gamma = 10^3$ and the bunches are supposed to consist of electrons all at one value of Ψ. In Figure 3, there are no laser fields, and it can be seen that collisions spread the beam way beyond the near-axial region. In Figure 4 the laser is on, and all electrons are taken to be at $\Psi = \pi/2$, giving maximum focusing but no acceleration; the distribution is dramatically narrowed, demonstrating the effectiveness of field focusing. In Figure 5 the electrons are at $\Psi = \frac{3\pi}{4}$ and are both accelerated and focused; their γ increases from 10^3 to 8×10^3 and the distribution is even tighter due to the effect of acceleration, although the focusing force is only 71% as large.

Emittance growth is another effect of collision scatter. It can be calculated for bunches undergoing both acceleration and focusing by utilizing the technique described in Reference 10. The betatron focusing function $\beta_\perp$ involved in the expressions found there is given for the plane geometry case by

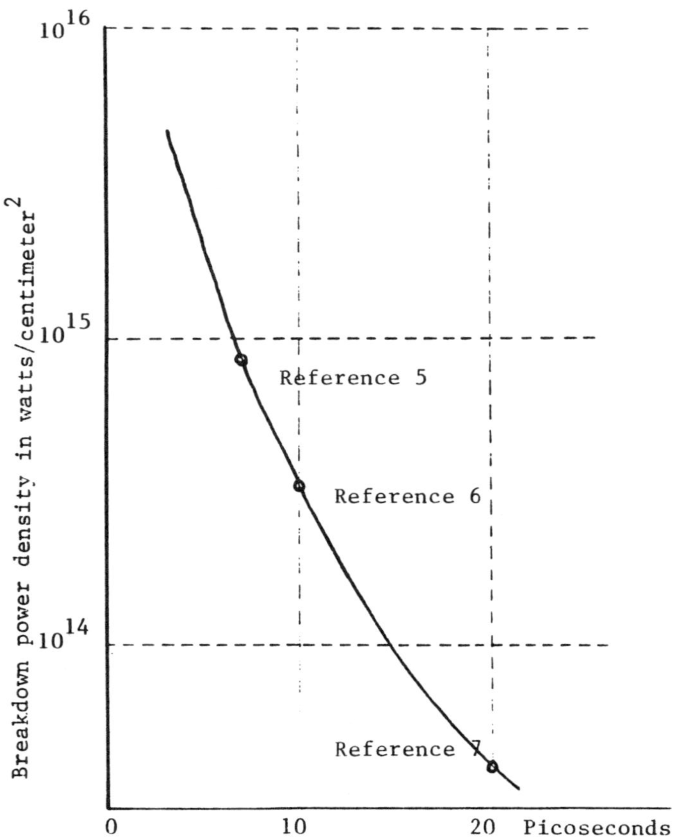

Fig. 2: Breakdown versus pulse length.

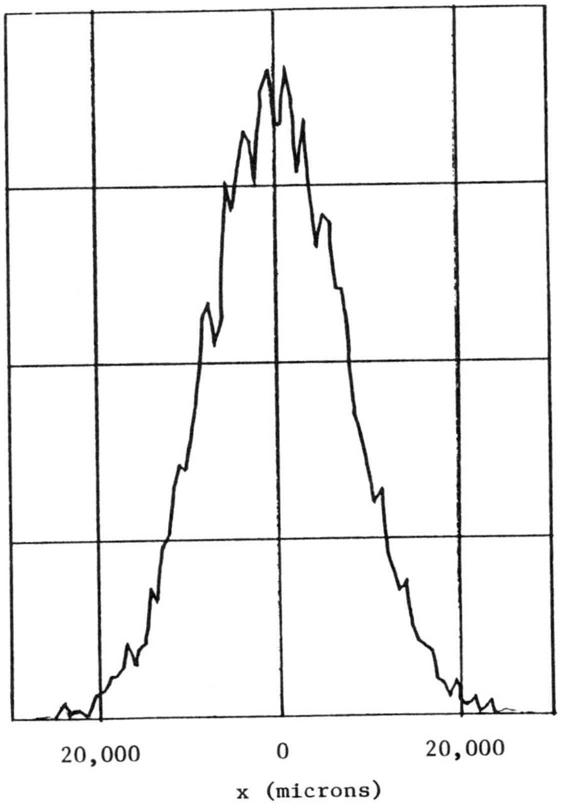

Fig. 3: Transverse electron distribution at z = 10 m without fields.

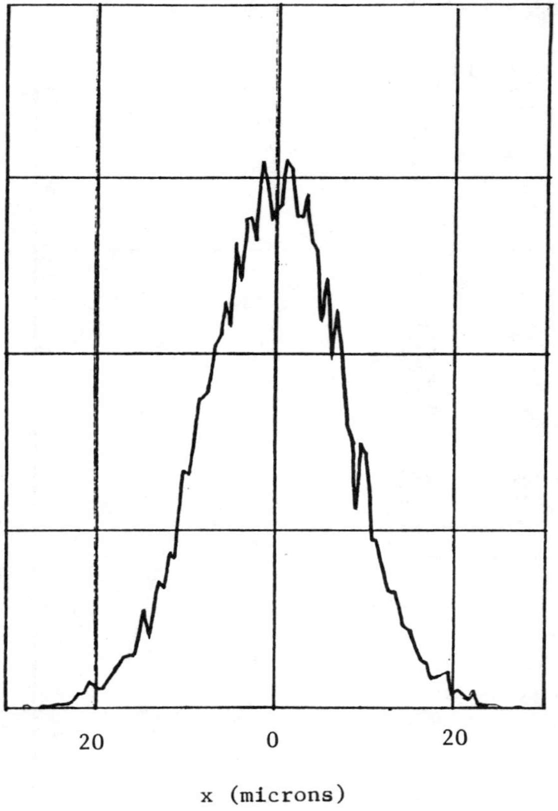

Fig. 4: Transverse electron distribution at $z = 10$ m and $\Psi = \pi/2$

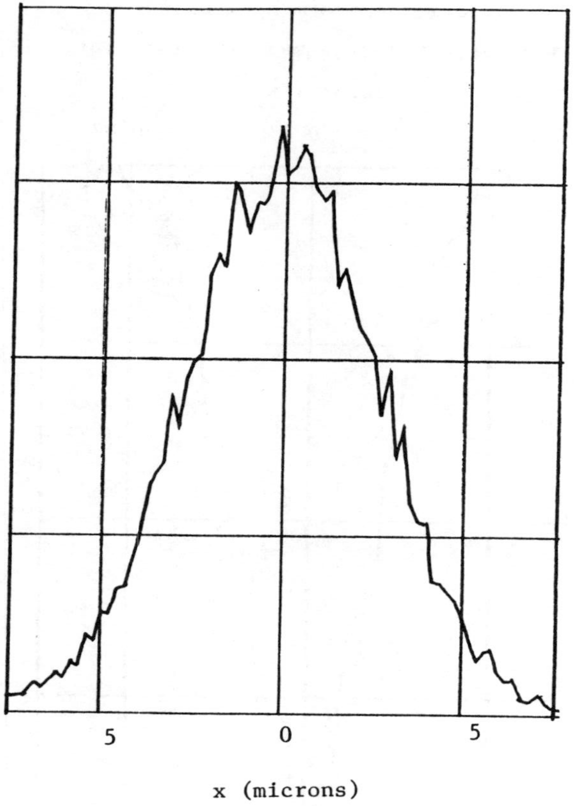

Fig. 5: Transverse electron distribution at $z = 10$ m and $\Psi = 3\pi/4$

$$\beta_{\perp}^2 = \frac{\gamma mc^2}{4eE_m\, \Theta^3\, (\pi/\lambda)\sin\Psi} \tag{14}$$

and for the cylindrical symmetry case by

$$\beta_{\perp}^2 = \frac{\gamma mc^2}{eE_0\, \Theta^2\, (\pi/\lambda)\sin\Psi} \tag{15}$$

The logarithmic term involving extreme values of collision angles has to be evaluated from scattering theory considerations.

ACCELERATOR SYSTEM CONSIDERATIONS

A significant aspect of user-oriented accelerators is overall energy efficiency. For any proposed scheme this entails a consideration of the closely related topics of laser power generation efficiency and of spent beam recovery. The inverse Cherenkov method appears comparatively well suited for both. It can use infrared wavelengths, such as from CO_2 lasers, where best efficiency has been demonstrated, and its spent beam consists of simply distributed plane wave components, which should facilitate recovery.

A simple method for reusing the spent beam is to use an optical waveguide. Suppose the incoming laser beam in Figure 1, instead of being allowed to diverge after interaction and be lost, is trapped by reflection on a cylindrical mirror surface of radius a. The fields inside the guide will still be expressed by Eqs. (7) to (9). Now the accelerating field on axis is

$$E_0 = \left[2\pi \sqrt{\frac{\varepsilon_0}{\mu_0}}\ \tan\Theta\ \frac{P}{\lambda a} \right]^{\frac{1}{2}} \sin\Theta \tag{16}$$

and, except for losses, it will occur over the whole length of the guide instead of being limited by the geometry. The radius a must be large enough to avoid damage to the guide surface by the field

$$E_r(a) \cong \frac{1}{a} \left[\frac{2}{\pi} \sqrt{\frac{\varepsilon_0}{\mu_0}}\ \tan\Theta\ P \right]^{\frac{1}{2}} \cong \frac{E_0}{\pi\Theta} \left[\frac{\lambda}{a}\right]^{\frac{1}{2}} \tag{17}$$

Apart from practical construction difficulties, waveguides present the following disadvantages:

(a) As the fields decay along their length due to beam loading and guide reflection losses, the average acceleration gradient is decreased;

(b) Reflection losses being irreversible, part of the power cannot be recovered; and

(c) If the interaction length is too long, the laser pulse will fall behind the electron bunches due to dispersion.

Another arrangement to reuse the beam is interaction inside a laser resonator. A resonator sustaining a recirculating, mode-locked pulse constitutes a system with built-in recycling of the spent power, successive passes of the laser pulse through the resonator being available for interaction with successive electron pulses. This mode of operation, instead of using a laser pulse with all its energy provided by an amplifier during the pulse time and shaped by suitable optics, would build up the pulse power in the course of several passes through the resonator gain medium, and then sustain it against beam loading and passive losses during the whole duration of the oscillation regime.

Interaction inside the resonator opens the possibility of accelerator devices based on a succession of optically bunched electron pulses passing through several laser resonators where oscillation takes place at the same frequency and with correlated phases. Mode-locked short laser pulse operation would be used in each resonator, the phase correlation being obtained by a common injected signal from a master oscillator and the fields being sustained by an amplifying section within each resonator. This overall accelerator configuration would resemble that of a microwave linac with successive sections powered by amplifiers fed from a common master oscillator. The relative phase of the fields could be maintained by passive components (phaseshifters) adjusted at the beginning by trial and error as in microwave machines.

Optimum-design resonators should produce a laser field configuration corresponding to a single interaction angle Θ for all plane wave components. The use of configurations which are easier to produce but involve a spread of values for Θ would be less efficient; compromises may however be practical.

REFERENCES

1. M. A. Piestrup, G. B. Rothbart, R. N. Fleming, and R. H. Pantell, Jour. Appl. Phys. 46, 132 (1975).
2. J. R. Fontana, and R. H. Pantell, Jour. Appl. Phys. 54 (8), 4285-4288, August 1983.
3. See the Table on page 4288 of Reference 2.
4. L. V. Keldish, Sov. Phys. JETP, 20, 1307-1314 (1965).
5. R. J. Dewhurst, J. Phys. D, 11, L 191-195 (1978).
6. A.J. Alcock, and M. C. Richardson, Phys. Rev. Lettrs 21, 667-670 (1968).
7. C. L. M. Ireland et al, Appl. Phys. Lettrs 24, (4), 15 February 1974.
8. N. M. Blachman, and E. D. Courant, Phys. Rev. 140 (1948).
9. See for example B. Rossi "High-Energy Particles", Prentice-Hall Inc., N. Y. 1952, Chapter 2.
10. B. W. Montague, CAS-ECFA-INFN Workshop on the Generation of High Fields for Particle Acceleration to Very High Energies, Frascati, 25 September to 4 October 1984.

THREE-WAVE ACCELERATOR AND HOW IT COMPARES WITH TWO-WAVE ACCELERATOR

M. J. Abedi
Department of Physics and Institute of Theoretical Science
University of Oregon, Eugene, OR 97403

ABSTRACT

A three-wave particle accelerator is briefly described and its energy gradient and efficiency are compared with those of two-wave accelerator.

INTRODUCTION

This paper describes an accelerator scheme which employs three coherent electromagnetic beams in vacuum. The acceleration mechanism is essentially equivalent to Inverse-Free-Electron-Laser (IFEL), and the three-wave accelerator may be identified as an IFEL accelerator in which the magnetic wiggler is replaced by two crossed coherent (laser or microwave) beams.

We assume that the particle density is sufficiently dilute so that the amplitudes of the external electromagnetic waves are not affected and space-charge forces can be neglected.

Thus, in the plane-wave approximation, we shall consider the motion of a single electron in the external electric and magnetic fields in the overlap region of three specified (monochromatic) electromagnetic plane waves intersecting each other at small angles. Also assumed is that the accelerating particle has an initial highly relativistic velocity in the forward direction which remains (approximately) constant during the acceleration period.

The main objective is to find an approximate expression for the particle's energy as a function of time and parameters of the waves (amplitudes and wavelengths) which may be used to examine the potential of the three-wave mechanism for particle acceleration. We will also compare the energy gradient and efficiency of this scheme with those of two-wave accelerator.[1]

SPECIFICATION OF THE ELECTROMAGNETIC WAVES

The i-th plane wave (for i = 1, 2, 3) of wavelength λ_i and angular frequency ω_i propagating in vacuum may be represented by the following equations:

$$\vec{E}_i\ (t, \vec{r}) = \vec{E}_{oi} \sin\ (\omega_i t - \vec{K}_i \cdot \vec{r} - \delta_i) \tag{1}$$

$$\vec{B}_i(t, \vec{r}) = \vec{B}_{oi} \sin\ (\omega_i t - \vec{K}_i \cdot \vec{r} - \delta_i) \tag{2}$$

$$\vec{B}_{oi} = \frac{1}{C} \vec{e}_{ki} \times \vec{E}_{oi} \tag{3}$$

where $\vec{E}_{oi}$ and $\vec{B}_{oi}$ are constant real electric and magnetic field amplitudes respectively, $\vec{K}_i$ is the propagation vector of magnitude $K_i = 2\pi/\lambda_i$ and direction of unit vector $\vec{e}_{ki}$, δ_i is a phase angle and C is the velocity of light in vacuum.

All the waves have their electric field and propagation vectors lie in the X-Z plane as shown in Fig. 1.

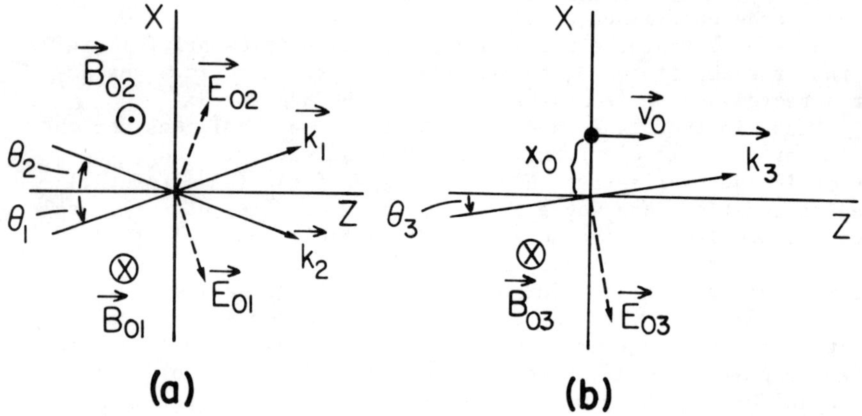

Figure 1. (a) Relative polarization and propagation of waves 1 and 2 which together form a dynamical wiggler analogous to the static wiggler of the IFEL accelerator. (b) Relative polarization and propagation of wave 3 as well as the initial position and velocity of the particle. (Symbol ⊙ or ⊗ indicates that the corresponding vector points out of or into the plane of the paper.)

Furthermore, waves 1 and 2 are propagating symmetrically about the Z-axis, i.e., $\theta_1 = -\theta_2$, $E_{01} = E_{02}$, $\lambda_1 = \lambda_2$ and $\delta_1 = \delta_2$; waves 1 and 3 have similar geometry but, in general, different parameters. All the angles will be small in practice but θ_1 and θ_2 must be (obviously) nonzero.

PARTICLE'S TRAJECTORY AND SYNCHRONISM CONDITION

The equations of motion of a particle of charge e and rest mass m in external fields $\vec{E}$ and $\vec{B}$ are the following:

$$\dot{\vec{P}} = e[\vec{E} + \vec{V} \times \vec{B}] \tag{4}$$

$$\dot{\varepsilon} = e\,\vec{V} \cdot \vec{E} \tag{5}$$

where $\vec{V}$, $\vec{P}$ and ε are the velocity, momentum and total energy of the particle respectively.

Starting with these equations and noting that $\vec{P} = \gamma\, m\, \vec{V}$ and $\varepsilon = \gamma m C^2$, where $\gamma = [1- |\vec{V}|^2/C^2]^{1/2}$, it is simple to show that:

$$\dot{\vec{V}} = \frac{e}{\gamma m} [\vec{E} + \vec{V} \times \vec{B} - (\vec{V} \cdot \vec{E}) \frac{\vec{V}}{C^2}] \tag{6}$$

We assume that $\dot{Z} \cong \dot{Z}_o \cong C$ where the subscript zero denotes values at $t = 0$. With this assumption which in turn implies $\dot{X} << \dot{Z}$, Eq. (6) for TWA field configuration of Fig. 1. is reduced to the following:

$$\ddot{X} = \frac{e}{\gamma m} [E_x - C\beta_Z B_y] \tag{7}$$

$$y = 0 \tag{8}$$

$$Z = \dot{Z}_o t \tag{9}$$

where $\beta_Z = \dot{Z}/C$ and we have set initial values Y_o and $\dot{Y}_o$ equal to zero.

Eq. (7) may be written as

$$\ddot{X} = \frac{1}{\gamma} [-2A_1 \cos\tilde{\omega}_1 t \sin K_{1x} X + A_3 \sin (\tilde{\omega}_3 t - \delta_3)] \tag{10}$$

where $$A_i = (e\, E_{oi}/m)\, (\beta_Z - \cos\theta_i) \tag{11}$$

$$\tilde{\omega}_i = \omega_i - K_{iz}\, \dot{Z},\ i = 1, 2, 3 \tag{12}$$

and we have set $\delta_1 = \theta_3 = 0$.
By defining a dimensionless quantity η through

$$X(t) = \tilde{X}_o + \eta(t)/K_{1x} \tag{13}$$

where $\tilde{X}_o$ is an arbitrary constant, one transforms Eq. (10) into

$$\ddot{\eta} = \cos\tilde{\omega}_1 t\, [A_s(t)\cos\eta + A_c(t)\sin\eta] + A_p(t)\sin(\tilde{\omega}_3 t - \delta_3) \tag{14}$$

where

$$A_s(t) = -2A_1 K_{1x} \sin(K_{1x}\tilde{X}_o)/\gamma \tag{15}$$

$$A_c(t) = -2A_1 K_{1x} \cos(K_{1x}\tilde{X}_o)/\gamma \tag{16}$$

$$A_p(t) = A_3 K_{1x}/\gamma \tag{17}$$

We choose $\tilde{X}_o$ such that

$$K_{1x}\tilde{X}_o = \pi/2 \tag{18}$$

in order to eliminate A_c. (This equation implies a non-essential constraint on $\tilde{X}_o$ which can be removed if needed.) Furthermore, we assume that $\eta << 1$ which implies $\sin\eta \simeq \eta$ and $\cos\eta \simeq 1$. Thus Eq. (14) reduces to

$$\ddot{\eta} = A_s(t)\cos\tilde{\omega}_1 t + A_p(t)\sin(\tilde{\omega}_3 t - \delta_3) \tag{19}$$

At this point we notice from the definition of A_s and A_p that the first term in Eq. (19) is contributed by waves 1 and 2, which together simulate the magnetic wiggler of an IFEL accelerator, and the other term is due to wave 3 with which the accelerating particle is supposed to stay in phase. Thus the synchronism condition is simply

$$\tilde{\omega}_1 = \tilde{\omega}_3 \tag{20}$$

which implies

$$\frac{\lambda_3}{\lambda_1} = \frac{1 - \beta_z}{1 - \beta_z \cos \theta_1} \tag{21}$$

This condition can also be obtained from geometrical consideration of the acceleration mechanism.

Returning to Eq. (19), we notice that coefficients A_s and A_p involve γ whose time dependence is not known and, hence the equation cannot be integrated. But, directed by our earlier numerical work[2] on the subject, we can write down an approximate solution

$$\eta = \alpha_s(t) - \alpha_p(t) \sin\delta_3 - \alpha_s(t) \cos\omega t - \alpha_p(t)\sin(\omega t - \delta_3) \tag{22}$$

where

$$\alpha_s(t) = A_s(t)/\omega^2 \tag{23}$$

$$\alpha_p(t) = A_p(t)/\omega^2 \tag{24}$$

$$\omega = \tilde{\omega}_1 = \tilde{\omega}_3 \tag{25}$$

and it is assumed that

$$\left| \frac{\alpha_s}{\alpha_p} \frac{\dot{\gamma}}{\gamma} \right| << \omega \tag{26}$$

and $|\alpha_p| < |\alpha_s|$. One can verify that Eq. (22) satisfies Eq. (19).

PARTICLE'S ENERGY

The particle's energy changes in time according to Eq. (5). From Eq. (13) we have

$$\vec{V} = \dot{\eta}/K_{1x} + \dot{Z}_o \vec{e}_z \tag{27}$$

Using Eqs. (22) and (27), we integrate Eq. (5) for $\gamma=\varepsilon/mc^2$ and obtain

$$\gamma = [\beta_1 t + S(t) - S(0) + \gamma_0^2]^{1/2} \tag{28}$$

where

$$\beta_1 = \frac{\cos\delta_3 \, e^2 \, E_{01} E_{03} \lambda_1}{\pi m^2 C^3} \, G\,(\beta_Z,\mu) \tag{29}$$

$$G(\beta_Z,\mu) = \beta_Z(1-\mu)(\beta_Z - \mu)/(1 - \beta_Z\mu)^2, \qquad \mu = \cos\theta_1 \tag{30}$$

and $S(t)$ is a periodic function of angular freqency ω whose explicit form is not of interest in the present discussion. $S(t)$ on the average will not contribute to the value of γ and, furthermore, it becomes negligible relative to the linear term as $t\to\infty$ which is the limit of interest. This motivates the definition of a "time averaged" γ by simply dropping $S(t) - S(0)$ from Eq. (28) to get,

$$\gamma_{avg}(Z) = \sqrt{bz + \gamma_0^2} \tag{31}$$

where $b = \beta_1/C$ and t has been expressed in term of Z. From Eq. (31), the (average) energy gradient in units of rest mass may be written as

$$\frac{d\gamma_{avg}}{dZ} = \cos\delta_3 \, \frac{e^2 E_{01} E_{03} \lambda_1}{2\pi m^2 C^4} \, \frac{1}{\gamma_{avg}} \, G\,(\beta_Z,\mu) \tag{32}$$

The physics is essentially contained in G which is a function of two variables β_z and μ. In our approximation, $\beta_z > \mu$ and we have

$$\lim_{\beta_z \to 1} G\,(\beta_z,\mu) = 1 \tag{33}$$

Thus, we may maximize the energy gradient with respect to δ_3 and G by setting

$$\delta_3 = 0; \; G(\beta_Z,\mu) = 1 \tag{34}$$

To get an idea about the energy gradient in the three-wave accelerator we resort to an example: Suppose the three waves are chosen such that $|e E_{01} \lambda_1| = 1$ MeV, $|e E_{03}| = 100$ GeV/m. Then, for a 50-MeV electron, the energy gradient is about 0.3 GeV/m.

Finally, a brief comparison between the three-wave accelerator presented here and the two-wave accelerator previously discussed in reference 1 seems to be both in place and necessary. In a two-wave accelerator only one wave serves as the wiggler, and one such scheme is obtained by eliminating wave 2 in Fig. 1. A similar calculation in the absence of wave 2 (or a proper comparison between Eq. (32) and the corresponding equation of reference 1) shows that,

$$[d\gamma/dZ]_{3WA} = 2[d\gamma/dZ]_{2WA} \tag{35}$$

Taking the extra wave employed in 3WA into account, we get,

$$\frac{[\text{Energy Efficiency}]_{3WA}}{[\text{Energy Efficiency}]_{2WA}} = \sqrt{2} \tag{36}$$

Before concluding, we note that two and three-wave accelerators have the advantages of being far-field and dispensing with the magnetic wiggler of the IFEL accelerator. But there is a limit on the maximum beam energy due to synchrotron radiation and the fact that energy gradient decreases as energy increases.

ACKNOWLEDGEMENTS

I am grateful to Professor Paul L. Csonka with whom I had frequent instructive discussions on this subject. The work is partially supported by the Department of Energy.

REFERENCES

1. R. H. Pantell, T. I. Smith, Appl. Phys. Lett. 40, 753 (1982).

2. M. J. Abedi, The Concept of a Three-Wave Laser-Driven Particle Accelerator-Preliminary Report, 1982. (Unpublished).

TRANSVERSE ELECTRON RESONANCE ACCELERATOR*

Paul L. Csonka
National Synchrotron Light Source
Brookhaven National Laboratory, Upton, N.Y. 11973

ABSTRACT

Transverse (to the velocity, $\bar{v}$, of the particles to be accelerated) electron oscillations are generated in high (e.g. solid) density plasmas by either an electromagnetic wave or by the field of charged particles traveling parallel to $\bar{v}$. The generating field oscillates with frequency $\omega = \omega_p$, where ω_p is the plasma frequency. The plasma is confined to a sequence of microstructures with typical dimensions of $d \approx 2\pi c/\omega_p$, allowing the generating fields to penetrate. Since ω_p is now high, the time scales, T, are correspondingly reduced. The microstructures are allowed to explode after t = T, until then they are confined by ion inertia. As a result of resonance, the electric field, E, inside the microstructures can exceed the generating field E_L. The generating force is proportional to E_L (as opposed to E_L^2). Phase matching of particles is possible by appropriate spacing of the microstructures or by a gas medium. The generating beam travels outside the plasma, filamentation is not a problem. The mechanism is relatively insensitive to the exact shape and position of the microstructures. This device contains features of various earlier proposed acceleration mechanisms and may be considered as the limiting case of several of those for small d, T and high E.

DISCUSSION

The underlying principles of a transverse electron resonance accelerator was originally described during the first workshop on the Laser Acceleration of Particles (1). It makes use of the large internal electric fields which can be generated in a dense plasma by resonantly oscillating electrons.

The accelerator contains a sequence of relatively high density objects, whose density may reach solid state densities. (2) The typical dimension of these objects is small, as explained below, therefore we will refer to them as "microstructures". The microstructures are to be located at appropriately chosen positions along, and close to, the trajectory of the particles to be accelerated (See Fig. 1).

Denote the free electron density inside the microstructures by ρ_e, and the plasma frequency by ω_p. Electron oscillations are generated in each microstructure by an oscillating generating electric field of amplitude E_g, and circular frequency ω. Choose

*Work supported by the U.S. Department of Energy.

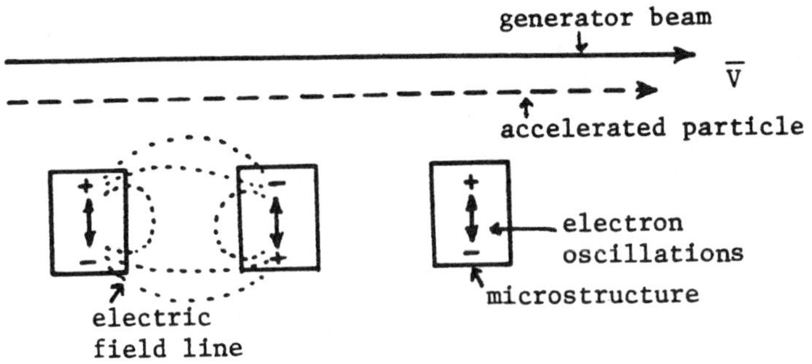

Fig. 1

$$\omega = \omega_p , \tag{1}$$

so that the generating field is in resonance with the plasma frequency. The oscillating electrons move along lines transverse to the velocity, $\bar{v}$, of the particle to be accelerated. As a result, a charge separation will be generated inside the microstructures, and those, in turn, create field lines which have a component parallel to $\bar{v}$ at the particle trajectory. This component will accelerate the particle. (Fig. 1 shows field lines corresponding to dipole type charge separation.) The oscillations are set up either by an electromagnetic wave or by bunches of charged particles which may move parallel to $\bar{v}$.

The typical diemsnion, d, of the microstructures is chosen to be of the order or less than λ_p, as defined below

$$d \lesssim 2\pi c/\omega_p = \lambda_r . \tag{2}$$

This condition insures that although ρ_e is high ($\lesssim 5 \cdot 10^{24}$ cm^{-3} $= \rho_{em}$), therefore so is ω_p ($\lesssim 1.3 \cdot 10^{17}$ sec^{-1} $= \omega_{pm}$ for nonrelativistic electrons) and although $\omega > \omega_p$, nevertheless the generating field can easily penetrate the microstructures from at least one, and perhaps several, directions.

As a result of resonant electron oscillations, large charge separations and high internal electric fields can be set up inside the microstructures. If the electrons can stay in resonance for N oscillations, where $N > 1$, the internal fields, E, can exceed the generating electric field amplitude E_g. For example, when $\rho_e = \rho_{em}$, one could reach $E = 3.3 \cdot 10^3$ GeV/cm $= E_m$ in about $N = 30$ to 10^3 oscillations, depending on the exact initial conditions. The nominally high value of E_m seems to justify further study of this mechanism.

We note that the electron oscillations being transverse, they are driven by a force proportional to E_g, which allows faster oscillation buildup than would be possible if the force were proportional to E_g^2.

The high internal fields will cause the microstructures to explode after a time T. For $t < T$ most electrons will be confined by the electrostatic force exerted by the ions, while those will be inertially confined. Since ω_p can now be high, even $N = 10^3$ oscillations will take place within a very short time interval (For $\rho_e = \rho_{em}$, this time is only $\approx 5 \cdot 10^{-14}$ sec), and confinement need to persist only during that interval.

The accelerating fields can be kept in phase with the accelerated particles in several ways:

a) Interrupt the sequence of microstructures in regions where dephasing would otherwise occur. Allow the particles to simply drift through these regions until the correct phasing is reestablished, and there resume acceleration.

b) If the generating field is due to an electromagnetic wave, that wave can be allowed to travel in a gas to insure phase matching. This is similar to "inverse Cherenkov acceleration", except that in contradistinction to that mechanism, in the present case the full transverse electric field in the wave <u>can</u> be utilized.

Thus we note with relief that to avoid dephasing of the oscillations, one does not have to resort to sophisticated devices, such as the proposed surfatron, or wave guiding structures. On the other hand, the present mechanism is flexible enough to allow "surfing", if one wishes to use that. In other words, the generating beam does not have to travel parallel to the sequence of microstructures, the generating fields need not penetrate the microstructures from the front, but may penetrate them obliquely, or from any other direction. If laser light is shone obliquely on the microstructures, then dephasing would tend to occur more often, but one can take care of that as described above. Since the generating laser light beam travels mostly outside the microstructures, no plasma related instabilities (e.g. filamentation) will develop in that beam.

Figure 2 a. illustrates a situation where tansverse resonant quadrupole electron oscillations are set up inside the microstructures. In this case the generating beam consists of a sequence of charged particle mcirobunches located at a distance λ_p from each other, and moving parallel to $\bar{v}$. (Microbunches can be generated in Transverse Optical Klystrons or in Free Electron Lasers. Values of λ in the visible range are feasible, and one expects eventually reach at least soft X-ray wavelengths[3]). The figure is drawn looking down on the microstructures, assuming they are located below that beam. The accelerated particles may move either above or below the microstructures. Quadrupole oscillations have the advantage of reducing collective radiation losses.

Figures 2b, 2c and 2d show schematic side views (unlike Fig. 2a) of certain alternative configurations.

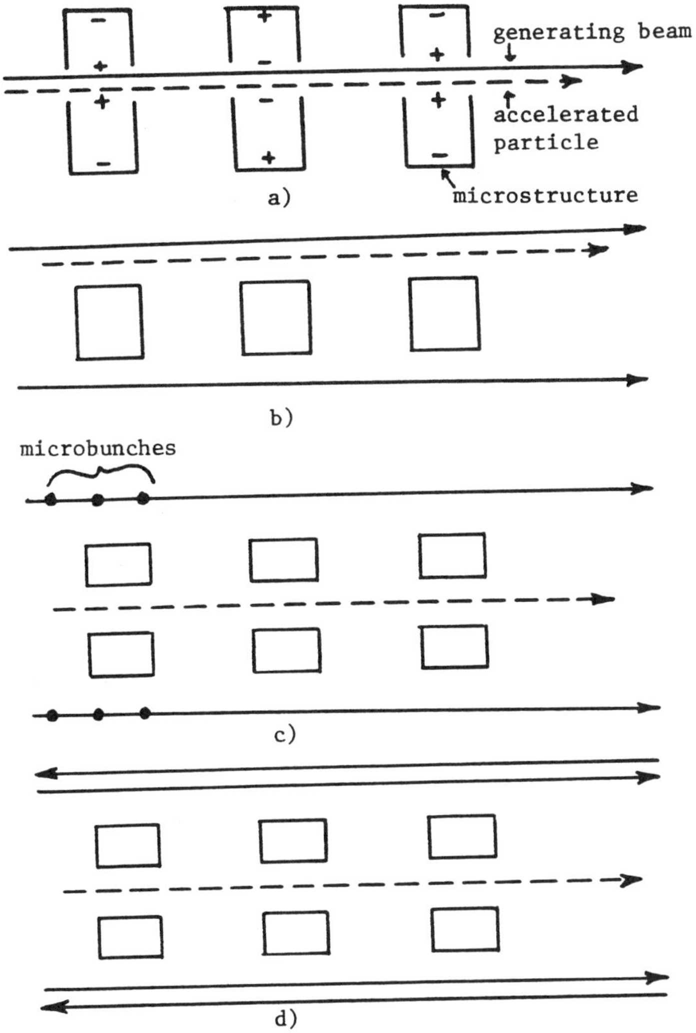
generating beam
accelerated
particle
microstructure
a)
b)
microbunches
c)
d)

Fig. 2

In all configurations shown, particle acceeration will be most effective if the accelerated beam passes the microstructures at a distance $\lesssim \lambda_p$. That underlines the importance of producing small crossection beams, including microbunched ones. For high energy colliders small beam crossections are required also by luminosity and power considerations.

If the oscillating electrons become relativistic, detuning of ω_p can be compensated in principle by changing λ_p for the generating beam. In microbunched beams that means changing the distance between microbunches (i.e. the corresponding particle density Fourier wavelength).

It is advantageous to induce electron oscillations with stronger electric fields, because the electron thermalization time increases with energy, so that a large initial amplitude will decrease randomization and increase N.

To illustrate, consider two cases. Parameters for the second case are more ambitious and are given in parentheses. Choose $\rho_e = 1.11 \cdot 10^{19}$ cm^{-3}, $\omega_p = 1.88 \cdot 10^{14}$ sec^{-1}. Let the generating field be due to a microbunched electron beam in which the particle density is 100% modulated with a Fourier wavelength 10 μm (1 μm). The beam passes the microstructures at a distance of 10 μm (1 μm), and the number of electrons per microbunch is 10^9 (10^8). Choose N = 1, a most pessimestic assumption. Then the internal field set up as a result of electron oscillations in the microstructures will be E ≈ 10 GeV/m (10^2 GeV/m). Of course, in reality one expects larger N, and therefore larger E. On should feel encouraged by these estimates.

The transverse electron resonance accelerator contains features from sevral distinct accelerating mechanisms proposed earlier. It is, of course, related to plasma accelerators, since plasma oscillations are induced. It is closely related to the acceleration by internal fields induced in a plasma focus, although there the charge separation is longitudinal. If a gas is used for phase matching, then in that respect it resembles an inverse Cherenkov accelerator. Since the generating beam travels mostly in vacuum, in that it is similar to far field accelerators. The presence of microstructures recalls the trend towards miniaturization in near field accelerators, such as when droplets are used as a wave guide, or in the two beam acceleration mechanism. One may argue with some conviction that in a certain sense the acceleration mechanism presented here is the limiting case for these various mechanisms, when the accelerating field is increased, breakdown of the materials is allowed, and sizes as well as times are reduced to what appears to be their minimum useful dimension. What one hopes to gain from this assortment of features, is a high accelerating field. Whereas that may not represent the supreme value in accelerator design, it is certainly a tempting goal.

Before the accelerator here described can hope to be of practical value, one will have to establish how large can N be for various realizable cases, how short microbunches can be created, and how close the microbunched beam can be allowed to pass near the microstructures.

Agenda for the near future: 1) Theoretically study electron oscillations, including numerical analysis. 2) Experimentally observe induced dipole amplitudes, and quadrupole amplitudes. 3) Study higher harmonic generation in TOKs and FELs.

ACKNOWLEDGEMENTS

I wish to thank several colleagues for interesting and occasionally animated discussions of this topic, especially Paul J. Channell, Shalom Eliezer, Heinrich Hora, Kenneth Lee, Michael J. Moravcsik and Andrew M. Sessler.

REFERENCE

1. Paul L. Csonka, in "Laser Acceleration of Particles", ed. Paul J. Channell. AIP, Conference Proceedings, Number 91, (1982), p. 216.
2. In principle, the density may exceed solid state density, but that case will not be considered here.
3. Paul L. Csonka in "SSRP Wiggler Workshop, (1977), and Particle Accelerators 7, 21 (1978), and 11, 45 (1980).

LASER FOCUS ACCELERATOR BY RELATIVISTIC SELF-FOCUSING AND HIGH ELECTRIC FIELDS IN DOUBLE LAYERS OF NONLINEAR FORCE PRODUCED CAVITONS

P.J. Clark, S. Eliezer*, F.J.M. Farley, M.P. Goldsworthy, F. Green, H. Hora, J.C. Kelly**, P. Lalousis***, B. Luther-Davies****, R.J. Stening and Wang Jin-Cheng
Department of Theoretical Physics, University of New South Wales,
Kensington 2033, Australia

ABSTRACT

The laser focus accelerator with relativistic self-focusing for achieving Z-separated heavy ions of energies beyond 10 GeV was studied experimentally, in detailed numerical work and estimations on intense muon sources, heavy nuclear collisions and generation of new isotopes are on the way. The recently detected inverted double layers in the nonlinear (ponderomotive) force produced cavitons with 10^9 V/cm nearly static field can be used for electron acceleration. An upgraded present days Antares system with 20 phase-optimized steps should arrive at TeV electrons. The spontaneous high magnetic fields should produce highly directed non-Z-separated ion bunches where the E × B mechanism of Forslund and Brackbill with thermally created electric fields can be improved drastically by nonlinear force generated fields. Further studies were on acceleration by relativistic Doppler shift and by the transverse free electron laser.

*Institute of Fusion Studies, Univ. Austin, Texas, USA
**Randwick Accelerator, School of Physics, UNSW, Sydney
***Euratom Awardee, Max-Planck-Institut f. Plasma Physics, Garching, West Germany
****Laser Physics Lab., R. Sch. Phys. Sc., Australian Nat. Univ., Canberra

1. INTRODUCTION

Since the first Workshop in 1982, an essentially new result are the nearly static electric fields of up to 10^{11}V/cm and more in the double layers of laser produced plasmas [1] [2] to be used in laser focus accelerators. The definition "laser focus accelerator" was given in the preceding workshop 1982 by A. Sessler to the cases where intensive laser beams irradiate solid targets in vacuum and cause the emission of ions up to 100 MeV energy [3]. This acceleration was not immediately applicable to electrons for same high energies as highly charged ions. Since then, the evaluation of double layers arrived at a result interesting for electron acceleration too.

Since 1982, the interest in proton acceleration has grown enormously with regard to planning the 20 TeV SSC accelerator. While it was questionable according to B. Richter [4] to look for higher energy proton collisions (these partons are a conglomerate, a "soup"), the aim of TeV electron colliders is more important. Richter's requirements are very hard: to build an accelerator with high efficiency of energy conversion to limit the power supply to few gigawatts, while the luminosity L of the TeV particles should be 10^{33} cm^{-2} sec^{-1} or more,

$$L = \frac{N_1 N_2 f}{2\pi r^2} = \frac{N^2 f}{A} \tag{1}$$

where f is the frequency of the particle bunches, $A = 2\pi r^2$ is the cross section of the colliding beams and $N = (N_1\ N_2)^{1/2}$ includes the numbers N_1 and N_2 of the particles in the colliding bunches.

Apart from these maximum requirements, there are a number of uses of laser focus accelerated protons, electrons or heavy ions, which are unique and can be achieved with present day laser technology after some upgrading towards the purposes of accelerators. These uses are listed in Table 1 [5].

The laser focus accelerator was described as being "in the enviable position" [3] that many properties have been realized and the theoretical interpretation has been developed to a high degree in agreement with the measurements and towards safe predictions for the next steps of developments.

2. RESULTS OF THE LASER FOCUS ACCELERATOR

The acceleration of protons or ions up to high charge number Z from solid state target in vacuum occurs after relativistic self-focusing of the laser beam to diameters between half and one vacuum wavelength, if the laser intensity is 1/1000 to 1/100 of the relativistic threshold [6] (above 3×10^{15} W/cm^2 for neodymium glass or 3×10^{13} W/cm^2 for CO_2 lasers) if the rise time of the laser pulses (or the pulsations of the beams or the pulsating reflection of the targets) is fast enough [7]. The ion energies and the Z-separation was measured in agreement with the formula for the ion energy ε_i

$$\varepsilon_i = Z\ P\ e^2/(6m\,\pi\,c^3) \tag{2}$$

in cgs units, or

$$\varepsilon_i\,[\mathrm{MeV}] = 3\ Z\,P \tag{3}$$

where the maximum laser power P is given in TW [7]. Experimental agreement is up to few hundred MeV, and detailed numerical evaluations for 10 TW laser pulses arrived at 6 GeV ion energies for 38 times ionized tin [8].

At present, measurements of the 100 MeV ions with neodymium glass laser pulses of few psec duration are under preparation [9]. A new technique for measuring the ion energy and angular distribution has been developed with plastic foils [10] of which one example of the etched holes due to each ion is shown in Fig. 1.

TABLE 1

Laser Accelerated Particle Pulses and Uses [5]

(Case) Particle	Energy	Number Per Pulse	Energy	Repetition Period	Uses
(a) Proton	600 MeV	10^{12}	100 J	1 - 5 sec	Pion, muon production; muon as probe for solid state, muon chemistry, some QED experiments. The short pulse would be useful.
(b) Proton	6 GeV	10^{11}	100 J	1 - 5 sec	Kaon, strange particles, antiproton production.
(c) Proton	60 GeV	10^{10}	100 J	1 - 5 sec	As (1), but higher intensity. Heavier unstable particles produced, many new experiments become possible.
(d) Proton	600 GeV	10^{10}	1000 J	1 - 5 sec	World class machine with many unexplored effects to be studied.
(e) Electron	1.2 GeV	10^{12}	200 J	1 - 5 sec	Could be a useful source of radiation for for far u.v. spectroscopy. Very little nuclear physics interest as many e^+e^- colliding beam machines have been built.
(f) Heavy ion	1-10 GeV	10^{11}	20-200 J	1 - 5 sec	Heavy ion beams are fairly new and there are a number of new effects to be explored.

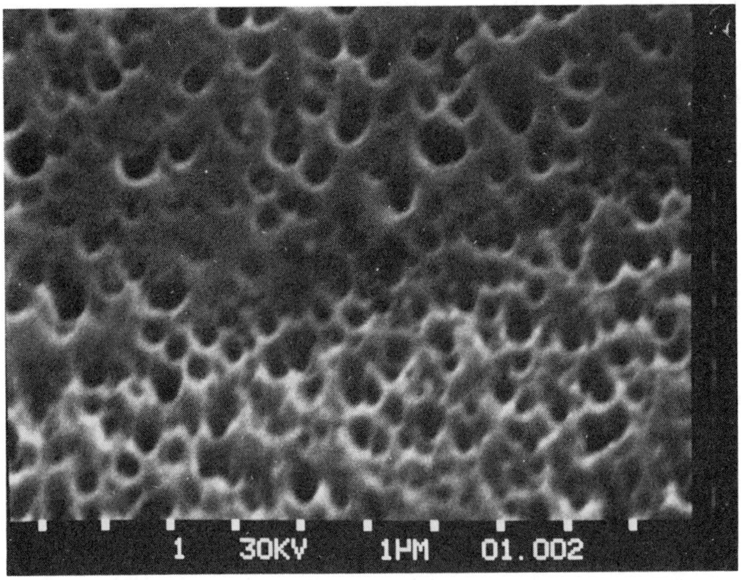

Figure 1 Scanning electron microscope picture (indicated scale length of 1 μm) of CB 39 plastic foil irradiated by laser produced ions of up to MeV energy. Upper part was covered with 0.75 μm thick aluminum foil.

3. ELECTRIC FIELDS AND DOUBLE LAYERS

Since about 1980, the generation of electric double layers in plasmas was developed on the basis of the ion-acoustic turbulence theory based on kinetic (Boltzmann) equations. Another way arrived at the realization of very high electric fields and double layers in plasmas on the basis of macroscopic hydrodynamic theory for laser produced plasmas [1] [2]. The force density in a plasma consists of the gasdynamic (thermokinetic) force given by the thermal pressure and by forces of electrodynamic origin (nonlinear force). The last one is given by

$$\underset{\sim}{f}_{NL} = \frac{1}{c}\,\underset{\sim}{j} \times \underset{\sim}{B} + \underset{\sim}{E}\,\underset{\sim}{\nabla} \cdot \underset{\sim}{E} + \underset{\sim}{\nabla}(\tilde{n}^2 - 1)\,\underset{\sim}{E}\,\underset{\sim}{E} \qquad (4)$$

where $\underset{\sim}{j}$ is the current density and $\underset{\sim}{E}$ and $\underset{\sim}{B}$ the electric and magnetic fields in the plasma. The complex refractive index $\tilde{n}$ is given in terms of the plasma frequency ω_p and collision frequency ν by

$$\tilde{n}^2 = 1 - \omega_p^{\,2}/\,[\omega^2(1 - i\nu/\omega] \qquad (5)$$

This general and complete result contains ponderomotive and non-ponderomotive parts [6]. Eq. (4) was derived for monochromatic laser radiation of frequency ω and is generally valid also for transient processes of laser pulses where Fourier spectra of monochromatic waves can be used.

While Eq. (4) was derived from a generalization of the theory of plasmas with space-charge quasi-neutrality [6], the model of quiver drift of electrons at laser-plasma-interaction indicated that there must be high electric fields between the optically (nonlinear force) driven electron clouds and the following ions. A numerical study of this process [1] with one-dimensional plasma (along) direction x) and perpendicular laser irradiation lead to use the plasma dynamics by the unrestricted electron and ion fluid equations plus Poisson's equation without the early restriction to space-charge quasi-neutrality. Only the realistic plasma with collisions, equipartition between electrons and ion temperature, complete nonlinear force, and the exact solution of Maxwell's equations for the propagation and reflection of laser radiation including the (nonlinear) dependence of the optical constants on the laser intensity, could be the basis for the treatment. Very short time steps and long integration at careful observation of appropriate boundary and numerical filtering procedure were necessary [1].

We arrived at the result that all plasmas with inhomogeneities in the electron density n_e or temperature T_e have an (internal) electric field E_x (which is longitudinal in the x-direction in our one-dimensional geometry) of [2]

$$E = \frac{m}{en_e}\left[\frac{\partial}{\partial x}\left(\frac{3n_e kT_e}{m}\right) + \frac{1}{m}\frac{\partial}{\partial x}(E_L^2 + H_L^2)/8\pi\right]\left[1 - \exp\left(-\frac{\nu}{2}t\right)\cos\omega_p t\right]$$

$$+ \frac{\omega_p^2 - 4\omega^2}{(\omega_p^2 - 4\omega^2)^2 + 16\nu^2\omega^2}\frac{e}{2m}\frac{\partial}{\partial x}(E_L^2 + H_L^2)\cos 2\omega t$$

$$+ \frac{2\nu\omega}{(\omega_p^2 - 4\omega^2) + 16\nu^2\omega^2}\frac{e}{m}\frac{\partial}{\partial x}(E_L^2 + H_L^2)\cos 2\omega t \qquad (6)$$

where E_L and H_L are the (external) transversal electric and magnetic fields of the laser irradiation. The first term shows how the electric fields are simply due to the gradient of density and/or temperature plus the gradient of the electromagnetic field density (nonlinear force) modified by collisional (ν) damped plasma (ω_p) oscillations. The second and third terms are of second harmonics of the laser radiation. The second term explains an observed density independent second harmonics emission [2] and the last term results in a new type of resonance mechanism at four times the critical density.

An evaluation of the resonance amplitude is shown in Fig. 2. For 10^{17} W/cm^2 neodymium laser irradiation, we expect an effective temperature above 10 keV with high profile steepening. In this case, the resonance arrives at longitudinal internal field amplitudes 10 to 100 times above the initial external laser field. Strong acceleration of electrons against the laser by quiver drift (similar to the usual resonance absorption working only at oblique incidence) [6] with little ion acceleration at this process can be expected.

4. ACCELERATION BY LASER MANIPULATED DOUBLE LAYERS

The very high electric fields in the double layers of the laser produced plasmas are demons-

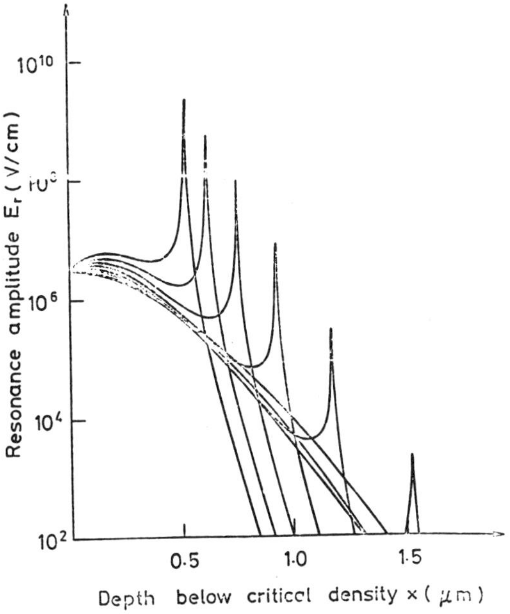

Figure 2 Amplitude E_r of the oscillating last term of Eq. (6) for neodymium laser irradiation of 10^{16} W/cm^2 into plasma of 1 keV temperature with a linear density increase from the critical density n_c at x = 0 to four times n_c at the maximum of the curves.

trated by one case of 10^{16} W/cm^2 neodymium glass laser irradiation. Fig. 3 [1] demonstrates the ion density at different times with a nonlinear force produced caviton at about 5 μm thickness. In Fig. 4 we see fields above 1.2×10^8 V/cm. The oscillation with the plasma frequency is damped much faster than the time steps indicate. The fields in Fig. 4 are then ("slowly") varying longitudinal fields with a generation of a very high amplitude fast (0.1 of speed of light) wave. As these fields are not the high frequency oscillations, as e.g. in the nodes of a standing wave to push the electrons towards the nodes but with a possible phase slip, but without these oscillations, they can directly be used for acceleration of particles.

The problem is only that the caviton dynamics in the plasma has to be manipulated by the laser with an appropriate pulse profile in such a way that the phasing of the accelerated electron cloud always follows the mentioned fast longitudinal wave maximum.

We have performed calculations of these non-conservative field accelerations of injected relativistic electrons and find just the expected magnitude of increase or decrease of electron energy.

For realistic cases, with laser beams of finite diameter, the one-dimensional calculations are a first approximation only as the realistic case has to include the varying magnetic field of high magnitude [12] surrounding the laser focus in the plasma. A further difference to the conditions of electrostatics of any (resting) plasma layer in vacuum without external fields where the collisionless transmission of a charged particle through all double layers of the plasma would result in neither gain nor loss of energy, is the dynamics of the plasma. The action of the laser causes a driving by the time dependence of the fields E(x,t) and of A(x,t) resulting in gains or losses of the transmitted charged particle.

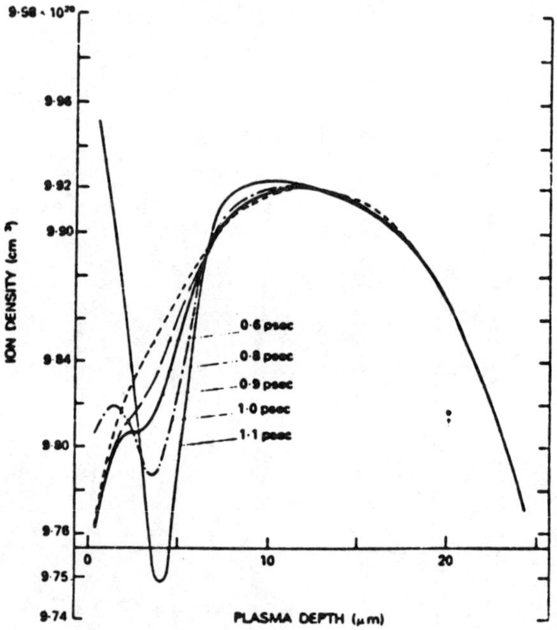

Figure 3 Ion density of a 25 μm thick initially symmetric plasma slab of 1 keV temperature irradiated from the left hand side by 10^{16} W/cm^2 neodymium glass laser. Curves are shown for different times. The generation of the density minimum shows the caviton as a typical nonlinear force mechanism (see, e.g. [6]).

The extrapolation of our computational result is possible for conditions of the present days CO_2 laser systems Antares. According to the presentation at the 1982 Workshop [13] , one beam of Antares (consisting of sub-beams) can provide 10^{14} Watts power if working at short 150 psec switched oscillator. Relativistic self-focusing [6] [7] in an irradiated target - would result in 10^{22} W/cm^2 intensities while unfocused plane wave dielectric swelling can well arrive at 10^{20} W/cm^2 . Longitudinal "slowly" varying fields and their fast waves - in analogy to the detailed computations [1] [2] for neodymium glass laser irradiation - can then arrive at values of 10^{11} to 10^{12} V/cm. While the fields in cavitons can spread over 5 to 10 wavelengths, the numerically observed spread of the fast wave mechanism is up to 50 wave lengths. For the CO_2 laser wave length of 10 μm, an "electrostatic" acceleration by a voltage of 5×10^{10} volts in one process is then possible. What is then feasible with today's technology at a synchronized amplification in a chain of 20 focuses of the described kind are then TeV electrons. A pre-amplification of the electron bunches to some GeV with usual accelerators is implicitly assumed when focusing to 10 μm, e.g. by using the ex-axial (space charge reduced) scheme [14] is realized. Working with 10^{12} electrons per bunch, and a frequency of 10^3 Hz, the luminosity is 10^{33} cm^{-2} sec^{-1} for symmetrically colliding TeV-electron beams.

5. FURTHER LASER FOCUS ACCELERATION SCHEMES

The laser focus acceleration of ions using relativistic self-focusing [7] and their detailed numerical evaluation [8] did not include the interaction with the high magnetic fields surrounding the laser focus in the plasma [12]. For the motion in axial direction in areas close to the axis, these

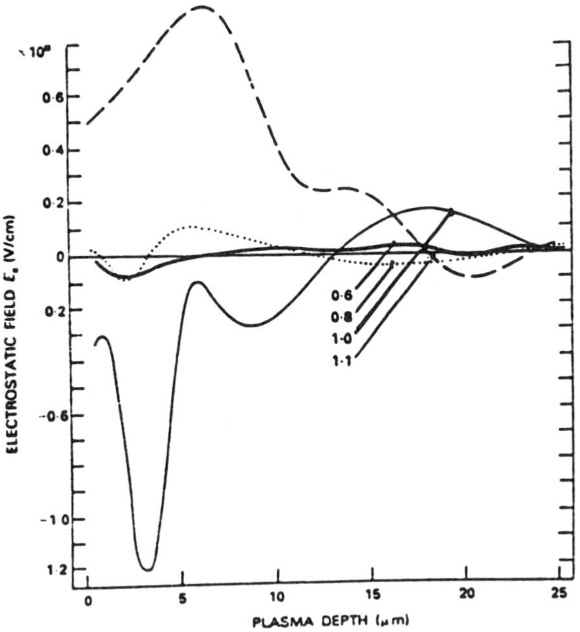

Figure 4 Electric field in the plasma at the corresponding times of Fig. 3.

fields are of no influence for which our evaluations [8] had been done. For other directions and at distances from the axis, the particle dynamics will be complicated. There is a direct proof from Doppler shifted x-ray lines of He-like phosphorus ions of 2 MeV energy at 5×10^{14} W/cm^2 neodymium glass laser irradiation [15] that at least an initially radial ion emission after relativistical self-focusing (even preferentially in radial direction) exist. In combination with the megagauss magnetic field in the plasma [12] solenoidally arranged to the laser beam axis, an interaction of the radially emitted ions with the magnetic field is evident.

Forslund and Brackbill [12] indicated the E × B-drift using the radial electric field of Eq. (6) without any oscillation

$$E = -\frac{1}{en_e}\frac{\partial}{\partial x}\left(\frac{3}{2}n_e kT_e\right) + \frac{1}{en_e}\frac{\partial}{\partial x}(E_L^2 + H_L^2)/8\pi \tag{7}$$

where only the first (thermal) term was considered. In this case [12] a simple calculation of the drift results in plasma velocities up to 10^8 cm/sec if (chaotic) plasma temperatures of 10 keV are assumed. The then resulting ions with energies in the multi 10^4 eV range have the advantage of high directivity, however not against the laser light as observed, but towards the interior of the target. The accelerated plasma has a constant block motion and no separation of ions by the charge number occurs (contrary to the Z-separation of the ions moving against the laser light [16] [17]). An open question in this model is that the $\tilde{E} \times \tilde{B}$ drift mechanism assumes a sufficiently large area of the interaction with the magnetic field and any transient process is ignored. The drift acceleration may therefore need a further modification where a reduction of the ion energies is possible.

A drastic extension of this drift mechanism happens, if not only the thermally produced radial electric field in the plasma of the focus region is considered (as Brackbill and Forslund did) but if the second nonlinear force term in Eq. (7) is considered too for the generation of the radial electric field E. For a laser beam of 10^{13} Watts and a radius of 15μm, the electric field is then near 10^9 V/cm, resulting in plasma drift velocities of 10^{10} cm/sec. The ion energies are then in the GeV range, therefore well competitive with the acceleration by the relativistic self-focusing, but with the difference of no charge separation and only with motion to the target interior, however with a very favourably small beam aperture.

If not the limitations of the model by restrictions of spatial inhomogeneities and transient mechanisms (e.g. with regard to the speed for the build up of the magnetic fields) are prohibitive, an extrapolation for 4 μm beam diameter and 10^{13} W laser pulses results in 500 MeV energy per nucleon for heavy ions, or in 0.5 TeV/nucleon with 10^{16} W laser pulses. The restrictions to the limitation of the modified Debye length [18] may arrive at a difficulty for the ion acceleration, but what remains then is the magnetic drift acceleration of plasma electrons.

Another acceleration scheme has been studied [5] where the laser light causes a direct acceleration of high density plasma slabs by the radiation pressure. While the simple acceleration by the radiation pressure is inefficient, the relativistic Doppler effect for the energy transfer results in interest on ion acceleration, if one can work with the ion plasma frequency. In this case, the just developed subpicosecond CO_2 laser pulses [19] of sufficient energy in the 10 to 100 j range result in small aperture ion bunches of GeV per nucleon with luminosity much above 10^{25} cm^{-2} sec^{-1}.

A further development for the free electron lasers should be mentioned since the inverse free electron laser [20] is very prospective as a laser accelerator. Inverting the radial emission of keV electrons from a 10^{16} W/cm^2 laser focus by the nonlinear force [18] into a laser amplifier [21], an increase of the otherwise low amplification has to be calculated by using nearly rectangular laser pulse profiles. When injecting precisely synchronized and energetically calibrated clusters of hydrogen (or neutral particle beams) into the laser pulses, a maximum global amplification per cluster of

$$A = 13.25/E_{ph}^2$$

has been derived where E_{ph} is the laser photon energy in eV. Amplifiers for the 100 Angstrom (or less) x-ray regime are then possible [22].

CONCLUSIONS

Defining the laser focus accelerator by mechanisms of particle acceleration at irradiation of high density targets by intensive laser pulses, the following results were reported.

1) Acceleration of ions after relativistic self-focusing is characterized by ion separation Z-dependence of the ion energy. Since experiments confirmed the theory up to the 100 MeV range, a detailed computation extended the mechanism to the 10 GeV ion energy range. Present days Antares laser should be able to produce energies of 100 MeV/nucleon. Upgrading to 600 MeV/nucleon and brightness of 10^{35} cm^{-2} sec^{-1} seem feasible and attractive for muon and pion sources and for the production of the needed 4000 new nuclei [23].

2) Acceleration of charged particles, especially of electrons, is possible in the (nearly static) electric fields in the double layers in the nonlinear force produced cavitons in plasmas at laser irradiation. The detailed numerical results of 10^9 V/cm fields for 10^{16} W/cm^2 Nd lasers, can be extended to 10^{11}..10^{12} V/cm fields for present day Antares-type lasers with very short pulse operation. Initially injected GeV electrons should gain 50 GeV per interaction with one caviton. 20 times repetition should produce TeV electrons of 10^{33} cm^{-2} sec^{-1} luminosity with present days systems. Speculations about very fast moving cavitons may simplify the mechanism.

3) The E × B drift plasma acceleration using the high magnetic fields in plasmas around the laser focus (Brackbill & Forslund [12]) results in thermally driven ions of some 10 keV energy without charge separation and highly directed emission. Extension of this scheme to driving by the nonlinear force produced electric fields, results in GeV ions if the assumptions of homogeneity, transient processes and Debye length limitations are not violated.

4) Acceleration by ordinary radiation pressure was improved by inclusion of the relativistic Doppler effect and by using the ion-plasma frequency for acceleration of thin plasma slabs by subpicosecond CO_2 laser pulses.

REFERENCES

[1] H. Hora, P. Lalousis and S. Eliezer, Phys. Rev. Letters **53**, 1650 (1984); S. Eliezer and A. Ludmirsky, **1**, 251 (1983).

[2] H. Hora, Laser and Particle Beams **3**, 57 (1985); P. Lalousis and H. Hora, Laser and Particle Beams **1**, 283 (1983).

[3] A. Sessler, First Workshop "Laser Acceleration of Particles", P. Channell ed. AIP Proceedings No. 91 (Am. Inst. Phys., New York, 1982) p. 10.

[4] B. Richter, this Workshop.

[5] F.J.M. Farley (unpublished).

[6] H. Hora, Physics of Laser Driven Plasmas (Wiley, New York, 1981).

[7] H. Hora, D.A. Jones, E.L. Kane, P. Lalousis, and B. Luther-Davies, First Workshop "Laser Acceleration of Particles" P. Channell ed. AIP Proceedings No. 91 (Am. Inst. Phys., New York, 1982) p. 120; H. Hora, J. Opt. Soc. Am. **65**, 882 (1975).

[8] D.A. Jones, P. Lalousis, E.L. Kane, and H. Hora, Phys. Fl. **25**, 2295 (1982).

[9] N. Nakano, M. Baba, Y. Tanaka, and H. Kuroda, Laser Interaction and Related Plasma Phenomena, H. Hora and G.H. Miley eds. (Plenum, New York, 1984) Vol. 6, p.127.

[10] J.C. Kelly and B. Luther-Davies (unpublished).

[11] R. Schrittwieser and G. Eder eds. Second Symposium on Electric Double Layers in Plasmas, Innsbruck, July 1984 (Theor. Phys., Uni. Innsbruck, 1984).

[12] D.W. Forslund and J.U. Brackbill, Phys. Rev. Lett **48**, 1614 (1982).

[13] Paul Channell ed. First Workshop "Laser Acceleration of Particle Beams" AIP Proceedings No. 91 (Am. Inst. Phys., New York, 1982).

[14] J.C. Kelly, IEEE Trans. Nucl. Sci. **NS-30**, 1476 (1983).

[15] A.V. Rode, Yu. A. Mikhailov, G.V. Sklizkov et al (unpublished).

[16] A.W. Ehler, J. Appl. Phys. **46**, 2464 (1975).

[17] B. Luther-Davies and J.L. Hughes, Opt. Comm. **18**, 605 (1976).

[18] B.W. Boreham and H. Hora, Phys. Rev. Letters **42**, 776 (1979).

[19] P. Corkum, this Workshop.

[20] P. Sprangle, private communication.

[21] H. Hora and G. Viera, Laser Interaction and Related Plasma Phenomena, H. Hora and G.H. Miley eds. (Plenum, New York, 1984) Vol. 6, p.201.

[22] H. Hora, P.J. Clark, and Wang Jin-Cheng (submitted for publication).

[23] E.W. Titterton, Lecture at the 40th Anniversary Celebration, Los Alamos, April 1984, Australian National University Report, April 1984.

INCREASING THE CENTER OF MASS ENERGY OF STORAGE RINGS AND COLLIDERS BY LASERS

R. Rossmanith
2000 Hamburg 52, W. Germany

ABSTRACT

It is demonstrated that a laser beam focussed into the interaction region of an electron-positron storage ring or a collider can, under certain conditions, substantially increase the center of mass energy of the colliding particles and reduce the costs of high energy accelerators.

INTRODUCTION

With increasing energy the size of electron-positron storage rings becomes extremely large. Therefore, linear colliders instead of storage rings were proposed for the next accelerator generation[1]. The length of these colliders is defined by the energy gain per unit length. In other words, the length is defined by the obtainable field strength. During the last years, several proposals were published where wake fields and laser fields instead of conventional microwave cavities and waveguides are used[2,3]. In the focus of an ordinary laser beam a field strength of 10E9 V/cm exists. But unfortunately, according to Maxwell's equations, the electric (and the magnetic) field aims in a direction perpendicular to the direction of the motion of the wave. Therefore, a particle travelling in such a field sees a field with different sign and, as a result, the net acceleration is zero. As a consequence, in order to gain a net acceleration it was proposed to modify the simple plane wave by different methods: e.g. by wave guide structures[4], by using media like gases[5] or by using plasma waves[6]. Also far fields modifications of the focus of a laser are proposed[7,8]. In principle, all these proposals would work out, but in practice they present manifold difficulties.

In this paper a different approach to this problem is discussed. It starts from the assumption that all kinds of the above mentioned light manipulations are (at the moment) impracticable. The only effect which is easy to obtain is the locally limited acceleration of the particle. Assume that in the interaction region of a storage ring both electrons and positrons are accelerated at the same time and collide, the center of mass energy of this storage ring is increased, even when the net acceleration across the whole focus is zero.

In the following, this effect is discussed. The advantages of such a system are evident: existing lasers and existing storage rings can be used together to investigate new and unexplored energy ranges.

ELEMENTARY DESCRIPTION OF THE LASER ENERGY BOOSTER ARRANGEMENT

The electric and the magnetic field of the laser beam is transversal (fig. 1). The peak power of a laser is usually expressed in terms of watts. The electric field strength E and the power density of a laser are related by the following relation[9]:

$$E\ (V/cm) = 27\sqrt{P\ (W/cm)} \qquad (1)$$

This relation will be explained in an example. A commercially available Nd-YAG laser with an oscillator-amplifier system achieves peak power of approximately 100 MW in 2 to 3 nanoseconds.

Assuming that the beam is focussed into a small area, e.g. into 10 times 10 microns the power density is 10E14 W/cm. The field strength according to formula (1) is 3E8 V/cm.
The energy gain of a particle in an electric field is given by the formula

$$E = e \int_a^b E(s)\, ds \qquad (2)$$

ΔE ...... energy gain of the particle between a and b
e .... electric charge
E(s) .... field strength along the particle trajectory

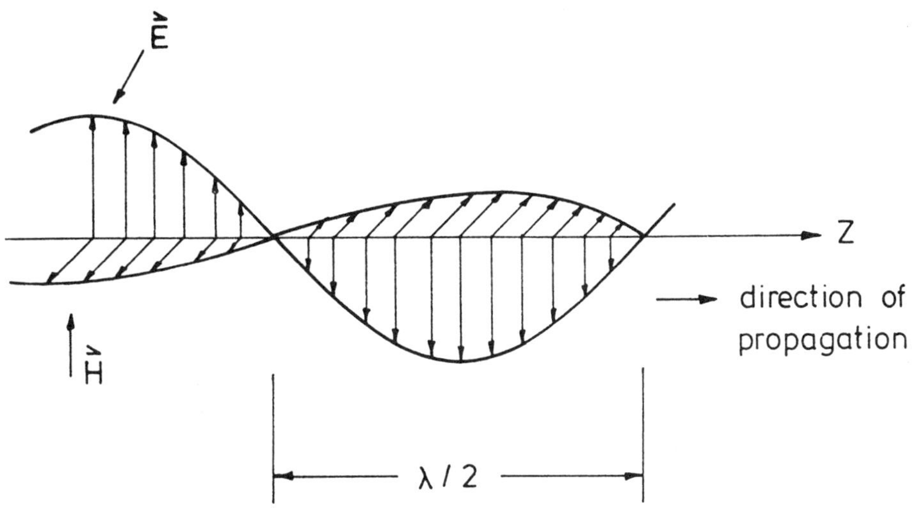

Fig. 1: Electric and magnetic field in a propagating wave.

Fig. 2 shows schematically the electric field strength at a certain moment in the focus of the laser beam. For the next considerations it will be assumed that there is only an electric field. Later, it will be shown how to reduce practically the influence of the magnetic field. In fig. 2 it is assumed that the laser operates in a TEM 00 mode and that the diffraction effects caused by the lens are neglected (the size of the lens is larger than the size of the incoming beam).
The field of the laser beam in the focus can be described by the following formula

$$E = A(x) \cdot \sin(\omega t - kz) \tag{3}$$

where A(x) describes the Gaussian shape of the focussed laser beam. The focal plane will be defined by z = 0 and the field acting on the particle can be calculated by replacing t in formula (3) by x/c and introducing a phase φ which takes into account the time by which the particle enters the field

$$E = A(x) \cdot \sin(\omega_x/c + \varphi) \tag{4}$$

The particle crossing the focal plane sees a sinosodial field and is therefore accelerated and decelerated as described before.

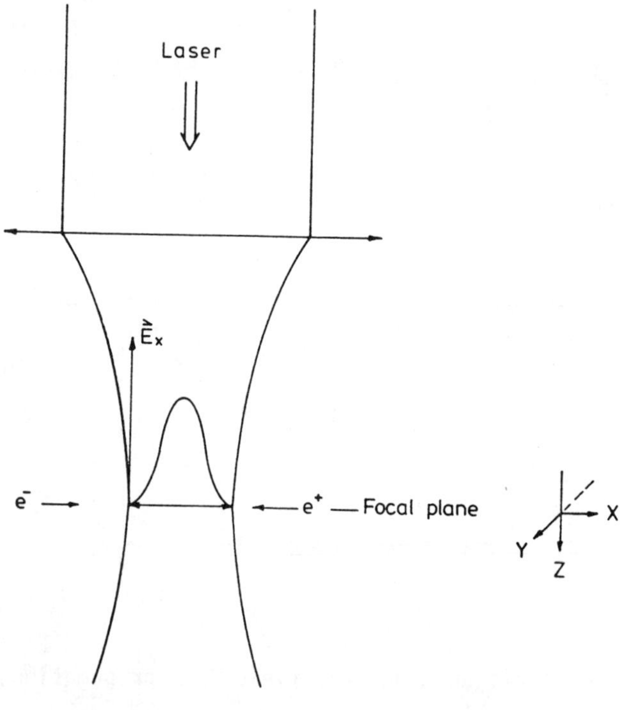

Fig. 2: The electric field in the focal plane of a laser beam

Before going into more detail, it will be explained how to minimize the influence of the magnetic field. This is done by using two laser beams instead of one as shown in fig. 3. The two laser beams or one splitted laser beam generate a standing wave. The electrical field in the focal plane is therefore

$$E_x = (A(x)/2)(\sin(\omega t-kz)+\sin(\omega t+kz))$$
$$= A(x)\cdot\sin(\omega t)\cdot\cos(kz)$$

For $z = 0$ $\quad E_x = A(x)\cdot\sin(\omega t)$ (5)

and the magnetic field

$$H_y = (A(x)/2)(\sin(\omega t-kz)-\sin(\omega t+kz))$$
$$= A(x)\cdot\cos(\omega t)\cdot\sin(kz)$$

For $z = 0$ $\quad H_y = 0$ (6)

The maximum of the electric field is separated by half a wavelength from the maximum of the magnetic field. Therefore, the assumption that only the electric field has to be taken into account is correct. The change of the particle energy is given by the following formula

$$E = e\int A(x)\cdot\sin(\omega x/c+\varphi)\,dx \qquad (7)$$

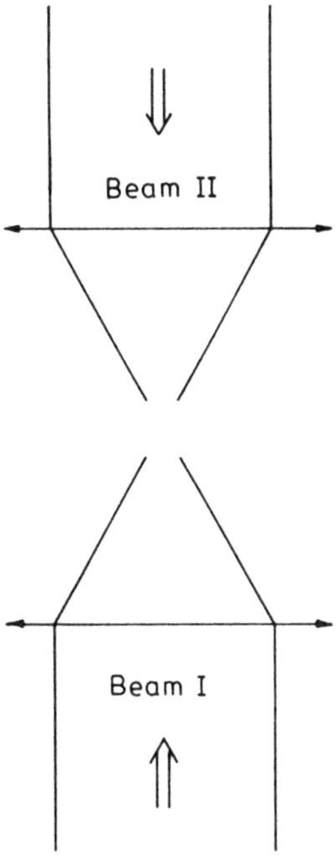

Fig.3: Separation of the electric and magnetic field in the focal plane by using the lasers.

To give an example, the function ΔE is calculated in fig. 4 for $A(x) = \exp(-(x^2/3\lambda^2))$

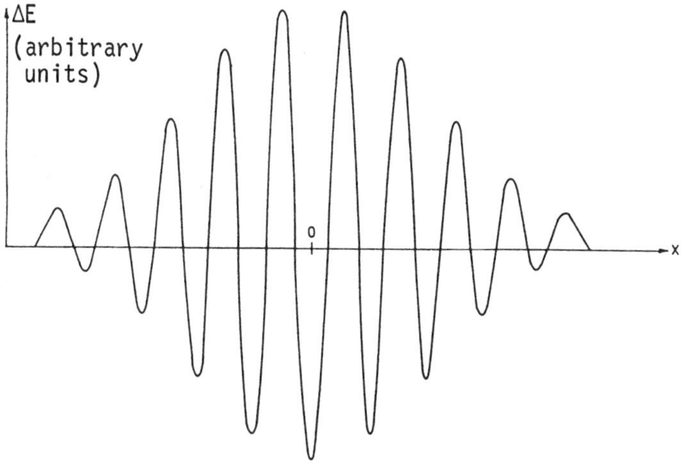

Fig. 4: The function E for a Gaussian shaped laser pulse.

Assume that both an electron and a positron approaches the focal plane as shown in fig. 2. For the two particles the charge and the direction of motion has opposite sign. Formula (7) is therefore the same both for electrons and positrons. As a result, there exist zones with high particle energy in the focal plane. The next question concerns the maximum possible energy gain and, connected with this question, the required laser configuration.

LAYOUT OF THE LASER SYSTEM FOR A 10 GeV BOOSTER

Let us assume that a laser booster only makes sense when an energy gain of something in between 1 to 100 GeV can be obtained. In this paper a 10 GeV booster will be discussed. The wavelength of the laser is assumed to 1 micron. In the following, we start from our well-known YAG-laser system and modify it in such a way that it can be used for an energy booster.

The required field strength for 10 GeV acceleration is 10E10 Volt per quarter wavelength. According to the assumed wavelength the required field strength is 4E10 V/micron or 4E14 V/cm. The simple focussed laser described in the previous chapter produces a field strength of 3E8 V/cm. The difference between the required field strength and the field strength obtained with this simple version of laser is about six orders of magnitude. The problem now is to modify the laser system in such a way that the required field strength is obtained. There are several possibilities how to proceed. In the following, these possibilities are discussed.

a) Increasing the power of the laser. This can only be used for small changes in the field strength according to the square root law of formula (1).

b) It is more effective to use more than one laser amplifier. This is the usual technique developed in fusion research. The master oscillator-amplifier combination is schematically drawn in fig. 5. The maximum number of amplifiers is somewhat up to 100. The above mentioned six orders of magnitude gap can be reduced by this method to four orders of magnitude.

c) The laser pulse has a length of 2 nsec. The bunch length is 20 times shorter. Therefore, the laser pulse can be folded in a Fabry-Perot type resonator with two lenses as shown in fig. 6. The field strength adds in the focus and therefore, the gap in the field strength becomes smaller by a factor of 20. The relation shrinks to 500.

d) The laser pulse can be made longer and at the same time the peak power reduced so that the total pulse power remains constant. According to c) the voltage is increased. By making the pulse e.g. 5 microseconds long with a peak power of only 20 kW we are only a factor of 7 away from our goal. We have now constructed a more than 1 GeV energy booster.

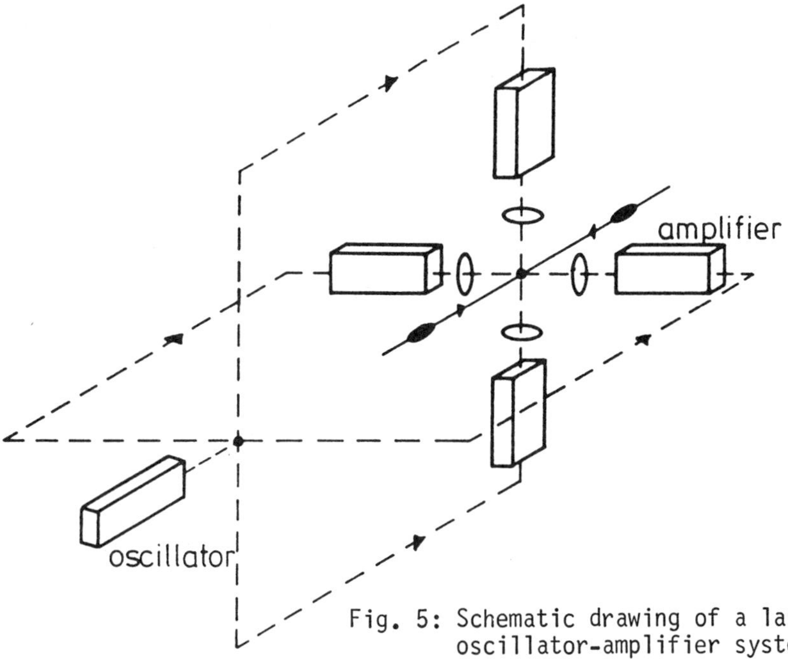

Fig. 5: Schematic drawing of a laser oscillator-amplifier system.

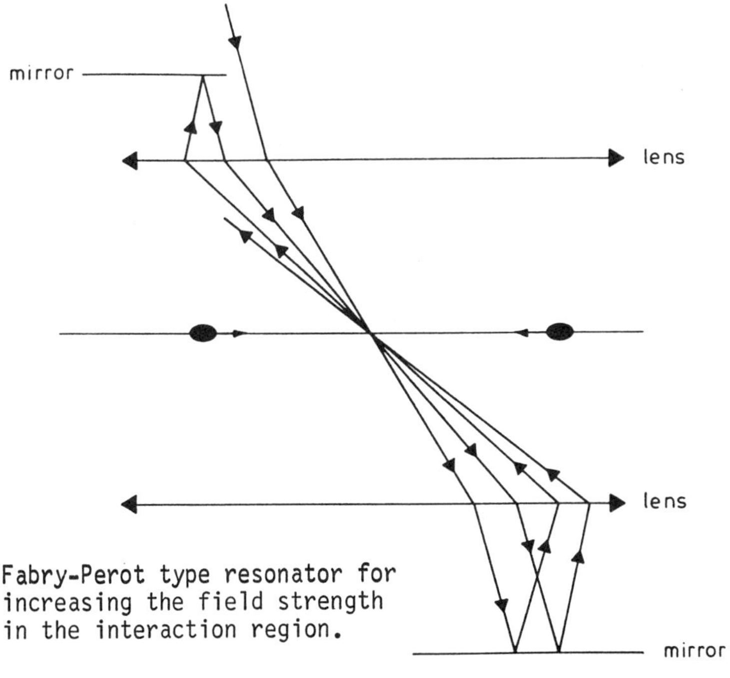

Fig. 6: Fabry-Perot type resonator for increasing the field strength in the interaction region.

The limit for method d) is evident: the maximum field strength is limited by the losses on the various surfaces similar to the limitations in microwave cavities.

e) To overcome this limitation the laser pulse can be splitted into two or more pieces as shown in fig. 7. One of the two pieces is delayed so that both of them enter the interaction region at the same time. This trick brings us into the 10 GeV region.

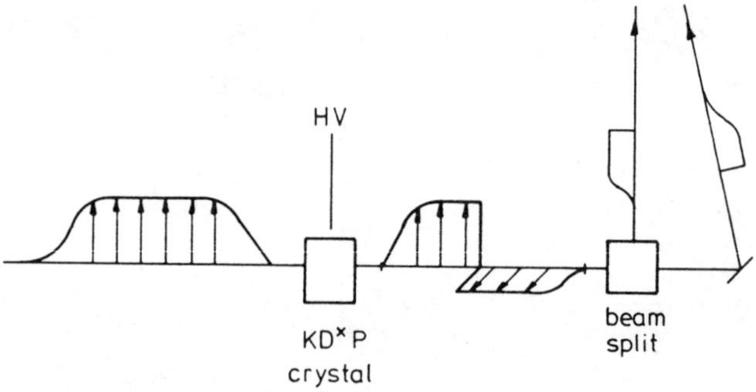

Fig. 7: Splitting of laser pulses with a KDP crystal.

It is evident that the first step b) (using several laser-amplifiers) is not the optimum solution. The number of amplifiers can be reduced by applying all the other methods.

Summarizing this chapter: from an optics point of view, it is not very difficult to increase the energy of the colliding beams by lasers. The method is rather uncomplicated and does not incorporate new optical techniques. The only remaining question to be answered is the following: does the whole method make sense from a luminosity point of view. This will be discussed in the next chapter.

LUMINOSITY

From the previous statements it is evident that acceleration is only possible in a small channel with a width of a tenth of a wavelength. Therefore, the incoming particle beam in the storage ring has to be focused into this channel. The technique for strong focusing is well-known from the SLC design[10]. Therefore, the laser booster does not influence the luminosity.

In conventional storage rings the situation is more complicated. This can be demonstrated by referring to a paper dealing with a microbeta interaction region in PETRA[11]. The basic idea was to use superconducting quadrupoles near to the interaction region. A beam with a horizontal emittance of 30E-6 m and a vertical emittance of 3E-6 m was focused by a 90 cm focal length quadrupole to 0.3 mm vertically and 4 mm horizontally. This is in terms of storage ring physics small but from the energy booster point of view intolerable high. The luminosity would be reduced by an enormous factor. Therefore, the beam has to be focused by additional quadrupoles. An example for a possible layout is shown in fig. 8. It is assumed that the optics can be changed in such a way that the horizontal beam dimension is less than 2 mm. The beam is focused by additional quadrupoles with a small focusing length. The two quadrupoles reduce the beam dimensions by a factor of 1000 and 100 respectively. They are now 4 microns and 3 microns. The nominal value of the luminosity increases by approximately 5 orders of magnitude but the high energy luminosity increases only by 2 orders of magnitude due to the still too big size of the beam. The numerical gain of luminosity is reduced or converted into a loss by taking into account the repetition rate of the laser. Assuming that the repetition rate is 30 Hz and the bunch repetition rate is 130 kHz, the high energy luminosity is reduced by the factor 2.5E-2. An upper limit for this method seems to be at the moment a luminosity of 10E30. Nevertheless, detailed studies have to be performed to discuss these figures in more detail.

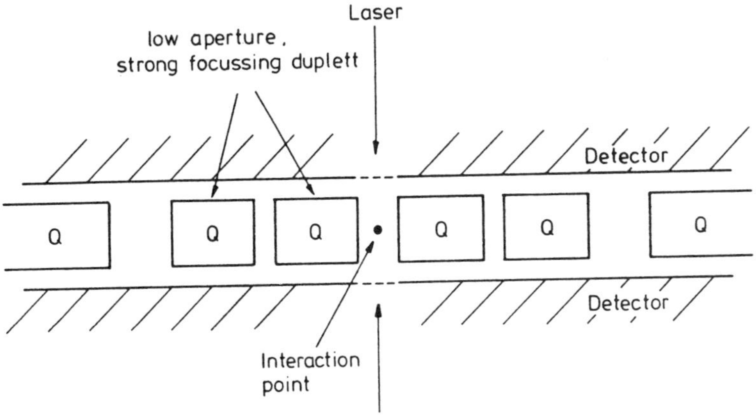

Fig. 8: Possible layout for the interaction region.

ACKNOWLEDGMENTS

The author wishes to thank many people for discussion and encouragement; Dipl. Phys. J. Kewisch and T. Limberg from DESY, Dr. Rüdiger Schmidt from CERN and Dr. Zhuang Jiejia from the Institute of High Energy Physics, Academia Sinica.

LITERATURE

1. SLAC Linear Collider Concept. Design Report. SLAC Report 229, June 1980
2. G.-A. Voss and T. Weiland, DESY 82-074, December 1982
3. For an overview see e.g.
   Israelyan, M.K.: Bibliography of papers on Laser Acceleration of charged particles with abstracts, Yerewan 1983
   Laser Acceleration of Particles, AIP Conference Proceedings No. 91, Los Alamos 1982
4. R.B. Palmer, in Laser Acceleration of Particles, see ref. 3, p. 179
5. R.H. Pantell et al., in Physics of Quantum Electronics, 9(1982)961
6. T. Tajima, in Laser Acceleration of Particles, see ref. 3, p. 69
7. R. Rossmanith, Nucl. Instr. Meth.154(1978)29
8. Peng Huangwu and Zhuang Jiejia, Scientia Sinica, Vol. XXIII, No. 2
9. L. Levi, Applied Optics, John Wiley and Sons, Inc.
10. B. Richter, The SLAC Linear Collider, Proc. XI Int. Conf. on High Energy Acc. Genua, 1980, p. 168
11. K. Steffen et al., DESY Internal Rep. M-81/20

RADIAL IMPLOSION ACCELERATION*

P. J. Channell, AT-6, MS 829
Los Alamos National Laboratory, Los Alamos, NM 87545

ABSTRACT

In this paper we propose a scheme to generate high accelerating gradients [approximately (∿) a few gigaelectron volts per meter]. The acceleration is nonresonant so that staging may be fairly easy, and the energy source is relativistic e-beams so that a relatively high overall efficiency may be achievable.

I. INTRODUCTION

A magnetic field of 100 kG is equivalent to an electric field of 3 GeV/m. The only way such a magnetic field can be made available for acceleration is through the equation

$$\vec{\nabla} \times \vec{E} = -\frac{1}{c}\frac{\partial \vec{B}}{\partial t} \quad ; \tag{1}$$

that is, we must move magnetic fields around or destroy them to generate electric fields. Letting $\nabla \sim 1/L$ (where L is a typical dimension) and $\partial/\partial t \sim 1/T$ (where T is a typical time) Eq. (1) is roughly

$$|E| \sim \frac{1}{c}\left(\frac{L}{T}\right) |B| \equiv \frac{V}{c} |B| \quad . \tag{2}$$

Thus, to get $|E| \sim |B|$, we must have $V/c \sim 1$; that is, we must move B-fields at nearly the velocity of light. An obvious technique for doing this is to push on the B-field with relativistic e-beams. The geometry we have in mind is shown in Figs. 1a and 1b.

Imploding the B-field requires a certain minimum e-beam pressure. We estimate the magnetic pressure as

$$P_B = \frac{B^2}{8\pi} \quad . \tag{3}$$

The e-beam pressure is approximately

$$P_e \simeq J\gamma m v \quad , \tag{4}$$

where J is the particle current/area ($J \sim cm^{-2}sec^{-1}$), γ is the relativistic mass factor, m is the mass, and v is the particles' velocity.

---

*Work supported by the US Department of Energy.

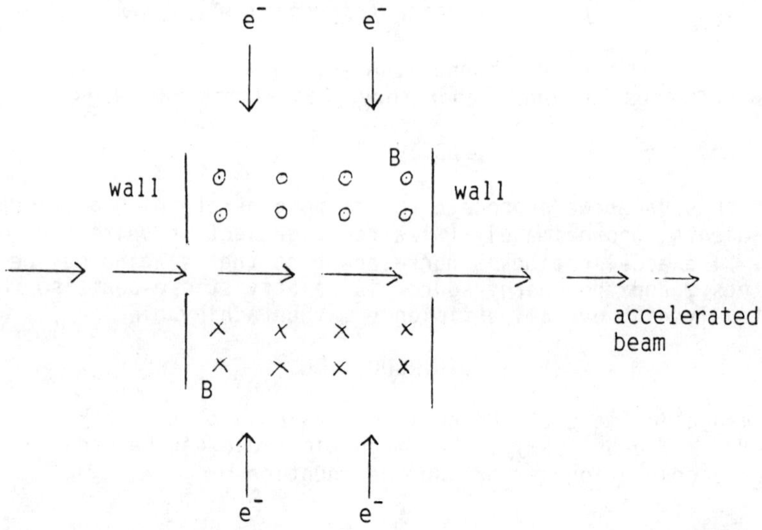

Fig. 1a. View from side of initial configuration.

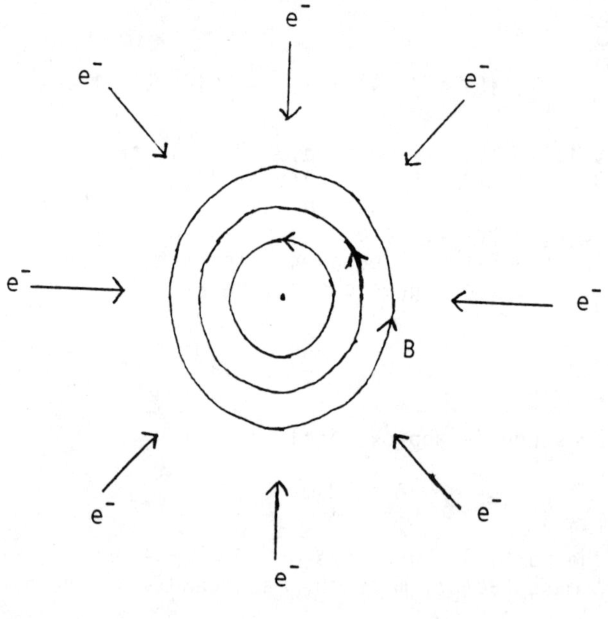

Fig. 1b. View along accelerated beam axis of initial configuration.

If we use electrons at 10 MeV and assume B = 100 kG, then we find

$$J \gtrsim 165 \text{ kA/cm}^2 \quad . \tag{5}$$

Such current densities have been achieved in a number of machines. The length of the e-beam pulse can be estimated to be the electron gyromagnetic radius divided by the implosion velocity. For the above example, this becomes

$$T_{pulse} \sim 10 \text{ ps} \quad . \tag{6}$$

Thus, the geometry and current density are typical of the Sandia proton beam fusion accelerator (PBFA) running with electrons; however, the pulse length is about 1000 times shorter, and the stored energy required is about 1000 times less.

Because the generated electric field is almost purely longitudinal, the coupling efficiency to accelerated particles can be quite high. In addition, relativistic e-beams can be generated with high efficiency, so that the overall efficiency can potentially be very good. How much of this potential can be realized depends on the details of the implosion process.

In the next section we present a one-dimensional snowplow model for the implosion process to elucidate the details and to show that the implosion velocity is approximately the speed of light. In Sec. III we discuss the validity of the one-dimensional assumption and point out the influence higher dimensional instabilities may have on tailoring the current and energy profile in the e-beams. In Sec. IV we present two schemes for setting up the initial B-field configuration. In the final section we discuss the advantages of our scheme and point out areas for future investigation.

II. SNOWPLOW MODEL

The implosion process not only allows us to move B-fields at the velocity of light, but also to begin with a modest B-field over a relatively large volume and to compress it to a large value. To investigate the implosion process let us use a Cartesian one-dimensional model.

Note from Fig. 1b that the field on axis (r=0) is zero. Letting the Cartesian variable x correspond to r, we use the magnetic field model shown in Fig. 2. This is a snowplow model in which we assume that the e-beams push all of the magnetic flux ahead of them and that the front is sharp. Analytically the magnetic field is

$$B_z = B(t) \frac{x}{L(t)} [H(x + L) - H(x - L)] \quad , \tag{7}$$

where $\pm L(t)$ are the positions of the fronts and where H is the Heaviside function. Using Eqs. (1) and (7) and recalling that

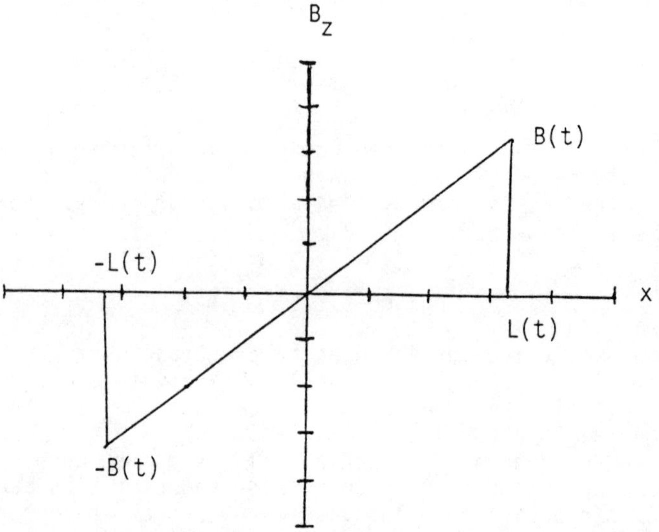

Fig. 2. Plot of magnetic field as a function of x at a fixed t.

B(t)L(t) = B(0)L(0) = constant, we find that the induced electric field is

$$E_y = B(t)\left(\frac{\dot{L}(t)}{c}\right)\left(\frac{x}{L(t)}\right)^2 [H(x + L) - H(x - L)] \quad , \qquad (8)$$

as shown in Fig. 3. Note that for $x = \pm L$, $E_y = B(\dot{L}/c)$; that is, the magnitude depends on the front velocity $\dot{L}$ as we expected.

To find an equation for L(t), we use momentum conservation. The momentum in the fields is

$$\vec{P}_{field} = \frac{1}{4\pi c} \vec{E} \times \vec{B} \quad . \qquad (9)$$

Note that $\vec{P}_{field}(x=0) = 0$, so that the momentum of the particle beam coming from the right (Fig. 2) is absorbed by the fields in $x > 0$. Using Eqs. (7) and (8) in the region $0 \leq x \leq L$, we find

$$\vec{P}_{field} = \frac{\hat{x}B^2}{4\pi c} \left(\frac{\dot{L}}{c}\right)\left(\frac{x}{L}\right)^3 \quad . \qquad (10)$$

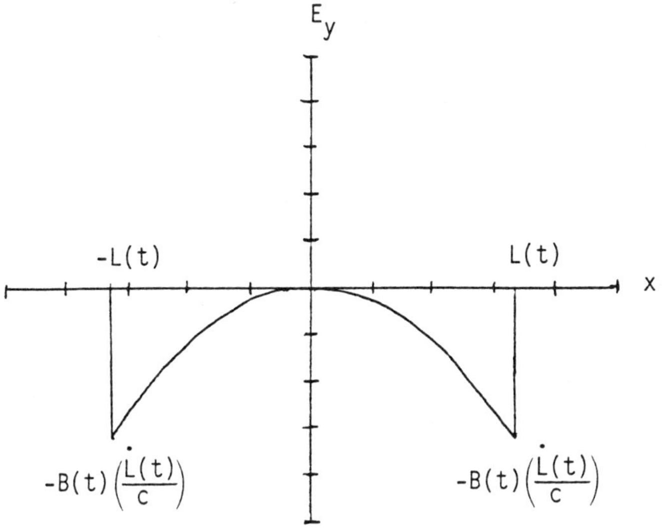

Fig. 3. Plot of electric field as a function of x at fixed t.

Integrating the momentum conservation equation

$$\frac{d\,\vec{p}_{field}}{dt} = - \frac{d\,\vec{p}_{particles}}{dt} \tag{11}$$

over the region x > 0, we find

$$\frac{B^2}{16\pi c^2}(L\ddot{L} - \dot{L}^2) = -J\gamma mv(1 - \frac{|\dot{L}|}{c}) \quad , \tag{12}$$

where the particle momentum transfer rate has been reduced to account for the electrons catching up to the front and where Eq. (4) has been used for the change in particle momentum.

Ideally, the particle beam pressure will be tailored to match, approximately, the magnetic field pressure so that excess energy use and instabilities can be minimized. We thus let

$$J\gamma mv \equiv \frac{\alpha(t)B^2}{8\pi} \quad , \tag{13}$$

where α is a, possibly time-dependent, factor of order 1. With this assumption, and noting that we want $\dot{L} < 0$, Eq. (12) becomes

$$L\ddot{L} - \dot{L}^2 = -2\alpha c^2(1 + \frac{\dot{L}}{c}) \quad . \tag{14}$$

A numerical solution of Eq. (14) for α = constant = 1 is shown in Fig. 4 where $\dot{L}$, the implosion velocity, is plotted as a function of time. It can be seen that fairly quickly the front starts moving at a uniform velocity. In this regime, we let $\ddot{L} = 0$, and find that Eq. (14) becomes

$$\dot{L} = \alpha c(1 - \sqrt{1 + \frac{2}{\alpha}}) \quad . \tag{15}$$

The quantity $|\dot{L}|$ achieves its maximum value when $\alpha = \infty$, where $\dot{L} = -c$. For $\alpha = 1$ we find

$$\dot{L} = -0.732\ c \quad . \tag{16}$$

We observe from Fig. 3 that the accelerating field is nonuniform transversely, and thus that one will want to accelerate hollow beams. The transverse nonuniformity will not be so dramatic as shown in Fig. 3 for two reasons: (1) the front will not be as sharp as we assumed in our simple model; (2) the front will move transversely while the accelerated particles traverse the accelerating region longitudinally, causing the accelerated particles to feel a transverse average of the accelerating gradient.

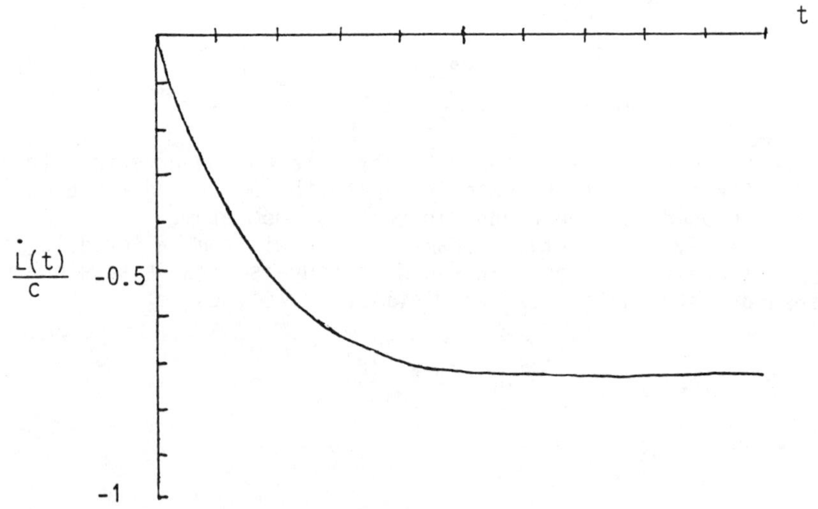

Fig. 4. Plot at $\dot{L}(t)/c$ as a function of t.

Although we have, for simplicity, assumed α = constant, it is clear that $\alpha(t)$ should be tailored in such a way to minimize instabilities and may also be chosen so as to produce more uniform accelerating fields. The actual $\alpha(t)$ will depend on a number of factors, including the extent to which J_Y can be controlled in practical e-beam machines. The detailed choice of the optimum $\alpha(t)$ probably will require extensive numerical computation.

III. RAYLEIGH-TAYLOR INSTABILITIES

In the previous section we assumed that the implosion was effectively one-dimensional. Higher dimensional effects will appear mainly through the onset of instabilities, principally the Rayleigh-Taylor instability. The implosion front is Rayleigh-Taylor unstable during the acceleration of the front and becomes neutrally stable during the uniform velocity and deceleration (maximum field) phases of the front.

Because we anticipate using the uniform velocity and deceleration phases of the front for particle acceleration, the principal concern arising from the Rayleigh-Taylor instability has to do with the amount of field flux actually captured by the front. The less flux captured, the higher will be the e-beam energy required for a given accelerating gradient and, thus, the lower the efficiency. By tailoring $\alpha(t)$ initially, it should be possible to minimize the effect of the Rayleigh-Taylor instability, because $\alpha(t)$ controls the acceleration rate. Finding the optimum $\alpha(t)$ probably will require numerical simulations.

IV. INITIAL CONFIGURATION

There are at least two ways that the initial B-field configuration can be obtained: exciting it with a high-current, low-energy e-beam along the accelerator axis, and running the gap as an rf cavity.

By exciting the gap with a preceding e-beam, the desired B-field can be set up. In this mode of operation, our scheme resembles a wake-field accelerator in which the wake function is dramatically increased by the imploding e-beams.

If we excite the gap with pulsed rf power in the TM_{010} mode, the B-field configuration will have the desired form for a half cycle of the rf. However, the vanishing of the azimuthal B-field at the walls seriously violates the 1-D geometry assumption. By adding a static axial B-field, we can prevent all the electrons from flowing along the wall. It is possible that such a configuration might develop into a roughly 1-D implosion, although probably this can only be investigated in numerical simulations.

It is, of course, also possible that other schemes can be devised to set up the initial B-field.

V. DISCUSSION

The scheme we have proposed may have several advantages to recommend it. First, the accelerating gradient can be quite high and might be pushed even higher than we have estimated if higher current-density e-beam machines become available. Next, the overall efficiency might be quite good because e-beams can be produced with high efficiency and the E-fields generated are almost purely longitudinal. Finally, the acceleration is nonresonant, which implies both easy staging of accelerating sections and the possibility of accelerating slower particles, such as heavy ions.

A number of issues clearly remain to be resolved. The best initiation scheme must be decided on. The influence on the Rayleigh-Taylor instability should be investigated in more sophisticated models, probably including numerical simulations. The acceleration process should also be studied in better models to determine both the best beam to accelerate and the energy spread induced by the transverse variation of field.

Once the theoretical tools have been devised to design and interpret an experiment, it will be appropriate to build such a machine. Note that such an experiment could be rather modest in scope because the implosion and acceleration processes can be tested in a single gap. The simultaneous promise of high accelerating gradient and high efficiency would certainly justify such an effort.

LASER FOCUSING OF PARTICLE BEAMS*

P. J. Channell, AT-6, MS H829, C. J. Elliott, X-1, MS E531
Los Alamos National Laboratory, Los Alamos, NM 87545

J. R. Fontana†
ECE Department, U. of California, Santa Barbara, CA 93106

ABSTRACT

We propose a scheme using the Inverse Cerenkov effect to focus particle beams with the potential of high focusing gradients ($\sim$200 kG/cm) and rapid, accurate control of the focusing element.

I. INTRODUCTION

The extremely high fields available in lasers have led various people to propose schemes to use those fields to accelerate particles. Many of these schemes continue to show great promise.

In general, particle beams must not only be accelerated longitudinally, but they must be focused in the transverse directions, sometimes to extremely small spot sizes. In this paper we propose a technique for using the high fields in lasers to do this focusing. The scheme we propose does, with fairly conservative assumptions, yield high focusing gradients in small focusing elements at modest laser power and with the potential for very fast, accurate control of focal placement and strength. With more optimistic assumptions on gas breakdown and laser power, the potential equivalent gradients approach the MG/cm level. However, even the gradients achievable with conservative assumptions ($\sim$200 kG/cm) are so large (and are combined with the virtue of fast accurate control) that we believe our proposed technique merits serious attention.

In the next section we describe our scheme qualitatively and discuss its potential. In Sec. III of this paper, we present a detailed analysis of the basic interaction in a 2-D planar model, starting from first principles (that is, Maxwell's equations and the Born approximation for the particle motion). This analysis illustrates the essential ingredients and provides quantitative estimates of the expected focusing strength, required gas density, and laser power. In Sec. IV the basic interaction is analyzed again in a cylindrical geometry using the paraxial ray equation. Essentially the same result as that of Sec. III is obtained. The Gaussian laser modes used in Secs. III and IV are clearly not the optimum modes; in Sec. V we build on the work done on Inverse Cerenkov Acceleration with axicon-focused laser beams to find yet

---

*Work supported by the US Department of Energy.

†Work supported by the US Department of Energy under contract DE-AT03-84ER40136.

another, more optimistic, estimate of the basic interaction. The reader can thus choose between a first-principles approach that may not be physically transparent (Sec. III), a more intuitive approach familiar to those in optics (Sec. IV), and an advanced treatment that offers the best results (Sec. V). In Sec. VI we briefly discuss the effect of gas scatter on the particle beam, and in the final section we summarize and discuss our results.

II. GENERAL DISCUSSION

The basic geometry we have in mind is shown in Fig. 1 where the laser electric field at the focal waist is primarily radial and is assumed to vanish on axis. A particle, as shown, traveling near the axis would indeed feel very strong focusing or defocusing forces from the electric field. However, the particle would also feel a force from the laser magnetic field, which would cancel the electric force to order $1/\gamma^2$, where γ is the relativistic mass factor. In this paper, we ignore all terms of order $1/\gamma^2$ as being too small to be of interest. Thus, in this approximation, the basic interaction of particle with laser focus vanishes.

Adding gas to the region of the laser focus has three principal effects. First, and most importantly, the electric and magnetic fields no longer balance to order $(\varepsilon-1)$, where ε is the dielectric constant of the gas; this implies that the interaction of a particle with the laser focus no longer vanishes. The next effect of the gas is to introduce a difference in velocity between the particle ($v = c$) and the light ($v < c$, if $\varepsilon > 1$). This effect causes a continuous phase shift and implies that the basic interaction must have finite range; that is, we cannot use plane waves.

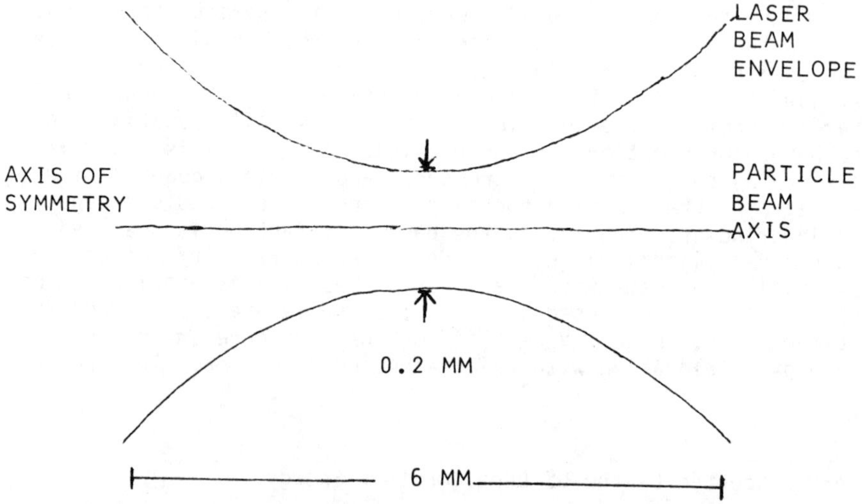

Fig. 1. General configuration and typical dimensions (not to scale).

The final effect of the gas is to limit the peak laser field to a value below breakdown. (It is possible that a plasma at the focus might also yield focusing, but we have not analyzed this case.) Breakdown values in gas depend strongly on laser pulse length and are not well known. We will use values appropriate to a 20-ps laser pulse ($\sim$20 GeV/m), even though actual pulse lengths may be shorter ($\sim$3 ps) and breakdown values presumably higher. Breakdown limits may also depend on pressure, although fairly weakly, but this dependence is not well understood at such short pulse lengths. Clearly, more experimental work needs to be done to explore the ultimate limits of our focusing technique.

In later sections, we present detailed models and explicit estimates for the focusing strength; here we summarize the results of the analysis qualitatively. The interaction of the particle with the light depends on the imbalance of E and B and is thus proportional to (ε-1). The interaction is maximized if the phase slip through the Rayleigh length (approximate length of laser focus) is a fraction ($\sim$1/3) of π. Finally, the interaction is optimized if the angle Θ, between the particle and the light [sin $\Theta \sim$ (waist width)/(Rayleigh length)] is the Cerenkov angle Θ_C. Our scheme can be described, essentially, as inverse Cerenkov focusing.

A difficulty that comes immediately to mind is that the interaction depends on the phase of the particle with respect to the light wave and can, depending on this phase, be either focusing or defocusing. There are two obvious solutions to this problem. The first is to accept the situation, that is, defocusing of half the particles. The remaining particles can be focused strongly enough to give a very high gain in luminosity. The second is to prebunch the particles at the correct phase. This might be possible, say with a free-electron laser.

A possible set of laser focus dimensions is shown in Fig. 1 ($\sim$0.2 mm x 6 mm) for a CO_2 laser ($\lambda \sim$10 μm). This is not obviously the optimum configuration and, in particular, the optimum sizes probably shrink as the laser wavelength decreases. Thus, for some applications, a glass laser or a laser with even shorter wavelengths, such as KrF, might be appropriate. In general, a complete optimization (including laser wavelength, breakdown level, pulse length, gas pressure, and laser configuration and parameters) has not been done. We will use a set of parameters that are fairly conservative, probably achievable, and yet that produce quite impressive results with very modest laser power.

III. PLANAR MODEL

In this section let us analyze a planar (that is, Cartesian) model assuming a Gaussian laser beam. We begin with Maxwell's equations and analyze the effect on particles using the Born approximation.

The geometry and coordinates are shown in Fig. 2. We assume a mode in which the laser magnetic field is purely z-directed ($B_x = B_y = 0$) and the electric field direction lies in the x,y

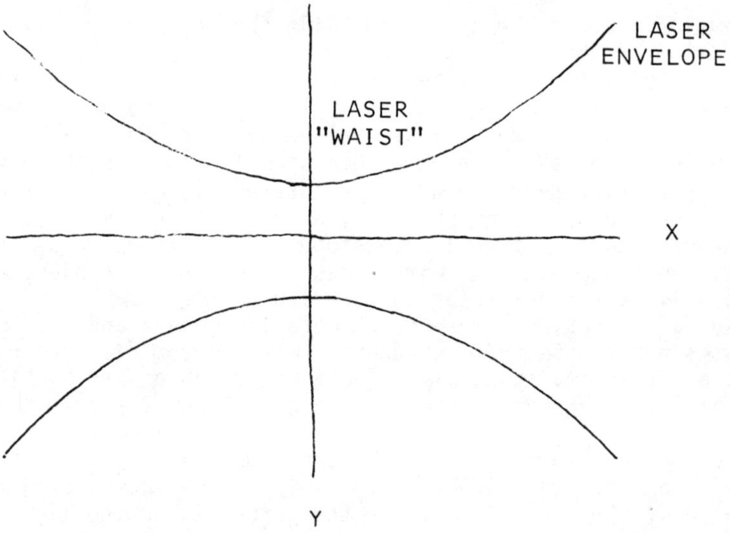

Fig. 2. Coordinates used in the analysis in Sec. III.

plane ($E_z = 0$). The relevant Maxwell's equations are then

$$\frac{\partial B_z}{\partial y} = \frac{\varepsilon}{c} \frac{\partial E_x}{\partial t} \quad , \tag{1}$$

$$\frac{\partial B_z}{\partial x} = \frac{-\varepsilon}{c} \frac{\partial E_y}{\partial t} \quad , \quad \text{and} \tag{2}$$

$$\frac{\partial E_y}{\partial x} - \frac{\partial E_x}{\partial y} = \frac{1}{c} \frac{\partial B_z}{\partial t} \quad , \tag{3}$$

where ε is the dielectric constant. From Eqs. (1)-(3) we can easily obtain the wave equation

$$\frac{\varepsilon}{c^2} \frac{\partial^2 E_y}{\partial t^2} = \frac{\partial^2 E_y}{\partial x^2} + \frac{\partial^2 E_y}{\partial y^2} \quad . \tag{4}$$

Writing the solution of this equation as a superposition of traveling waves we find

$$E_y = \int_0^\infty dk \; \overline{E}(k) \sin Ky \cos(kx - \omega t) \quad , \tag{5}$$

where

$$K \equiv \sqrt{\frac{\varepsilon\omega^2}{c^2} - k^2} \quad , \tag{6}$$

and where

$$\overline{E}(k) = 0 \qquad \text{if} \qquad |k| > \frac{\omega}{c}\,\varepsilon \quad . \tag{7}$$

To determine $\overline{E}(k)$, we impose the initial-boundary-value condition

$$E_y(x=0,\ y,\ t=0) = \sqrt{2e}\ E_p \cdot \left(\frac{y}{L}\right) e^{-\left(\frac{y}{L}\right)^2} \quad , \tag{8}$$

where E_p is the peak electric field at the waist. We then find that

$$\overline{E}(k) \simeq \frac{1}{2}\sqrt{\frac{2e}{\pi}}\, kL^2E_p \exp\left[-\frac{L^2}{4}\left(\frac{e\omega^2}{c^2} - k^2\right)\right] \quad . \tag{9}$$

The electric and magnetic fields are thus, near $y = 0$,

$$E_y \simeq \frac{1}{2}\sqrt{\frac{2e}{\pi}}\, L^2E_p y \int_0^{\frac{\omega}{c}\sqrt{\varepsilon}} dkkK \exp\left[-\frac{L^2}{4}\left(\frac{\varepsilon\omega^2}{c^2} - k^2\right)\right]$$

$$\cos(kx - \omega t) \quad , \quad \text{and} \tag{10}$$

$$B_z \simeq \frac{1}{2}\sqrt{\frac{2e}{\pi}}\, L^2E_p y\left(\frac{\varepsilon\omega}{c}\right) \int_0^{\frac{\omega}{c}\sqrt{\varepsilon}} dkK \exp\left[-\frac{L^2}{4}\left(\frac{\varepsilon\omega^2}{c^2} - k^2\right)\right]$$

$$\cos(kx - \omega t) \quad . \tag{11}$$

To evaluate the effect of these fields on the particles, we use the Born approximation, that is,

$$\Delta P_y = \int_{-\infty}^{\infty} dt\ F_y \quad , \tag{12}$$

where ΔP_y is the change in y-momentum, where $F_y = q(E_y - B_z)$ is the y-component of the force, and where we use the trajectory

$$y = \text{const} \quad , \quad \text{and} \tag{13}$$

$$x = c(t - t_0) \quad , \tag{14}$$

where t_0 is the time of arrival of the particle at $x = 0$ (assumed arbitrary). Using Eqs. (10), (11), (13), and (14) in Eq. (12), we find

$$\Delta P_y = \frac{1}{2}\sqrt{\frac{2e}{\pi}}\, L^2 (qE_p) y \left\{ \int_0^{\frac{\omega}{c}\sqrt{\varepsilon}} dkK \exp\left[-\frac{L^2}{4}\left(\frac{\varepsilon\omega^2}{c^2} - k^2\right)\right] \right.$$

$$\left. \cdot \left(k - \frac{\varepsilon\omega}{c}\right) \int_{-\infty}^{\infty} dt \cos(kct - \omega t - kct_0) \right\} \quad . \tag{15}$$

Using the fact that

$$\int_{-\infty}^{\infty} dt \cos \alpha t = 2\pi\delta(\alpha) \quad , \tag{16}$$

where δ is the Dirac delta function, Eq. (15) reduces to

$$\Delta P_y = -\sqrt{2\pi e}\,(\cos\varphi_0)(qE_p) y \frac{L^2\omega^2}{c^3}(\varepsilon - 1)^{3/2} \exp\left[-\frac{L^2\omega^2(\varepsilon - 1)}{4c^2}\right] \quad , \tag{17}$$

where φ_0 is the wave phase at the time of the particle's arrival at $x = 0$. We choose the gas pressure, and thus ε, so as to maximize Eq. (17). We thus choose

$$\varepsilon = 1 + \frac{6\lambda^2}{4\pi^2 L^2} \quad , \tag{18}$$

where λ is the wavelength of the laser. With this optimization, Eq. (17) reduces to

$$\Delta P_y = - \cos \varphi_0 \sqrt{2\pi e} \left(\frac{6}{e}\right)^{3/2} (qE_p) \frac{c}{\omega} \left(\frac{y}{L}\right) \quad . \tag{19}$$

The focal length L_F is

$$L_F = \frac{P_x y}{\Delta P_y} \tag{20}$$

and is given by

$$L_F = 2\sqrt{\frac{\pi}{2e}} \left(\frac{e}{6}\right)^{3/2} \left(\frac{L}{\lambda}\right) \left(\frac{\gamma mc^2}{qE_p}\right) \left(\frac{1}{\cos \varphi_0}\right) \quad . \tag{21}$$

Equation (21) is our basic result. Note that by decreasing L/λ, we can decrease the focal length; however, from Eq. (18) we see that the price we pay is that we must increase the pressure. The peak laser power required, P, can be estimated to be

$$P \simeq \frac{(\pi L^2) E_p^2 c}{4\pi} \quad . \tag{22}$$

As an example, let us assume

$$\gamma mc^2 = 50 \text{ GeV} \quad ,$$

$$qE_p = 20 \text{ GeV/m} \quad ,$$

$$L/\lambda = 10 \ (\rightarrow \text{pressure} \sim 10 \text{ atm}) \quad , \quad \text{and}$$

$$\varphi_0 = 0 \ (\text{maximum focus}) \quad . \tag{23}$$

We then find

$$L_F = 11.25 \text{ m} \quad , \quad \text{and}$$

$$P = 10^{10} \text{ W} \quad . \tag{24}$$

For a magnet of the same length (equal to $2\pi L^2/\lambda$) to produce the same deflection, it would have to have a field gradient $qB' \sim 220$ kG/cm, a quite high value.

Thus, with modest peak power (10^{10} W) and reasonable gas pressure (∿10 atm), we can produce very strong focusing with a fairly conservative value of the breakdown limit. Optimistic assumptions about laser pulse lengths and gas breakdown would allow us to push the equivalent gradients into the range of 1 MG/cm, albeit at the expense of greater laser power. The ability to accurately adjust the focal position, phase, and strength on a fast time scale makes this scheme very attractive.

IV. PARAXIAL EXPRESSION FOR A SINGLE FOCAL ELEMENT

Electrons that travel at the speed of light experience no focusing force because the electric field and the induction forces exactly balance. This cancellation does not occur in a dielectric medium of refractive index n where the net force is proportional to $(n^2 - 1) \cong 2(n - 1)$. Coincident with focusing, the electron sees a phase modulation $(n - 1)\omega t$, where $\omega = 2\pi/\lambda$, λ being the wavelength and t the time. When the electrons travel through a TEM_{01} Gaussian beam mode, this dielectric modulation is chosen roughly to match one-third its period to a Rayleigh range Z_r.

Consider the following y component of the exact Gaussian electric-field mode, represented as an integral over the transverse components of the wave number $\underline{k}$ at location $\underline{x}$:

$$E_y = C_o \int_{-\infty}^{\infty} dk_y dk_x \; i\, k_y \exp\left[-\frac{1}{4}(k_x^2 + k_y^2)\omega_o^2 + i\,\underline{k}\cdot\underline{x} - i\omega t\right] \quad , \quad (25)$$

where we use the paraxial approximation, k_x, $k_y << k$, $\underline{k} = [k_x, k_y, k_z]$, $\underline{x} = [x,y,z]$, and

$$k_z = [k^2 - k_x^2 - k_y^2]^{1/2} \simeq k - (k_x^2 + k_y^2)/(2k) \quad (26)$$

to compute the fields to lowest significant order. The convention we use for the field is that the right-hand side of the equation is the real part of what is written. The quadrature requires evaluation of

$$\int_{-\infty}^{\infty} dt \begin{bmatrix} 1 \\ t \end{bmatrix} \exp(-At^2 + iBt) = \begin{bmatrix} 1 \\ \frac{iB}{2A} \end{bmatrix} \sqrt{\frac{\pi}{2A}} \exp(-B^2/4A) \quad , \quad (27)$$

giving

$$E_y = \sqrt{2e}\, E_B \frac{y}{\omega_0} \frac{\exp - \left[\frac{x^2 + y^2}{\omega_0^2(1 + iz/z_r)} + i\,kz - i\omega t \right]}{(1 + iz/z_r)^2} \quad , \qquad (28)$$

where $z_r = k\omega_0^2/2$, and we have renormalized the field to its peak value E_B at the focal plane $z = 0$. The x-component of the induction is given by

$$B_x = -\, kE_y/\omega\, [1 + \Theta(k_y/K)^2] \quad . \qquad (29)$$

Integration of the transverse force $q(E_y - cB_x)$ over the electron particle trajectory

$$\underline{x} = [0,\, y,\, ct + \psi/k] \quad , \qquad (30)$$

in the spirit of the Born approximation, gives the transverse impulse. The ratio of the electron impact parameter y to the focal length is evaluated on axis by dividing the impulse by the momentum

$$\frac{y}{f} = \frac{q}{mc\gamma} \int_{-\infty}^{\infty} (E_y - cB_x)dt \quad . \qquad (31)$$

Using the expressions for the fields, this gives

$$\frac{W_0}{f} = 2^{3/2}\, e^{1/2}\, \frac{(1 - n)\, qE_B}{mc\gamma}\, e^{i\psi} \int_{-\infty}^{\infty} \frac{e^{i(n - 1)\omega t}}{(1 + iz/z_r)^2}\, dt \quad , \qquad (32)$$

where q is the electron charge, m its mass, and γ its total energy measured in rest mass units. Completing the contour in the lower half-plane by an infinite semicircle and using the residue of the integrand, we obtain

$$\frac{W_0}{f} = -2^{5/2}\, e^{1/2}\, \pi\, \frac{qE_B}{\omega mc\gamma}\, \delta^2 e^{-\delta} \cos\psi \quad , \qquad (33)$$

where δ is a measure of the refractive index,

$$\delta = (n - 1)\omega z_r/c \quad . \tag{34}$$

Optimization with respect to δ gives

$$\frac{W_o}{f} = -2^{7/2}\, e^{-3/2}\, \frac{qE_B\lambda}{mc^2\gamma} \cos\psi \quad , \tag{35}$$

which, except for a factor near 1, is the inverse of Eq. (21).

V. FOCUSING WITH SINGLE-Θ LASER BEAMS IN A GAS

A laser configuration that has been proposed for acceleration[1] consists of a continuous set of plane waves in a gas medium, all converging on the z-axis at the same angle Θ, related to the index n of the medium and the speed βc of the particles by the Cerenkov relation $n\beta \cos\Theta = 1$.

This configuration has also focusing properties, given for an electron of charge -q by the radial force[2]

$$F_r(r,\psi) = qE_0 \tan\Theta J_1(kr \tan\Theta/\beta)(-\sin\psi) \quad , \tag{36}$$

where

$\psi = kct - nkz \cos\Theta$ is the (constant) phase angle of a particle,

E_0 = the peak field on the z-axis, and

$k = 2\pi/\lambda$ is for the vacuum wavelength.

For $r << \beta\lambda/\tan\Theta$, $\beta \simeq 1$, and $\tan\Theta \simeq \Theta$:

$$F_r(r,\psi) \simeq [-qE_0\, \Theta^2(\pi/\lambda) \sin\psi]r \quad . \tag{37}$$

Thus, an electron traveling in the general +z-direction at a distance r from the axis will perform betatron oscillations

$$r(z) \propto \sin(z/\beta_\perp) \quad , \tag{38}$$

where

$$\beta_\perp^2 = \frac{\gamma mc^2}{qE_0(\pi/\lambda)\,\Theta^2 \sin\psi} \quad . \tag{39}$$

There is focusing in the range $0 < \psi < \pi$ and defocusing otherwise.

When the length L of the laser interaction region is small compared with $\beta_\perp$ there is still a focusing (or defocusing) effect that can be calculated by considering the angle α by which a particle entering the region parallel to the z-axis and with a given r is deflected away from the axis. If p_z is the axial momentum and Δp_r the acquired radial momentum over L,

$$\alpha \simeq \tan \Theta = \frac{F_r L/\beta c}{\gamma m \beta c} \simeq \left[\frac{-qE_0 \Theta^2 L(\pi/\lambda) \sin \psi}{\gamma mc^2}\right] r \qquad (40)$$

for $\beta \simeq 1$.

The focal length of this lens is

$$f = \frac{r}{-\tan \alpha} \simeq \frac{\gamma mc^2}{qE_0 \Theta^2 L(\pi/\lambda) \sin \psi} \quad . \qquad (41)$$

Substituting E_0 as a function of L, λ and the total power P in the laser beam,[3] we obtain

$$f \simeq 2.3 \times 10^3 \frac{\gamma \lambda^{3/2}}{\Theta^3 L^{1/2} P^{1/2} \sin \psi} \text{ (mks units)} \quad . \qquad (42)$$

For example, with Θ = 52 mrad (corresponding to hydrogen at 10 atm), $\lambda = 10^{-5}$ m, $\gamma = 10^5$ (that is, 50 GeV), $L = 10^{-2}$ m, and $P = 10^{10}$ W, we have

$$f \simeq 5.3/\sin \psi \quad , \qquad (43)$$

which is about a factor of 2 better than the result [Eq. (24)] obtained for a Gaussian laser beam.

VI. GAS SCATTER

A beam propagating through a gas will show an increase in emittance caused by molecular collisions. B. W. Montague has given a method[4] to calculate this increase in the presence of a focusing force. For a given betatron focusing function $\beta_\perp$, the normalized (that is, invariant) emittance growth is[5]

$$\Delta\varepsilon = \frac{2F}{\gamma'} [\gamma_f^{1/2} - \gamma_i^{1/2}] \quad ,$$

where γ' is the rate of increase in γ, and the subscripts f and i indicate the final and initial values. The function F is given as[6]

$$F = 2\pi r_e^2 N \ln\left(\frac{\Theta_{max}}{\Theta_{min}}\right)\frac{\beta_\perp}{\gamma^{1/2}} \quad ,$$

where $r_e = 2.8 \times 10^{-15}$ m is the classical electron radius, and Θ_{max} and Θ_{min} are the extreme atom collision deflection angles for the particles.

We now use these expressions to calculate $\Delta\varepsilon$ for the short-lens case discussed above, taking $\psi = \pi/2$ for no acceleration and maximum focusing. With $\ln(\Theta_{max}/\Theta_{min}) = 10$ and using $\beta_\perp = 4\ f/2\pi$, we have

$$\Delta\varepsilon \simeq 8.5 \times 10^{-3}\left[\frac{fLp}{\gamma}\right] \quad ,$$

where the emittance, focal length, and interaction length are in meters, and p is the gas pressure in atmospheres.

For the example above, with $f = 5.3$ m, $L = 10^{-2}$ m, $\varepsilon\gamma = 10^5$, and $p = 10$ atm we obtain

$$\Delta\varepsilon \simeq 4.2 \times 10^{-8} \text{ mrad} \quad .$$

The focusing (positive or negative) properties discussed above apply to particles with a given phase ψ relative to the laser fields; because of the Cerenkov condition this phase is conserved during the interaction when a sufficiently large β is assumed. For beams unbunched at the laser wavelength, a continuous variation in ψ occurs along the beam, resulting in all the possible positive and negative values of the focal length f given by Eq. (7). Nevertheless, a positive focusing effect can be obtained for all ψ's by using a succession of lenses phased so the ψ of any particle increases by $\pi/2$ between lenses. Four such lenses provide overall positive focusing for any particle in the beam.

VII. DISCUSSION

We have presented analyses of simple models starting from either Maxwell's equations or the paraxial ray equation, which show that with reasonable assumptions on laser power, gas breakdown, and gas pressure we obtain very high focusing forces. Using more sophisticated models developed for inverse Cerenkov accelerators, we have seen that we can do even better. Optimistic scaling of breakdown with laser pulse length would imply that even higher focusing forces ($\sim$1 MG/cm) might be available.

Our scheme has a number of obvious advantages that merit attention. With conservative assumptions on breakdown, we obtain very high focusing fields: with the potential (with favorable breakdown scaling) of extremely high focusing forces. Perhaps more intriguing is the possibility that such strong focusing can be very rapidly and accurately controlled by optical techniques. This fast and accurate control is partially because our scheme requires very little solid material near of the particle beam; the focusing element is the laser focus and requires only gas to be effective. Finally, the laser power required is modest and thus readily available, although high repetition rates at short pulse lengths may still require some development.

Essentially, what we have done is to turn the Inverse Cerenkov Acceleration scheme into a focusing scheme. It is probable that other laser acceleration schemes, for example droplet or beat-wave schemes, can also be used for focusing. Our scheme has the advantage of simplicity; namely, the medium (gas) is passive and is not required to have any coherent structure, with the result that fast accurate control depends primarily on the optical elements. On the other hand, it is likely that other schemes, in particular the beat-wave scheme, may give even higher focusing forces than our scheme. The droplet scheme is intermediate between the two schemes in that it may provide easy control, if the droplet formation can be done accurately, and yet have higher focusing forces than our scheme. Thus, our scheme has the relative advantage of simplicity, but it is clear that all the techniques should be investigated for their focusing potential.

A number of topics still need investigation. The limits of our scheme depend on gas breakdown at short pulse lengths and can probably be best studied experimentally, although adequate focusing seems to be available even with conservative breakdown assumptions. Clearly the most important task is to create an optimized design for some particular purpose, such as, for example, final focusing in the Stanford Linear Collider. The examples we have presented have not been optimized in any sense. Finally, it would be useful to have a detailed design, including laser "architecture," to see how such a system might actually fit into accelerator systems and to determine whether any relevant technologies require further development. The potential great advantages of laser focusing would certainly seem to justify such studies.

ACKNOWLEDGMENTS

The authors are grateful to A. M. Sessler and B. W. Montague for many suggestions and discussions.

REFERENCES

1. J. R. Fontana and R. H. Pantell, J. Appl. Phys. 54 (8), August 1983.

2. Eq. (18) of Ref. 1.

3. Eq. (13) of Ref. 1.

4. B. W. Montague, "Emittance Growth from Multiple Scattering in the Plasma Beat Accelerator," presented at the CAS-ECFA-INFN Workshop at Frascati, 25 September to 4 October 1984.

5. Eq. (25)] of Ref. 4.

6. Eqs. (21) and (23) of Ref. 4. The range of collision deflection angles has been readjusted to fit the case of a neutral gas rather than that of a plasma.

SWITCHED POWER LINAC

W. Willis
CERN
1211 Geneva 23
Switzerland

ABSTRACT

The proposed linac is powered by switched pulses instead of radio frequency power, using distributed photodiodes driven by short pulses of light. If the required photodiode and light source requirements can be met, this device will have the following advantages;

1) Very short pulses can be produced by the high pervaence photodiodes, leading to a small stored energy.
2. The distribution of the power source over the whole accelerator surface allows a high density which can produce a large gradient.
3. The short pulse reduces the heating of accelerator structure and other gradient limitations.
4. The power switching can be highly efficient.
5. It is straightforward to recover most of the energy in the pulse, raising the overall efficiency or allowing the beam loading to be lowered for fixed efficiency.
6. The high gradient and low beam loading give low wake field effects, and good output emittance.

This note is an updated version of one submitted to the Frascati Conference.

The proposal can be understood by reference to Figs. 1 to 3. The accelerating structure consists of copper disks of radius R and thickness d, with a central aperture of radius r, spaced by a distance s. Near the outer radius is a wire photocathode of radius r_c, at a distance of g from the nearest copper surface. These wires are supported to form an octagon, and charged through an "external" photodiode as shown in Fig. 3.

The operation proceeds by charging the wire photocathode through the external photodiode in a moderately short time $\sim$ 0.5 ns. (See the waveforms in Fig. 2.) The charge then has time to distribute uniformly on the wire. Next a short light pulse discharges the wire photodiode in a much shorter time onto the disk structure, generating a pulse which travels inward at increasing voltage until it is reflected from the central aperture. At that time the particle bunch passes through the gap and feels the accelerating field. Most of the energy in the pulse is reflected and travels back out to the outer radius of the structure, where the energy is recovered by another photodiode switch.

An approximate quantitative analysis of this device will now be given. One has to put the analysis in a parametric form to realize certain matching conditions, but for brevity only a numerical example which satisfies these conditions is shown, with parameters given in Table 1. Derived quantities are given in Table 2. The scale was chosen to match the stored energy to the required value and to push the accelerating field toward the limits given by field emission. The plate thickness is a structural choice. The diode characteristics were chosen to match the impedance of the structure, for efficient power transfer.

The copper disk structure driven at the edge looks like a transmission line of impedance 1.3 ohms. It is essential that the photodiode current into that produce a potential drop matched to that on the photocathode, V, so that the photoelectrons give up all the stored energy in their passage across g and arrive at the anode with small energy. For an accurate calculation of the photodiode performance one must consider the motion of the electrons taking into account their space charge and magnetic field, and the radiation in the structure. Since ours is a two dimensional problem, a numerical solution adapting existing programmes is perhaps possible, but for an approximate solution we ignore space charge and use the transmission line approximation, then check the current we obtain against the space charge limit.

It is important to keep the time constant of the electron motion short, in order to keep the power high and stored energy low. We may assume the light pulse itself is sufficiently short. The cylindrical geometry of the cathode wire is favourable in this respect, since the electrons acquire most of their energy in a short distance. The voltage assumed gives a surface field on the wire of 400 MV/m, well below the point where field emission would be a problem. This voltage is maintained for less than 1ns, so that breakdown processes which depend on ion motion cannot be effective. A breakdown process dependent on photon feedback is not favoured because of the small solid angle subtended by the photocathode.

The risetime of the photocurrent pulse based on free electron motion is about 2.5 ps. The completion of the current pulse is dependent on the reaction of the electromagnetic field in the structure. One way to see a qualitative analysis is to note that in the first phase of charging the copper structure, its transmission line character is not relevant, and we may consider that we are charging a capacitor large compared to that of the photocathode. This is our justification for computing the risetime slope using the static field. Next, the structure looks like two transmission lines, one going inward and one outward, with a total impedance of 1.3/2 ohms. Then the reflection from the open outward line arrives, several ps later. By then, the electrons have fallen through most of the potential drop, the energy stored has been transfered, and the motion of the electrons will be stopped on the average or slightly reversed. A number of parameters are available for detailed timing of this condition. It seems that one should not lose more than a few percent of the stored energy in this process.

On the basis of this discussion, we take the full width of the current pulse to be 10ps. The average value of the current is then 3.2×10^4 amps, which develops a voltage in the transmission line equal to that originally on the photocathode. We note that this current is considerably smaller than that of a space charge limited diode, again noting that magnetic field is not considered.

As the pulse travels inward, the impedance of the line rises with the inverse of the radius. Near the centre, when the inner aperture is reached and the pulse is reflected, 1/r has increased by a factor 120. We assume an increase in voltage by a factor of 40. (This is probably optimistic, but numerical calculations are in progress). The resulting acceleration field in the gap is 1.6 GeV/m, close to the field emission limit, though the computed field emission current is small. Note that the field is only maintained for 10ps at any given point, eliminating breakdown mechanisms based on regeneration. The heating due to the current at the inner edge of the copper disk, is not sufficient to cause a problem.

The beam bunch length must be smaller than the length of the field pulse, probably about 1 mm. There is an opportunity to shape the electric field pulse, within the constraints determined by keeping the photodiode losses small, in order to compensate for longitudinal wake field.

The reflected pulse arriving back at the outer radius of the structure has lost 24% of its energy in ohmic dissipation and about 10% to the beam. About 2/3 remains to be recovered, well worthwhile. This is done by illuminating the other side of the photocathode, on a ridge provided on the copper disk, with a photocathode surface. The charge is transfered back to the wire and through the external diode into the power supply.

The use of the external photodiode solves two problems. It is essential to keep the total capacitance of the primary photocathode low, to match the stored energy to the desired value. With the coupling only through the very low capacitance of the external photodiode, the capacitance of the primary photocathode is determined essentially by its active portion. Also, we wish to charge or discharge the photocathode line very quickly, in order to prevent breakdown. It would be difficult to find another charging mechanism on this timescale, and to transport pulses of this length involves undesirable ohmic losses. In this system, the short pulses travel a very small distance. Further local filtering of the reactive power associated with the repetition rate ($\sim 10^4$Hz for example) means that the power distribution in the system is essentially continuous current, with very low losses. If the switch efficiency is high, the overall efficiency of the accelerator could be 25%, for 10% beam loading, with the important exception of the light source power.

The optics must focus light on the 100 μm wire, not a difficult requirement for the short wavelengths which will be used. The light wave front must have cylindrical symmetry to preserve the isochrony, but the photocathodes can be polygonal, as in Fig. 5. The alignment should be within 10 μm, not too difficult with a short focal length cylindrical lens or mirror mounted on the accelerator structure. The pathlength tolerance is ∿ 0.1 mm, not very severe. The most difficult part seems to be the distribution of the light over the whole surface of a long accelerator, and the initial set-up.

The light source has to provide very large power in ∿ 1 ps pulses at the repetition rate of the machine. One gap of this machine needs, for 4 eV photons and 10% cathode efficiency, about 10^6w, or ∿ 1/2 GW/m, 1/2 Tw/km. This is large, but not unreasonable. Presumably an FEL is the way to do it.

The photocathode is the element which presents the fundamental uncertainty. All photocathodes used in current practical devices are based on semiconducting layers, and usually on very special surface conditions. The surfaces are very easily damaged, and even the bulk properties of the semiconductors can easily change as the defects alter under high current loads or bombardment by energetic particles, unless they are already in an amorphous highly doped condition. Worse, in the particular design described here, the diode is driven both ways, so that each electrode receives electron bombardment as well as emitting electrons. (In fact, a design with two wire photocathodes can avoid this particular aspect.) It seems that we would prefer a photocathode which is a nominaly pure, intrinsic conductor with a surface which is not special. Then its properties can be expected to remain stable. Work functions of decent metals are all more than 3.5 ev. Consequently, they are photo-emitters only in the far UV. Their quantum efficiency is found experimentally and theoretically to be of the form

$$QE = \frac{5 \times 15^{-4}}{\cos\theta} (h\nu - w)^2$$

where θ is the incident angle with respect to the normal. For W = 3.5 (thorium) hν = 9.5 ev (CaF_2 or MgF_2 optics) and θ = 75°, the QE = 7.2%. This is a solution without fundamental uncertainties, though the use of floride optics (albeit in tiny pieces) or mirrors is not too pleasant, and the efficiency is a bit low. Quartz optics are much nicer, restricting use to $h\nu \leq 7$ eV. Then we should consider surfaces of heavily doped semiconductors, still without special surfaces. We may expect initial QE > 20%, and experiments must show if they are adequately rugged.

We must also note the effect of the strong electrical field on the primary photodiode. The effect of this field is to lower the effective work function by more than 0.5 v, from the Schottky formula.

Last, an aside on the transverse wake field. This effect and its damaging effects on emittance growth, is the source of limits on the allowed beam loading. We note that the open structure used in this accelerator allows an interesting attack on this problem, which can be reduced if the alignment of the beam with the inner aperture in the structure is improved. As shown in Fig. 3, antennae are provided in four orthogonal directions outside the accelerator structure, to pick up the wake field radiation escaping from the slots at the outer radius. This is done on each accelerator pulse in 1% of the sections which are left unpowered for that purpose. The radiation from opposed pairs of antennae is subtracted and then detected, to obtain a check of the symmetry of the radiation. This method looks at the effect to be suppressed directly, so that no separate alignement is necessary. These signals are Fourier analyzed and give, and every 100 machine pulses, an indication of steering corrections. There can be feed back with a time constant of $\sim$ 0.01s, so that even earth motion can be servoed out. To first order, then, the transverse wake field problem can be suppressed.

These ideas were developed in collaboration with R. Palmer, J. Claus, V. Radeka and I. Stumer.

Table 1 - Parameters

R	Outer radius	60 mm
S	Disk spacing	1 mm
d	Disk thickness	0.6 mm
r	inner radius	0.5 mm
g	photodiode gap	0.5 mm
r_C	photodiode radius	50 μm
V	Charging voltage	40 kV
τ_e	Light pulse length	1 ps
C_x	External photodiodes cap.	1 pF
τ_x	Charging time	400 ps
E_γ	Light photon energy (wavelength)	4 eV (320 nm).

Table 2 - Derived Quantities

Photodiode capacitance	8 pF
photodiode charge	0.32 μC
stored energy	6.4 mJ
discharge time	10 ps
power	0.64 Gw
outer radius impedance Z_R	1.3 ohm
inner radius impedance Z_I	∿ 30 ohm
current	32 KA
current/mm² at photocathode	2000A/mm²
space charge limit (Child's law)	4.2×10^5 A
line voltage at external feed	40 kV
voltage at reflection	∿ 1.6 MV
accelerating field	∿ 1.6 GV/m
average gradient	∿ 1 GV/m
surface resistivity of Cu	0.06 ohm
ohmic energy dissipation	1.5 mJ
ohmic loss at inner radius	200 Mw/cm²
energy deposit in skin volume	40 J/cm³
no. particles to exhaust stored energy	2.5×10^{10}
accelerator efficiency for 10% beam load	.64/2.14 = 30%
light power, 4 eV and 10% photocathode eff.	1.3 MW
power density	100 KW/mm²
light energy	13 μJ
length for 1 TeV	1 km
beam stored energy/pulse of 2.5×10^9c	400 J
wasted energy lost in accelerator	933 J
power required for 10^4 Hz	13 Mw
beam power	4 Mw
light power	26 KW
light source power, if 1% eff.	2.6 MW.

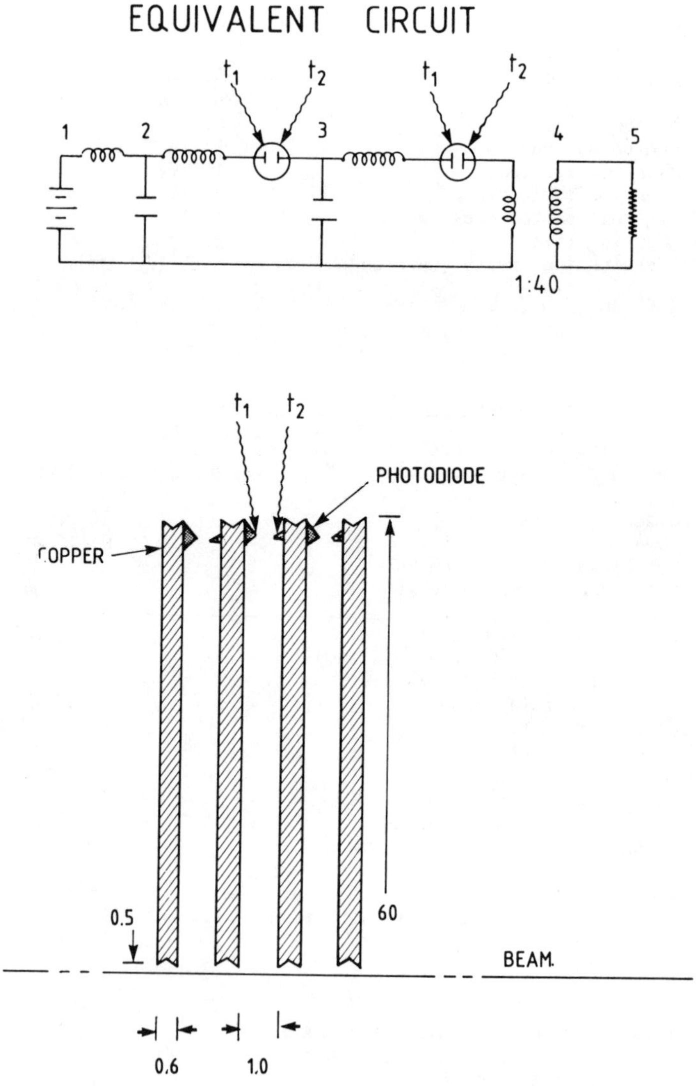
EQUIVALENT CIRCUIT
t_1
t_2
t_1
t_2
1
2
3
4
5
1:40
t_1
t_2
PHOTODIODE
COPPER
0.5
60
BEAM.
0.6
1.0

Figure 1

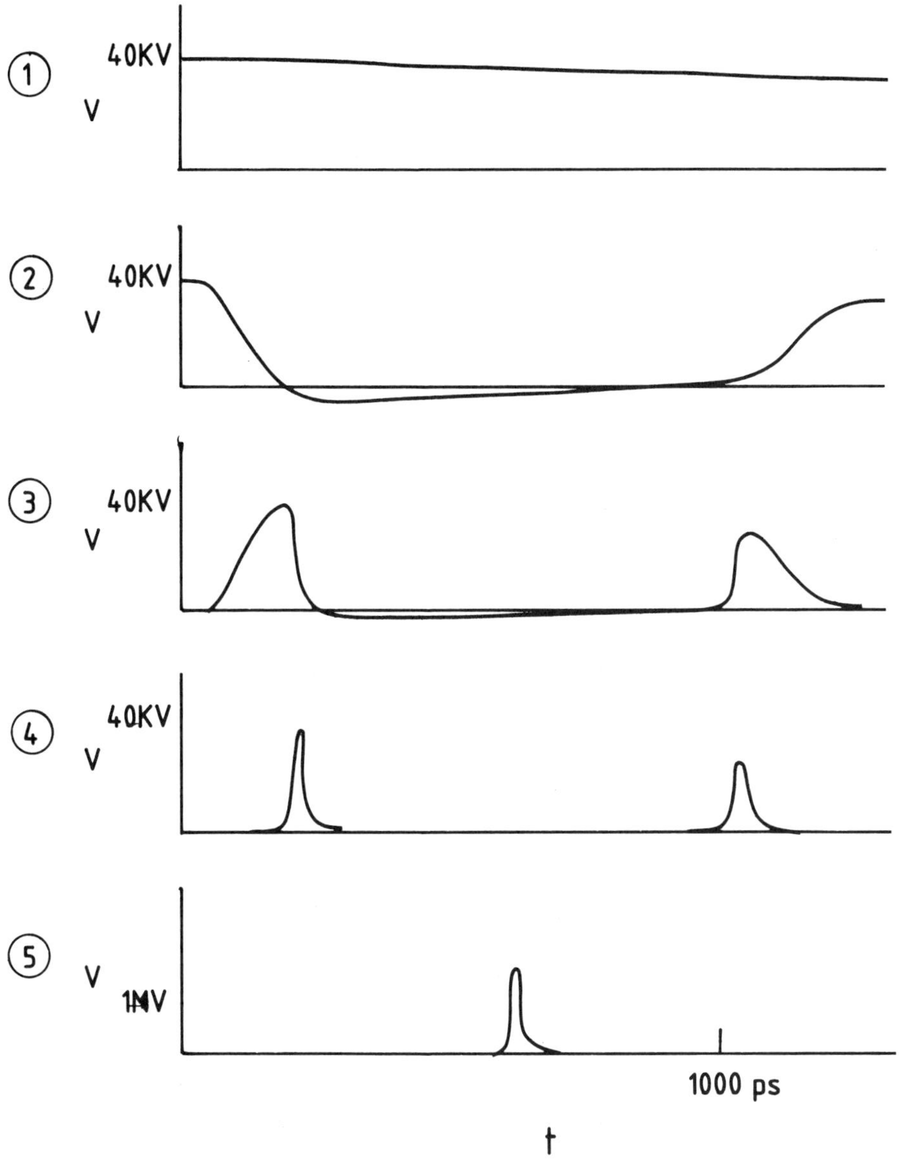

Figure 2(a)

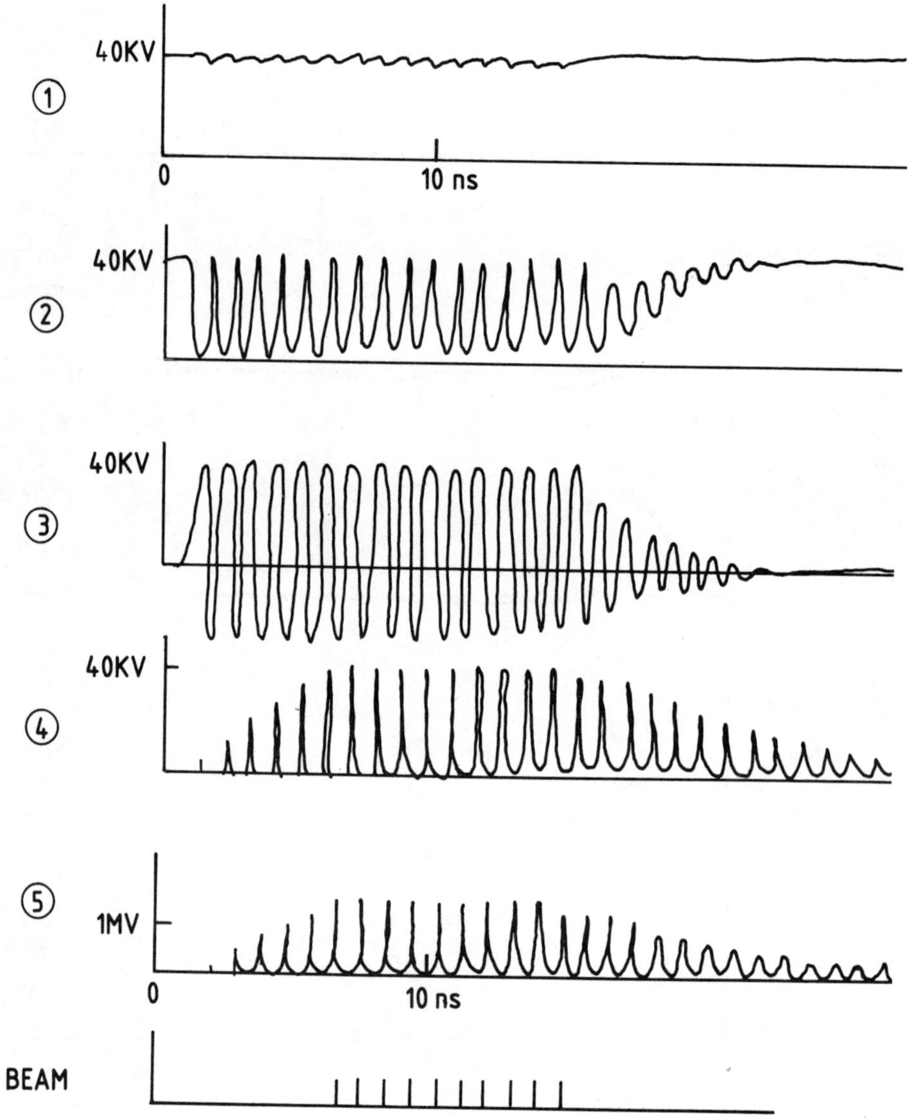

Figure 2(b)

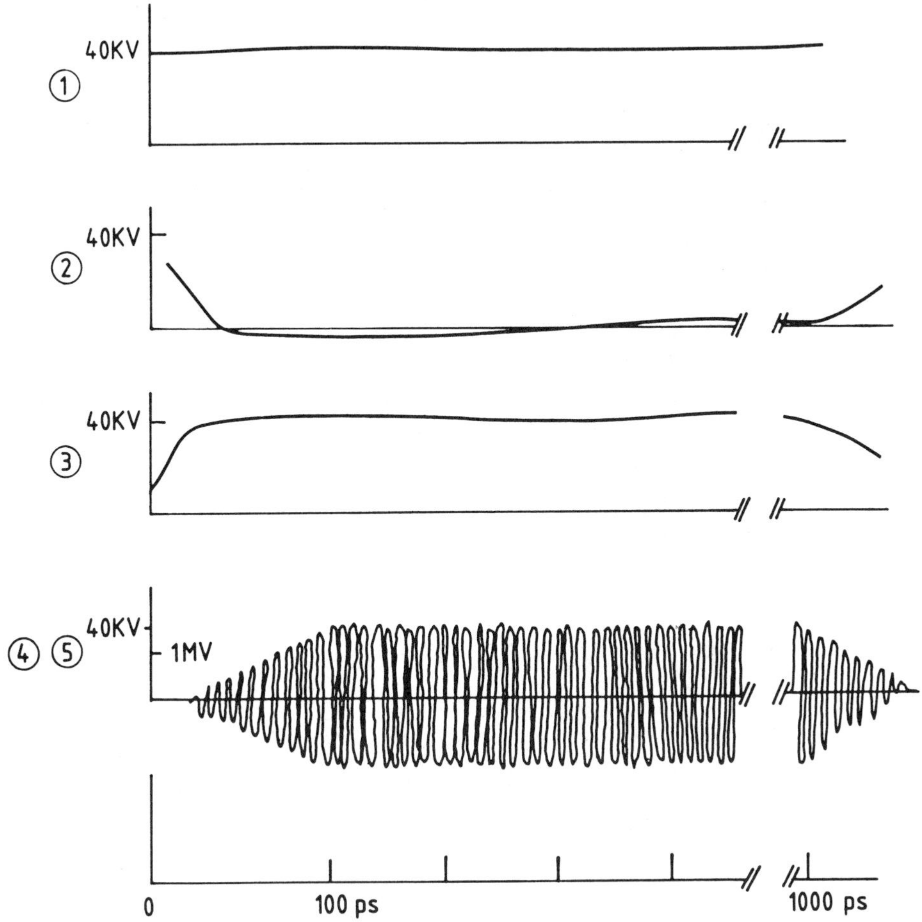

Figure 2(c)

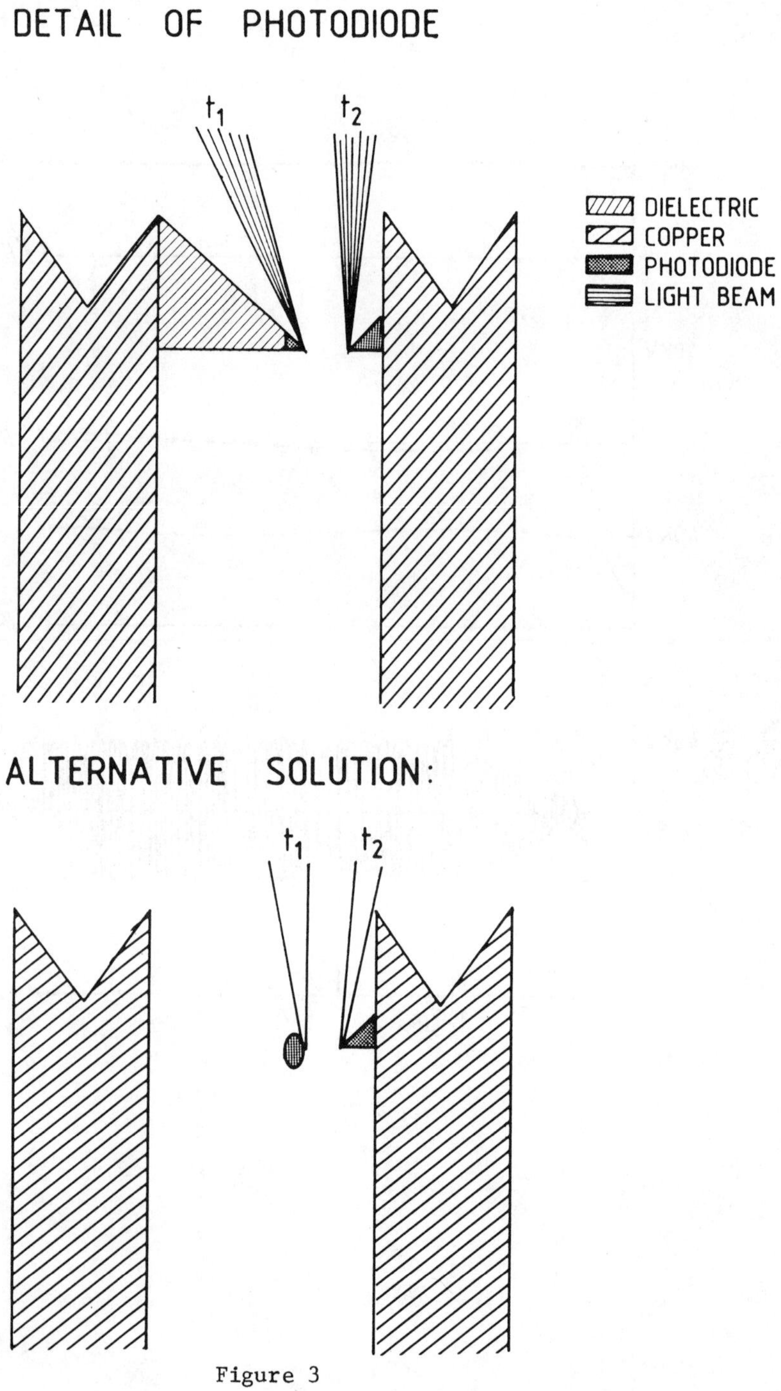
DETAIL OF PHOTODIODE
t_1
t_2
DIELECTRIC
COPPER
PHOTODIODE
LIGHT BEAM
ALTERNATIVE SOLUTION:
t_1
t_2

Figure 3

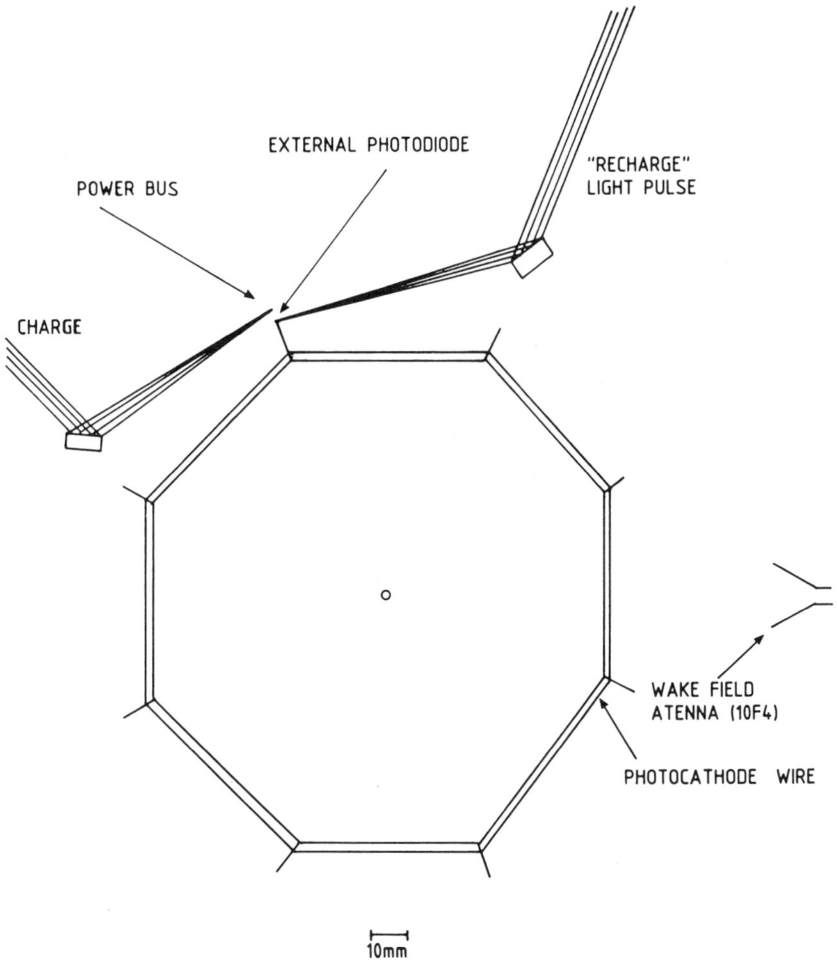
EXTERNAL PHOTODIODE
"RECHARGE"
LIGHT PULSE
POWER BUS
CHARGE
WAKE FIELD
ATENNA (10F4)
PHOTOCATHODE WIRE
10mm

Figure 4

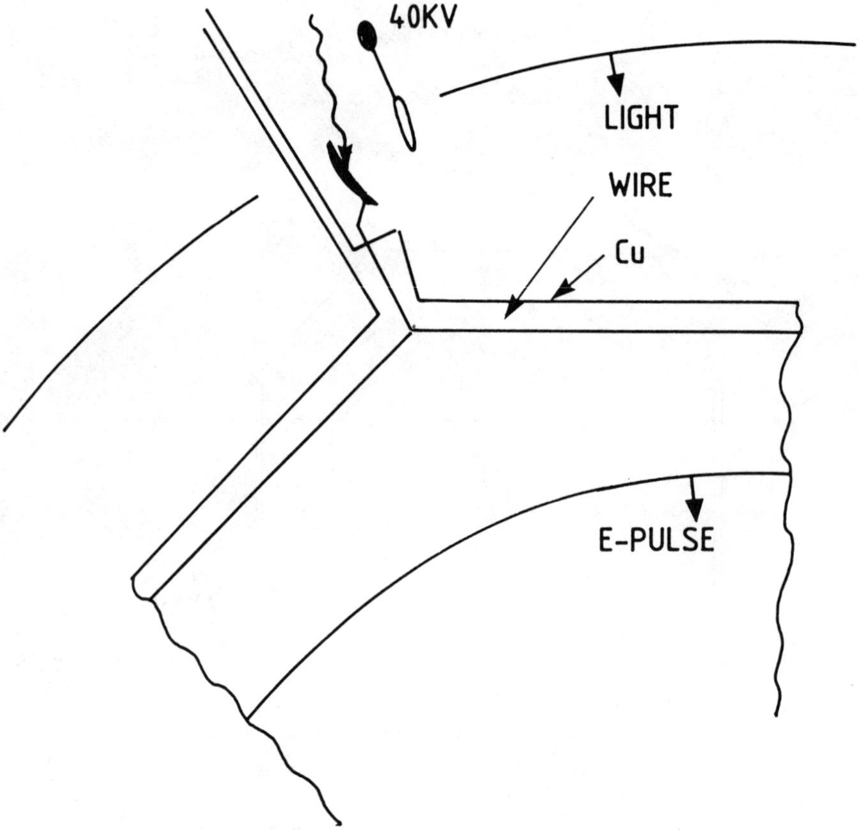

Figure 5

A PERIODIC PLASMA WAVEGUIDE

F.T.Cole*
Argonne National Laboratory **
January 18,1985

INTRODUCTION

It is proposed in this paper that a low-energy electron beam (a non-neutral plasma) with periodic density variation be used to perform the functions of a disc-loaded waveguide linear-accelerator structure. An accelerator is envisaged as a sequence of such periodic plasma waveguides, each a few meters long and each with an electron gun to produce the low-energy beam and an external rf power source at the upstream end. The particles being accelerated, either electrons or ions, are a separate second beam that passes sequentially through the waveguides, as in a conventional travelling-wave linear accelerator. A sketch of the device is given in Fig. 1.

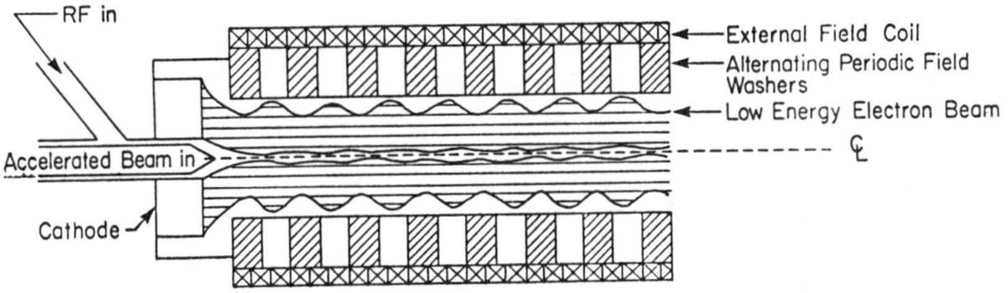

Fig. 1

The function of the low-energy beam is to provide a low-loss slow-wave structure for propagation of TM electromagnetic waves. The low-energy beam is not accelerated, nor is it a source of power. One can envisage it as a solid beam, a periodic medium, or a hollow beam where the medium provides periodic boundary conditions. In this paper, we treat the case of a solid beam.

The low-energy beam moves in an external periodic longitudinal magnetic field and the periodic density variations are produced by the standing waves of this motion. An externally produced electromagnetic wave will propagate in this medium with space harmonics corresponding to slow waves. The TM wave in synchronism with the second beam will accelerate that beam.

The most severe limit on the accelerating gradient arises from trapping the low-energy electrons in the accelerating bucket. Within this limit, accelerating fields of the order of 1 GV/m appear to be possible at millimeter wavelengths. The major advantage of the device over a conventional linac is that power lost to the wall can

* On leave from Fermi National Accelerator Laboratory
** Supported by the US Department of Energy

be much smaller, because the electron density in the beam is many orders of magnitude smaller than in a metal.

The device proposed here is different from many other devices using an electron beam in a periodic longitudinal field[1,2,3] in that those devices utilize coherent transverse motion ("sinuous" motion in Rayleigh's terminology), as a source of power, while the present proposed device utilizes "varicose" motion only to create a medium. It is different from other proposed plasma accelerators [4,5] in that it utilizes electromagnetic waves rather than plasma or cyclotron waves. Tajima[6] has proposed a somewhat similar device, a hollow-beam periodic plasma waveguide at laser frequency created by a laser beam. It should be noted that the system proposed here can in principle also be used as a travelling-wave tube microwave generator or amplifier.

In the following sections, we discuss the motion of the electron beam, the dispersion relation for electromagnetic waves, and the trapping limit. We consider only axially symmetric waves and beams, so there is no dependence on the azimuthal angle about the longitudinal z axis.

MOTION OF THE ELECTRON BEAM

We discuss the motion of the slow-energy electron beam. Our primary interest is in the forced oscillations of the beam envelope as it is acted only by both the external longitudinal field and the transverse self fields. We treat this motion under the assumptions that transverse displacements are small enough that we can neglect nonlinear terms in the displacement, that the amplitude of variation of the external field is small compared with the constant field, and that higher harmonics of the external field are small compared with the first harmonic. That is, we take the field on axis to be

$$B_z = B_0 + B_1 \cos k_0 z \tag{2.1}$$

$$B_r = B_\Theta = 0,$$

where k0 is the wave number of the periodic field. Off axis, there are transverse field components, but these do not affect the motion in the linear approximation and we omit them here. We write the field in terms of the relativistic cyclotron frequency $\omega_c = -eB_z/m\gamma c$, where, as usual, e is the electron charge, m its rest mass, γ its total energy in units of rest energy and c is the speed of light, as

$$\omega_c = \omega_{c0} + \omega_{c1} \cos k_0 z \tag{2.2}$$

The field of Eqs. (2.1) or (2.2) can be derived from a vector potential

$$A_\Theta = - \frac{m\gamma c}{e} (1/2\omega_c r) \quad (2.3)$$

The radial space-charge force is

$$1/2\ m\gamma \frac{\omega_p^2}{\gamma^2} f(r) \tag{2.4}$$

$$f(r) = r : r<r_b$$
$$= \frac{r_b^2}{r} : r \geq r_b)$$

where $\omega_p^2 = 4\pi ne^2/m\gamma$, n is the electron density and r_b is the radius of the beam. The single-particle equation of radial motion is

$$\ddot{r} = r\dot{\Theta}^2 - r\dot{\Theta}\omega_c + 1/2 \frac{\omega_p^2}{\gamma^2} f(r) \tag{2.5}$$

The canonical angular momentum

$$p_\Theta = m\gamma r^2\dot{\Theta} + \frac{e}{c} rA\Theta \tag{2.6}$$
$$= m\gamma r^2(\dot{\Theta} - 1/2\omega_c)$$

is a constant of the motion. Then

$$\ddot{r} = (\frac{p_\Theta}{m\gamma})^2 \frac{1}{r^3} - \frac{1}{4} \omega_c^2 r + 1/2 \frac{\omega_p^2}{\gamma^2} f(r) \tag{2.7}$$

If there were no longitudinal variation, the beam would have a uniform radius r_0 and

$$\frac{p_\Theta}{m\gamma r_0^2} = \pm \sqrt{\omega_{c0}^2 - \omega_{p0}^2} \tag{2.8}$$

We define $x = r - r_0$ and keep only terms linear in x in Eq. (2.7). We use the external form of f(r). Then

$$x'' + (\omega_c^2 - \omega_p^2)\ x = 0,$$

a well-known equation of motion for a beam electron.[7] Here ω_c^2 and ω_p^2 are periodic functions of z. We change the independent variable to $z = v_0 t$, where v_0 is the beam velocity. Then

$$\frac{d^2x}{dz^2} = x'' = - \frac{1}{v_0^2} (\omega_c^2 - \omega_p^2)x \tag{2.9}$$

Individual electrons will oscillate with wave numbers approximately equal to

$$(\sqrt{\omega_{c_0}^2 - \omega_{p_0}^2}\ ) /v_0,$$

which is not in general the same as k_0, the period of the field. The periodic part of the coefficient will give rise to Floquet stop bands when

$$\sqrt{\omega_c^2 - \omega_p^2} = nk_0 v_0,$$

where n is an integer. This is the same behavior found experimentally by Mahaffey and Trivelpiece[7] and explained by them.

At the same time, the beam envelope will oscillate with the wave number k_0 of the external field, just as the amplitude function or envelope in a strong-focusing lattice has the period of the structure, although individual particles have wave numbers different from k_0. We take

$$x = A(z)\cos\Phi(s) \tag{2.10}$$

and find by substitution in Eq. (2.9) and harmonic balance

$$\Phi' = \frac{E}{A^2}$$

$$A'' - \frac{E}{A^3} + \frac{1}{v_0^2}(\omega_{c0}^2 - \omega_{p0}^2)A = 2\frac{\omega_{c1}\omega_{c0}}{v_0^2} A\cos k_0 z.$$

We take

$$A = A_0 + A_1 \cos k_0 z \tag{2.11}$$

(A_0 refers to the envelope, while r_0 is for an individual electron)

and find

$$E = \frac{\omega_{c0}}{v_0} A_0^2$$

$$A_1 = \frac{2\omega_{c1}\omega_{c0}A_0}{\omega_{c0}^2 - \omega_{p0}^2 - k_0^2 v_0^2}, \tag{2.12}$$

demonstrating one of the Floguet resonances posited above. Finally, the plasma frequency varies periodically as

$$\omega_p^2 = \omega_{p0}^2 - 2\omega_{p0}^2 \frac{A_1}{A_0} \cos k_0 z \tag{2.13}$$

DISPERSION RELATION

The external magnetic field has only a small direct effect on the propagation of electromagnetic waves. In this discussion, we therefore neglect the external field and with it the coupling between electric and magnetic waves. Assuming that all quantities vary with time as $e^{i\omega t}$, the wave equation for the axially symmetric TM wave is

$$\frac{\partial^2 E_z}{\partial z^2} + \frac{1}{r}\frac{2}{\partial r}(r\frac{\partial E_z}{\partial r}) + \frac{1}{c^2}(\omega^2 - \omega_p^2)E_z = 0. \quad (3.1)$$

Here ω_p^2 is a periodic function of z with period $2\pi/k_0$. Note the difference from the conventional disc-loaded waveguide case or from the hollow-beam case. There the wave equation does not have periodically varying coeffcients, but periodic boundary conditions. In the case treated here, we can separate variables

$$E_z = R(r)\, E(z) \quad (3.2)$$

$$\frac{1}{r}\frac{2}{\partial r}(r\frac{\partial R}{\partial r}) + k_r^2 R = 0$$

$$\frac{\partial^2 E}{\partial z^2} + \frac{1}{{}^2 c}(\omega^2 - \omega_p^2 - c^2 k_r^2)\, E = 0, \quad (3.3)$$

with k_r the separation constant. Boundary conditions on a surface of relatively large radius can be satisfied by this simple solution.

The radial equation hs the solution

$$R = J_0(k_r r) \quad (3.4)$$

and the longitudinal equation satisfies Floguet's theorem. Its solution is therefore of the form

$$E = \Sigma E_n e^{ik_n z}$$

$$k_n = k_z + nk_0 \quad (3.5)$$

Substitution of this solution in Eq. (3.1) gives the set of recursion relations

$$D_n E_n = \sum_{m \neq 0} \omega_{pm}^2 E_{n-m}$$

$$D_n = \omega^2 - \omega_{p0}^2 - c^2(k_r^2 + k_n^2) \quad (3.6)$$

When ω_{pm} = 0 for m =0, the right hand side vanishes and Eq. (3.6) reduces to the usual dispersion relation for electromagnetic waves in a plasma

$$D_0 = \omega^2 - \omega_{p0}^2 - c^2(k_r^2 + k_z^2) = 0 \quad (3.7)$$

If D_0 = 0 at the point (ω, k_z), then D_n = 0 at the point (ω, k_n), a new set of waves. Let us consider two cases:

(1) Far from D_n = 0.

Since $\omega_{pn} \ll \omega_{p0}$, we assume that $E_n \ll E_0$. Then the only significant term of the right hand side of Eq. (3.6) is the term containing E_0 and

$$E_n = \omega_{pn}^2 E_0/D_n.$$

Then for n = 0,

$$(D_0 - \sum_{m\neq 0} \frac{\omega_{pm}^2 \omega_{p-m}^2}{D_m}) E_0 = 0, \tag{3.8}$$

which is a dispersion relation for this case. Of more interest is the case close to a zero of the denominator, when the assumptions leading to Eq. (3.8) are not valid.

(2) Near $D_1 = 0$.

Now E_0 and E_1 can be comparable. We neglect all other harmonics and have

$$D_0 E_0 - \omega_{p-1}^2 E_1 = 0$$

$$D_1 E_1 - \omega_{p1}^2 E_0 = 0 ,$$

from which the dispersion relation is

$$D_0 D_1 - \omega_{p1}^2 \omega_{p-1}^2 = 0 \tag{3.9}$$

There are two branches of the solution, approximately parabolas centered at $k_z = 0$ and $k_z = k_0$, respectively. At $k_z = 1/2\ k_0$,

$$(\omega^2 - \omega_p^2 - c^2 k_r^2) = c^2 \frac{k_0^2}{4} \pm (\omega_{p1}^2 \omega_{p-1}^2), \tag{3.10}$$

so there is a stopband between the two branches of width

$$\delta = 2\omega_{p1}\omega_{p-1} \tag{3.11}$$

and each branch has slope zero at this point. A slow wave can be propagated along one of these branches, as in a conventional travelling-wave structure.

FIELD LIMITATION

Consider the motion of a particle in the travelling wave. The calculation is carried out in the spirit of Chao and Ruth's work.[8] The Hamiltonian of the motion is

$$H = c\sqrt{p^2 + m^2 c^2} - \frac{eE_0}{k} \cos(\omega_0 t - kz). \tag{4.1}$$

We make a canonical transformation to the wave system by the generating function

$$F(P,z,t) = P(z - v_0 t), \tag{4.2}$$

where

$$v_0 = \omega_0/k \tag{4.3}$$

The new Hamiltonian is

$$\bar{H} = c\sqrt{P^2 + m^2c^2} - \frac{eE_0}{k} \cos kF - v_0 P, \tag{4.4}$$

where

$$Z = z = v_0 t. \tag{4.5}$$

There are fixed points at $P = P_0 = m\gamma_0 v_0$, $\gamma_0 = (1 - v_0^2/c^2$ $Z = 0$ (stable) and $Z = \pm\pi$ (unstable). The value of the Hamiltonian on the separatrix is

$$\bar{H} = mc^2 (\frac{1}{\gamma_0} + \delta), \tag{4.6}$$

where

$$\delta = \frac{eE_0}{k\ mc^2}, \tag{4.7}$$

The separatrix is given by

$$P = mc[\beta_0 \gamma_0 A \pm \gamma_0 \sqrt{A^2 \gamma_0^2 -1}] \tag{4.8}$$

$$A = \frac{1}{\gamma_0} + \delta\ (1 + \cos k\ Z) \tag{4.9}$$

The bucket is centered on the high-energy second beam. The low-energy first beam will be outside the bucket and be untrapped if the bucket half-height is smaller than the height of the bucket center, that is, if

$$\gamma_0 \sqrt{A^2 \gamma_0^2 - 1} < \beta_0 \gamma_0^2 A,$$

or

$$A^2 \gamma_0^2 - 1 < \beta_0^2 \gamma_0^2 A,$$

or

$$A^2 < 1.$$

At the bucket center, $Z = 0$,

$$\frac{2eE_0}{k\ mc^2} < 1 - \frac{1}{\gamma_0}. \tag{4.9}$$

Thus if the accelerated beam is at low energy $(\gamma_0 \cong 1)$, the maximum accelerating field is very small. For $\gamma_0 >> 1$, and $\lambda = 1$ mm,

$$eE_0 < \frac{\pi}{2} \cdot 10^9 \text{ eV/m} \tag{4.10}$$

CONCLUSIONS

In the sections above, we have proposed a periodic plasma waveguide and presented heuristic discussions of some of the accelerator-physics problems of the device. We have not discussed here motion of the accelerated beam or the electrical properties of the medium as a waveguide.

We envisage such a device as operating in the millimeter or submillimeter wavelength regime as a linear collider with a pulse length of approximately 1 nanosecond and a repetition period of approximately 1 microsecond. These parameters have not been optimized. The accelerating field is approximately 1 GV/m. The low-energy electron beam is of the order of 1 keV in energy and 1 kA in peak current.

We think of this device as a more conventional plasma wave accelerator, bridging the gap between conventional linear accelerators and future plasma and laser accelerators that will achieve still higher accelerating fields.

REFERENCES

1. R.M. Phillips, IRE Trans. Electron Devices 231 (1960). C.E. Enderby and R.M. Phillips, Proc. IEEE 53, 1648 (1965).
2. M. Friedman and N. Herndon, P.R.L. 28, 210 (1972); PRL 29, 55 (1972).
3. W.A. McMullin and G. Bekefi, Phys. Rev. A, 25, 1826 (1982).
4. Y.B. Fainberg, Proc. 1956 CERN Symposium on High-Energy Accelerators (CERN, Geneva, 1956) Vol. I, p. 80.
5. M.L. Sloan and W.E. Drummond, P.R.L. 31, 1234 (1973).
6. T. Tajima, Proc. 12th International Conference on High Energy Accelerators (Fermilab, 1983) p. 470.
7. R.A. Mahaffey and A.W. Trivelpiece, P.R.L. 36 1044 (1976).
8. R.D. Ruth and A.W. Chao in "Laser Acceleration of Particles" AIP Conf. Proc. 91 (AIP, New York, 1982) p. 94.

IONIZATION FRONT ACCELERATOR: HIGH GRADIENTS, DEMONSTRATED PARTICLE ACCELERATION, AND A PROPOSED RELATIVISTIC ACCELERATOR*

C. L. Olson, C. A. Frost, E. L. Patterson,
J. P. Anthes, and J. W. Poukey
Sandia National Laboratories, Albuquerque, NM 87185

ABSTRACT

The Ionization Front Accelerator (IFA) is a collective ion accelerator for which high-gradient particle acceleration has now been demonstrated. In the IFA, the space charge field at the front of an intense relativistic electron beam is controlled by a laser and used to accelerate an ion bunch. Two complete IFA systems have been built (IFA-1 and IFA-2). Here we present initial IFA-2 ion results that demonstrate that ions have been accelerated with controlled accelerating fields of 33 MV/m over 30 cm.

Space charge fields of accelerators like the IFA and the plasma beat wave accelerator are compared, and both are shown to be capable of producing fields 1 GV/m and higher. The IFA systems are discussed and initial IFA-2 ion results are presented. Lastly, a relativistic IFA is proposed that should, in principle, permit the attainment of virtually unlimited ion energies.

INTRODUCTION

Collective accelerators are high-gradient accelerators that use the collective fields of one particle species (usually electrons) to accelerate a select group of charged particles (usually of a different species). Collective ion accelerators typically use the fields of an electron beam or electron ring to accelerate ions.[1,2] In the Ionization Front Accelerator (IFA), the space charge field at the front of an intense relativistic electron beam (IREB) is controlled by a laser, and used to accelerate an ion bunch.[3,4] Two complete IFA systems have been built (IFA-1 and IFA-2).[5-8] Here we present some initial IFA-2 ion results. These results demonstrate that ions have been accelerated with controlled accelerating fields of 33 MV/m over 30 cm.

First, we will compare the space charge accelerating fields possible for two classes of collective accelerators--(1) those that use an IREB like the IFA, and (2) those that use a plasma wave like the plasma beat wave accelerator. Both have accelerating fields created by net space charge fields. We will show that both are capable of creating accelerating fields of

*Supported by Division of Advanced Energy Projects, U.S. Department of Energy.

~1 GV/m and higher. We will then briefly describe the IFA concept, the IFA-1 system, and the IFA-2 system. IFA-2 ion results are then discussed. Lastly, a relativistic IFA concept is proposed that should, in principle, permit the attainment of virtually unlimited ion energies.

SPACE CHARGE FIELDS

Consider the space charge field at the edge of an IREB. For a long beam of uniform electron density n_e and radius r_b, the beam appears as a rod of charge and the radial electric field at the edge of the beam is

$$E_r = 2\pi n_e e r_b \quad , \tag{1}$$

where e is the charge of an electron. For a steep axial charge neutralization gradient, the maximum axial electric field possible at the end of the rod of charge is[1]

$$E_z = E_r \quad . \tag{2}$$

From (1) and (2) we note the important scaling

$$E_z \sim n_e \quad . \tag{3}$$

In Table I, some examples are given of these results. Here, ε_e is the electron energy, I_e is the electron current, and J_e is the electron current density. The parameters in the first line are those as used in the IFA-2 experiment, and they represent a modest-sized IREB with $J_e = 10^4$ A/cm^2. The parameters in the second line show a scale up of one order of magnitude in n_e, and for this case $J_e = 10^5$ A/cm^2 and $E_z \approx 1.8$ GV/m. A current density of $J_e \gtrsim 3 \times 10^4$ A/cm^2 was routinely achieved with IFA-1, and current densities up to 10^7 A/cm^2 have been produced in pinched electron beam experiments,[9,10] so it is not at all unrealistic to consider current densities of 10^5 A/cm^2 or more. It should be noted that these large space charge fields are automatically present with IREBs. If steps are taken to neutralize the IREB in a controlled manner (as in the IFA), then these large fields may be utilized to accelerate particles.

Now consider a neutral plasma with electron density n_e and ion density n_i so $n_e = n_i$. The plasma is automatically neutral, with essentially negligible space charge fields. However, if a plasma wave is driven (as in the plasma beat wave accelerator), then the characteristic wave number k is given by

$$k^{-1} = c/\omega_{pe} \quad , \tag{4}$$

where $\omega_{pe} = (4\pi n_e e^2/m)^{1/2}$ is the electron plasma frequency, and m is the mass of an electron. The peak axial electric field associated with the plasma wave is

$$E_z \approx \frac{\Delta n_e}{n_e} 4\pi n_e e(c/\omega_{pe}) \quad , \tag{5}$$

where $\Delta n_e/n_e$ represents the magnitude of the density modulation. From (5), we note the important scaling

$$E_z \sim \sqrt{n_e} \quad . \tag{6}$$

In Table II, some examples are given of these results for a 10 percent density modulation. Note that for a plasma density of $n_e = 10^{17}$ cm^{-3}, we have $E_z = 3.2$ GV/m.

In Fig. 1, a comparison of the above space charge results is given. The IREB result is a plot of the result (1)-(2), and it shows the scaling $E_z \sim n_e$. The plasma wave result is a plot of result (5), and it shows the scaling $E_z \sim \sqrt{n_e}$. The IFA-2 particle acceleration results presented in this paper are represented by the heavy dot, and the dashed line represents the fields possible with IFA-2. The plasma beat wave results of Joshi et al. presented at this conference are represented by the dashed line marked BWA--this line represents the wave fields inferred from their data.[11] Note, e.g., that to achieve 1 GV/m requires an IREB density of $n_e \approx 10^{13}$ cm^{-3}, or a plasma density of $n_e \approx 10^{16}$ cm^{-3}.

In the Proceedings of the First Workshop on Laser Acceleration of Particles,[12] it was noted that IREB densities typically are $\lesssim 10^{14}$ cm^{-3}, but then it was assumed that the scaling $E_z \sim \sqrt{n_e}$ of (5) applies, and this led to the (incorrect) conclusion that IREBs were not interesting for high gradient accelerators. In fact, the correct scaling for IREBs is $E_z \sim n_e$ as in (3), and very large fields can be produced as shown in Fig. 1. The important conclusion is that with either IREBs or plasma waves, fields as high as 1 GV/m or higher should be attainable.

Table I. Examples of Space Charge Fields for IREBs.

$n_e(cm^{-3})$	$\mathcal{E}_e$ (MeV)	I_e (kA)	r_b (cm)	J_e (A/cm^2)	E_z
2.0×10^{12}	1	30	1	1.0×10^4	180 MV/m
2.0×10^{13}	10	300	1	1.0×10^5	1.8 GV/m

Table II. Examples of Space Charge Fields for Plasma Waves.

$n_e(cm^{-3})$	ω_{pe} (sec^{-1})	c/ω_{pe} (cm)	E_z for 10% modulation
1.0×10^{15}	1.8×10^{12}	0.017	320 MV/m
1.0×10^{17}	1.8×10^{13}	0.0017	3.2 GV/m

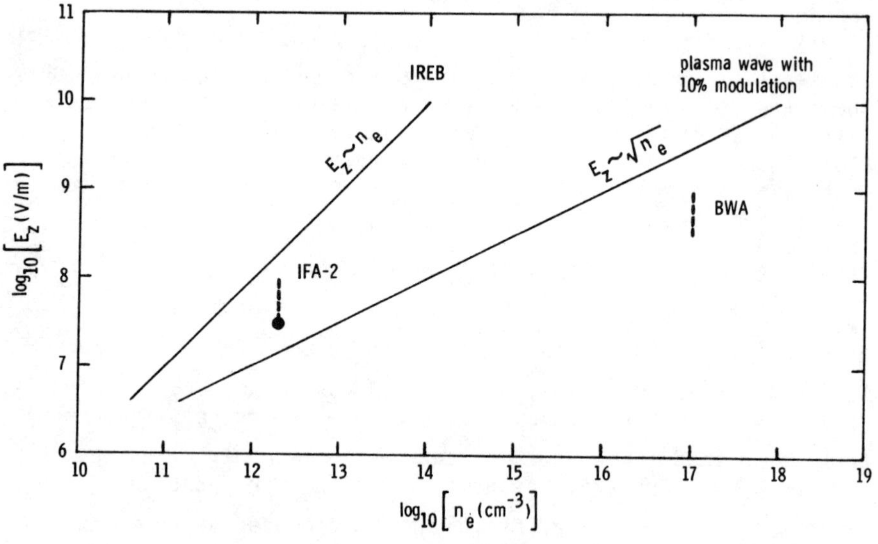

Figure 1. Space Charge Fields.

IONIZATION FRONT ACCELERATOR

The IFA is a high-gradient collective ion accelerator in which ions are trapped and accelerated in a strong-potential well at the head of an IREB.[1,4] The IFA is a direct extension of the acceleration process that occurs when an IREB is injected into low pressure neutral gas. In the IFA concept, as shown in Fig. 2, the potential well at the IREB head is programmed to move with the desired phase velocity by actively controlling the ionization of a suitable background working gas with a laser. The working gas pressure is chosen low enough so that IREB-induced ionization processes can not substantially ionize the gas before the laser does. Typically, the working gas ions have a very low charge-to-mass ratio and are not trapped. Ions from a separate source with a very high charge-to-mass ratio (e.g., H^+ or He_4^{+2}) are trapped and accelerated in the IREB potential well.

The laser parameters are chosen so that as the laser is swept, a background plasma of density n_p is created with $n_p \approx n_b$, where n_b is the IREB electron density. The IREB parameters are typically chosen with $I_e \approx I_\ell$, where I_e is the injected IREB current and I_ℓ is the space charge limiting current. With these parameters, the IREB will initially enter the drift tube, be stopped by its own space charge, and create a strong potential well. As the laser is swept, a charge-neutralizing channel is created, and the IREB will propagate through it. In the channel, plasma electrons are expelled radially until the plasma ion background density n_{pi} is about equal to n_b, giving charge neutrality to the IREB. In the channel, the IREB is held together by its self-magnetic field, and individual electrons undergo betatron oscillations as they travel to the head of the channel. Beyond the head of the channel, there is no neutralization; there the IREB sees its own space charge and quickly diverges radially. At all times, a strong potential well is created just past the end of the channel by the space charge field of the IREB head. As the laser is swept, the potential well synchronously follows it.

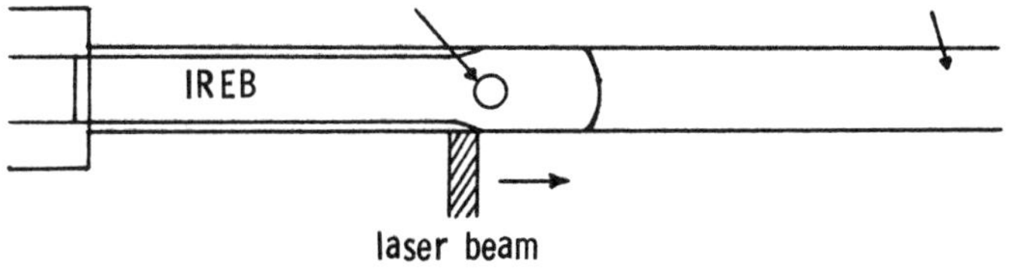

Figure 2. Ionization Front Accelerator (IFA).

It should be noted that the IREB supplies the power and the fields to accelerate the ions trapped in the potential well; the laser supplies only a small amount of energy to control the motion of the IREB. The distribution of the laser energy, the ion energy, and the IREB energy is, respectively,

$$\varepsilon_{laser} << \varepsilon_{ions} < \varepsilon_{IREB} \quad . \tag{7}$$

This means that a very small amount of (expensive) laser photon energy is used to control a very large amount of (inexpensive) pulsed power IREB energy.

A summary of design parameters for three IFA examples is given in Table III. The IFA-1 system was constructed and operated during 1977-1979. The IFA-2 system was constructed and operated during 1981-1984. Our main goal has been to develop a compact ion accelerator with ion energies in the range of a few 100 MeV to a few GeV for several applications, as discussed, e.g., by the U.S.D.O.E. Study on Collective Accelerators.[13] In addition, with the relativistic IFA concept discussed later in this paper, the IFA could be used to reach virtually any ion energy desired.

Table III. IFA Design Parameters.

	IREB	acceleration region	protons
IFA-1	0.6 MeV 20 kA 0.5 cm radius 10 nsec 0.01 TW 0.1 kJ	50 MV/m 1.2 cm diameter 10 cm length 6.5 nsec of IREB used to accelerate protons	5 MeV 0.8 kA 0.25 cm radius 0.17 nsec 0.004 TW 0.7 J
IFA-2	1.2 MeV 30 kA 1 cm radius 30 nsec 0.04 TW 1 kJ	100 MV/m 2.2 cm diameter 1 m length 16 nsec of IREB used to accelerate protons	100 MeV 5 kA 0.5 cm radius 0.08 nsec 0.5 TW 40 J
1 GeV example	3 MeV 30 kA 1 cm radius 40 nsec 0.09 TW 3.6 kJ	100 MV/m 2.2 cm diameter 10 m length 40 nsec of IREB used to accelerate protons	1 GeV 10 kA 0.5 cm radius 0.04 nsec 10 TW 400 J

The IFA-1 system is shown in Fig. 3. For these experiments, Cs was the working gas, and two-step photoionization was used with a dye laser (852.1 nm) for Cs excitation and a frequency-doubled ruby laser (347 nm) for photoionization of Cs from the excited state. The advantage in using two-step photoionization of Cs is that the excited state photoionization cross section is about 50 times larger than the ground state photoionization cross section. In IFA-1, the "exciter" laser (dye laser) was swept and the "kicker" laser (frequency-doubled ruby laser) was "on" during the experiment. The dye laser was accurately swept quadratically in time by using transit time delays in a programmed light pipe array. The key results from the IFA-1 experiments[5,6] were: (1) Cs was demonstrated to be a feasible working gas at a density of 10^{15} cm^{-3}. Since the IREB density is $\sim 10^{12}$ cm^{-3}, this meant that the Cs only had to be ionized ~0.1 percent for the IFA to work. (2) Accurately-controlled motion of the front of the IREB was demonstrated with three different programmed sweep rates. These results demonstrated that IFA-controlled motion of the potential well at the IREB head had been achieved. (3) Three different ion data sets were obtained (H^+, D^+, and He^{+2}) that implied that controlled accelerating fields of 50 MV/m were achieved over an acceleration length of 10 cm. More ion data was needed, but the data acquisition rate was severely limited due to jitter between the lasers and the IREB caused by jitter in the self-breakdown oil switches on the Blumlein. A new system (IFA-2) was conceived to alleviate these problems.[7]

The IFA-2 system is shown in Fig. 4. Originally, IFA-2 was to use a single laser with two-photon ionization of a new room temperature working gas (diethyl anilene). However, upon discovery that the photoionization occurred by a two-step process with a much slower temporal dependence,[8] we were forced to return to using Cs as the working gas. However the final IFA-2 system has several differences from IFA-1, as follows. In IFA-2, a dye laser (852.1 nm) is used to pre-excite the entire Cs volume, and the "kicker" laser is an XeCl laser (308 nm) that is swept. An electro-optic crystal deflector is used to sweep the beam in time, typically 30 cm in 20 nsec, with a quadratic temporal dependence. Laser-triggered Blumlein switches are used to provide low-jitter synchronization of the IREB with the lasers. The IREB parameters are larger (1 MV, 30 kA, 30 ns) and the IFA drift section is larger (30 cm long, 1.1 cm radius).

Although analysis of the IFA-2 ion results is still in progress, we present here some initial results. First, we performed experiments to verify that the Cs photoionization scheme was working as planned. In these experiments the lasers were used to pre-ionize a 10 cm section of the Cs. Then by observing the IREB propagation behavior with a streak camera, we were able to observe the threshold for charge neutralization at which the IREB would propagate quickly through only the first 10 cm. From these

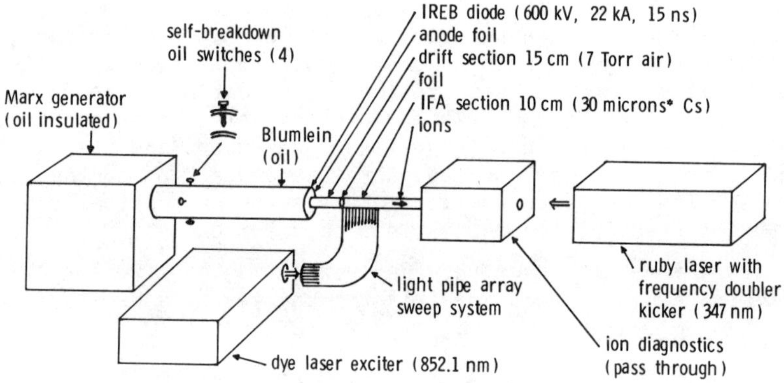

Figure 3. IFA-1.

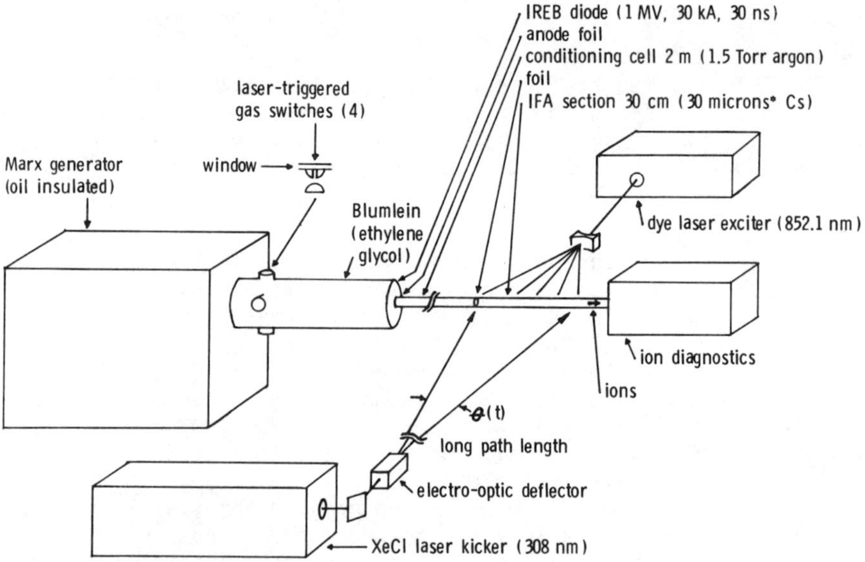

Figure 4. IFA-2.

experiments, we found, in agreement with theory, that only 1 mJ/cm^2 of dye laser and 0.5 mJ/cm^2 of swept XeCl laser are needed for IFA control of the IREB (which has about 1 kJ of energy). Thus we emphasize that the scaling relation (7) holds well for the IFA.

Some of the IFA-2 ion results are as follows. The ion energy spectra are produced by ion tracks in a solid state nuclear track detector (CR-39) placed inside a magnetic spectrometer. After etching in NaOH, the individual track pits are visible and can be counted with the aid of a microscope. In Fig. 5a, we show the proton spectrum produced for the IFA-2 system at room temperature, filled with 100 microns H_2, and the lasers blocked. This result gives the natural collective acceleration result for our system and shows a peak at ~1 MeV. In Fig. 5b, we show the proton spectrum for the full IFA-2 system, heated, with 50 microns Cs and 50 microns H_2, and lasers fully operational. Here, the sweep length was 10 cm and the final laser sweep speed βc had $\beta = 0.058$, which corresponds to a 1.7 MeV proton. The proton spectrum is peaked very close to 1.7 MeV indicating that ion trapping and ion acceleration occurred as programmed. Note that a shift in the main ion energy from Fig. 5a to Fig. 5b is clearly demonstrated.

In Fig. 6a, we show the He^{+2} ion spectrum produced for the IFA-2 system at room temperature, filled with 100 microns He, and with the lasers blocked. The natural collective acceleration process gives a peak at ~2 MeV. In Fig. 6b, we show the He^{+2} spectrum for the full IFA-2 system, heated, with 50 microns Cs and 50 microns He, and the lasers fully operational. Here the sweep length was 30 cm and the programmed final laser sweep speed βc had $\beta = 0.1$, which corresponds to a 20 MeV He^{+2} ion. The He^{+2} ion spectrum shown in Fig. 6b has a clear peak at the programmed ion energy. The ion number is low in Fig. 6b, which is why the noise level of the CR-39 is visible. In any case, Fig. 6b shows that the ions trapped and accelerated have a peak energy very close to the programmed sweep energy. This result (20 MeV He^{+2} in 30 cm) means controlled accelerating fields of 33 MV/m over 30 cm have been achieved.

The number of ions N that can be trapped under optimum conditions should be about $NZ \approx n_b r_b^3$, where Z is the charge of the ions.[1] For IFA-2 parameters, this is $NZ \approx 2.0 \times 10^{12}$. Based on the estimated phase space distribution of the final ion bunch, and assuming that the bunch is charge neutral and drifts ballistically, then it is possible to estimate N based on the small number of ions in the phase space accepted by the magnetic spectrometer. The rough estimates for N corresponding to the data in Figs. 5a, 5b, 6a, and 6b are, respectively, 5×10^{12}, 1×10^{12}, 1×10^{11}, and 3×10^7. The last number is low presumably because

(a)

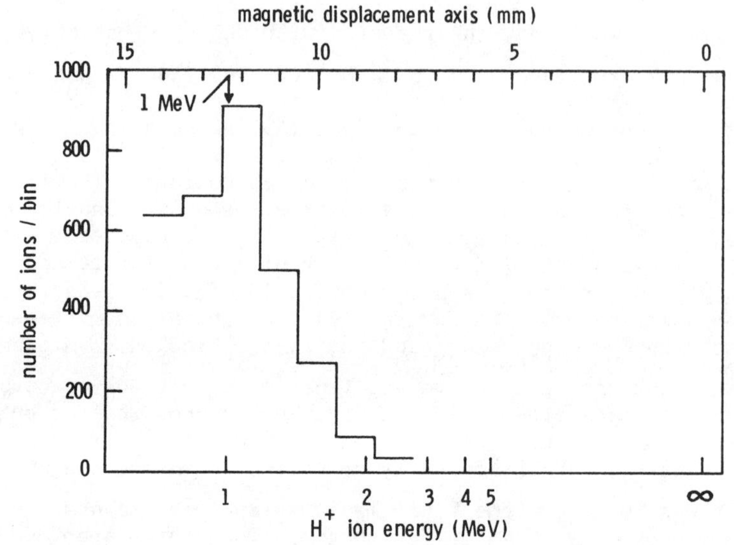

(b)

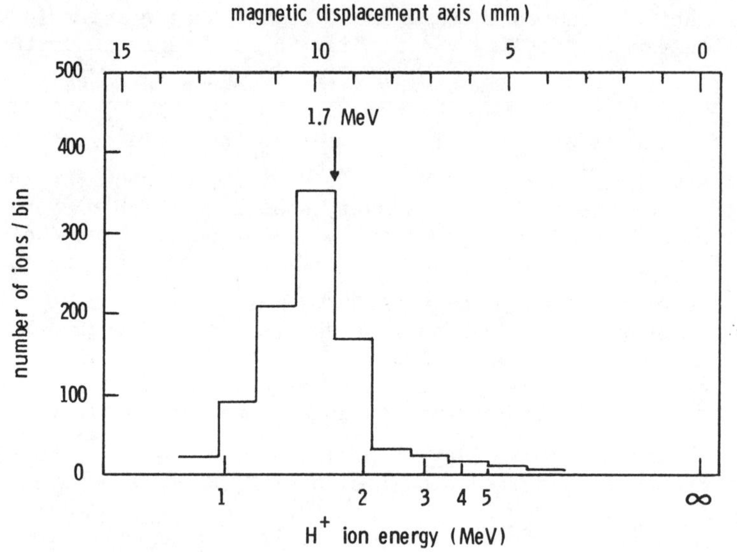

Figure 5. Proton Acceleration.

(a) Natural collective ion acceleration process for IREB injection into 100 microns hydrogen at room temperature. (Bin size: 0.1 mm x 0.1 mm.)

(b) Full IFA-2 system with 10 cm sweep and an ion source of 50 microns hydrogen. The final programmed sweep speed βc had $\beta = 0.058$ which corresponds to a 1.7 MeV proton. (Bin size: 0.1 mm x 0.1 mm.)

(a)

(b)

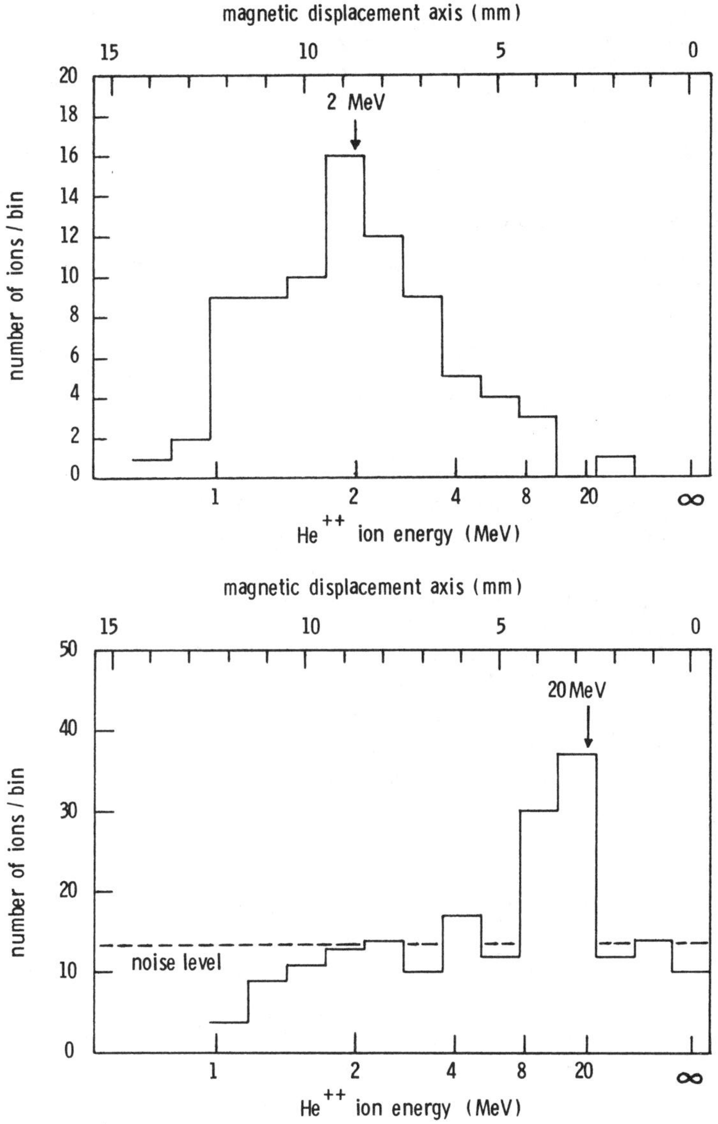

Figure 6. Helium Ion Acceleration.

(a) Natural collective ion acceleration process for IREB injection into 100 microns helium at room temperature. (Bin size: 0.1 mm x 0.1 mm.)

(b) Full IFA-2 system with 30 cm sweep and an ion source of 50 microns helium. The final programmed sweep speed βc had $\beta = 0.1$ which corresponds to a 20 MeV He^{++} ion. (Bin size: 1.0 mm x 4.0 mm.)

the swept XeCl laser intensity was just marginal for this case. We needed 127 mJ in the swept XeCl laser beam; with a DC bias on the deflector, the actual laser energy transmitted varied from 65 mJ to 195 mJ. Also, there was no opportunity to optimize the laser timing and ion source gas pressure to maximize N. For the 10 cm IFA-2 case of Fig. 5b, the laser intensity was ample, and we have $N \approx 1 \times 10^{12}$ without any optimization of laser timing or ion source gas pressure. Thus with proper laser intensity, and ion source optimization, we expect that $N \gtrsim 10^{12}$ should be routinely attainable.

We would like to emphasize that our present results are limited by the low swept laser intensity. With the appropriate laser intensity, the same experiment should control 100 MV/m accelerating fields over $\gtrsim$ 1 meter.

RELATIVISTIC IFA

Although our present interest with the IFA has been in accelerating particles from $\varepsilon_i/A = 0$ to $\sim$1 GeV (i.e., from $\beta = 0$ to $\beta = 0.87$), there is also the possibility of creating a relativistic IFA that has a phase velocity essentially equal to c. In the relativistic IFA, it is necessary to have the un-neutralized (or at most, partially neutralized) IREB propagate first, and then to launch the ionization front along the already propagating beam.[1] As indicated in Fig. 7, the initial IREB propagation can be accomplished by (1) injecting the IREB into a low density preformed plasma with

$$\gamma_e^{-2} < f_e \ll 1 \qquad (8)$$

(where γ_e is the IREB relativistic factor, and f_e is the fractional space charge neutralization), or (2) by injecting the IREB along an externally-applied axial magnetic field B_z. The first case could be easily accomplished by barely ionizing the working gas with a laser pulse just before the IREB is injected. For either case, the IREB front speed will be less than c, and a propagating IREB will be created that is very weakly neutralized ($f_e \ll 1$) or completely un-neutralized. A short time after the IREB is injected, an ionizing laser pulse is injected on axis, and it propagates co-linearly with the IREB. The laser pulse must be short (i.e., $\lesssim$ 30 psec if the ionization length is to be $\lesssim$ 1 cm) and have high enough intensity to create a plasma density equal to the IREB density. As the laser pulse passes by, the usual IFA sequence of events occurs. A plasma is created, the plasma electrons are expelled to the conducting walls, and the remaining plasma ion background provides charge neutralization for the IREB. A steep potential well back is created synchronously with the end of the laser pulse.

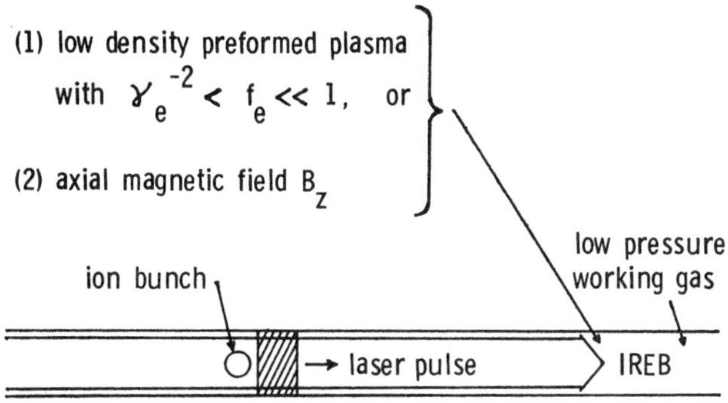

Figure 7. Relativistic IFA.

Note that the scheme is actually simpler than the non-relativistic IFA in that here no sweep method is required--the laser pulse is already moving at the desired phase velocity of $c/n \approx c$, where n is the index of refraction of the working gas. In addition, the phase velocity can be finely tuned by selection of the working gas type, pressure, and temperature, and by choice of the kicker laser wavelength.

It is of interest to consider what ion injection energies are needed to prevent ion slip-out over a reasonable accelerator section length, if the laser pulse phase velocity is taken to be equal to c. For example, a 20 GeV proton has $\beta = 0.999$ and would slip 1 cm (relative to a laser pulse which has $\beta = 1$) over a length of 10 m. This means that for a potential well back length of 1 cm, and an injected ion energy $\varepsilon_i/A \gtrsim 20$ GeV, we can envision a relativistic IFA accelerator producing an energy gain of ~10 GeV per 10 m section without ion slip-out.

The relativistic IFA should permit, in principle, the attainment of any desired ion energy. Further analysis of this and related relativistic IFA particle accelerator concepts is planned.

CONCLUSIONS

We have shown that large space charge fields of 1 GV/m or more can be produced with IREBs, as in the IFA. A key feature of the IFA is that a small amount of (expensive) laser energy is used to control a large amount of (inexpensive) pulsed power IREB energy. Since the ion energy can be a sizeable fraction of the IREB energy,[1] and since pulsed power IREB machines can have a high wall-plug efficiency, the IFA can be an efficient accelerator.

Two complete IFA systems have been built and operated. In both IFA-1 and IFA-2 we have demonstrated laser two-step photoionization of Cs. In IFA-1, the exciter laser was swept, and the kicker laser was fixed; in IFA-2, the exciter laser was fixed and the kicker laser was swept. Both schemes work. In both IFA-1 and IFA-2 we have demonstrated a fast laser sweep with a quadratic temporal dependence. In IFA-1 a short pulse laser and a light pipe array were used; in IFA-2 a long pulse laser and an electro-optic deflector were used. Both schemes work. In both IFA-1 and IFA-2, we have demonstrated laser-controlled motion of the potential well at the head of an IREB. In IFA-1 we obtained limited ion data. In IFA-2 we have obtained ion data with both a 10 cm sweep length and a 30 cm sweep length. Our helium ion data demonstrates that we have produced controlled accelerating fields of 33 MV/m over 30 cm.

Lastly, we have proposed a relativistic IFA with phase velocity essentially equal to c. This scheme opens the possibility of having accelerating fields of ~1 GV/m in 10 m acceleration sections, and, in principle, the attainment of unlimited ion energies.

ACKNOWLEDGMENTS

The assistance of W. Jaramillo and G. Samlin with the IFA-2 experiments is gratefully acknowledged.

The IFA-1 experiments (1977-1979) were supported by the Division of Nuclear Physics, U.S. Department of Energy, and the Air Force Office of Scientific Research. The IFA-2 experiments (1981-1984) were supported by the Division of Advanced Energy Projects, U.S. Department of Energy, and by Defense Programs, U.S. Department of Energy.

REFERENCES

1. C. L. Olson and U. Schumacher, Collective Ion Acceleration--Springer Tracts in Modern Physics, Vol. 84 (Springer-Verlag, Heidelberg, 1979).
2. Collective Methods of Acceleration, Papers presented at the Third International Conference on Collective Methods of Acceleration, edited by N. Rostoker and M. Reiser (Harwood, N.Y., 1979).
3. C. L. Olson, in Proceedings of the IX International Conference High Energy Accel., SLAC, Stanford, CA, May 2-7, 1974 (National Tech. Inf. Service, Springfield, VA, 1974), p. 272.
4. C. L. Olson, in Proceedings of the Int. Top. Conf. E-Beam Res. and Tech., Albuquerque, NM, Nov. 3-5, 1975 (National Tech. Inf. Service, Springfield, VA, 1976) Vol. 2, p. 312.
5. C. L. Olson, J. W. Poukey, J. P. VanDevender, A. Owyoung, and J. S. Pearlman, IEEE Trans. Nucl. Sci. NS-24 No. 3, 1659 (1977).

6. C. L. Olson, IEEE Trans. Nucl. Sci. NS-26 No. 3, 4231 (1979).
7. C. L. Olson, J. R. Woodworth, C. A. Frost, and R. A. Gerber, IEEE Trans. Nucl. Sci. NS-28 No. 3, 3349 (1981).
8. C. L. Olson, C. A. Frost, E. L. Patterson, and J. W. Poukey, IEEE Trans. Nucl. Sci. NS-30 No. 4, 3189 (1983).
9. G. Yonas, K. R. Prestwich, J. W. Poukey, and J. R. Freeman, Phys. Rev. Lett. 30, 164 (1973).
10. I. P. Afonin, M. V. Babykin, K. A. Baigarin, N. U. Barinov, A. V. Bartov, A. G. Bel'tsevich, A. V. Gubarev, V. N. Kiselev, V. I. Mizhiritskii and V. A. Treshchin, Proc. of the Second All-Union Conference on Engineering Problems of Thermonuclear Reactors, Leningrad, USSR, June 23-25, 1981, Vol. 3, p. 249.
11. C. Joshi et al., this conference (1985).
12. Laser Acceleration of Particles (Los Alamos, 1982), edited by P. Channell, AIP Conference Proc. No. 91 (Series Editor, H. C. Wolfe) (AIP, N.Y., 1982), p. 6 and p. 16.
13. Collective Accelerators, Report of U.S.D.O.E. Study Group (F. Cole, Chairman), Fermi National Accelerator Laboratory Report FN-355 (FNAL, Batavia, IL, 1981).

THE WAKEATRON: ACCELERATION OF ELECTRONS ON THE WAKE FIELD OF A PROTON BUNCH

A. G. Ruggiero
Argonne National Laboratory, Argonne, IL 60439

ABSTRACT

We explore in this note the idea of accelerating a low intensity electron or positron bunch, travelling through a linear rf structure, following at a short distance an intense proton bunch which leaves behind a wake field. This device acts like a transformer where two beams are involved: one made of protons at high current and low energy, the other made of either electrons or positrons, at low current and high energy. The two beams are coupled electromagnetically to each other by a specially designed rf structure made of a long sequence of cavities.

INTRODUCTION

We follow the approach already outlined in a previous paper[1]. This note is divided into six sections. The first two sections deal with a theorem on the maximum energy gain, the transformation efficiency and related considerations. The other four sections describe the Wakeatron as a linear collider for electrons and positrons from 1 TeV to 10 TeV per beam and luminosity 10^{32}-10^{33} cm^{-2} s^{-1}.

The wake field accelerators are devices with two beams of charged particles that we can label beams #1 and #2. The first is a leading bunch moving in a structure made of a long sequence of rf cells and leaves behind a wake field to which loses a considerable amount of energy. The second beam (#2) travels behind, and it is accelerated by the wake field of the first. In order to have a positive net acceleration the second beam is of much weaker intensity than the first. A theorem has been established recently[2] which under rather stringent conditions establishes that a particle of beam #2 cannot gain more than twice the initial energy of a particle in beam #1. Therefore it seems that in order to accelerate electrons (beam #2) to say 1 TeV one needs a primary beam of particles, of so far unspecified nature, at 500 GeV. There are several ways to go around this theorem by not satisfying the preassumed conditions. It seems possible to have the two beams moving in two different parts of the same rf structure, for instance a hollow beam moving along the periphery of an axially symmetric structure which then generates a higher field gradient in the center of the structure where the second beam is located[3]. It is also possible to take the leading bunches with asymmetric particle distribution[4], or as we shall propose here with a longitudinal extension large enough to cover several periods of the rf structure. With these techniques the original factor of two for the transformation ratio can be exceeded. In this note we estimate a possible ratio of 10. Thus to produce 1 TeV electrons now one requires a leading intense bunch of 100 GeV particles. But so far there has been no need to say anything about the nature of the leading particles, they could have been electrons as well as protons or any others that can be easily produced with an energy of 100 GeV. We would like to point out that it is considerably easier, cheaper, and it requires less power to produce a 100 GeV proton beam than an electron beam of the same energy and intensity. Therefore in this note, as in the original one[1], we propose to use proton bunches to generate wake fields by which accelerate electrons. In other schemes where electrons are also used as leading bunches[5], they start with an initial energy of 1 GeV, lose their total energy to the rf structure and are then accelerated again with conventional means to 1 GeV in parallel before they can be used again. This is equivalent to a sequence of 100 conventional linacs at 1 GeV each. If on the other hand protons are used, they can be generated initially at the required energy of 100 GeV in one single acceleration cycle in a conventional synchrotron.

Finally, we also point out that because of their larger mass, it is possible with protons to rearrange the longitudinal sequential order of the particles in the leading bunch since the amount of energy they lose or gain depends on their position within the bunch and this can be easily shifted compared to one another. It is because of this shuffling mechanism that it is actually possible to achieve a higher transformation ratio in a co-linear structure.

A MORE GENERAL THEOREM ON THE MAXIMUM ENERGY GAIN FOR THE WAKE FIELD CO-LINEAR ACCELERATORS (WAKEATRONS)

In a recent paper[2] a theorem was established about the maximum energy gain that a particle can receive under the accelerating effect of the wake field of an intense bunch of leading particles. According to this theorem the maximum energy gain cannot exceed twice the initial energy of a leading particle. This limitation is due to the fact that the primary bunch can travel only some distance L before it comes to a stop after having lost all its energy to the wake field. The derivation of the theorem mentioned above applies to the case where both bunches, the one that leaves the wake field and the one that is being accelerated, have zero longitudinal length. In this case the energy lost to the wake field is the largest and therefore the distance L travelled before coming to a stop is the shortest. Also in this case the acceleration gradient is the largest for a given number of leading particles.

In this section we want to derive a more generalized theorem where the leading bunch has a finite but not vanishing longitudinal length. In this case the rate of energy lost to the wake field is lower and the leading bunch can travel a longer distance. Although the acceleration gradient is reduced, the maximum energy that the second beam can gain is no longer limited to only twice the initial energy of the primary particles.

The energy change of the leading bunch per unit length due to its own wake is given by

$$\frac{d(N_1E_1)}{dz} = - N_1^2 e^2 W(0) \exp(-\sigma^2/2g^2) \tag{1}$$

where all the symbols have the same meaning as in reference [2]. What has been added is the reduction of the amount of coherent radiation due to the length of the bunch. We are assuming gaussian distribution with rms length σ. Equation (1) is clearly model dependent and applies in particular to the case of wake fields left behind by a bunch travelling along the axis of an rf structure with periodicity of length g. Eq. (1) was derived a long time ago[6] and it seems to be correct for the so-called "optical resonator" model made of a sequence of parallel planes of infinite extension. This case is shown in Fig. 1 and Eq. (1) applies more correctly when the conditions

$$a \ll g \ll b \tag{2}$$

are fulfilled, where, assuming cylindrical geometry, b is the outer radius of the structure and a is the radius of the opening on the axis. Also it is known[6] that for the optical resonator model

$$W(0) = 1/2a^2 \tag{3}$$

Observe that Eq. (1) gives the total amount of the energy lost by the total charge in the bunch. Individual particles may lose a different amount of energy per unit length depending on their location within the bunch.

The energy change of the trailing bunch, which we assume without any longitudinal extension, is given by

$$\frac{d(N_2E_2)}{dz} = -N_2^2e^2W(0) - N_1N_2W(y)\exp(-\sigma^2/4g^2)\,e^2 \tag{4}$$

The first term is of obvious origin, since the bunch has no longitudinal width. The second represents the positive energy gain provided $W(y) < 0$. Observe that the dependence on the rms bunch length σ of the first beam is weaker compared to the one shown in Eq. (1) due to the presence of a different coefficient in the exponential factor. A single explanation for it is that the energy loss is given by the integral of the square of the wake field on the rf structure volume[6].

The energy conservation requires

$$[N_1^2\exp(-\sigma^2/2g^2) + N_2^2]W(0) + N_1N_2W(y)\exp(-\sigma^2/4g^2) \geq 0 \tag{5}$$

Define

$$\bar{N}_1 = N_1\exp(-\sigma^2/4g^2) \tag{6}$$

as the effective numbers of particles that can radiate coherently when the bunch length is included. Then Eq. (5) becomes

$$(\bar{N}_1^2 + N_2^2)W(0) + \bar{N}_1N_2W(y) \geq 0 \tag{7}$$

Since this relation is valid for any values of $\bar{N}_1$ and N_2, we derive the following inequality

$$[-W(y)] \leq 2W(0) \tag{8}$$

which is the same result as obtained in reference [2]. Observe that in analogy to Eq. (3) we can also write

$$[-W(y)] = \varepsilon/a^2 \tag{9}$$

where the form factor $\varepsilon < 1$ as a consequence of (8). It can be thought to be made of the product of two quantities

$$\varepsilon = \varepsilon_{geom} \cdot \varepsilon_{loss} \tag{10}$$

The first would represent the maximum accelerating gradient a trailing particle can receive even assuming a perfectly conductive material with no resistive or thermal losses. It depends on the distance y from the leading bunch and on the geometry (a,b,g) of the cavity structure. The second factor takes into account the resistivity of the material involved and the losses of field energy through this material that cannot possibly be recovered by the trailing particle.

We can now derive three important equations: the acceleration gradient for the trailing particles, the total length of the acceleration structure, given by the drift the leading particles can travel before they come to a stop or close to it, and the total energy gain per trailing particle which is given by the product of the former two quantities.

Acceleration Gradient

The rate of energy gain per particle is obviously given by Eq. (4) when both sides are divided by N_2. We will make use of the more explicit definitions (3) and (9) for W(0) and W(y). Observe that Eq. (4) is made of two terms; the first is the rate of energy lost to its own wake field and the second the rate of gain from the wake field of the leading bunch. Let us define a loss-to-gain energy ratio for the trailing particle,

$$\eta = \frac{\text{Energy Lost to Own Wake Field}}{\text{Energy Gain from Leading Bunch Wake Field}}$$

that is

$$\eta = N_2 \exp(\sigma^2/4g^2)/2\varepsilon N_1 \tag{11}$$

then we can write for the acceleration gradient

$$G \equiv dE_2/dz = \varepsilon N_1 \frac{r_o E_o}{a^2} (1 - \eta) \exp(-\sigma^2/4g^2) \tag{12}$$

where $r_oE_o = e^2$ and, for instance,

$r_o = 1.535 \times 10^{-18}$ m, classical radius of a proton
$E_o = 0.93826$ GeV , rest energy of a proton

Total Acceleration Length

This can be derived from Eq. (1) which gives the total energy loss rate from all the particles of the leading bunch. As we have already observed the actual loss rate of an individual particle depends on its location within the bunch and on the dependence of the form factor ε on the distance between particles. It is obvious that the particles at the head of the leading bunch do not suffer appreciable energy loss. In reference [2] it was assumed that ε is practically constant (or the equivalent of it) so that the energy loss rate for a particle increases steadily with the distance from the head of the bunch. In reality for the structure we are considering an oscillatory behaviour of ε with a period closely related to the gap width g is expected, and if the bunch has an extension larger than g then the energy loss will also have an oscillatory behaviour along the length of the bunch. In particular there will be other locations within the bunch where there are no energy losses, and some sections of the bunch will actually gain energy instead of losing it. It is clear that the estimate and understanding of the factor ε, which we do not claim here, is crucial to estimate the spread of energy loss rates. For the moment we will assume that all the particles have an average energy loss given by Eq. (1) after having divided both sides by N_1. Then the total acceleration length is

$$L = 2\Delta E_1 a^2 \exp(\sigma^2/2g^2)/N_1 E_o r_o \tag{13}$$

where ΔE_1 is the average energy lost per particle in the leading bunch. It is convenient to dump the primary beam before the velocity of the particles is slightly reduced below the velocity of light. Indeed an energy spread ΔE in the beam due to a gradient of energy loss rate would correspond to a velocity spread Δv which in turn would give the following increase in length

$$\Delta s = \frac{e}{\beta\gamma^3} \frac{\Delta E}{E_o}$$

after travelling a distance ℓ short enough so that γ is about constant. This might cause a change from the originally assumed gaussian distribution, but also a mixing mechanism by which the longitudinal position of the particle is continuously changed so that we can indeed speak of an average energy loss per particle, a quantity which is then common to all the particles. A convenient parameter to describe this effect is the mixing rate Δ which we can define as the change per unit length of the distance between a particle which suffers the maximum rate of energy loss and a particle which does not suffer any at all. The mixing rate is a function of the distance z travelled, it is zero at the beginning and it will constantly increase since then. It is convenient to estimate Δ half-the-way of the rf structure, that is for $z = L/2$, in which case one has simply

$$\Delta = E_0^2/E_1^2 \tag{14}$$

This quantity depends only on the initial energy value E_1 and on the particle next energy E_0. It is clearly seen that because of their mass, protons would allow considerably more mixing than electrons.

Total Energy Gain

This is given by the product of Eq. (12) with Eq. (13). One obtains

$$\Delta E_2 \equiv G \cdot L = 2\varepsilon\Delta E_1(1-\eta)\exp(\sigma^2/4g^2) \tag{15}$$

As one can see, by making the length of the primary bunch longer, the total amount of energy gain can increase well beyond the factor of two proven in the theorem of reference [2]. Of course the distance to travel increases also and at a faster rate as one can observe by inspecting Eq. (13). Similarly the acceleration gradient reduces. Observe that if one desires $\Delta E_2 \to \infty$ than also $L \to \infty$ and $G \to 0$. For the special case the primary bunch acts like a point charge, that is $\sigma = 0$, we recover the results of the theorem of reference [2].

Finally it should be noted that only two of the three equations (12, 13, 15) are essential since any one of them can be derived from the other.

TRANSFORMATION EFFICIENCY AND RELATED CONSIDERATIONS

The device we describe here can be thought of as a transformer where high energy particles at low intensity are obtained from a larger number of particles at lower energy. A transformation efficiency can therefore be defined as follows:

$$\alpha = \frac{\text{Total Energy Gain for Trailing Bunch}}{\text{Total Energy Lost from Leading Bunch}}$$

that is

$$\alpha = N_2\Delta E_2/N_1\Delta E_1 \tag{16}$$

By making use of Eq. (15), Eq. (16) becomes

$$\alpha = 4\varepsilon^2\eta(1 - \eta) \tag{17}$$

where we have made use of the definition (11) for the loss-to-gain ratio η.

In order for α to be positive, one has to require $\eta < 1$ which condition sets a limit on the maximum number of trailing particles that can be accelerated, that is

$$N_2 < 2\varepsilon N_1 \exp(-\sigma^2/4g^2) \quad (18)$$

At this limit, $\alpha = 0$, there is no acceleration for the trailing bunch. In this case the amount of energy gained from the wake field of the leading bunch is offset by the energy lost to its own wake field. If the number of trailing particles is increased beyond the limit (18) so that $\eta > 1$, then a net energy loss (deceleration) will result also for the second bunch.

The optimum transformation conditions are obtained, as one can see by inspecting (17) for $\eta = 1/2$, in which case $\alpha = \varepsilon^2$, the best of the total energy transformation possible, as it is possible to see in the special case $\varepsilon = 1$ in which case also $\alpha = 1$.

The optimum condition $\eta = 1/2$ corresponds to the following distribution of the numbers of particles between the two bunches

$$N_2 = \varepsilon N_1 \exp(-\sigma^2/4g^2) \quad (19)$$

Consider the special case $\sigma = 0$ and $\varepsilon = 1$, with maximum energy transformation, that is the total energy lost by the first beam is gained in all of the second beam. Of course since the two beams have the same intensity, the second one will end up with the same initial energy of the first.

One more useful parameter which describes the performance of the proposed device is the <u>transformer ratio</u> that is the ratio <u>r</u> of the rate of the energy gain per particle of the trailing bunch to the rate of energy loss per particle of the leading bunch. From Eq. (1), after having divided both sides by N_1, and from (12) we obtain

$$r = 2\varepsilon(1 - \eta) \exp(\sigma^2/4g^2) \quad (20)$$

Once more, for the case of point distribution ($\sigma = 0$) and no resistive losses (ε - 1), we recover the result of the theorem of Ref. [2] which states that the ratio of energy gain cannot exceed twice the rate of energy loss per particle. Nevertheless it is pointed out here that by lengthening the primary bunch to several widths g of the rf gaps it is possible to enhance the transformer ratio.

OUTLINE OF THE DEVICE

The Wakeatron which we describe next is intended as a linear collider for electron and positron beams each with energy of 1-10 TeV and luminosity of 10^{32}-10^{33} cm^{-2}s^{-1}. In this device electrons and positrons are accelerated on the wake field of intense and relatively short proton bunches which these days we know how to produce in efficient and reliable ways. All the bunches involved travel along the axis of an RF structure as shown in Fig. 1. Therefore the analysis in the previous two sections above apply to this case where the leading bunch is made of protons and the trailing bunch of either electrons or positrons.

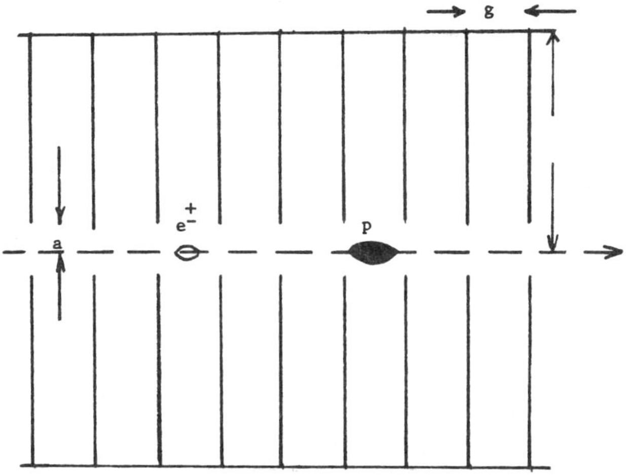

Fig. 1 CONCEPTUAL DESIGN OF AN RF STRUCTURE TO ACCELERATE ELECTRONS OR POSITRONS ON THE WAKE FIELD OF AN INTENSE PROTON BUNCH (PROTON KLYSTRON, WAKETRON).

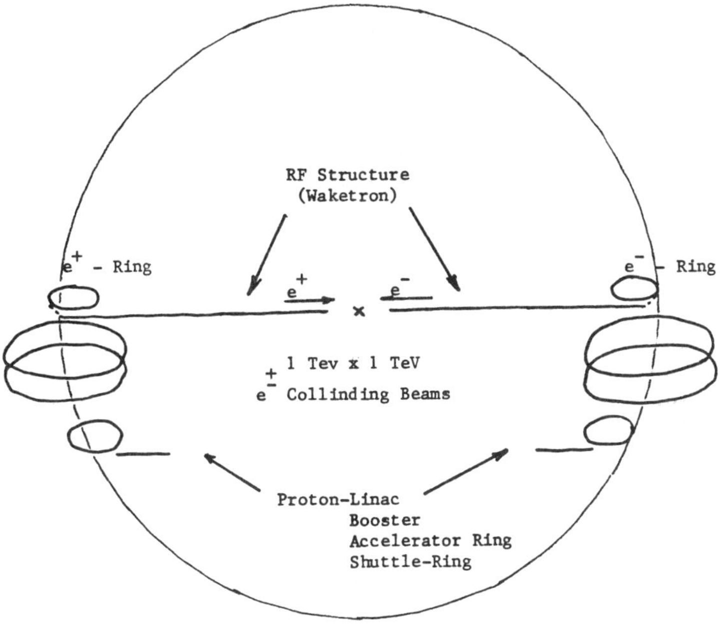

Fig. 2 LAY-OUT OF AN ELECTRON-POSITRON LINEAR COLLIDER WHICH MAKES USE OF THE IDEA OF ACCELERATING ELECTRONS AND POSITRONS ON THE WAKE FIELD OF INTENSE PROTON BUNCHES.

A possible lay-out of the Wakeatron is given in Fig. 2. It is made of two parts which are identical to each other but arranged symmetrically to each other around the crossing point where the two beams collide. One part is to accelerate electrons and the other positrons. Each part is made of a proton source which generates tight bunches in a conventional way. There is an electron beam source at one side and a positron beam source at the other. The acceleration of electrons and positrons takes place in the two sections of the Wakeatron itself which are identical to each other and of the same rf structure. The mode of operation we conceive is that one proton bunch is extracted from each side and injected into their respective sections of rf structure immediately followed by either an electron or a positron bunch. This will occur at some repetition rate at which all sources are to be adjusted to.

As a comparison, we show in Fig. 2 a circle which represents an SSC ring, a hadron collider for energies of 20 TeV per beam and luminosity of 10^{33} $cm^{-2}s^{-1}$. With today's superconducting magnet technology, this ring is expected to have a circumference of about 100 Km, that is a diameter of 30 Km. One can presume, and we shall here, that given the performance of the two possible colliders, circular and linear, they can occupy roughly the same side of real estate, as shown in Fig. 2. Thus we can take 30 Km for the total length of the linear collider, that is an acceleration length L = 15 Km for each RF structure that makes the Wakeatron.

Table I gives the summary for the choice of the proton beam energy and intensity for three energy cases: 1, 3 and 10 TeV per beams. These values correspond to L = 15 Km as previously stated, and have been derived from Eqs. (13 and 15). For reasons also explained above we have assumed a low efficiency, that is $\eta \ll 1$. Finally we can see from both Eqs. (12) and (13) that a very small opening radius a is desirable, on the other hand the opening must be large enough to accommodate both beams dimensions and we have chosen a = 1 mm. The ratio σ/g is taken as a variable parameter in Table 1 as well as the form factor ε, for which we have considered only the two values $\varepsilon = 1$ and $\varepsilon = 0.5$.

It can be seen from Table 1 that the proton beam total energy $(N_1 \Delta E_1)$ does not depend on σ/g for a fixed acceleration length L but increases with the square of both ε and the electron/positron final energy ΔE_2. Therefore for a given ε and ΔE_2, the number of protons per bunch increases with the inverse of the initial proton beam energy ΔE_1. An interesting case is the one circled in by a dashed line in Table I, which corresponds to $\exp(\sigma^2/4g^2) = 5$. In the following we will continue our analysis for this special case.

The performance of the collider is described by the parameters listed in Table II. These parameters are consistent with those agreed upon by the community at the 1984 Frascati Meeting.[7] They are based upon the repetition rate of 4000 bunch-bunch encounters. For a luminosity of 10^{32} cm^{-2} s^{-1} this corresponds to 3×10^9 electrons or positrons per bunch and to a normalized rms emittance of 10^{-6} mrad for the case of 1 TeV per beam. These parameters have been chosen so that the disruption parameters D is about unit and the energy spread due to Beamstrahlung is not too large, in this case 3%. These conditions are satisfied for the same beam dimensions and intensity at all three values of the collider energy we have chosen. The luminosity increases from 10^{32} cm^{-2} s^{-1} at 1 TeV to 10^{33} cm^{-2} s^{-1} at 10 TeV.

It is our understanding that the performance of the linear collider described here is equal or even superior to that of the circular collider for the hadron beams. The SSC performance lies somewhere between the 1 TeV and 3 TeV case shown in Table II. The high energy 10 TeV case, is certainly a project for the very far future although it does not quite make the level of luminosity that would be desired[8] (around 10^{34} cm^{-2} s^{-1}) at least in the context of this paper.

As we have already said, the Wakeatron is made of a long sequence of RF cells which closely resemble the structure of a conventional linac. The energy of the proton bunches is used to power the device and generate the accelerating field, rather than klystrons or other power sources. There is a basic difference between a conventional linac and the device we describe here. The linac operates at one single frequency mode, whereas the Wakeatron we propose is capable of generating many modes in the wake field of the proton bunches, and the sum of all the modes together can generate a higher accelerating gradient. It is then necessary that the particles which are to be accelerated follow the primary bunch at close distance, likely to be no more than a few times the period length g. In fact most of the higher order mode will damp rather fast and at a longer range, only the fundamental mode will predominate.

We observe that the bunch rms length σ and the gap width g enter only as a ratio σ/g in all our considerations so far. The only other quantity that might depend on σ and g independently is of course the form factor ε but this still remains a subject of further study. Therefore we are free to chose the gap width g which is large enough to allow the proton bunches to be long enough to support a considerably large number of particles. The dimension of the rf cavities we have chosen are shown in Table III. It is worth noting that proton bunches, each with 1.3×10^{11} particles, have been obtained in the CERN SPS at 270 GeV and with a rms bunch length of 10 cm. We have therefore chosen a bunch length and a gap width g to roughly match this configuration for a proton beam energy of 100 GeV (1 TeV for electrons and positrons). Since the larger energies also have higher intensity, we have chosen to increase the bunch length σ and the gap width g accordingly. We believe that condition (2) needs to be satisfied and we have chosen the outer radius of the cavities $b = 5g$. The thickness of the walls can be set to 1 or 2 mm, large enough compared to skin depth in the material at any frequency of interest.

All other parameters can then be easily calculated and are also shown in Table 3. In particular we give the acceleration rate which is ideally 200 MeV/m at 3 TeV. The energy lost per unit length for the protons is 10 times smaller. The dependence of these parameters with the beam energy is just a consequence of our rule of the exercise by which the total acceleration length, L has been fixed for all cases to 15 Km. We observe that indeed the efficiency factor η is very small for all cases, and that the transformation coefficient α is about 12%. This is to produce for example a 2 MW electron beam as required for the collider at 1 x 1 TeV^2, one needs 16 MW in the proton beam.

THE SOURCES FOR ELECTRONS AND POSITRONS

The electron and positron sources are each made of a 1 GeV linac and a damping ring of the same energy. To generate positrons a target is inserted after the early stages of the linac. Parameters are given in Table IV and are intended as just as example. A more careful analysis is required to optimize the design of the two sources. Anyway, in principle, the linac will generate bunches at the rate f = 4000 per second and could be made operable in the S-band mode like SLAC which would given an overall length of about 100 m at a gradient of 10 MV/m. We have found that the energy of 1 GeV is an optimum which provides fast radiation damping and small natural emittance at the same time.

The electron/positron bunches are transferred in the damping ring where they are kept circulating for a while. The primary function of the damping ring is to hold the beam until it is "cooled" effectively by synchrotron radiation. We propose that there be 100 bunches at any time circulating in the ring; since bunches are to be extracted at a rate of f = 4000 per second, the time each bunch will spend in the ring is 100/f = 25 msec. For the radiation effects to take over effectively we require that the typical radiation damping time correspond to a fraction of the circulating time, for example 1/3 which is 8 msec. The parameters given in Table IV would roughly yield an equilibrium rms emittance $\varepsilon = \sigma^2/\beta_L = (10^{-6}\text{m})/\gamma$ similar to the proton beam. For the electron/positron bunch length, we have taken an rms value σ_e = 1mm smaller than the length of the proton bunches. The design of these sources should be rather straight-forward and conventional.

THE PROTON SOURCE

Each proton source is made of a Linac, a Booster, Accelerator and Shuttle Ring. The energy and size for the components are given in Table V. Table VI summarizes the beam parameters for the Shuttle Ring. The Accelerator Ring has a cycle of one ramp per second and will deliver 4000 proton bunches to the Shuttle Ring where they will be stored for a period of one second during which they will be extracted one by one until the Shuttle Ring is empty and ready to receive a new load of 4000 bunches from the Accelerator Ring. This mode of operation will provide a constant rate of f = 4000 encounters/second. To fill up the Accelerator Ring with 10 consecutive pulses over a short front porch of one second cycles, we require the Booster to operate at 30 cycles per second and to have dimensions about one tenth of the Accelerator and Shuttle Rings. For this last mode or operation we impose a maximum limit bending field of 5 KG. However, the separation of bunches in the Shuttle

Ring will be 5 nsec at 1 TeV and 50 nsec at 10 TeV, obviously of some concern in designing the extraction kicker. The proton sources have to deliver, for obvious reasons, very intense, short bunches. The short bunch length can be achieved by raising the transition energy and choosing large accelerating rf frequency and voltage. The design of the entire proton source should clearly await a more detailed study which will require some of these parameters to change in order to meet the beam specifications. In particular we believe that an individual bunch area of 0.5 eV-sec for 1 TeV and correspondingly higher at larger energies is adequate for the beam to stabilize against microwave coherent excitation, assuming a coupling impedance $|Z/n|$ -1 ohm. Also the rms emittance, the same in both planes, $\varepsilon = \sigma^2/\beta_L = (10^{-6})/\gamma$ is consistent with the assumptions being made for other current projects (SSC, LHC).

Finally, observe that because of the assumed maximum field of 5 KG the dimension of the proton rings increases linearly with the electron beam energy i.e., a radius of 1 km for 1 TeV and 10 km for 10 TeV. We are confident that with more careful analysis we will find means to optimize these parameters in future.

POWER REQUIREMENTS, OVERALL EFFICIENCY AND COST ESTIMATES

We can make an estimate of the power requirements scaling from well known proton accelerators like those existing in CERN or Fermilab. A summary is given in Table VII. There are two major contributions; one is the power required to operate the magnets, and this increases with the size of the ring, and the other is the power which goes into the rf accelerating system. This is about twice the power of each proton beam since the same amount is dissipated in the rf cavities themselves. The operation of the electron linacs and damping rings as well as of the focussing systems along the length of the Wakeatron do not take a large amount of power. In the end we estimate about 100 MW to operate the entire facility for a 1 x 1 TeV^2 collider; this requirement increases almost quadratically with the collider energy. We can then estimate the overall efficiency, that is the ratio between the power in the electron beam and the total power required to operate the facility. We expect an overall efficiency of 4% at 1 TeV which then decreases linearly with the beam energy. We believe that an overall efficiency of a few percent is a significant figure and deserves careful consideration.

It is relatively easy to make an approximate cost estimate of the device from existing proton facilities. There are few items like the electron source (linac and damping ring) and the accelerating structure itself (Wakeatron) that have a cost independent of the choice of the collider energy. On the other hand because of the different sizes, the proton source has a cost that should increase linearly with the collider energy. One should also add the cost for beam transfer and focussing along the accelerating structure; we believe this cost also increases with the energy. The summary of a very rough cost estimate is given in Table VIII. Although a more detailed study is required before one can reach a more definite conclusion, it nevertheless seems that from a cost and performance point of view the device we have described here has some advantages over a circular collider like the SSC.

REFERENCES

1. A. G. Ruggiero, "The Wakeatron: Acceleration of Electrons on the Wake Field of a Proton Bunch." Paper submitted to the Proceedings of the Workshop on the Generation of High Fields, Frascati, Italy, Sept. 1984.
2. R. D. Ruth *et al.* "A Plasma Wake Field Accelerator". SLAC-PUB-3374, July, 1984.
3. T. Weiland, "High Gradient Acceleration of Particles by Wake Field Transformation" KEK Report 84-1, April 1984.
4. P. Morton, R. Ruth Private Communication, Malibu Beach, California, January 1985.
5. G. Voss and T. Weiland, DESY M-82-10, April 1982 and DESY 82-074, Nov. 1982.
6. E. Keil, Proc. Intern. Conf. High Energy Accelerators, II (1969), p. 551. Also Nuclear Instr. and Methods, 100 (1972), p. 419.
7. L. Hand and J. Rees, Proceedings of the Workshop on the Generation of High Fields, Frascati, Italy, Sept. 1984.
8. B. Richter, Invited Talk to this Workshop.

Table I Proton Beam Parameters vs. Electron Beam Energy

		$\Delta E_2 = 1$ TeV		$\Delta E_2 = 3$ TeV		$\Delta E_2 = 10$ TeV	
r	σ/g	ΔE_1	N_1	ΔE_1	N_1	ΔE_1	N_1
2.	0	0.5 TeV	$0.45.10^{11}$	1.5 TeV	$1.4.10^{11}$	5 TeV	$4.6.10^{11}$
		1.0	0.93	3.0	2.8	10	9.3
4.	1.665	0.25	0.93	0.75	2.8	2.5	9.3
		0.5	1.9	1.5	5.6	5	19.
10.	2.537	0.1	2.3	0.3	6.9	1.	23.
		0.2	4.6	0.6	14.	2.	46.
20.	3.035	0.05	4.6	0.15	14.	0.5	46.
		0.1	9.3	0.3	28.	1.	93.
first line:		$\varepsilon = 1$		$\alpha = 1$ mm		$L = 15$Km	
second line		$\varepsilon = 0.5$		$\eta << 1$			

Table II Collider Parameters

Energy/Beam	1 TeV	3 TeV	10 TeV
Repetition Rate		4,000 encounters/sec	
No. of $e^{\pm}$/Bunch		3×10^9	
β^* (H and V)		5mm	
β-emittance σ^2/β_L (H and V)		10^{-6}m/γ	
Rms Bunch Length		1mm	
Rms Beam Spot, $\sigma^*_{H,V}$	500 A°	300 A°	150 A°
Disruption Parameters, D		1	
Energy Spread due to Beamstrahlung		3%	
Luminosity, $cm^{-2}s^{-1}$	1.0×10^{32}	3.0×10^{32}	1.0×10^{33}

Table III The Wakeatron RF Structure

Energy/Beam	1 TeV	3 TeV	10 TeV
No. of p's/Bunch	2.3×10^{11}	6.9×10^{11}	2.3×10^{12}
p-Beam Energy	100 GeV	300 GeV	1000 GeV
σ/g, ($r = 10$)		2.537	
Rms p-Bunch Length	12.7 cm	38.1 cm	127 cm
Gap Width, g	5 cm	15 cm	50 cm
Iris Radius, a	1 mm	1 mm	1 mm
Outer Radius, b	0.5 m	1.5 m	5 m
Wall Thickness	1-2 mm	5 mm	1-2 cm
Total Length		2 x 15 Km	
Accelerat. Rate	67 MeV/m	200 MeV/m	667 MeV/m
Energy Loss (p)	6.7 MeV/m	20 MeV/m	67 MeV/m
Energy Deposition	0.25 J/m	2.25 J/m	25 J/m
Form Factor, ϵ		1.	
Loss-to-Gain Ratio, η	0.03	0.01	0.003
Transformation Coefficient, α	0.12	0.04	0.012
Power/Beam: electrons	2 MW	6 MW	20 MW
protons	16 MW	144 MW	1600 MW

Table IV Electron/Positron Source

Linac:	
Output Energy	1 GeV
RF	3 GHZ
Repetition Rate	4 KHZ
Damping Ring:	
Energy	1 GeV
Average Radius	10m
Packing Factor	50%
Dipole Field	8 KG
Betatron Tune, (H,V)	15
No. of Bunches	100
No. of $e^{\pm}$/Bunch	3×10^9
Radiat. Damping Time	8 msec
Time between two consecutive bunch extractions	25msec
σ^2/β_2 (H and V)	10^{-6}m/γ
Rms Bunch Length	1mm
Energy Loss	20 KeV/turn
RF:	
Frequency	500 KHZ
Voltage	100 KV

Table V Parameters for the Proton Source

Energy/Beam		1 GeV	3 GeV	10 GeV
Components:				
Linac (200 MHz)		1 GeV	3 GeV	10 GeV
Booster		10 GeV/100m	30 GeV/300m	100 GeV/1km
Accel. Ring		100 GeV/1Km	300 GeV/3Km	1 TeV/10Km
Shuttle Ring		100 GeV/1Km	300 GeV/3Km	1 TeV/10Km
Cycle Rate:				
Booster			30Hz	
Acceler. Ring			1Hz	
RF:				
Booster		175-200 MHz/2MV	67MHz/1MV	20 MHz/0.2MV
Accel. Ring	(h = 4000)	200 MHz/7MV	67MHz/2MV	20 MHz/0.7MV
Shuttle Ring	(h = 4000)	200 MHz/10MV	67MHz/3MV	20 MHz/1.0MV
Bending Field:				
Booster			0.5-5KG	
Accel. Ring			0.5-5KG	
Shuttle Ring			5KG	

Table VI Beam Parameters in the Shuttle Ring

Energy/Beam	1 TeV	3 TeV	10 TeV
Average Radius	1Km	3Km	10Km
Bending Field		5KG	
RF: Frequency	200 MHz	67 MHz	20 MHz
Harmonic No.		4000	
Voltage	10 MV	3MV	1MV
No. of Bunches		4000	
No. of p's/bunch	2.3×10^{11}	6.9×10^{11}	2.3×10^{12}
Rms Bunch Length	12.7cm	38.1cm	127cm
Bunch Separation	5nsec	15nsec	50nsec
Total No. of Protons	9.2×10^{14}	2.8×10^{15}	9.2×10^{15}
Longit. Phase Space Area(95%)	0.5 eV.s	1.5 eV.s	5 eV.s
β-Emittance σ^2/β_L(H and V)		10^{-6}m/γ	

Table VII Power Requirements

Energy/Beam	1 TeV	3 TeV	10 TeV
Power/Beam: electrons	2MW	6MW	20 MW
protons	16MW	144MW	1600MW
α	12%	4%	1.2%
Each p-Source:			
Magnet Power	10MW	30MW	100MW
RF Power (x2)	32MW	288MW	3200MW
Each $e^{\pm}$-Source	6 MW	6MW	6MW
Transv. Focus/Section	2MW	6MW	20MW
Total Power for Entire Facility	100MW	660MW	6.65GW
Overall Efficiency	4%	1.8%	0.6%

Table VIII Cost Estimate (approximate)

Energy/Beam	1 TeV	3 TeV	10 TeV
Each Proton Source	200 M$	600 M$	2000 M$
Each Electron Source	25	25	25
Each Section of RF Structure	100	100	100
Transport, Transfer and Focussing for each side	25	75	250
Total each side	350	800	2375
	x 2	x 2	x 2
Total Project(*)	0.7 B$	1.6 B$	4.75 B$

(*)1985 US dollars. No contingency and no escalation added.

REPORT OF THE WORKING GROUP ON LASER TECHNOLOGY

S. Singer
Los Alamos National Laboratory, Los Alamos, NM 87545

At the Second Workshop on Laser Acceleration of Particles, the Laser Technology Working Group examined basic issues of feasibility, realizability, and cost, in addition to the specific technical issues relevant to the various acceleration concepts. We agree with others that the opportunities are considerable, that the prospects are promising, and that more work definitely needs to be done. We also observe that because of considerations of efficiency, the cost of "laser power supplies" is great, and work is needed in the area of recovery and reusability of laser light in ways relevant to the particular accelerator concept. These conclusions are based on accelerator requirements and performance attributes defined at the Workshop and should be re-evaluated if those attitudes change.

The members of the Working Group are listed in Table I; the representation from industry, universities, and national laboratories, and the expertise in laser-matter interactions, laser physics and system design, optics, pulsed power engineering and accelerators contributed greatly to the activities of the working group.

In a particle accelerator powered by a laser, it is evident that the laser subsystem must deliver the right beam, generated at high efficiency, to the right place at the right time at a high repetition rate, and all at a cost that is perceived to be acceptable. The Working Group spent much of its time determining the meaning of these terms and assessing the capabilities of present day laser-optical technology. Our conclusions, assessments and recommendations for future work are the subjects of the following discussion.

The "right" laser beam is viewed in terms of its time, space, phase and energy attributes; Table II summarizes our assessments, concerns and observations. Almost all advanced concepts desire ultra short optical pulses, <1-50 ps. This seems to be true for several reasons: microstructure of electron beams from the injector, a desire to forestall the effects of materials damage and plasma production; and optimization of the physical process producing the accelerating electric field (i.e., for the beat wave process, avoiding energy coupling to the plasma ions or to non-useful instabilities). Details of the laser pulse shape best suited for each accelerator concept are not yet available, but we expect that as the accelerator concepts become more refined, the appreciation of a well-shaped laser pulse that optimizes the acceleration efficiency will grow. Optical Dispersion is a property of the accelerator "structure" material that affects wave propagation; it could be, for example, an issue for the Beat Wave Accelerator (BWA) (discussed below). Pulse chirp refers to a time dependent frequency-changing process in the laser subsystem that produces a specific frequency modulation in the optical pulse. Many high power laser devices produce some chirp, and frequency modulation is not difficult to produce. What remains to be determined is how much is desirable. For example, in the BWA chirping can reduce the undesirable effects of relativistic detuning in the BWA, but it may conceivably lead to filamentary self-focussing of the light beam, an undesirable consequence. Phase front distortion affects the spatial properties of the optical beam electric field and intensity in both the near field and far field. In a real-world optical system, delivery of a "diffraction-limited" (or nearly so) beam to an application device can be complex, difficult, and expensive, particularly for short optical wavelengths. The near field accelerator (NFA) (i.e., the "grating" concept) requires a programmed phase/spatial angle/electric field in the optical pulse; it is not yet clear how implementation of such conceptual requirements impact efficiency. The issue of laser energy can be appreciated by the following consideration. For a 5 TeV beam of 10^9 particles per macropulse with a repetition rate of 1 kHz, the average beam power is 800 kW. If the conversion efficiency of laser light to beam energy is 4 percent, the total laser beam power required is 20 MW per particle beam. The laser energy per pulse per beam is 20 kJ and the total peak power per particle beam is 2×10^{15} watt (assuming a 10 ps pulse duration). Generation of 10s of kilojoules of light with short pulse duration is not new; both the Nova system (Nd:glass) and Polaris (multiplexed KrF concept) will generate ~100 kJ of light; doing so for 10 ps pulses will be challenging. More relevant is the light energy per acceleration stage. Extensive staging is almost inherent in the NFA and some staging may be inescapable in the BWA. The energy per state will, of course, depend on efficiency and staging details, but may be ~1-1000 joules. This is an important point because it may permit the used laser devices whose individual size is comparable to the length of an accelerator stage and may thus permit integration of the laser into the accelerator module. On the other hand, generation of such pulse energies for ultra-short pulses at an acceptable efficiency may require the use of power multiplying schemes such as optical multiplexing or nonlinear optical pulse compression. While conceptually simple, these processes can lead to greater laser system size and complexity.

Efficiency is the single most important laser issue relevant to their use for an optical power supply to accelerate particles. Current laser efficiency varies from ~.5-5 percent (see Table III) and projections of overall efficiency >10 percent are speculative. Using the example above, a 10 percent efficiency implies a 0.2 GW electricity supply to the laser per beam and (allowing for the general facility use of electricity) a 1 GW electrical power supply. In today's coal-fired generating plant market, this implies a cost of $2-3 B, a substantial number.

At the present level of understanding, beam timing and beam positioning do not appear to be beset with fundamental problems. The side illumination scheme of the NFA is easiest to accomplish; transverse and axial positioning accuracies of 2-10 μm seem achievable. Angular positioning of a few microradians is also feasible. Considerably greater accuracy is possible, but if long (many meters) atmospheric propagation paths are contained in the laser system architecture and if high power pulses are to be propagated, careful consideration must be given to the refractive and scattering effects of atmospheric turbulence and thermal gradients. Translated to time differences, these physical spatial tolerances become time "differences" of ~10 femtoseconds, and suitably small value. Pulse to pulse jitter and jitter between the laser and electron beam pulses (i.e., synchronization) are a cause for concern. The FEL may be a particularly useful combined laser-electron beam sourcefor which synchronization is "automatic." The issue of "staging," i.e, the ability to inject the electron beam into the next stage of the accelerator with the proper phase (so that it is accelerated rather than decelerated) is an important one requiring more analysis. Although interstage positioning accuracy of 1/10 λ (optical) and phasing resolutions of 1/100 λ are possible, jitter between the two beams would be a problem as would jitter induced by earth motions. These can cause position changes of, say, mirrors to be ~10 μm at frequencies up to ~60 Hz. Dynamic compensation for such ground motion (due mostly to vehicle traffic and electrically operated equipment) is difficult and expensive, for frequencies >15 Hz.

For gas lasers, repetition rates of ~100 Hz have been accomplished and considerable attention is being given to increasing that rate and increasing the average laser power at high repetition rates. The issues associated with this effort are typically: maintenance of the optical quality of the laser gain medium, suitable lifetime of the thin vacuum window of the laser's electron gun, and adequate lifetime of the electron gun cathode. Considerable progress is expected in the next few years. For Nd:glass, high average power has been difficult to achieve. However, recent advances in architecture (e.g., use of helium-cooled slabs) and high efficiency diode "flash-lamps" indicate that relatively high power operation at tens of hertz with good optical quality can be achieved. Progress will depend on the development of large scale diode flash lamp arrays, a task requiring considerable cost.

Efficient interaction of the light pulse with the accelerator structure, while essential, poses requirements on the light beam which are not well understood. An efficient interaction could impose difficult conditions on the optical phase or intensity; compensation may be required for distortions produced by the accelerator medium due to the interaction itself; special conditions may be required to insure formation of the long self-focussed beam channels apparently required by the BWA. In that context, we note that conventional (gaussian) in-line optics may not be usable for the BWA because the focal length of the optical focussing device is so much longer than the region of acceleration that the accelerator's overall length may simply be too great (see Appendix A). Additionally, the interaction efficiency may depend on the properties of the accelerator itself. For example, the two beam overlap required for the BWA can be affected by optical dispersion in the plasma (see Appendix B).

The cost of average power in a laser device is, at present, high. The working group estimates that in the near term, that cost will be in the range 50-few x 100 dollars/watt. Using the example of the discussion above where the laser average power is 20 MW, the cost at $100/watt is a very impressive $2B. Equally impressive is the electrical operating cost for the facility, which is ~$1M per day (1 GW at $.10 per kw hr). It is interesting to contrast the high cost of average power with the low cost of peak power. We estimate that, from published statements of peak power and cost for Nova (~100 TW) and Antares (~50 TW) the cost of peak power is microdollars/watt. The cost impact of using lasers for particle acceleration would be much ameliorated if laser peak power were the needed commodity instead of average power energy. These considerations emphasize the need for improving laser efficiency and for finding ways to reuse or recycle laser light.

Table III shows the compilation of attributes of various laser devices; the data speak for themselves. We can summarize our perceptions of the status and future possibilities of relevant (i.e., potentially efficient) laser systems as follows:

<u>CO_2 Lasers</u>

- Efficiency ~ 10 percent possible
- High repetition rates possible

- Short pulses possible
- Best for NFA
- Not necessarily best for BWA
- Optical quality OK

Nd: Glass Lasers

- Efficiency ~10 percent possible, much effort required to develop diode pumps
- Repetition rates ~10's of Hz possible
- Optical quality OK
- Short pulses OK

Excimer Lasers

- Efficiency ~7 percent possible
- Efficiency ~15 percent theoretically possible in modest sized devices
- Repetition rates 100's/sec probable
- ps pulses OK
- Optical quality OK

The technology development issues, broadly stated, may be summarized as follows:

- Improve processes and understanding of ultra-short pulse generation and amplification (all lasers)
- Obtain damage data for ps pulses (all lasers)
- Develop new solid state laser materials (glass lasers)
- Promote development of diode flash lamps (glass laser)
- Develop methods of phase/intensity control (CO_2 for NFA.

Another way to summarize our deliberations is to assess the difficulties in accomplishing desired laser performances. Table IV shows that assessment for gas (G) and solid-state (S) lasers. The assessment that development of high average power solid state lasers borders on very difficult is primarily due to the cost associated with that development, not its technical possibility. The table reemphasizes the difficulty of achieving the essential quality of overall efficiency by any process.

In conclusion, we recommend that prospects for laser powered accelerators would be much improved if laser efficiency were significantly increased, or - more likely - efficient ways could be found to recycle or reuse the laser light not coupled into the particle beam. Additionally, attention should be given to acceleration concepts which use high laser peak power and derive their bulk energy (or high average power) from more traditional souces such as capacitors or RF devices. Of course, if the accelerator parameters such as the number of particles per macropulse and repetition rate decrease considerably, the concern over laser efficiency and cost diminish accordingly.

APPENDIX A

Use of Conventional Optics

In the beat wave acceleration process, the electrons being accelerated move slightly faster than the phase velocity of the Langmuir wave that is the source of the accelerating electric field. Eventually, the electron phase change relative to the wave is large enough that acceleration ceases and the interaction region must be terminated to prevent de-acceleration. The length (in the laboratory frame of reference) L_D of the interaction region is taken from the work of Chen, Lawson, Tajima to be

$$L_D = \frac{c}{\omega_p} \left(\frac{\omega}{\omega_p}\right)^2 \qquad \text{(cm)}$$

In the basic concept of the BWA, the light beam must be focussed to a beam waist of length L_D. If this is to be done with traditional (i.e., linear) optical elements such as lenses and mirrors, the length R of the beam waist is

$$R = 4\pi \left(\frac{f}{D}\right)^2 \frac{c}{\omega} \qquad \text{(cm)} \qquad \begin{bmatrix} F = \text{focal length} \\ D = \text{diameter of lens} \end{bmatrix}$$

R is taken from the "depth of field" expression of Born and Wolf (Optics, Third Edition). At the ends of the beam waist, the light intensity is 90 percent of its value at the nominal focus.

Equating these two expressions, we obtain

$$f^2 = \left(\frac{D}{4\pi}\right)^2 \left(\frac{\omega}{\omega_p}\right)^3$$

Other relevant expressions are:

G_B = beam energy gain per "stage" = $200\ mc^2\ \left(N/10^9\right)\left(\omega/\omega_p\right)^2$ (joules).

ϵ_L = laser energy per stage = G_B/ϵ (joules)

D^2 = E_L/F (cm^2)

where

N = number of electrons per macropulse in multiples of 10^9

ϵ = conversion efficiency of laser light to electron energy (assumed to be independent of wavelength of light), assumed to be 1 percent.

F = permissible optical flux handled by focussing element, in joules/cm^2 (assumed to be 1 joule/cm^2 from work of Corkum).

Combination of these results gives:

$$f = \frac{.006}{\sqrt{\epsilon}} \left(\frac{\omega}{\omega_p}\right)^{5/2} \quad (cm)$$

Table V gives some numerical values for ω/ω_p = 1, 10, 100 (and for λ = 1/4 μm). The table gives the lens diameter D, the focal length f, the lens F , the diameter of the beam waist, the accelerating length L_D and the ratio f/L_D of the focal length to the accelerating length. These results suggest that for any reasonable value of ω/ω_p, the length of the accelerator (which is 2f x No. of stages) is very much longer than the actual accelerating length. Also, the beam waist diameter is generally much larger than what one guesses the electron beam diameter to be, a circumstance which could make an efficient interaction less likely.

We conclude that for the range of parameters used by the workshop, conventional optics leads to a design (BWA) that is not necessarily the most efficient or the least expensive. While the matter may be ameliorated for accelerator parameter values less demanding, an exciting alternative is to somehow cause the light beam to undergo relativistic self-focussing. Then, the lens focal lengths can be short compared to the acceleratating length, and the beam diameter is several plasma wavelengths, a situation likely to be much more compatible with a low cost accelerator.

APPENDIX B

Effects of Optical Dispersion

The BWA requires that beams of somewhat different frequency overlap in a plasma characterized by a plasma (angular) frequency w_p. The index of refraction of the plasma is

$$M = \left(1 - \left(\frac{\omega_p}{\omega}\right)^2\right)^{1/2}$$

If the beams are co-propogating in the same direction, they will eventually separate if the propagation path is long enough or if the optical frequency or plasma frequency is small enough. To estimate the nature of this issue, we require that the optical pulses at ω_1 and ω_2 ($\omega_1 - \omega_2 = \omega_p$) and pulse duration τ must not separate by more than 10 percent of their length while traversing an acceleration length L_D. It is easy to show that this condition translates to

$$\omega \tau > 10$$

with the assumption that $\omega_1 + \omega_2 \sim 2\omega$, $\omega/\omega_p \gg 1$. Thus, dispersion effects are an issue for the long optical wavelengths (e.g., >10 μm), low plasma frequency and very short pulses. At 10 μm ($\omega \sim 2 \times 10^{14}$) and ~10 ps, the condition appears to be well satisfied.

TABLE I

WORKING GROUP MEMBERS

JAMES MORGAN	*ITEK*
RICH SHEFFIELD	*LANL*
AL SAXMAN	*LANL*
DENNIS LOWENTHAL	*SPECTRA TECH*
WALLY LELAND	*LANL*
SIDNEY SINGER	*LANL*
IRVING BIGIO	*LANL*
JONAH JACOB	*SCIENCE RESEARCH*
J J EWING	*SPECTRA TECH*
JOHN SOURES	*ROCHESTER*
NORM KURNIT	*LANL*
DEAN JUDD	*LANL*
CHARLES FENSTERMACHER	*LANL*
BARBARA SCHAAD	*DE PAUL*
BOB UMSTADTER	*UCLA*

TABLE II

THE RIGHT LASER PROPERTIES:

• PULSE WIDTH:	5-50 ps ALL WAVELENGTHS ~ 0.1 ps FOR NEAR FIELD MAY BE DESIREABLE
• PULSE SHAPE:	NOT WELL DEFINED PROBABLY IMPORTANT
• OPTICAL DISPERSION :	IMPORTANT ISSUE FOR BWA
• PULSE "CHIRP":	EASILY PRODUCED EXISTS NATURALLY FOR SOME LASER DEVICES MAY BE DESIRABLE FOR BWA MAY LEAD TO SELF FOCUSSING IN A PLASMA
• PULSE ENERGY:	1-1000 JOULES NOT BELIEVED TO BE A PROBLEM (MULTIPLEXING)
• PHASE DISTORTION:	NEAR-DIFFRACTION-LIMIT DIFFICULT SPATIAL VARIATIONS IN FOCAL PLANE AN ISSUE PHASE FRONT CONTROL NEEDED FOR NEAR FIELD SCHEME

TABLE III

THE STATUS OF LASER TECHNOLOGY

	FEL	CO_2	I	Nd: Glass	EXCIMER	HF	CO
λ (μ)	.5–200 .1 event.	9–10	1.3	1.0	.24–.4	2–3	5
Δλ (lines)	yes	yes	yes	yes	yes	yes	yes
POWER/ ENERGY	5NW/ 10MJ	20 TW	IGW	5.7TW	6KJ (500ns)	kilojoules	1KW (~μsec)
PULSE LENGTH	30 ps	2 ps	500ps	ps–ηsec	ps	1ηsec	
EXPECTED EFFICIENCY%	20	10	1	1–10	5–10	200?	15–20
REP RATE	100MHz	1 KHz	1 Hz	100 Hz low power	.1– 1 KHz	SINGLE SHOT	100 Hz
FOCUSSABILITY	DL	DL	DL	DL	1–2xDL	2–3xDL	DL
DAMAGE (J/cm^2)	λ dependent	6–8 mirrors 5 windows	10–20	10	2–10	4–5	4–5

TABLE IV

GRAND SUMMARY

ATTRIBUTE	OK	HARD	TOUGH
EFFICIENCY > 15%			G, S
ENERGY RECOVERY			G, S
REP RATE > 1 KHZ		G	S
HIGH AVE. POWER		G	S
BEAM QUALITY		G, S	
PS PULSES		G, S	
BEAM HANDLING	G, S		
ARCHITECTURE	G, S		

OK: NO BIG PROBLEMS
HARD: LOTS OF WORK NEEDED
TOUGH: VERY DIFFICULT INDEED

TABLE V

EXAMPLE

ω/ω_p	D CM	f CM	$E_{\#}$	d CM	L_D CM	f/L_D
10	2	20	9	5 μm	.004	4775
100	21	6000	282	.014	4	1500
1000	213	10^6	9000	.5	4000	475

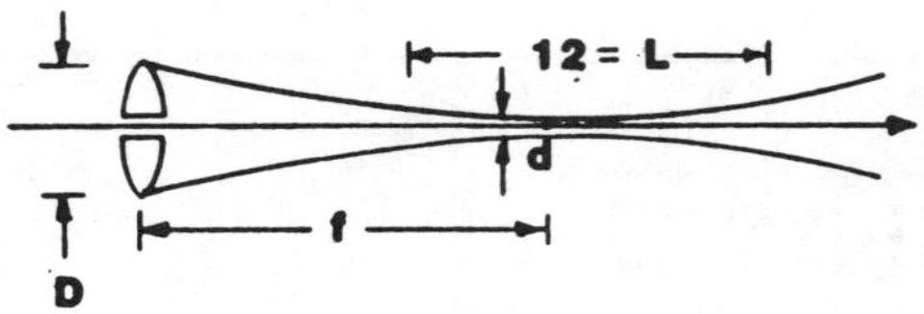

$\lambda = 249$ nm
$\epsilon = .01$

VERY HIGH POWER LASER PULSES

P.B. Corkum
Division of Physics
National Research Council of Canada
Ottawa, Ontario, Canada K1A 0R6

ABSTRACT

Techniques appropriate for the generation of very high power laser pulses are reviewed with particular emphasis on CO_2 lasers. In keeping with the speculative nature of this workshop, I suggest future developments in CO_2 and other lasers.

I. INTRODUCTION

The acceleration of charged particles under the influence of laser radiation ultimately relies on the strong electric fields lasers can create.

In principle, stimulated emission allows us to release whatever energy is stored in a laser over either a short or long time interval. If we wish to obtain the maximum power from a laser it is clear that we must develop methods for extracting the stored energy in the shortest possible time. Typically, pulses on the order of 100 fsec to 1 psec are consistent with the bandwidth limit of the important energy storage lasers. Thus we should expect the highest power lasers to use picosecond or femtosecond (ultrashort) pulses.

Ultrashort pulse techniques have been extensively developed over the last decade or more.[1] It seems odd that the adaptation of these techniques for the generation of very high power pulses has, as yet, hardly been addressed.

Aside from the strong electric fields, there are other reasons for using short pulses for laser-driven particle acceleration:

(i) With picosecond pulses, higher field intensities can be obtained without damage to nearby material surfaces. This has been extensively studied for semiconductors. Even with ~ 70 fsec pulses, damage to semiconductors begins with surface melting at an energy density ~ 100 mJ/cm^2 (that is, $\underset{\sim}{E}$ ~ 2x10^{7}V/cm).

(ii) The breakdown threshold in gases increases dramatically with picosecond pulses.[3] For infrared radiation, with intensity below the threshold for Keldish ionization[4] significant plasma formation is impossible. (For example, in the case of He, even at pressures of ~ 10 atm., the plasma growth rate due to avalanche ionization cannot exceed (300 fsec)$^{-1}$.)

(iii) In plasmas, picosecond pulses should simplify the otherwise wide array of instabilities that characterize laser plasma interactions by minimizing the effects of ion motion. Both pondermotive and thermal self-focusing are virtually impossible on the time scale of a few optical cycles. Similarly, ion waves (including the surface waves on a laser-grating accelerator) probably cannot grow or decay on this time scale.

There are three obvious candidate lasers for the generation of very high power radiation: CO_2 lasers, Nd:glass lasers and excimer lasers. Multi-atmosphere CO_2 lasers have a bandwidth sufficient for high gain amplification of pulses as short as 2 psec on each of the four main spectral features. For Nd:glass lasers, the gain bandwidth should allow the amplification of pulses as short as 100 fsec, while for excimer lasers, both XeCl and KrF have bandwidths consistent with amplification of pulses of ~ 200 fsec.

All three lasers can be constructed with very large apertures and, therefore, in preliminary experiments, we should aim to extract the maximum power density from the gain medium. In general, however, lasers operated at very high power density can be expected to be particularly sensitive to nonlinear effects. These can be either multiphoton absorption, which limits the maximum intensity of the radiation, or self-focusing (a severe limitation for Nd:glass lasers) which affects the wave front. In general, infrared radiation should be much less sensitive to these effects because (i) the photon energy is low, making multiphoton absorption less probable, and (ii) the wavelength is long, making self-focusing and other phase-related processes more difficult than at shorter wavelengths. (In Section IV, I will show that there are cases where the latter assertion is not correct).

With these considerations in mind, it seems wise to begin the development of very high power lasers with CO_2. Historically, however, picosecond techniques were developed on Nd:glass lasers[5] and, because of the relatively easy availability of picosecond sources, excimer lasers[6] were the next to be investigated. Picosecond CO_2 lasers are only now being explored.

This paper illustrates some of the techniques appropriate for very high power pulse generation. Because CO_2 lasers appear to be the most exciting for this application, I will use CO_2 as the primary example. However, I will discuss some ideas for generating very high power pulses in glass lasers. I think glass lasers may have the long term potential to surpass CO_2. (Recent work on excimer lasers is discussed elsewhere.[6,7])

II. THE SOURCE

The development of ultrashort pulses over the last decade has been largely dominated by improvements in c.w. mode-locked dye lasers. These are highly reliable, very stable sources which can now reach pulse durations as short as 70 fsec.[8] Although pulses typically contain less than 1 nJ of energy, they can be amplified in dye laser amplifers to an energy of ~ 1 mJ.[6] Such pulses are sufficiently intense that a variety of nonlinear techniques can be used to generate pulses of a similar duration over all of the visible spectrum and over much of the infrared and U.V spectrum as well. Thus experiments can now be performed on all three of the above mentioned lasers without undue concern for the source of an ultrashort input pulse. For completeness, I will briefly outline the approach that I have used to generate picosecond 10 μm pulses.([3,9,10])

Fig. 1 is a schematic of the experimental arrangement. The 616 nm visible pulse was generated by a c.w. mode-locked dye laser which produced a train of 2 psec, 616 nm pulses at an 80 MHz repetition rate. A single pulse from the dye laser output was amplified in a XeCl pumped dye amplifier chain[6] to a maximum energy of 1 mJ. The s-polarized pulse illuminated a two-component CdTe reflection switch[10] at an incident energy density of ~5 mJ/cm^2 near Brewster's angle. The slab was simultaneously illuminated by a transform-limited, 1 MW, 100 nsec pulse of p-polarized 10 μm radiation.

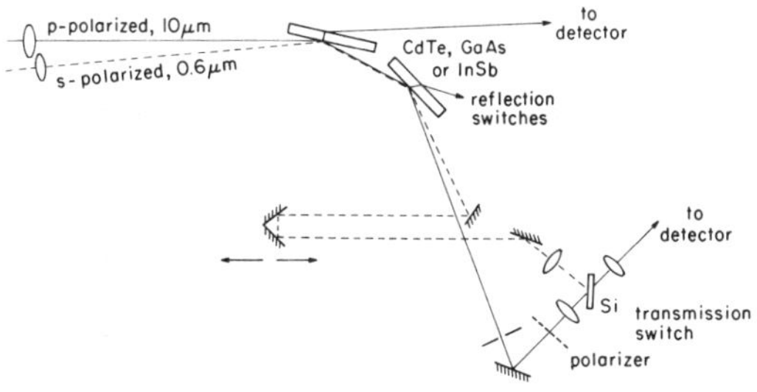

Fig. 1 Experimental arrangement used to generate picosecond 10 μm pulses.

The semiconductor plasma generated by the visible radiation produced a 2 psec rise-time, ~ 60 psec duration, 10 μm pulse. A 10^6:1 signal-to-background power contrast ratio was obtained.

To decrease the duration of the 10 μm pulse, a Si wafer was placed in the path of the 10 μm beam. Illuminated with the remaining 616 nm radiation, the transmission of the Si could be switched to T <10^{-4}. Thus the system consisted of a two-element reflection switch to "turn on" the 10 μm pulse, followed by a Si transmission switch to turn it off. A similar configuration was used previously.[11] Pulses having a duration variable from 2-40 psec could be generated.

III AMPLIFICATION

The small signal gain spectrum of low pressure CO_2 lasers is strongly modulated by the individual vibrational-rotational transitions of the CO_2 molecules. Near the 10.6 μm P(20) transition, adjacent rotational lines are separated by 55 x 10^9 Hz.

As the operating pressure of a CO_2 laser is increased, the rotational transitions broaden at a rate of ~ 3 GHz/atm. By 8 atmospheres, the adjacent lines overlap sufficiently to allow continuous tuning over 4 broad frequency bands. Thus, at high pressures, we obtained nearly unmodulated bands, wide enough for the amplification of picosecond pulses.

To illustrate the amplification characteristics of high pressure CO_2 lasers, I will describe an experiment that we performed to determine both the maximum power density (characterized by a saturation energy) and the minimum pulse duration that can be obtained from a high-gain laser.

The 2-40 psec pulse was injected into a regenerative amplifier (Fig. 2) where it was amplified through a total gain of ~ 10^{10}. (The regenerative amplifier is not fundamental to picosecond pulse amplification, but allows us to obtain large gains from a single gain module). The injected pulse, coupled into the slave cavity by reflection from a wedged NaCl beam-splitter, had a peak power of ~ 100W. An injected power of only ~10^{-8}W is sufficient to dominate the spontaneous emission of the gain medium.[12]

Radiation coupled from the cavity by the NaCl beam-splitter provides a measure of the regeneratively amplified pulse on each round-trip cavity transit. Fig. 3 is an oscilloscope trace of the output observed from one surface of the NaCl beam-splitter. The regeneratively amplified 10 μm pulse reaches a maximum energy density of ~ 1J/cm².

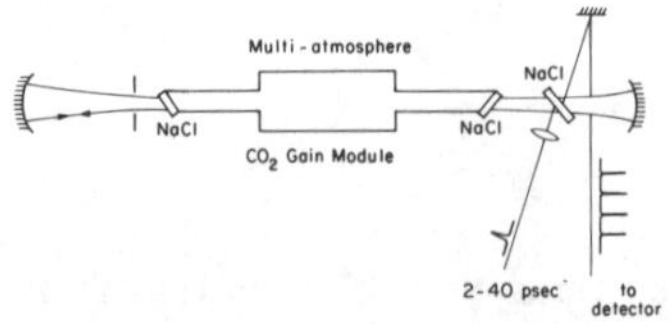

Fig. 2 Schematic of the regenerative amplifier used investigate amplification picosecond 10 μm pulses.

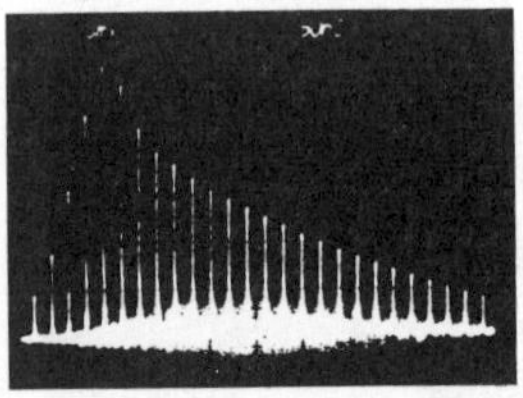

Fig. 3 Oscilloscope trace of the output observed from a single surface of the NaCl beam-splitter. The injected pulse duration was 2 psec and the laser operating pressure was 10 atmospheres.

Figure 4 shows the peak pulse in the mode-locked train obtained with a 3 psec injected pulse. The laser was operated on both the 9 µm R branch (left trace, 10 atmospheres laser operating pressure) and the 10 µm P branch (right trace, 14 atmospheres laser pressure).

Both traces show evidence of satellite pulses. These are due to the residual gain modulation resulting from insufficient overlap of the pressure broadened rotational lines. The 18 psec pulse separation in Fig. 4 (right trace) and the 25 psec pulse separation in Fig. 4 (left trace) are characteristic of the rotational line separation on the P and R branch respectively. Two important conclusions emerge from these picosecond regenerative amplification studies:

(i) Multi-atmosphere CO_2 lasers are capable of amplifying pulses as short as ~ 2.5 psec through a total gain of 10^{10} without significant pulse broadening, and

(ii) Nearly all of the stored energy (≧ 1 J/cm^2 for our device) in a multi-atmosphere CO_2 laser within the mode volume of our resonator (0.01 cm^2 x 40 cm) can be concentrated in a single picosecond, regeneratively amplified pulse.

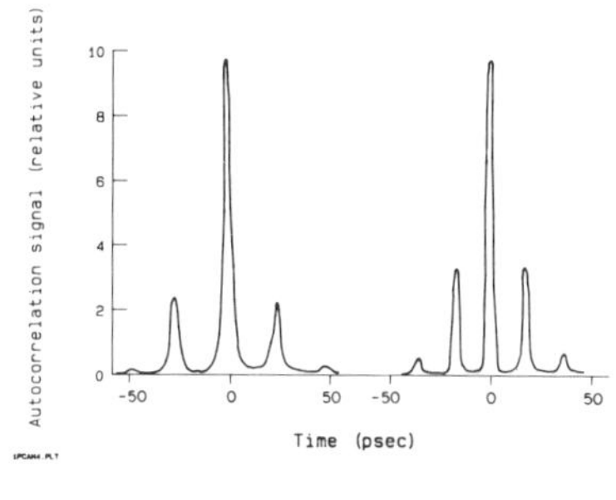

Fig. 4 Autocorrelation of the regeneratively amplified pulse. Left trace: 9 µm R branch, Right trace: 10 µm P branch.

Although single pass amplification studies of multi-atomsphere CO_2 lasers are required to accurately determine the energy extraction efficiency under scalable conditions, it is possible to roughly estimate the saturation energy from the time dependence of the regeneratively amplified output train. Within the limitations of the procedure, we obtain a saturation energy in the range of 200 $mJ/cm^2 \leqq E_{sat} \leqq 600\ mJ/cm^2$. Even the smallest value is well in excess of the saturation energy of any other picosecond amplifying medium, with the exception of Nd:glass where the nonlinear index of refraction imposes an ~ 8 x $10^9 W/cm^2$ limit on the instantaneous output power for a well designed laser.

IV. SPECULATION CONCERNING VERY HIGH POWER CO_2 AMPLIFIERS

If we accept a value of 400 mJ/cm^2 as the saturation energy of a 10 atm CO_2 amplifier for 2.5 psec pulses, it is possible to calculate the output energy density that could be obtained from a well designed amplifier chain. Energy densities of 2 J/cm^2 appear feasible, corresponding to power densities of ~ 10^{12}W/cm^2.

I have described the amplification of a picosecond pulse through a total gain of ~ 10^{10}. For such high gains, a pressure of ~ 10 atm was necessary to minimize the energy in the satellite pulse (a trailing pulse). If less gain is necessary from an amplifier module, a stronger modulation of the gain spectrum can be tolerated for a given fraction of energy in the satellite pulse. A lower pressure amplifier can then be used. A high power picosecond amplifier chain could consist of a high pressure ($\geq$ 10 atm) preamplifier, with a maximum output pulse energy in the 1-20 J range followed by a lower pressure (~ 5 atm) larger aperture power amplifier.

Single amplifier modules are now constructed with a discharge aperture of ~ 100 cm-atmospheres. This would correspond to a 20 cm discharge at 5 atm. A 20 cm x 20 cm aperture amplifier should be capable of output pulse energies approaching one petawatt while modules with rectangular apertures appear feasible.

Pulses of still higher power can in all likelihood be obtained from power broadened amplifiers. Such an amplifier could be any currently available, few-atmosphere CO_2 amplifier provided it is operated at a power density level of $\geq 10^{11}$W/cm^2. At such an intensity, the Rabi frequency ω_r is

$$\omega_r = \frac{\mu E}{h} \sim 4\text{cm}^{-1} \qquad (1)$$

where μ is the dipole moment of the CO_2 laser transition, and E is the electric field of the laser radiation. (To obtain the value of 4 cm^{-1} I used a value of μ appropriate for the vibrational transition (μ = 0.035D). Of course the rotational and orientational quantum numbers also affect the value of the dipole moment.) Thus at I $\geq 10^{11}$W/cm^2, the power broadening is greater than the separation between rotational levels. It is, in fact, as large as the bandwidth of the incident 10 μm pulse.

V. OPTICAL PULSE COMPRESSION

Although it has been clear for some time that short pulses are appropriate for the acceleration of charged particles, it has not been explicitly noted that there can be a "trade-off" between shorter pulse durations and resonant enhancement required from the accelerating mode. For example, by further decreasing the pulse duration, we can make the design of high-Q cavities in the laser-driven driven grating accelerator[13] less important by supplying higher incident fields.

After reaching the bandwidth limits of laser amplifiers, a further increase in the power of ultrashort optical pulses is possible by using pulse compression. I will illustrate pulse compression for CO_2 lasers using plasma nonlinearities, but the concept was developed in the visible and near infrared using electronic nonlinearities. Advances in pulse compression have been extremely important for ultrashort pulse technology in recent years. It has, however, been applied almost exclusively to relatively low power pulses which can be propagated in optical fibers. For small compression ratios ($\lesssim 5$), however, high power, larger aperture compression should be possible.

Optical pulse compression consists of two steps: (i) The optical pulse is frequency chirped in a nonlinear medium. For 10 μm radiation, the nonlinearity can be plasma production and (ii) a dispersing medium to allow the trailing portions of the pulse (now colour coded) to "catch up" with the front of the pulse. For 10 μm radiation we have used linear dispersion in NaCl. I will outline the physics behind the chirping method that we used. This is of more than academic interest, for it will demonstrate that plasmas can be extremely non-linear media for infrared radiation. As with other nonlinearities, this can be useful or deleterious, depending on the circumstances. Consequently, it is an area that will require considerable care when using high power infrared pulses.

Consider a plasma of density N_e. The refractive index is given by

$$\eta = \eta_0 \, (1 - N_e/N_{cr})^{1/2} \tag{2}$$

where η_0 is the index of refraction in the absence of a plasma and N_{cr} is the critical density:

$$N_{cr} = \frac{\omega^2 m}{2\pi e^2} \, \eta_0 \tag{3}$$

In Eq. 3, $\omega = 2\pi c/\lambda$ where λ is the vacuum wavelength of light, m is the free-carrier reduced mass, and e is the electron charge.

For $N_e < N_{cr}$, Eq. 2 can be expanded to give

$$\eta = \eta_0 \left\{1 - \frac{1}{2}\frac{N_e}{N_{cr}} + \ldots\right\} \tag{4}$$

Plasma production, therefore, results in a change in the refractive index of

$$\delta\eta = \frac{1}{2}\frac{N_e\lambda^2 e^2}{2\pi\eta_o^2 mc^2} \qquad (5)$$

The important point to note in Eq. 5 is the λ^2 dependence of $\delta\eta$. This says that refractive index changes, and therefore wavefront changes, can be produced with less plasma in the infrared than in the visible. (The refractive index can also be very sensitive to carrier mass changes for long wavelength radiation).

Frequency chirping follows from Eq. 5 in a straightforward way. Any process that leads to plasma production during a pulse will produce an increase in the phase velocity of the back of the pulse compared to the front. Thus, as the pulse propagates, it becomes increasingly blue chirped. The frequency chirp can be expressed as follows

$$W_{ch}(T) = \frac{\pi\eta_o}{\lambda}\frac{1}{N_{cr}}\int\frac{dN_e(T)dl}{dT} \qquad (6)$$

where the integral is over the path followed by the optical pulse and T is the time measured in a frame moving with the optical pulse. Pulse compression is optimized for a linearly chirped pulse.

Plasma production is possible under external control or, for example, by avalanche ionization in an appropriately preionized gas or in a semiconductor. Experimentally, the CO_2 gain medium itself is a weakly pre-ionized plasma. I have plotted in Fig. 5 the frequency spectrum of the amplified pulse as the path integral accumulates, that is, for various pulses in the output train.

The laser resonator also contains a dispersing medium, NaCl, which has a higher group velocity for the blue spectral components of the pulse than for the red components. Thus, as the chirped pulse makes successive passes through the NaCl windows and the beam-splitter, the pulse duration should decrease. This is shown in Fig. 6 where the autocorrelation pulse width (fwhm) is plotted after each successive double pass transit of the gain medium.

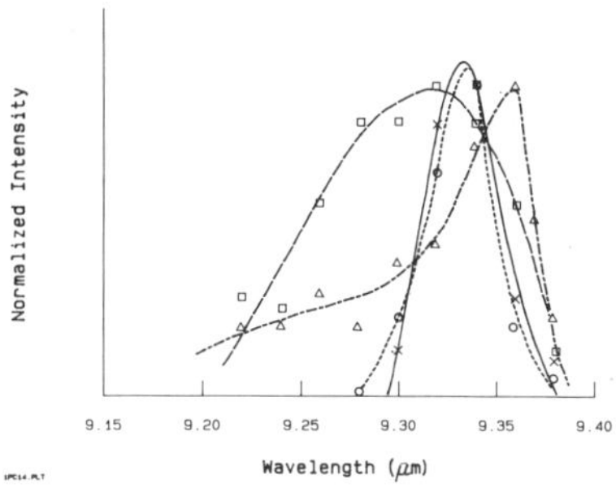

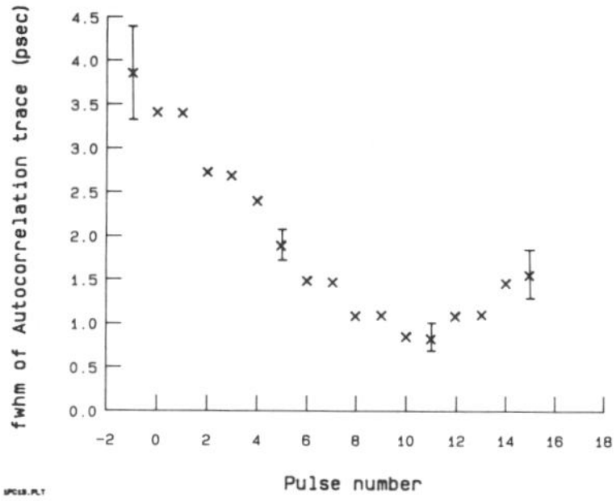

Fig. 5 Spectrum of selected pulses in the mode-locked train. The hybrid oscillator was tuned to 9 R(20) and the multi-atmosphere laser was operated at 10 atm. The injected pulse width was 2 psec. The experimental points are labeled by (x) - pulse No. -3 (O) - peak pulse, (□) - pulse No. +6, (Δ) - pulse No. +12. (The peak energy pulse in the output train is numbered 0 while those after the peak are given positive numbers).

Fig. 6 Plot of the autocorrelation fwhm as a function of the pulse number in the mode-locked train.

VI. CONCLUSIONS

I have speculated about the design of very high power CO_2 lasers. It seems that other advances will permit power densities even higher than 10^{12} W/cm^2. For example, (1) a pulse compressor, located part way along the CO_2 amplifier chain should allow output pulses as short as 1 psec, even in the absence of power broadening. (2) Isotopic mixtures of CO_2 can be used to fill the gap between the P and R branches in the CO_2 spectrum, allowing, in principle, much shorter pulses. (3) The implications of power broadening are only beginning to be considered. In the extreme, it may be possible, by using both isotopic mixtures and power broadening, to amplify pulses overlapping both vibrational bands generating shorter pulses still. I would not be surprised if, ultimately, power densities in excess of 10^{13} W/cm^2 can be obtained at the output of CO_2 lasers.

Can higher power density pulses be generated by other lasers? I began this article by mentioning excimer lasers and Nd:glass lasers as potentially important picosecond amplifiers.

In excimer lasers, the maximum output power density will be limited by the energy storage time of the gain medium (typically ~ 5 nsec). In Reference 6, we estimate the likely maximum power density of an XeCl laser to be ~ 10^{11} W/cm^2 in the absence of pulse compression. KrF lasers should have similar parameters.[7]

Nd:glass is the other candidate laser mentioned previously. Nd:glass lasers are characterized by a very high saturation energy[14] (E_S ~ 4 J/cm^2) and an inhomogenously broadened gain profile with a bandwidth[15] of ≥ 200Å. If a bandwidth limited pulse could be amplified to saturation in Nd:glass, the output power density could reach ~ 10^{14} W/cm^2. Unfortunately self-focusing in the laser medium limits the practical intensity in a near-diffraction limited beam. The maximum intensity[14] is ~ 8 x 10^9 W/cm^2.

It should be possible to "side-step" this nonlinear limit. Consider the following: a 100 fsec pulse is sent through a dispersive delay line and is thereby frequency chirped. The much longer pulse that emerges can then be amplified to the nonlinear (or saturation) limit. The pulse is then compressed by a linear medium of opposite dispersion.

All of the above steps can be performed now! Pulses of less than 100 fsec can be generated in colliding pulse mode-locked dye laser amplifiers. Nonlinear mixing techniques can be used to generate 100 fsec, 1.06 μm pulses. These pulses can be dispersed in long fiber optic delay lines. (It may be more feasible to disperse the dye pulse before nonlinear conversion). The resultant 1.06 μm chirped pulse can be amplified and recompressed with grating pairs.

No one has performed an experiment such as this, but as it becomes more widely known that very high power lasers can now be conceived, and indeed that they can even be quite simple when judged on the scale of current high power (fusion) lasers, experiments of this type will surely be performed.

REFERENCES

(1) see for example Ultrafast Phenomena IV, D.H. Auston and K.B. Eisenthal, editors, Springer-Verlag, New York, 1984.

(2) M.C. Downer, R.L. Fork and C.V. Shank, "Imaging with Femtosecond Optical Pulses", published in Ultrafast Phenomena IV, D.H. Auston and K.B. Eisenthal, editors, Springer-Verlag, New York, 1984, pp 106-110.

(3) P.B. Corkum, "Amplification of Picosecond 10 μm Pulses in Multiatmosphere CO_2 Lasers" accepted for publication in IEEE J. Quantum Electron. March 1985.

(4) L.V. Keldysh, "Ionization in the Field of a Strong Electromagnetic Wave", Sov. Phys. JETP, Vol. 20, pp 1307-1314, 1965.

(5) A.J. De Maria, D.A. Stetser and M. Heynau, "Self-Mode-Locking of Lasers with Saturable Absorbers", Appl. Phys. Lett. 8, pp 174-176 (1966).

(6) P.B. Corkum and R.S. Taylor, "Picosecond Amplification and Kinetic Studies of XeCl", IEEE J. Quantum Electron, QE-18, 1962-1975 (1982).

(7) S. Szatmari and F.P. Schafer, "Picosecond Gain Dynamics of KrF", Appl. Phys. B 33, pp 219-223 (1984).

(8) C.V. Shank, R.L. Fork and R.T. Yen, "Moving from the Picosecond to the Femtosecond Time Regime", published in Picosecond Phenomena III, K.B. Eisenthal, R.M. Hochstrasser, W. Kaiser and A. Laubereau, editors, Springer-Verlag, New York, 1982, pp 2-5.

(9) P.B. Corkum, "High-Power, Subpicosecond 10 μm Pulse Generation", Opt. Lett., 8, pp 514-516, (1983).

(10) A.J. Alcock and P.B. Corkum, "Ultra-fast Switching of Infrared Radiation by Laser-Produced Carriers in Semiconductors", Can. J. Phys. 57, 1280-1290, (1979).

(11) S.A. Jamison and A.V. Nurmikko, "Generation of Picosecond Pulses of Variable Duration at 10.6 μm", Appl. Phys. Lett., 33, pp 598-600, (1978).

(12) P.B. Corkum, A.J. Alcock, D.J. James, K.J. Andrews, K.E. Leopold, D.F. Rollin and J.C. Samson, "Recent Development in High Power CO_2 Laser Mode-Locking and Pulse Selection", in Laser Interaction and Related Plasma Phenomena Vol. 4A, edited by H.J. Schwartz and H. Hora, Plenum Publishing Corp. 1977, pp. 143-160.

(13) R.B. Palmer, "A Laser Driven Grating Linac", Particle Accelerators, 11, pp 81-90 (1980)

(14) W. Seka, J. Soures, O. Lewis, J. Bunkenburg, D. Brown, S. Jacobs, G. Mourou and J. Zimmermann, "High-Power Phosphate-Glass Laser System: Design and Performance Characteristics", Appl. Opt., 19, pp. 409-419, (1980).

(15) A. Yariv, Quantum Electronics, second edition, J. Wiley and Sons , Inc. New York (1976) p. 205.

AN FEL-POWERED PARTICLE ACCELERATOR?

Jack Slater
Spectra Technology Inc., Bellevue, Washington 98004

ABSTRACT

This paper is a condensed FEL tutorial and consideration of the FEL as a driver for a laser particle accelerator. The basic FEL mechanism is described and the commonality of the various FEL schemes is identified. It is seen that the FEL has many characteristics suitable for the laser particle accelerator including high power, high efficiency, and wavelength tuneability. The FEL peak power is limited by the accelerator that drives it, and this peak power may be lower than desirable for the laser particle accelerator.

1. INTRODUCTION

The free-electron laser (FEL) operates by means of an energy exchange betwen free electrons and an electromagnetic wave. Unlike conventional lasers, transition between fixed atomic or molecular energy levels is not involved, so the FEL is inherently wavelength tunable. A similar energy exchange has been utilized for years in conventional microwave devices such as klystrons or magnetrons, but this discussion addresses the newer short wavelength devices using a relativistic electron beam and wiggler, or undulator, magnet. The wiggler magnet is typically an array of permanent magnets which produce a spacially periodic transverse magnetic field that gives a small transverse velocity to an electron beam passing through. This transverse velocity couples to the transverse E-field of the photon to provide a mechanism for energy exchange. Useful interaction occurs at any frequency for which the transverse velocity and E-field oscillate in phase. This resonant condition implies a particular relationship between the e-beam energy, the photon wavelength, the wiggler wavelength, and, to some extent, the wiggler field strength. This high voltage, short wavelength FEL has several characteristics which separate it from the longer wavelength, lower voltage devices. First, the photon wavelength is not tied to the cavity dimensions and the electromagnetic wave typically propagates in free-space fashion. The phase matching necessary to achieve synchronism between the electron transverse velocity and photon E-field occurs by control of the axial electron velocity rather than through a slow wave structure for the photons. As a result, no metal or dielectric surface is required in close proximity to the interaction; therefore, the device can operate at short wavelength and high power simultaneously. The electron energy not converted to photon energy on a single pass through the wiggler magnet remains in a highly collimated beam and is largely recoverable.

With electron beam energy recovery, high power systems can be envisioned which operate efficiently at ultraviolet or longer wavelengths. Such systems would obtain nearly diffraction limited beam quality as a result of the narrow bore wiggler magnet and well controlled gain medium, i.e., the electron beam. Therefore, it is natural that FELs be considered as candidate drivers for laser-powered particle accelerators. To this end, the remainder of this paper contains an overview of the FEL physics and system configurations, a review of some existing devices, and comments on the scalability of the existing concepts and technology to regimes of interest for laser particle accelerators.

In Section 2, the basic equations of motion describing the FEL are reviewed. A classical treatment is employed with a point electron interacting with the combined wiggler magnetic field and photon electric field. This classical equivalent to the quantum-mechanical treatment(1) was first described by Colson(2) and allows a remarkably simple visualization of the interaction. Various methods of extracting energy from the electron beam are discussed in Section 3. It is shown that, in general, high electron energy extraction and high photon gain are mutually exclusive. Furthermore, a wide range of interaction schemes lead to approximately the same product of fractional gain and fractional extraction. Section 4 highlights several existing FEL devices. These existing devices are limited by the available electron beams and are uniformly low power devices. Even so, they demonstrate that the basic science of the FEL is well in hand and that extrapolation to higher power requires primarily technology development of e-beams and optical systems. Included is a discussion of the scalability of FELs to levels of interest for laser particle accelerators. It is a general conclusion that the FEL has excellent potential to be a high average power device but that the peak powers available may be lower than desirable.

2. BASIC EQUATIONS OF MOTION

A classical picture of the FEL involves electron motion under the influence of the ponderomotive potential formed by the combined action of the magnetic field, the wiggler, and the electric field of the photon. This potential has a sinusoidal variation for movement parallel to the electron or photon beam. For appropriate choice of wiggler parameters, photon wavelength, and e-beam energy, the system is resonant and this potential surface moves at nearly the same speed as the electron, thereby allowing significant energy exchange. What follows is a review of the basic equations necessary for determination of the electron behavior under the influence of the ponderomotive potential. An in-depth derivation is given in Reference 3. Equations of motion can be developed by starting with an assumed form for the photon E-field, E_L, in wiggler B-field, B_w, and the geometry of Figure 1.

$$B_w(z) = \hat{y}\, B_o \cos k_w z \qquad [1]$$

$$E_L(z) = \hat{x}\, E_o \cos \left[k_L z - w_L t\right] \qquad [2]$$

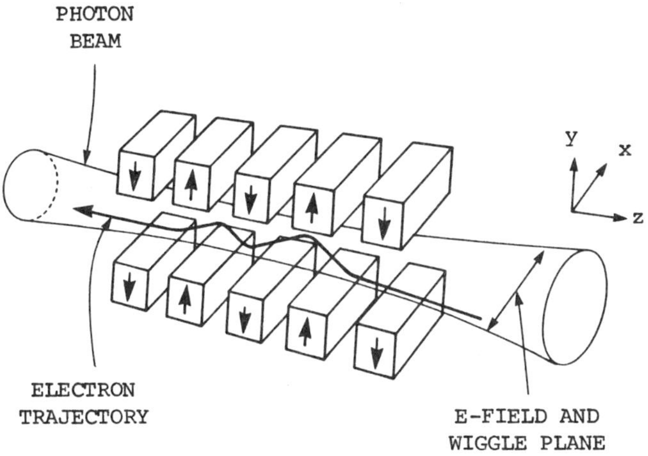

Figure 1. FEL geometry for plane polarized wiggler with permanent magnets.

Here B_o is the peak magnetic field of the wiggler and E_o is the peak E-field of the photon. The wiggler and photon wave numbers are k_w and k_L, respectively. For a relativistic electron moving in the same direction as the photon, the electric and magnetic components of the Lorentz force due to the photon nearly cancel and, as a result, the transverse motion of the electron is established almost entirely by the action of the wiggler B-field. This force results in a small perpendicular velocity component, $v_\perp$, given by

$$v_{\perp} = c\ (2)^{1/2}\ \frac{a_w}{\gamma} \sin k_w z \qquad [3]$$

where

$$a_w = \frac{1}{(2)^{1/2}}\ \frac{e\ B_o}{mc^2\ k_w} \qquad [4]$$

The value of a_w is typically of order unity for a realistic system optimized for gain.(4) This perpendicular velocity couples to the transverse E-field of the photon. The equation of motion for the electron energy follows from examining $v_{\perp} \cdot E$ and can be shown to be

$$\frac{d\gamma}{dz} = \frac{e_L a_w}{\gamma} \sin \psi \qquad [5]$$

where

$$e_L = \frac{1}{(2)^{1/2}}\ \frac{eE_o}{mc^2} \qquad [6]$$

Here the phase parameter, ψ, is defined as $(k_L + k_w)\ z - w_L t$, with z being the electron position at time t. It is simply the phase angle of the electron with respect to the sinusoidal ponderomotive potential well. The equation of motion for ψ is

$$\frac{d\psi}{dz} = k_w - \frac{k_L}{2\gamma^2}\ [1 + a_w{}^2] \qquad [7]$$

Equations 5 and 7 together define the FEL interaction for a single electron as long as the E- and B-fields are specified. A self-consistent model of the E-field and electron dynamics has been fertile ground for theoretical study but is not discussed here.(5) A significant exchange of energy between the electron and photons can occur whenever ψ is not rapidly varying. This condition of resonance is then $d\psi/dz \approx 0$, which from Equation 7 implies

$$\gamma_r^2 = \frac{\lambda_w}{2\lambda_L}\left[1 + a_w^2\right] \qquad [8]$$

Above some critical electron density, the effect of space charge within the e-beam becomes increasingly important, both from the standpoint of electron transport and the physics of the FEL interaction. At high enough densities, the electrostatic forces of the e-beam completely dominate the ponderomotive potential and, in this case, the interaction is cast in terms of a three-wave Raman process involving the photon signal wave, the wiggler or pump wave (the static B-field appears as an electromagnetic wave in the electron frame), and a longitudinal plasma wave. The Raman regime is not generally accessible in experiments operating on IR or shorter wavelengths, and it is not considered further here. Examples of unified descriptions including various density and gain regimes can be found in References 6 and 7.

The single particle equations of motion 5 and 7 can be combined to give a second order equation for the phase angle ψ, which is the pendulum equation, first studied by Colson.[8] This pendulum-like motion describes the evolution of the phase angle ψ, with ψ marking the electron position in the ponderomotive potential. The rate of change, $d\psi/dz$, is proportional to the difference between the electron energy and resonant energy given by Equation 8. We define this detuning by $\Delta\gamma = \gamma - \gamma_r$. Depending on the initial phase, the detuning, and the height of the potential well, particle motion is either trapped within one valley of the potential or untrapped. The height of the potential in energy is the function of the electric field and photon wavelength and is given by[3]

$$\frac{\Delta\gamma}{\gamma} = \left[\frac{2\lambda_L e_L a_w}{\pi\left[1 + a_w^2\right]}\right]^{1/2} \qquad [9]$$

A useful pictorial means of understanding this trapping can be seen in the γ-θ phase diagram of Figure 2. Trapped particles have energies near resonance and make rotations about the point $\psi_r = 0$, $\gamma = \gamma_r$, while untrapped particles have off-resonant energies and experience ever-increasing ψ values. Their energy oscillates as they ride over the crests and valleys of the ponderomotive potential.

The trapped particles make one orbit of the well in a distance ℓ_s along the wiggler given by

$$\ell_s = \frac{\pi}{k_w}\left[\frac{k_L\left[1 + a_w^2\right]}{a_w e_L \cos\psi_r}\right]^{1/2} \qquad [10]$$

The length ℓ_s is called the synchrotron period.

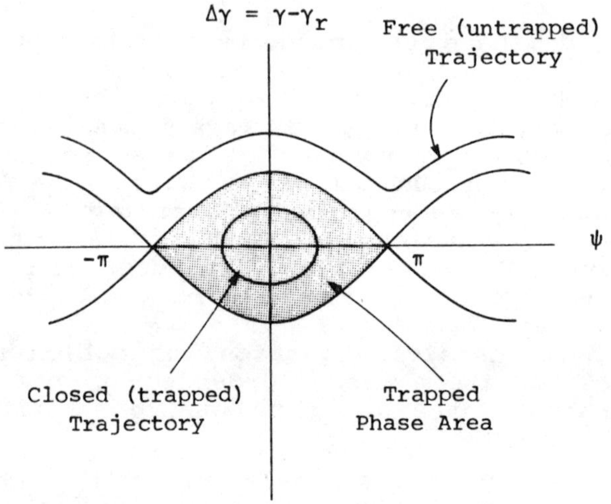

Figure 2. Electron motion in γ-ψ space.

Equations 9 and 10 together characterize the strength of the FEL interaction. Together they show what fractional energy change can occur over what length. The two are plotted as a function of intensity and photon wavelength in Figure 3. Here a_w has been set equal to unity, which is a typical value resulting from optimization exercises.[4] As the photon wavelength decreases, the bucket size also decreases and the synchrotron wavelength increases. This loss of interaction strength at short wavelength shows why the FEL has been thus far demonstrated only at visible and longer wavelengths.

3. EXTRACTION OF ENERGY

There are several different ways in which the "bucket" of Figure 2 can be used to extract energy. The various schemes differ widely in such basic parameters as photon gain, maximum electron energy extraction, and allowable e-beam energy spread and emittance. The common schemes and their most notable features, listed in relative fashion, are given in Table I. These schemes are described in only the briefest detail here, as they are united by a common feature described further below. Details of the physics of each scheme can be found in Reference 9.

The optical klystron employs two untapered wigglers separated by an energy dispersive section. The dispersive section leads to bunching of the electrons following the velocity modulation induced in the front

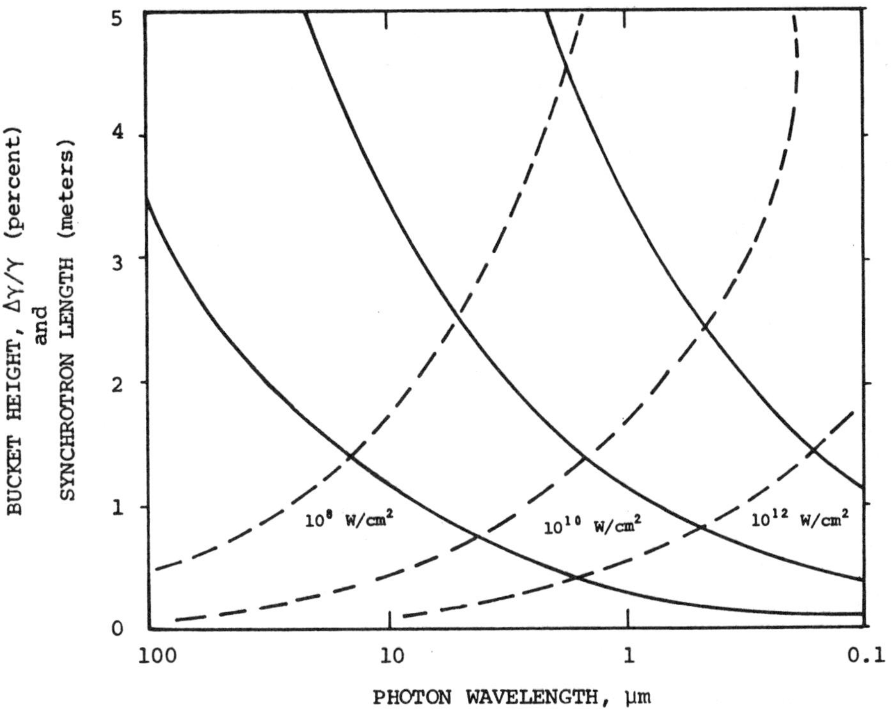

Figure 3. The bucket height and synchrotron wavelength are indicated by the solid and dashed lines, respectively, for a range of photon intensities. These lines apply to a wiggler with a wavelength of 2 cm and an a_w value of unity.

Table I. Energy Extraction Schemes

Scheme	**Extraction**	**Gain**	**Bandwidth**
Optical Klystron	Very low	Very high	Very low
Untapered	Low	High	Low
Tapered	High	Low	High
Phase Displacement	Moderate	Low	High

wiggler. The rear wiggler then acts as an energy extractor. The system exhibits very high gain because of the efficient bunching. For the same reason, the electrons are easily overbunched and the gain saturates at low optical power. The extraction and bandwidth (energy spread acceptance) are low.

The untapered wiggler has less efficient bunching and, therefore, lower gain, but it can tolerate higher energy spread and thus allows larger electron energy extraction before gain saturation occurs. Still higher extraction, but lower gain, is possible with the tapered wiggler. In this scheme, electrons are trapped in the ponderomotive well, and the well is decelerated by adjusting λ_w, B_w, or both along the wiggler length.

The phase displacement scheme involves a tapered wiggler but the ponderomotive well is accelerated rather than decelerated. To provide deceleration of the electrons, the well is created at an energy well below the input electron energy and then accelerated through the electron distribution to a point well above the input electron energy. If the acceleration is adiabatic, electrons avoid being captured in the well and, in the process, end up with lower mean energy.

With the exception of the phase displacement technique, Table I shows a clear correlation between gain and extraction. When the fractional gain and extraction are both small, it can be shown that for all three cases (i.e., the optical klystron, the untapered wiggler, and the tapered wiggler) the product of fractional gain and fractional extraction depends only on several physical parameters and is independent of the extraction scheme. Thus, details of the individual mechanism are less important in a sense; the various schemes are simply ways to access a wide range of gain values, with the extraction always inversely related. With the aid of Reference 4, it is relatively simple to show that, for the tapered wiggler, the gain and extraction are related as

$$(\text{gain})(\text{extraction}) = 0.04\ P_{eb}\ B^2\ \lambda_p\ L_w \qquad [11]$$

Here P_{eb} is the e-beam power in MW, B is the RMS B-field value in kg, λ_p is the photon wavelength in cm, and L_w is the wiggler wavelength in cm. The gain and extraction are both expressed in percentages. In Reference 4, it was assumed that $\delta \sin \psi = 0.3$ and $q = 1.5$. The constant in Equation 11 is correct for $a_w \approx 1$. For other a_w values, the constant should be replaced by $0.16/(1 + a_w^2)^2$.

A similar value for the gain-extraction product is found for the untapered wiggler. The standard equation for the small signal gain and maximum extraction yields[10] a product which exceeds Equation 11 by roughly a factor of 10. However, to achieve maximum extraction, the gain saturates significantly and the maximum value of self-consistent gain extraction product is close to that of Equation 11.

For the optical klystron, Ellaume has shown[11] that the gain is boosted over that for an equal length untapered wiggler by a factor dependent on the degree of energy dispersion between the front and rear sections. However, the maximum extraction is reduced by the same factor so that the maximum self-consistent gain extraction product is near that given by Equation 11.

4. FEL CAPABILITY

Numerous FELS have been operated successfully over the last several years. Table II contains a representative sample of these devices. Some of these

Table II. Representative Existing Free-Electron Lasers

Location	Type	Wave Length	Beam Energy	Peak Power	Macro Power	Macro Length	Efficiency
LANL[12]	RF Linac untapered wiggler	10 μm	20 MeV	7 MW	4 KW	100 μs	1%
LLNL[13]	Induction Linac untapered wiggler	8.7 mm	3.3 MeV	80 MW	80 MW	10 ns	5%
NRL[14] (VEBA)	Induction Linac untapered wiggler	2.5 cm	1.2 MeV	35 MW	35 MW	30 ns	2.5%
Orsay[15]	Storage Ring optical klystron	0.65 μm	160 MeV	60 mW	75 μW	Continuous	2.4×10^{-3}%
Stanford[16] (Madey)	Superconducting RF Accelerator, untapered wiggler	3.3 μm	43 MeV	130 KW	5 W	1-20 ms	0.2%
Stanford[17] (TRW)	Superconducting RF Accelerator, tapered wiggler	1.6 μm	66 MeV	1.2 MW	4 W	5 ms	1.1%
UCSB[18]	Electrostatic w/recovery, untapered wiggler	400 μm	2.5 MeV	3 KW	3 KW	50 μs	2%

devices have demonstrated additional capability to that shown. Note that several accelerator types and wiggler schemes are represented. The NRL device listed differs from the others in that plasma effects must be taken into account to understand the basic interaction. These effects become unimportant at shorter wavelength and high e-beam energy. With the exception of two cases, the efficiencies lifted refer to the percent of electron energy given to the optical beam in a single pass through the wiggler. For these systems, the e-beam energy remaining after the interaction is not recovered. The two exceptions are the Orsay and UCSB work. In the Orsay storage ring experiment, the e-beam is continually recycled and the efficiency listed is the ratio of laser power to e-beam synchrotron power. In the UCSB experiment, the e-beam energy is electrostatically recovered and the efficiency listed refers to the laser power divided by the fraction of e-beam power which is not recovered. All of these devices operate at low peak and average powers when compared to possible laser particle accelerator requirements. While the potential of high power and high efficiency has motivated these FEL programs in general, the FEL is sufficiently new that achieving oscillation is an overriding concern (except for the long wavelength devices) and parameters such as efficiency, peak power, and average power have not been pushed to their limits.

An idea of the near-term potential of the FEL can be gained by examining the accelerator. The FEL performance is obviously limited by the e-beam, and the accelerator technology is quite mature in comparison to the FEL itself. In this spirit, Table III shows what the near-term FEL capability would be for several different classes of accelerators. The generic accelerator parameters listed are those familiar to the author and are not intended to represent fundamental limitations. However, significant improvement in our accelerator technology presumably would be required to exceed these parameters by large factors.

Table III. Representative Near-Term FEL Capability

Device	Pulse Length	E-beam Peak Power (W)	Laser Peak Power (W)	Laser Average Power (W)
Diode (amplifier)	ns-μs	10^{10} (10 KA, 1 MeV)	2×10^9	2×10^5 (100 ns, 1 kHz)
Induction Linac (amplifier)	ns-μs	10^{12} (10 KA, 100 MeV)	2×10^{11}	10^7 (50 ns, 1 kHz)
RF Linac (oscillator)	ps	4×10^{10} (200 A, 200 MeV)	4×10^9	10^7 (20 ps, 100 MHz)
Storage Ring (oscillator)	ps-ns	3×10^{11} (300 A, 1 GeV)	2×10^5	600 (1 ns, 3 MHz)

The peak laser power for the diode and induction LINAC accelerators is based on the assumption that 20 percent e-beam energy extraction is possible. This limits the laser peak and average power to 20 percent of the e-beam power. Pulse compression techniques may allow higher peak power. The RF LINAC will have less gain-extraction product available and the laser power in this case is based on 10 percent e-beam energy extraction.

The storage ring parameters are based on a representative high peak current ring designed for high energy spread acceptance, namely the Stanford X-Ray Research Center presently under construction. For the ring, Reniere has shown[19] that the laser power P_L is related to synchrotron power by

$$P_L \leq P_s \frac{\Delta E}{E}$$

where $\Delta E/E$ is the allowable energy spread induced in the e-beam. This limits both the efficiency and power. For this example, a synchrotron loss time constant of 5 msec and 5 percent energy acceptance were chosen. The allowable energy spread is roughly a factor of five less than the ring acceptance if beam loss is to be avoided.

5. CONCLUSIONS

The variety of possible schemes for laser particle accelerators and the lack of systems studies pertaining to these schemes precludes specification of precise laser requirements at present. However, there are some key parameters that can be estimated by examining pulse formats which give maximum luminosity at minimum average laser power. These considerations lead to scenarios with short pulses, high rep rates, and generally high instantaneous powers. They are discussed elsewhere in this volume. The FEL has several characteristics which match with the possible requirements. Among these are

1) Scalability to high average power
2) High potential efficiency
3) Wavelength at UV or longer
4) Picosecond pulse lengths
5) Excellent beam quality

A characteristic which may not be well suited to the laser particle accelerator is the peak power which, as shown in the rough estimates of Table III, is only 2×10^{11} Watts for the long pulse induction LINAC and only 4×10^{9} for the short pulse RF LINAC. Pulse compression techniques could allow some increase in peak power, or the photon transport system may be devised to expose a single electron bunch to more than one photon bunch. However, it should be noted that the FEL is probably not compatible with angular multiplexing, so that compression techniques or smart transport systems must function by means of nonlinear effects or other active optics. Thus, one challenge for the laser particle accelerator design studies is to find an accelerator mechanism and compatible photon transport system which is suited to the format of the FEL.

REFERENCES

1. J.M.J. Madey, J. Appl. Phys. 42: 1906 (1971).

2. W.B. Colson, Phys. Lett. 64A: 190 (1977).

3. N.M. Kroll, P.L. Moryton, and M.N. Rosenbluth, "Variable Parameter Free-Electron Laser," Physics of Quantum Electronics, Vol. 7, p. 89 (1980).

4. J.M. Slater, IEEE J. Quant. Electr. QE-17: 1476 (1981).

5. See the following references: W.B. Colson, "Optical Pulse Evolution in The Stanford Free-Electron Laser and in a Tapered Wiggler," Physics of Quantum Electronics, Vol. 8, p. 457 (1982); W.B. Colson and J.L. Richardson, Phys. Rev. Lett. 50: 1050 (1983); H. Al-Abawi, J.K. McIver, G.T. Moore, and M.O. Scully, "Pulse Propagation in the Tapered Wiggler," Physics of Quantum Electronics, Vol. 8, p. 415 (1982); and P. Sprangle, C.M. Tang, and I.B. Bernstein, NRL Memorandum Report 5011, April 1983; see also Report 5110.

6. N.M. Kroll and W.A. McMullin, Phys. Rev. A 17: 300 (1978).

7. P. Sprangle, R.A. Smith, and V.L. Granatstein, "Free Electron Lasers and Stimulated Scattering from Relativistic Electron Beams" in *Infrared and Millimeter Waves*, Vol. 1, K.J. Button, Editor (Academic Press, New York, 1979) p. 279.

8. W.B. Colson, Physics Letters 64A: 190 (1977).

9. Physics of Quantum Electronics, Vol. 7 (1980) and Volumes 8 and 9 (1982).

10. D.C. Quimby, unpublished calculations.

11. P. Elleaume, "Optical Klystron Emission and Gain," Physics of Quantum Electronics, Vol. 8, p. 199 (1982).

12. B.E. Newnam, presented at 1984 FEL Conference, Castelgandolfo, Italy, 3-7 September 1984; and R.L. Sheffield, presented at Lasers '84, San Francisco, 26-30 November 1984.

13. A. Sessler, presented at Lasers '84, San Francisco, 26-30 November 1984.

14. S. Gold, presented at Lasers '84, San Francisco, 26-30 November 1984.

15. P. Elleaume, J.M. Ortega, M. Billardon, C. Bazin, M. Bergher, M. Velghe, Y. Petroff, D.A.G. Deacon, K.E. Robinson, and J.M.J. Madey, J. Physique 45: 989 (1984).

16. S. Benson, D.A.G. Deacon, J.N. Eckstein, J.M.J. Madey, K. Robinson, T.I. Smith, and R. Taber, Journal De Physique 44(2): C1-353 (1983).

17. J.A. Edighoffer, G.R. Neil, C.E. Hess, T.I. Smith, S.W. Fornaca, and H.A. Schwettman, Physical Review Letters 52: 344 (1984).

18. L. Elias, presented at Lasers '84, San Francisco, 26-30 November 1984.

19. A. Renieri, Il Nuovo Cimento 53B: 160 (1979).

LASER TECHNOLOGIES FOR LASER ACCELERATORS

Dennis Lowenthal and Jack Slater
Spectra Technology Inc., Bellevue, Washington 98004

ABSTRACT

This paper presents a partial summary of work completed on a DOE contract entitled "Laser Technologies for Laser Accelerators." Luminosities of 10^{32} cm^{-2} sec^{-1} force 1 TeV laser accelerators to be driven by ~10 MW of output laser power. Consequently, laser system efficiency, cost, and transport geometries for the optical beams are of critical importance. In this paper the laser requirements are reviewed in terms of the laser luminosity requirements. Transport systems for far field geometries are considered, and it is concluded that glancing incidence optics will be required for all accelerator mechanisms unless the coupling strength (ratio of acceleration gradient to vacuum laser electric field) is very large, $\geqslant 1.0$. Both CO_2 and KrF lasers are reviewed from the point of view of high efficiency and pulse format. For CO_2 lasers, it is shown that one must extract with a burst of 10-30 pulses to achieve wall plug efficiencies in the range of 10-15 percent and for KrF lasers, one must extract with bursts of several hundred pulses to achieve wall plug efficiencies of 7-10 percent. In addition, the gas flow systems for these lasers are considered and shown to set upper bounds on the laser apertures for a given rep rate. These results indicate that acceleration mechanisms will be favored that can easily employ a large number of "small" lasers.

I. INTRODUCTION

During the past year, Spectra Technology, Inc. (STI), formerly Mathematical Sciences Northwest, has carried out a program for DOE on the evaluation of laser technologies for laser accelerators. The main goal of the program was the evaluation of laser system characteristics set by final electron energy and luminosity requirements. This work entailed the familarization of STI staff with the various accelerator mechanisms (grating linac, beat wave, inverse FEL, and inverse Cherenkov) being proposed; an evaluation of some transport schemes and scaling; and the consideration of laser types that may fulfill the needs of high average power coupled with high wall plug efficiency. In the course of this work, several adjacent technologies such as Raman beam combination, which may be a component of a full accelerator transport system, were reviewed and will be discussed in our final report to DOE.

When this work was initiated, it was thought that a full system conceptual design might be one possible end product. It became clear that the complexity of this problem is enormous and the optimum parameters of such a conceptual design are far from being available for most of the proposed accelerator mechanisms. Consequently, it is virtually impossible to propose end-to-end conceptual designs for any of the accelerator mechanisms. As a result, our work has developed around specific technologies that are likely to play a role in an eventual accelerator design. We begin with a brief discussion of laser systems requirements based on luminosity, followed by an analysis of transport systems for far-field geometries. This paper concludes with a discussion of CO_2 and KrF laser technologies and the conditions under which these technologies can be applied to the laser accelerator endeavor.

II. LASER POWER REQUIREMENTS

In this section we examined the laser system requirements that are driven by the final electron energy and the desired luminosity. Laser particle accelerators must achieve high aceleration gradients (>200 MeV/m) if they are to have an impact on the development of accelerators of the future. However, an equally important factor is the luminosity that can be obtained with a particular design. The luminosity defines the maximum frequency of events that an experimentor may be able to detect for a process of a given cross section.

The luminosity for a beam-beam collider machine may be expressed as

$$\mathcal{L} = \frac{hN^2 f}{4\pi\sigma_{\perp}^{*2}} \qquad [1]$$

where N is the number of electrons per bunch, $\sigma_{\perp}$ is the diameter of the bunch, f is the bunch rep rate, and h is the enhancement factor. The quantities shown with an asterisk are to be taken at the beam-beam intersection point. Other related quantities include the beam emittance

$$\epsilon = \frac{\pi\gamma\sigma_{\perp}^{*2}}{\beta_s} \qquad [2]$$

which relates the betatron function β_s and cross section of the beam bunch; the synchrotron radiation emitted by electrons when their trajectories are deflected during the collision process has been coined as "beamstrahlung."[1]

$$\frac{\Delta\xi}{\xi} \cong \frac{2}{9} r_o^3 \left[\frac{\gamma N^2}{\sigma_{\perp}^{*2} \sigma_{11}}\right] \qquad [3]$$

Here $\Delta\xi/\xi$ is the fractional energy spread introduced in the beam by beamstrahlung. The beam brightness, b, has been found to be nearly constant over a wide range of accelerator parameters;[2,3] therefore, it becomes a useful parameter

$$b = \frac{\langle I_d \rangle}{\epsilon^2} \qquad [4]$$

where $\langle I_d \rangle$ is the average current in the accelerator driver. A final parameter is the disruption, D, which determines the magnitude of beam pinch and the resulting enhancement factor h.[4]

$$D = \frac{r_o \sigma_{11} N}{\gamma \sigma_{\perp}^{*2}} \qquad [5]$$

In our initial work, the beamstrahlung was set equal to ~3 percent and assumed to be a hard limit; during the Malibu meeting, however, it became clear that for ≳1 TeV accelerators and large luminosities, the beamstrahlung condition based on a classical calculation does not apply because of quantum effects, so that the beamstrahlung equation can be removed from consideration. Since this is a recent development, we present results for both cases where the following assumptions are used:

$$b = \text{const} = 10^4 \ \text{amps/cm}^2\text{rad}^2$$

$$\sigma_{\parallel} = \sqrt{3}\ \beta_s$$

$$\langle I_d \rangle = Ne/\tau_d$$

$$\tau_d \cong 10^{-9} \ \text{sec}$$

Setting $b = 10^4$ is just the Lawson-Penner[2] scaling for existing machines. Choosing the beam bunch length $\sigma_{\parallel}$ approximately equal to the betatron function β_s can be shown to lead to a maximum luminosity. τ_d is the period at which electron bunches are delivered to the accelerator. With these assumptions, the beam brightness equation can be expressed as $\epsilon^2 = Ne/b\tau_d$ and then combined with the emittance definition to yield

$$\frac{\sigma_{\perp}^{*2}}{\sigma_{11}} = \frac{1}{\pi\gamma}\left[\frac{Ne}{3b\tau_d}\right]^{1/2} \qquad [6]$$

Case A: With Beamstrahlung Limit Applied

By combining Equation [6] with the definition of beamstrahlung, one obtains results for $\sigma_{\perp}^{*2}$, $\sigma_{\parallel}$, and the luminosity $\mathcal{L}$ in terms of the number of particles per bunch.

$$\sigma_{11} = \frac{2}{3} \gamma N^{3/4} \left[\frac{3\pi^2 b \tau_d r_o^6}{4e\left[\frac{\Delta\xi}{\xi}\right]^2} \right]^{1/4} \equiv K_1 N^{3/4} \gamma$$

$$\left[\sigma_\perp^*\right]^2 = \frac{1}{3} N^{5/4} \left[\frac{4e r_o^6}{3b\tau_d \pi^2 \left[\frac{\Delta\xi}{\xi}\right]^2} \right]^{1/4} \equiv K_2 N^{5/4} \qquad [7]$$

$$\mathscr{L} = \frac{hfN^{3/4}}{8\pi K_2}$$

These results can be expressed in terms of the laser energy per pulse, E_ℓ, and the laser average output power, P_{av}

$$E_\ell = \frac{\gamma(mc^2)N}{\eta_t} \qquad [8]$$

$$P_{av} = E_\ell f$$

Here η_t is the efficiency with which laser energy is transferred to the electrons. These expressions yield the interesting result

$$P_{av} = 16\pi K_2^{4/5} \quad \frac{\gamma mc^2}{h\eta_t} \left[\sigma_\perp^*\right]^{2/5} \qquad [9]$$

which implies that for constant luminosities, the average laser power required is reduced as the beam bunch cross section is lowered. This result is shown in Figure 1 for a 1 TeV accelerator. Notice that for luminosities of 10^{32} cm^{-2} sec^{-1}, the average laser power is several megawatts with beam bunch diameters in the 1 μ range. The laser power requirements are lowered slightly if beam bunch diameters of $\leqslant 10^{-2}$ μ can be achieved. The laser energy per pulse is related to the rep rate f and is found to be a few kilojoules at a kilohertz.

Case B: Without Beamstrahlung Limit Applied

It became clear at the Malibu meeting that application of the beamstrahlung limit may not be appropriate. Instead, we apply the beam pinch condition that maximizes the luminosity when the beam disruption D is of order 2-10. However, both Equation [6] and the definition of disruption gives the ratio $\sigma_\perp^{*\,2}/\sigma_\parallel$ so that the beam bunch values $\sigma_\perp^*$ and $\sigma_\parallel$ cannot be found directly. Instead, we combine these two expressions to find the number of electrons per bunch as a function of D

$$N = K_3 D^2 \qquad [10]$$

where

$$K_3 = \frac{e}{3b\tau_d \pi^2 r_o^2}$$

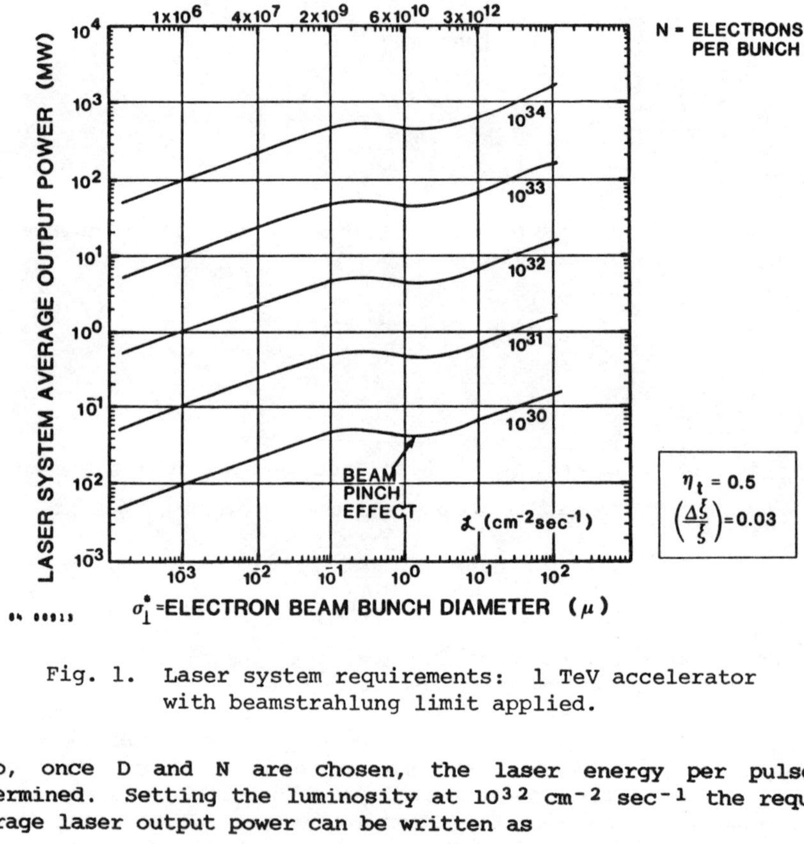

Fig. 1. Laser system requirements: 1 TeV accelerator with beamstrahlung limit applied.

Also, once D and N are chosen, the laser energy per pulse is determined. Setting the luminosity at 10^{32} cm^{-2} sec^{-1} the required average laser output power can be written as

$$P_{av} = \frac{1.3 \times 10^{12}\, \sigma_{\perp}^{*\,2} \text{ (microns)}}{E_{\ell}(J)h}, \qquad [11]$$

where $b = 10^4$, $\eta_t = 0.5$, and $\tau_d = 10^{-9}$. A plot of this function is presented in Figure 2, where lines of constant laser energy per pulse and laser rep rate are illustrated.

It is clear that lowering the beam bunch cross section remains an important goal in terms of minimizing the laser average output power. In fact, it has a much stronger effect than in Case A where beamstrahlung was taken as a limiting quantity. Nevertheless, we

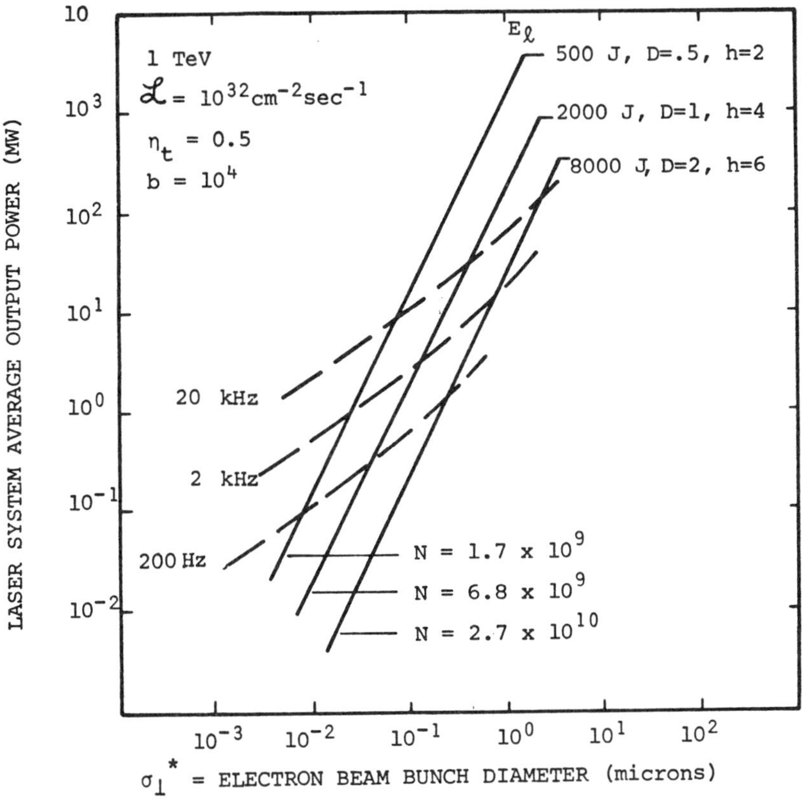

85 09222 Fig. 2. Laser System Requirements: 1 TeV accelerator without beamstrahlung limit applied.

note that laser average power levels of megawatt proportions will be required for 1 TeV accelerators operating at luminosities of $\sim 10^{32}$ cm^{-2} sec^{-1}. Further, this level of power can easily increase by an order of magnitude, depending on the actual coupling or transfer of laser energy to the electrons. In Figures 1 and 2, we have assumed a 50 percent transfer, but this could easily be much lower depending on the details of the accelerating mechanism and the transport system employed.

Comparison of Average Laser Power Requirements

It is important to understand the difference between the predictions of Figures 1 and 2. This can best be understood by picking an electron number per bunch which then sets the laser energy. For this example, we choose $N \approx 6 \times 10^9$ and E_ℓ becomes ≈ 2 kJ. In Figure 1, this condition corresponds to a single point at a luminosity of 10^{32} cm^{-2} sec^{-1}. Here the beam bunch diameter is ~10^{-1} μ, the required laser average power is ~5 MW, and the rep rate is 2.5 kHz. On the other hand, Figure 2 is not so constraining. At this electron number per bunch and laser energy, there is a whole range of beam bunch diameters and laser average power levels available resulting from the removal of the beamstrahlung limit. In fact, by lowering $\sigma_{\perp}^{*}$ to 10^{-2}, the required laser average power drops to ~2×10^{-2} MW and the laser rep rate is only ~10 Hz. Just how much advantage one can achieve depends on the minimum size of $\sigma_{\perp}^{*}$ achievable and on a correct formulation of beamstrahlung radiation, which is bound to become important again once $\sigma_{\perp}^{*}$ is made very small.

Throughout this discussion, the pulse length of the laser has not been mentioned, but the electron bunch length $\sigma_{\parallel}$ will be specified by the luminosity conditions. The laser pulse length can be longer than $\sigma_{\parallel}$ as long as there is slip between the photon pulse and the electron bunch so that the entire energy per pulse can be used or transferred to the electrons. However, the pulse length will have an upper bound set by the acceleration gradients desired and the nature of the transport system, as outlined in the next section.

III. ACCELERATION GRADIENTS AND OPTICAL TRANSPORT

An attempt has been made to provide a framework within which accelerator staging can be examined without reference to specific accelerator mechanism details. Both far- and near-field geometries were considered in this work; here only the far-field scaling is presented, but the near-field will be discussed in the DOE program final report. In order to remove the details of specific accelerator mechanisms, they are modeled using a coupling strength, g_c, that relates the local acceleration gradient to the optical power density

$$G_{local}\,(\text{GeV/met}) = g_c K\left[P_{av}(\text{W/cm}^2)\right]^{1/2} \qquad [12]$$

where K is a constant equal to 2.75×10^{-6}. When $g_c = 1$, the local acceleration gradient equals the vacuum electric field of the focused laser beam. The various accelerator mechanisms tend to divide into two categories:

(1) Strong Coupling Strengths ($g_c \cong 1$)

Beat Wave
Grating Linac

(2) Weak Coupling Strengths ($g_c \cong 10^{-3}$)

Inverse Free-Electron Laser
Inverse Cherenkov

The average acceleration gradient for an accelerator will always be less than the local gradient because of the need to expand beams prior to reflection from optical elements.

Far-Field Geometries

Figure 3 shows a generic far-field geometry constructed from N_s stages. We assume each stage is driven with $1/N_s$ of the required laser energy for acceleration, E_ℓ. This is the simplest arrangement; a more complicated scheme transfers unused laser energy from one stage to the next but does not change the basic results of this discussion. Within each stage, there can be one or more refocusings; we assume these refocusings can be described according to ideal Gaussian beam propagation. The length of each refocusing section is L_R and the Rayleigh range associated with each refocusing is Z_R. Significant acceleration is taken to occur only over twice the Rayleigh range so that the average acceleration gradient becomes

$$G_{av} = G_{local} \left[\frac{2Z_R}{L_R}\right] \qquad [13]$$

The equations that describe the Gaussian[5] beam propagation are

$$\omega^2 = \omega_o^2 \left[1 + \left[\frac{L_R}{2Z_R}\right]^2\right]^{1/2}$$

$$Z_R = \frac{\pi\omega_o^2}{\lambda} \qquad [14]$$

$$L_R = 2Z_R \left[\frac{\omega^2 - \omega_o^2}{\omega_o^2}\right]^{1/2}$$

where 2ω is the beam size at the refocusing optical elements.

The local gradient at each refocusing can now be expressed as

$$G_{local} = g_c K \left[\frac{E_s}{\pi\omega_o^2 \tau_p}\right]^{1/2} \qquad [15]$$

where $2\,\omega_o$ is the minimum beam size, E_s is the laser energy in each

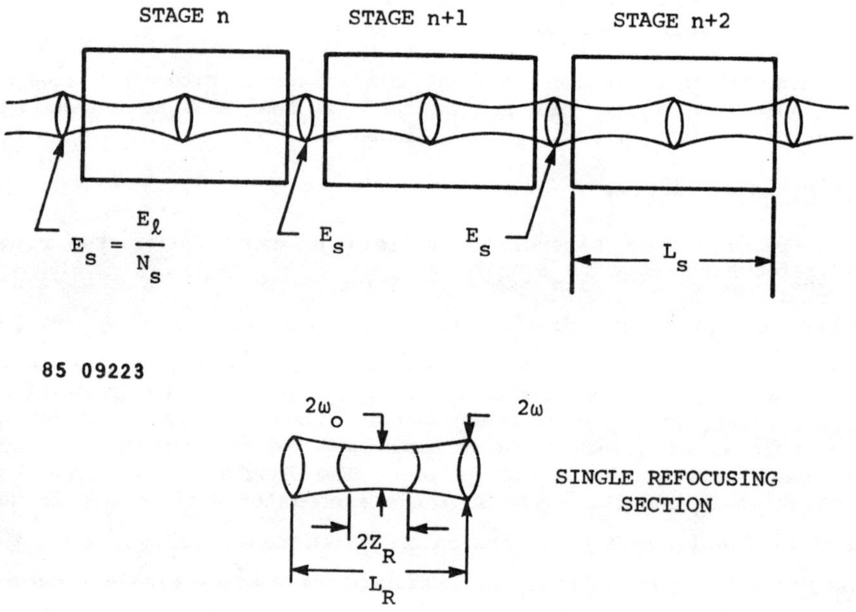

Fig. 3. Far field transport geometry.

stage, and τ_p is the laser pulse length. The ratio 2 Z_R/L_R goes to (ω_o/ω) when $\omega_o << \omega$. Under this condition, the average acceleration gradient becomes

$$G_{av} = g_c K \left[\frac{E_s}{\pi\omega^2\tau_p}\right]^{1/2} \qquad [16]$$

which is independent of ω_o and only depends on the flux at the refocusing elements. It is useful to compute the power density at the refocusing elements for an accelerator with an average acceleration gradient of 1 GeV/m and different coupling strengths. Typical values are shown below.

$G_{av} = 1$ GeV/m

g_c	**Power Density at Refocusing Elements (W/cm²)**
1	1.3×10^{11}
10^{-1}	1.3×10^{13}
10^{-2}	1.3×10^{15}
10^{-3}	1.3×10^{17}
10^{-4}	1.3×10^{19}

The refocusing elements must operate at glancing angles of incidence to handle these flux levels unless the coupling strength is $\geqslant 1$.

<u>Exact Solution</u>

Combining the Gaussian beam propagation equation without making approximations yields

$$\omega_o^2 = \frac{\omega^2}{2}\left\{1 \pm \left[1 - \left[\frac{L_R\lambda}{\pi\omega^2}\right]^2\right]^{1/2}\right\} \qquad [17]$$

and

$$G_{av}(\text{GeV/Met}) = g_c k\left[\frac{\ell_{int}}{2Z_R}\right]\left[\frac{E_s}{\pi\omega^2\tau_p}\right]^{1/2} \cdot \qquad [18]$$

$$\cdot \left[\frac{2}{1 \mp \left[1 - \left[\frac{L_R\lambda}{\pi\omega^2}\right]^2\right]^{1/2}}\right]^{1/2}$$

The factor $(\ell_{int}/2\,Z_R)$ is included because, for some cases, the acceleration region is less than twice the Rayleigh range; this occurs when the refocusing optics are inside twice the Rayleigh range. Table I shows the three possible cases that result from these equations. Notice that the average acceleration gradient is the same for all three to within a factor of $\sqrt{2}$. The first case is the least interesting because the refocusing elements are placed inside twice the Rayleigh range and the length between refocusing elements is small ($L_R << \pi\omega^2/\lambda$), leading to large numbers of elements for a fixed accelerator length. Case 2 has the largest acceleration gradient by the factor $\sqrt{2}$ and also the largest distance between refocusing elements ($L_R = \pi\omega^2/\lambda$). Case 3 could be useful because it allows the photon beam to be tightly focused and then expanded rapidly. The short Rayleigh range and high intensity focus could be useful for some acceleration mechanisms. However, this flexibility is paid for in very short distances between refocusing elements ($L_R << \pi\omega^2/\lambda$).

Therefore, for the present discussion, we concentrate on Case 2 and estimate the number N_R of refocusing elements required. N_R is given by

$$N_R = \frac{L_{acc}}{L_R} = \frac{E_f(\text{GeV})}{L_R\ G_{av}(\text{GeV/met})} \quad [19]$$

For the Case 2 transport, one substitutes for L_R, G_{av}(GeV/m) and uses $E_s = E_\ell/N_s$ to find

$$\frac{N_R}{N_s} = \frac{5 \times 10^{-3}\ \lambda(\mu)\ \tau_p(\text{sec})\ E_f(\text{GeV})\ G_{av}(\text{GeV/met})}{E_\ell(\text{J})\ \left[Kg_c\right]^2} \quad [20]$$

where $N_R \geqslant N_s$.

Table I

POSSIBLE SOLUTIONS TO THE GAUSSIAN BEAM TRANSPORT GEOMETRY

CASE	CONDITIONS	GEOMETRY	AVERAGE ACCELERATION GRADIENT NORMALIZED TO: $Kg_c\left[\frac{E_s}{\pi\omega^2\tau_p}\right]^{1/2}$
1	$L_R << \frac{\pi\omega^2}{\lambda}$; +, − ; $\omega_o \rightarrow \omega$, $L_R << 2Z_R$, $\ell_{int} = L_R$	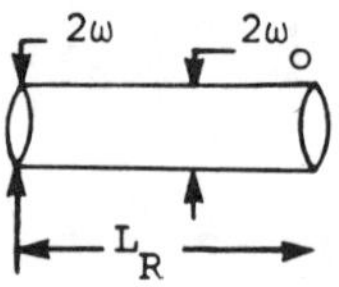	1
2	$L_R = \frac{\pi\omega^2}{\lambda}$; $\omega_o = \omega/\sqrt{2}$, $\frac{L_R}{2} = \frac{\pi\omega_o^2}{\lambda} = Z_R$, $\ell_{int} = 2Z_R$	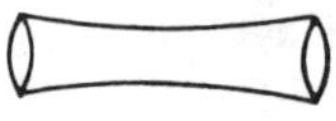	$\sqrt{2}$
3	$L_R << \frac{\pi\omega^2}{\lambda}$; −, + ; $\omega_o = \frac{\omega}{2}\left[\frac{L_R}{\frac{\pi\omega^2}{\lambda}}\right]$, $L_R >> 2Z_R$, $\ell_{int} = 2Z_R$		1

These results are plotted in Figure 4, where the number of refocusing elements and the diameter of the refocusing elements are

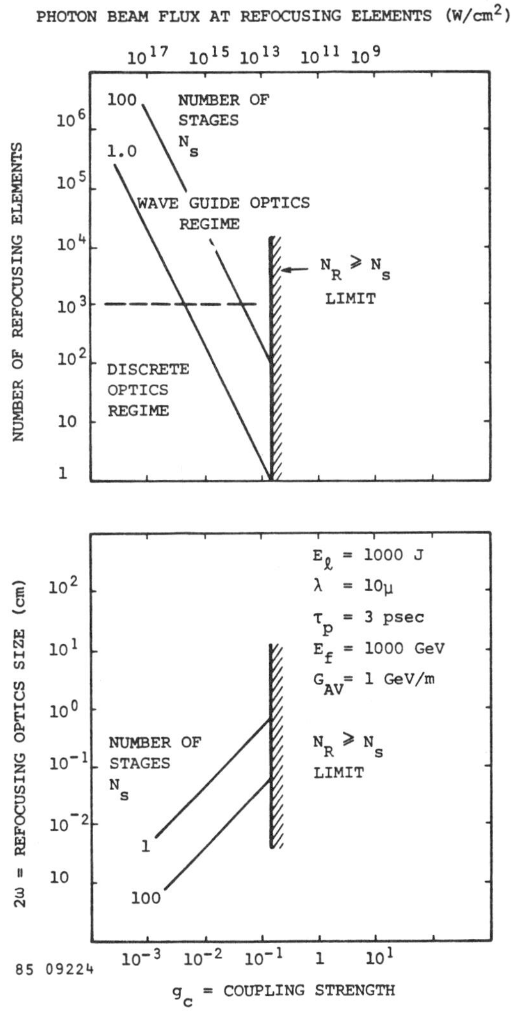

Fig. 4. Required number of refocusing elements vs. coupling strength.

shown versus the coupling strength. For these plots, we have arbitrarily taken the laser energy, E_{ℓ}, to be 1000 J and the pulse length as 3 psec. The actual values would need to be chosen based on the luminosity requirements and specific details of the mechanism employed. The limit on the right occurs because we demand $N_R \geqslant N_s$. It does not mean that coupling strengths above 10^{-1} are not accessible but, instead, that coupling strengths above 10^{-1} are not needed to achieve the assumed acceleration gradient of 1 GeV/m.

These plots also illustrate the regime of waveguide optics and discrete optics based on the number of refocusing elements. Note also that the price one pays for a larger number of stages (a large number of small lasers, each delivering energy E_s such that $N_s E_s = E_\ell$) is an increased number of refocusing elements. At the top of each plot, the normal incidence flux in each stage is indicated. For all but the highest coupling strengths, the optical elements would have to be used at glancing angles of incidence to avoid damage.

Table II lists these results for an accelerator design using both high and low coupling strengths to achieve 1 GeV/m gradients. The number of stages equals the number of laser units required to deliver the full 1 kJ. The parameter E_ℓ^1 is the individual laser unit output energy per pulse. In the next section, it will be shown that the requirements of high efficiency, when gas lasers are employed, will force the use of at least 10-100 stages of individual laser units.

IV. EVALUATION OF LASER TECHNOLOGY

From the previous two sections, it is apparent that total laser output power levels for laser accelerators will be in the megawatt range and that rep rates are likely to fall between 10^2 to 10^4 Hz. Additionally, the individual pulse lengths must be short and probably must be less than 10 psec if acceptable acceleration gradients are to be achieved. Other factors specific to the proposed accelerator mechanisms, such as plasma formation at droplets and plasma instabilities, also drive the pulse length into the picosecond range.

Besides the requirements of high average power, high rep rates, and short pulse length, the laser system must be efficient in order to minimize wall plug power and capital costs. These general constraints force one to consider both CO_2 and KrF lasers as strong candidates for the laser accelerator application. Both of these laser systems have demonstrated relatively high efficiencies, generation of picosecond pulses has been accomplished, and the gas medium lends itself to rep rate operation. Both solid state and free-electron lasers seem less desirable at this time. Solid state lasers currently have pulse repetition limits set by heating effects, efficiency limits set by the current flash lamp pumping technology, and maximum energy fluence limits set by the onset of non-linear effects. While all of these limitations are currently receiving study and improvements are likely, we have temporarily set solid state laser systems aside in favor of KrF and CO_2. The free-electron laser has very high efficiency potential, produces short pulses naturally, and will deliver high average output power levels. However, the pulse format of the free-electron laser is difficult to interface with the requirements of a laser accelerator in terms of rep rate and energy per pulse. Here again, we have chosen to set aside the free-electron laser in favor of the gas laser options of KrF and CO_2.

TABLE II

Far-Field Accelerator Parameters
for Low and High Mechanism Coupling Strengths

$\mathcal{L} = 10^{32}\ cm^{-2}\ sec^{-1}$

$E_f = 1$ TeV

$\lambda = 10\ \mu$

$\tau_p = 3$ psec

$E_\ell = 1$ KJ, $f = 1$ kHz

$\eta_t = 0.5$

$\sigma_\perp^* = 6 \times 10^{-2}\ \mu$

N_s		N_R	E_ℓ^1(J)	$2\,\omega$(cm)
$g_c=10^{-1}$	1	2	1000	5
	10	20	100	1.6
	100	200	10	0.5
$g_c=10^{-3}$	1	2×10^4	1000	5×10^{-2}
	10	2×10^5	100	1.6×10^{-2}
	100	2×10^6	10	5×10^{-3}

N_s = number of stages = number of individual laser units

E_ℓ^1 = individual laser unit output energy per pulse

$2\,\omega$ = optical beam diameter at refocusing elements

During the Malibu workshop, excellent reviews of excimer, CO_2, solid state, and free-electron lasers were presented. As a result, there is no need to review these laser technologies in any detail here. Instead, we have chosen to examine the limitations imposed on the use of CO_2 and KrF lasers by the rep rate and efficiency requirements. These limitations result from the necessary use of gas flow systems and specific pulse format arrangements that result in high efficiency operation.

Limits to High Average Power Gas Lasers

Figure 5 illustrates two flow loop geometries currently used with rep rated gas laser systems. The first is a "conventional" flow loop and the second is called a "compact" flow loop system. Both of these flow loop geometries are currently used at STI. For example,

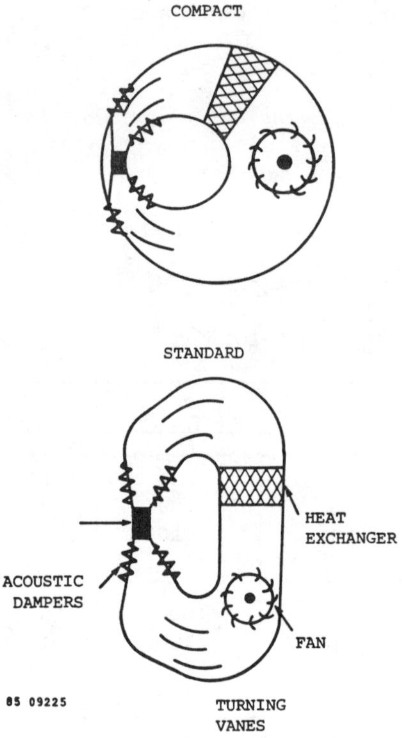

Fig. 5. Typical Gas Flow Loop Geometries

the 200 W excimer Mistral laser operates at 1 kHz and uses the conventional flow arrangement; the Windvan 100 W CO_2 laser built for NOAA operates at 50 Hz and uses a compact flow loop design. Both of these flow arrangements have the purpose of removing waste heat from the gas and providing an optically uniform gas medium for each laser pulse. To achieve these goals, each flow loop system must include acoustic damping materials to attenuate acoustic waves; heat exchangers to remove unwanted heat; driving fans or turning vanes; and, most important, very detailed and careful engineering design and analysis to assure efficient operation while maintaining acceptable medium homogeneity. Just to indicate the magnitude of the problem, we note that the medium density fluctuations must be less than

$$\left[\frac{\Delta\rho}{\rho}\right] \leqslant \frac{2 \times 10^{-3}}{P(atm)} \qquad [21]$$

to assure wave front distortions of less than ~0.1 wave. The flow loop must maintain this type of fractional density uniformity on each laser shot.

Flow Loop Power Requirements

Besides maintaining the medium homogeneity, the flow loop must remove waste heat and use as little power as possible if the laser wall plug efficiency is not to be seriously degraded. This condition will set an upper limit on the product of rep rate and laser aperture. Figure 6 shows the general geometry. The power required by the flow loop will be proportional to the round-trip pressure drop ΔP and the volume flow rate Γ.

$$P_{flow} = \frac{1}{\eta_{fan}} (\Delta P)\, \Gamma \qquad [22]$$

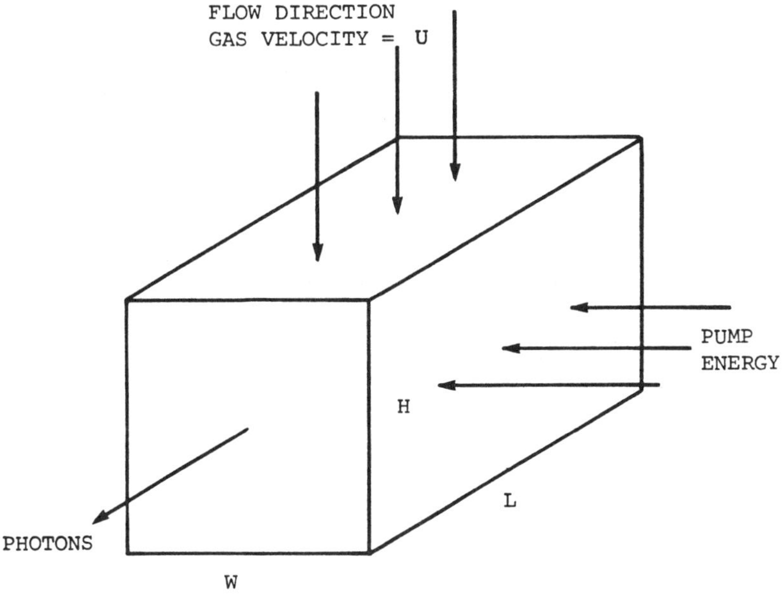

Fig. 6. Laser flow loop geometry.

85 09226

Here η_{fan} is the electrical efficiency of the driving fan. The flow loop pressure drop can be conveniently related to the cavity dynamic head pressure drop through a constant, K.

$$\Delta P = K\left[\frac{1}{2}\rho U^2\right] \qquad [23]$$

where U is the gas flow velocity in the laser cavity. The volume flow rate Γ is UWL. The laser gas must be completely removed from the laser cavity before each successive shot; in fact, one generally must remove more than the cavity volume before each successive shot in order to maintain the desired homogeneity. The number of cavity volumes removed is normally called the clearing factor C_f and is simply related to the rep rate and laser cavity dimension

$$U = HfC_f \qquad [24]$$

One can now express the flow loop power in terms of the cavity dimension, the flow speed, the clearing factor, and the constant K. However, a more useful quantity is the ratio of the flow loop power to the power deposited in the active medium. This ratio is defined as g_f and can be expressed as

$$g_f = 6 \times 10^{-10}\,K\,\frac{(Mwt)\,P(atm)\,H^2(cm)\,f^2(Hz)\,C_f^3}{\eta_{fan}\;T(°K)\,Pump(J/cc)} \qquad [25]$$

where pump is the energy density supplied to the active medium and Mwt is the gas molecular weight in grams. This ratio is useful because it can be simply related to the laser wall plug efficiency. If g_f were equal to unity, the wall plug efficiency would drop by approximately a factor of two from the case where no flow loop power was required. Consequently, we generally require the ratio g_f to be less than ~0.25 for reasonably efficient operation. The fan efficiency is a big question, but one can generally assume 25 percent is obtainable.

Typical values for a KrF laser are

Pump (J/cc)	≈0.1	(100 J/ℓ)
T (°K)	≈273	
Mwt	≈80	(Rich Kr mixture)
P (atm)	≈1.0	

Under these conditions, one finds

$$H^2(cm)\,f^2(Hz)\,C_f^3 K \leqslant 3.6 \times 10^7 \qquad [26]$$

Clearly, the clearing factor plays a strong role in setting this limit. Typically, the clearing factor is ~2.5 for self-sustained gas discharges and somewhat smaller, say 1.5, for e-beam sustained discharges. The factor K can be held to ~1.5 for very well designed flow systems; this means the pressure drop around the loop can be held to 1.5 times the cavity dynamic head pressure. With these

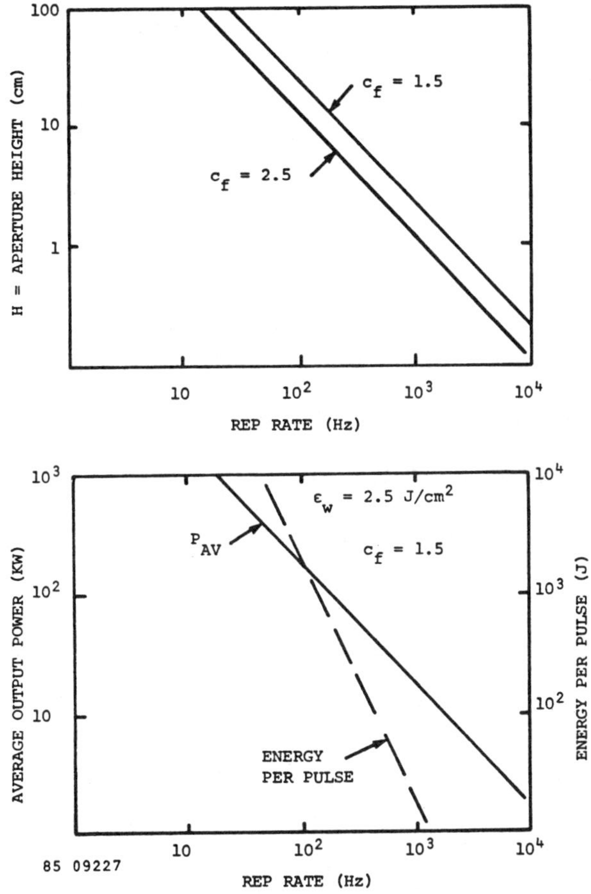

Fig. 7. Laser aperture rep rate trade-off based on flow loop power consideration. These plots were constructed for KrF lasers.

numbers, Figure 7 has been constructed. The first plot shows the laser aperture height versus laser rep rate; notice that at 1.0 kHz the aperture height is only a few cm, and at 10–20 Hz the aperture may be ~50 cm. One can convert this aperture height into average power and energy per pulse by assuming the laser window fluence must be less than some value and that the aperture width is approximately equal to its height, $W \cong H$. For KrF, the window fluence is taken as 2.5 J/cm^2 and leads to the second plot in Figure 7. This window fluence is for a long pulse (microseconds); in reality, the output from the laser will occur in a large number of psec-wide pulses, but the burst of picosecond pulses will occur over ≈500–1000 μsec. It is the sum of energy in each burst of picosecond pulses that must be held below ~2.5 J/cm^2. This topic is discussed later in this section. The message of this plot is that when rep rates of 1.0 kHz are used, the average output power per laser is low (~10 kW) compared to the 1-10 MW needed for the entire accelerator. As a consequence, a large number of such lasers (~500) will be needed to drive a 1 TeV accelerator structure based on these simple flow loop power considerations.

Accelerator mechanisms that can easily interface with a large number of rep rated lasers will be favored by these results. In general, this will include mechanisms that couple strongly to the laser electric field (beat wave and grating linac); otherwise, acceptable acceleration gradients will not be achievable.

Specific Laser Systems

As mentioned earlier, only the KrF and CO_2 gas lasers are considered in this paper because of their demonstrated high efficiency and rep rate capability. In this section, the efficiency and energy extraction conditions required by these lasers are presented. Detailed information about the specific molecular and atomic kinetics will not be presented, as several excellent tutorial[6,7] talks on these lasers were given at the workshop and additional reviews will be presented in our report to DOE.

KrF: A Non-Storage Laser

KrF is considered a non-storage laser because the upper state lifetime is quite short (≈5 nsec) compared to the required pumping time of 100-1000 nsec. Consequently, to achieve efficient operation, extraction must occur throughout the pumping period. The formation efficiency will depend on the detailed excimer kinetics and pumping technique but, with KrF, this efficiency is found to be ~25-32 percent for Kr-rich mixtures. However, the efficiency with which this stored power density can be extracted is not unity for excimers; this is because of non-saturable absorption processes that are characteristic of this laser type.

For excimer lasers, the extraction efficiency may reach ~60 percent if the gain-to-loss ratio is maintained above ~10. Therefore, intrinsic efficiencies (product of extraction and formation) near 15-20 percent are possible. If pulse power efficiencies are held to ~60 percent, then wall plug efficiencies near 9-12 percent will be obtainable before power is allocated to auxillary equipment and the flow loop. Including estimates for these components leads to wall plug efficiencies goals of order 7-10 percent.

Short Pulse Extraction

KrF does, in fact, store energy over a time that is short compared to the upper state lifetime. Therefore, a train of short picosecond pulses spaced a few nanoseconds apart and continuing over the length of the pump time (100-500 nsecs) may extract the full laser energy with nearly the extraction efficiency of a single long pulse. In this case, the extractable stored energy density is given by $g_o E_{sat}$ where E_{sat} is the medium saturation fluence expressed as

$E_{sat} = h\nu/\sigma$.

Figure 8 shows the suggested arrangement for a KrF laser that is extracted with a burst of picosecond pulses spaced by the interval

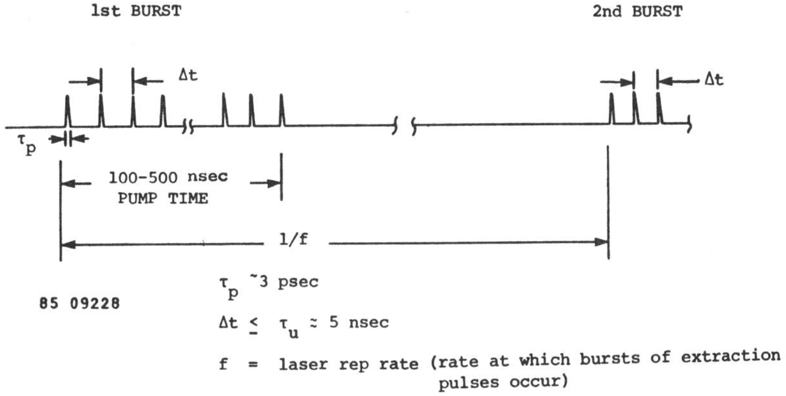

Fig. 8. KrF Pulse Extraction Format

Δt. This burst is repeated at the laser rep rate f and contains a number of pulses given by (pump time)/Δt. The number of pulses are likely to be several hundred to maintain high efficiency. For KrF, the saturation fluence is ~2.5 mJ/cm^2, so that the output fluence from the amplifier is on the order of 7.5 mJ/cm^2 or 2.5 x 10^9 W/cm^2 for a 3 psec pulse length. This flux level is probably OK in terms of non-linear problems in the windows. Assuming that a few hundred of these pulses are used, the total burst window fluence becomes 200 x 7.5 mJ/cm^2 = 1.5 J/cm^2, which is below the long pulse window damage limit of 2-3 J/cm^2. We define this multi-pulse extraction scheme as short pulse temporal multiplexing.

There are a number of uncertainties in these predictions. First, just how many picosecond pulses are really needed (what is their spacing) to achieve acceptable efficiency? Second, are there kinetic processes that become important or unimportant when extraction pulses of a few picoseconds are used (for example, the upper state manifold vibrational relaxation rates)? Third, what effect does ASE[8] have on the extraction efficiency when temporal multiplexing is exploited? The fact that the flow loop will restrict the aperture and force a long aspect ratio (length-to-width) on the individual laser units will help mediate this possible problem with ASE.

CO_2: A Storage Laser

The CO_2 laser has a distinct advantage over the KrF system because it is an excellent storage medium. One might expect that the medium could be pumped and then fully extracted with a single pulse instead of the several hundred required by KrF. However, this is only true if the extraction pulse is very long (1-10 microseconds). Charlie Fenstermacher[7] presented an excellent review of CO_2 laser kinetics at the workshop and pointed out that to make the CO_2 laser efficient, a large fraction of the gas mix must be nitrogen. This is because the efficiency of electrically pumping energy into nitrogen and then transferring this energy to CO_2 through collisions is much higher than electrically pumping the CO_2 molecule directly. Unfortunately, the transfer time between nitrogen and CO_2 is roughly 250 nanosecond-atm and 5-10 times as much energy is stored in the nitrogen as compared to the CO_2. Therefore, to extract all of the energy stored in the medium, one must use long extraction pulses or a burst of sequentially spaced short pulses. Here the spacing between the sequential pulses is set by the transfer time between nitrogen and CO_2.

To complicate matters, each vibrational level is subdivided into a number of closely spaced rotational levels that share the stored energy in CO_2. The relaxation time (the time it takes to transfer energy from one rotational level to the others) is of order 0.2 nsec-atm. This means that if we extract energy from a CO_2 laser amplifier with a sub-nanosecond pulse that has a bandwidth covering a single rotational line, then only energy stored in that rotational level will be extracted. Since there are about 16 participating rotational levels, only ~1/16 of the energy stored in CO_2 will be removed on any given pulse.

The need for picosecond pulses solves this problem automatically. The spectral bandwidth of a 3 psec pulse encompasses the entire rotational band structure so that all rotational lines will be extracted simultaneously. However, because the laser gain width is not uniform but has a structure made up of ~16 peaks, the 3 psec pulse will break up into a number of short pulses. It turns out that the rotational lines are pressure broadened so that raising the pressure above 1 atm tends to smooth the gain spectrum. Corkum[9] has demonstrated that at 10 atm, the rotational lines are broadened sufficiently such that adjacent lines begin to overlap and a fairly uniform gain spectrum extends over the full rotational bandwidth. Under these conditions, 3 psec pulse amplification is possible.

Energy Extraction

Measurements at NRC[9] show the saturation fluence for 3 psec extraction pulses in a 10 atm CO_2 mixture to be 200-600 mJ/cm². This means that with a single 3 psec pulse, one can remove nearly all the energy stored in CO_2 with an extraction fluence of ~1.5 J/cm². In 3 psec, this is a flux of 5×10^{11} W/cm². Measurements indicate that NaCl windows will withstand only 0.5 J/cm² in 3 psec so that the output beams must be expanded to avoid damage. An alternative would be to use a lower level of extraction flux, but this would lead to a lower extraction efficiency on each pulse.

CO_2 Laser Efficiency

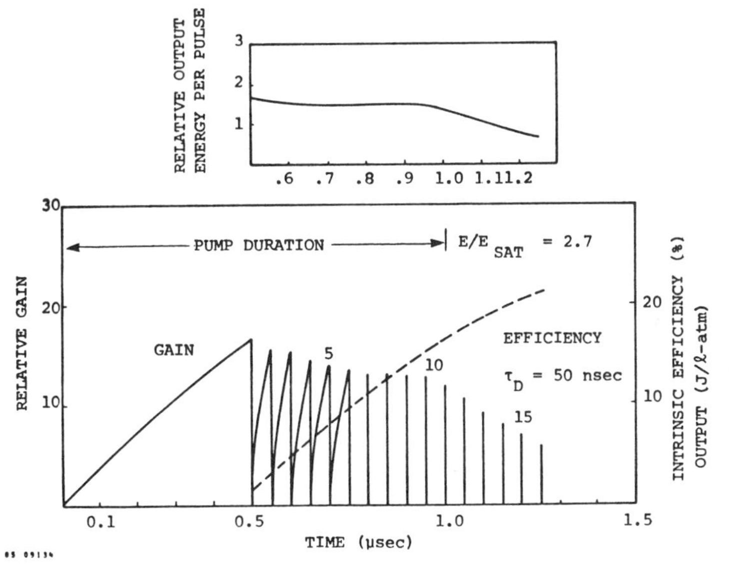

Fig. 9. CO_2 kinetics code calculation illustrating the transfer of energy from N_2 to CO_2 in a 10 atm 3/2/1 ($He/N_2/CO_2$) mixture.

Figure 9 shows the results of a CO_2 kinetic code calculation completed for BNL by Spectra Technology, Inc. This calculation illustrates how energy is extracted from the medium on a number of successive short pulses spaced ~50 nsec apart. The figure shows the gain as a function of time and the sum of the extracted energy after each extraction pulse. Also shown is the laser amplifier intrinsic efficiency as a function of the number of extraction pulses. Notice that 20 percent intrinsic efficiencies are obtained after ~15 pulses. To construct this plot, the extraction efficiency was essentially unity; this is possible in CO_2 because there is no non-saturable absorption, as there is with KrF, but requires that the extraction fluence be several saturation fluence levels. If the extraction fluence is lower, then some fraction of the energy stored in CO_2 is left behind but can be removed on a subsequent pulse. However, a larger number of extraction pulses are required to achieve the same intrinsic efficiency. This effect is shown in Table III where the number of extraction pulses are listed versus desired intrinsic efficiency and extraction fluence level.

Table III

NUMBER OF EXTRACTION PULSES REQUIRED

$g_oL = 2.5$

Multiple Pulse Intrinsic Efficiency	E_{OUT}/E_{sat} 1.0	1.75	1.5	∞
10%	20	12	9	7
15%	30	18	14	10
20%	40	24	19	14
Extraction Efficiency/Pulse	35%	57%	75%	100%

The pulse extraction arrangement for a CO_2 is similar to that for KrF. Here a burst of picosecond pulses spaced 50 nsecs apart is employed. This burst is repeated at the laser rep rate f and contains 15–30 pulses. The CO_2 saturation fluence is ~0.5 J/cm^2 so that the output fluence per pulse is ~1.5 J/cm^2. The energy fluence in the burst is just the number of pulses times the extraction fluence or 20–50 J/cm^2. Probably 20 J/cm^2 is the maximum the windows will tolerate over a few microseconds, so the beam may need to be expanded before exiting through the window.

As with the KrF laser, there are a number of uncertainties about these predictions for CO_2. First, the number of extraction pulses that are actually required must be more accurately determined, and demonstration experiments should be carried out to show high efficiencies are possible. Also, more measurements with picosecond pulses need to be carried out in order to determine saturation fluence levels for 10 atm operation using mixtures of interest. Second, there may be kinetic processes that are important to understand on the 3 psec timescale and that impact the overall design. For example, energy stored in CO_2 levels above the lasing state will not be available during the extraction pulse.

Interface Between Laser System and Accelerator Structure

The interface between the laser system and the accelerator structure will be a strong function of the accelerator mechanism chosen. In any case, the multiple output pulse format from the laser must be matched to the number of stages (near or far field) and the electron bunch temporal format. The simplest configuration would be one where the temporal spacing of photon pulses just matches the electron bunch spacing; in addition, the energy in each photon pulse should be completely transferred to the associated electron bunch after traversing each transport stage. This simplest arrangement may not be feasible when the conditions such as phase matching, energy transfer to electrons, available electron bunch temporal spacing, pulse length and slip requirements, as well as optical beam quality requirements are applied to each mechanism. As a result, angular pulse multiplexing may be necessary, and this greatly complicates the interface geometry. The interface question becomes even more difficult if laser energy from one stage must be carried along into the next stage because it is not fully transferred to the electrons. In this case, photons from one stage must be combined coherently with photons supplied by the next laser unit. Clearly, the interface problem must receive much of our attention in future work and will require different solutions for each of the various accelerator mechanisms.

V. SUMMARY

The accelerator luminosity requirements were shown to set minimum bounds on the laser system average output power of 1-10 MW for a 1 TeV accelerator. Rep rates are in the kilohertz regime and impose limits on the individual laser apertures through consideration of the power required to drive the gas flow systems (KrF and CO_2 lasers). Individual lasers operating at these rep rates are likely to have apertures in the <2-5 cm range. This constraint merged with maximum window fluence levels of 2-5 J/cm^2 yields upper limits on the average individual laser output power of ~10-100 kW. As this is far below the necessary 1-10 MW for a 1 TeV accelerator, one must use a large number of laser units (>100) to drive the accelerator structure. Therefore, mechanisms that can interface easily with a large number of lasers will be favored.

Transport systems for far-field geometries were studied briefly. The results show that accelerator mechanisms with low coupling strength require waveguide-like optical systems in order to maintain acceleration gradients near 1 GeV/m. For high coupling strength mechanisms, the transport system can be constructed from discrete optical components (1-100 refocusing stages). In all cases, the optical components must be used in a glancing incidence configuration. Another interesting feature occurs when we try to break the accelerator into a number of stages, with each stage being driven by a separate laser. In this case, the total laser energy required per electron bunch is the sum of the energy injected into each stage, i.e., $E_\ell = E_s N_s$, but we find the number of refocusing elements increases linearly with the number of stages employed.

Finally, the operating conditions and pulse format for efficient KrF and CO_2 lasers was presented. It was estimated that the wall plug efficiency goals for these lasers should be 7-10 percent for KrF and 9-14 percent for CO_2. A number of design verification tests need to be carried out to demonstrate or show this feasibility. An advantage of CO_2 over KrF is higher efficiency and a lower number of extraction pulses (10–20 for CO_2 vs. 100–500 for KrF). This could be a significant advantage if the pulses must be angularly multiplexed as well as time multiplexed.

A major stumbling block during this work and during the workshop was finding a correct interface between the laser output pulse format and the acceleration mechanisms. Solving this problem is a formidable task because it requires an optimum merging of all the discussed technologies including the laser, its multiplexing system, the transport system, and the luminosity requirements. In addition, there are specific problems concerning the efficient transfer of laser energy to the electrons, the maintaining of required laser beam quality, and the possible need for nonlinear processes to combine beams and to improve beam quality. To attack this question, one must choose a specific accelerator mechanism and work through a complete self-consistent conceptual design that will undoubtedly require many iterations. But before this can occur, there are many details of the various mechanisms that must be more fully understood.

For the near-term, it seems more worthwhile for the laser community to concentrate on individual technology issues such as laser extraction, wall plug efficiency, short pulse production and amplification, as well as possible beam clean-up and beam combination processes that may become essential components of a final design.

REFERENCES

1. H. Wiedemann, "Linear Collider", SLAC-PUB-2849, November 1981.

2. J.D. Lawson, *The Physics of Charged Particle Beams*, (Clarendon, Oxford, 1978) p. 178.

3. G. Gattoli, T. Letardi, J.M.J. Madey, and A. Renieri, IEEE J. of Quantum Electronics QE-20, 637 (1984).

4. R. Hollebeck, NIM 184, 1981, p. 333.

5. A.E. Siegman, *An Introduction to Lasers and Masers*, (McGraw Hill, New York, 1971) p. 293.

6. For other published Reviews see: J.J. Ewing, "Rare-gas Halide Lasers", Physics Today 32 (May 1978); J.J. Ewing, "Excimer Lasers", *Laser Handbook*, (North-Holland Publishing Company, 1979).

7. For other published Reviews see: C. Fenstermacher, "Laser Systems for High Peak-Power Applications", SPIE Proceedings 61, (1975); B.J. Feldman, "Short-Pulse Multiline and Multiband Energy Extraction in High-Pressure CO_2 Laser Amplifiers," IEEE J. Quantum Electronics, QE-9, 1070 (1973); G.T. Schappert, "Totational Relaxation Effects in Short-Pulse CO_2 Amplifiers," Appl. Phys. Lett. 23, 319 (1973).

8. D.D. Lowenthal, J.J. Ewing, R.E. Center, P.B. Mumola, W.M. Grossman, N.T. Olson, and J.P. Shannon, IEEE J. Quantum Electronics QE-17, 861 (1981); D.D. Lowenthal and J.M. Eggleston, "ASE Effects in Small Aspect Ratio Laser Oscillators and Amplifiers with Non-Saturable Absorption", submitted to JQE, January 1985.

9. P.B. Corkum, "High-power, Sub-psec 10 μm Pulse Generation," Optics Letters _8, 514 (1983).

THIN LAYERED INTERFERENCE MIRRORS TO REDUCE RADIATION DAMAGE

Paul L. Csonka
National Synchrotron Light Source
Brookhaven National Laboratory, Upton, N.Y. 11973

ABSTRACT

Calculations have been performed which show that damage induced by radiation of wavelength λ, incident on mirrors (and other optical elements) can be significantly reduced by a multilayered construction, in which the layers most sensitive to radiation damage have thickness $d \ll \lambda$. The layers may or may not be free-standing. The threshold to radiation damage in certain cases is calculated to increase by a factor $\approx 10^2$.

INTRODUCTION

Thin layered interference mirrors consist of a sequence of material layers designed so as to increase the radiation damage threshold.

The basic idea on which the design is based has been described earlier[1,2] and can be outlined as follows:

Consider a monochromatic electromagnetic plane wave with vacuum wavelength λ normally incident on the plane surface of a thin material layer. When the wave has high enough intensity, it will damage the layer. Assume now that a second wave of the same kind is simultaneously incident on the back surface of the layer. In general one expects even more damage, since now more intensity is incident on the layer. However, it turns out that if the relative phases of the two plane waves reaching the layer are correctly chosen, the damage will be less, instead of more, than the damage caused by either incident wave alone. In other words, more incident energy flux can actually help to reduce damage. The reason is, of course, that if the phases are such that the two plane waves tend to cancel each other inside the layer, then damage will be reduced. The reduction is most dramatic when the thickness of the layer, d, is $\ll \lambda$.

We use this simple phenomenon to design layered optical elements so constructed that when an electromagnetic wave is incident, the element automatically sets up an interference pattern within itself which is such that it has a minimum at all those layers which are most sensitive to radiation damage. One expects that the most dramatic damage threshold improvement can be achieved for a collection of free standing layers, i.e. structures in which each material layer is separated from its neighbors by a layer of vacuum. Denote by f_r the factor by which the damage threshold is

*Work supported by the U.S. Department of Energy.

improved. For a free standing layer as $d \to 0$, one would have $f_r \to \infty$. In fact, d is limited from below by practical considerations such as the uniformity and stability of the layer. One the other hand, structures in which the layers are not free standing are easier to make, but for these f_r will be governed by the properties of those materials which are located where the interference pattern inside the structure has its local maxima.

We have performed calculations[3] to evaluate f_r and the overall reflectivity R_{tm} for a plane electromagnetic wave normally incident on an interference mirror containing (N + 1) free standing plane layers. It was assumed that damage to the layers is proportional to the square of the local electric field amplitude (as it is for damage caused by induced currents in good conductors). The calculations show that for N = 10, on paper at least, $R_{tm} \approx 100\%$ and $f_r \approx 10^2$ can be simultaneously achieved.

CALCULATION

Figure 1 shows schematically a mirror which consists of N + 1 layers. Radiation is incident from the left, and is partly transmitted and partly reflected by each layer, as well as the whole assembly. The rightmost layer is called the "zeroth" layer, the leftmost is the "N^{th}" layer. The layers are separated by a medium (e.g. vacuum) whose complex index of refraction for radiation with wavelength λ is $n_1 = n_{1r} + in_{1i}$. The mth layer ($m = 0,1,\ldots,N$) has complex index of refraction $n_{2,m}$. We denote $n_{2,m}/n_1$ by $n^m = n_r^m + in_i^m$. The mth layer has thickness d_m, and is separated from the $(m+1)^{th}$ one by a distance ℓ_m. The amplitude reflectance of the mth layer will be denoted by R_{am}. The reflectance of the assembly of layers containing the zeroth , first, ..., and including the mth one, is R_{tam}.

For purposes of mirror design (one is interested in the reflected as opposed to the transmitted wave) the interesting parameter characterizing the interaction of the zeroth layer with the other layers is $R_{ao} = R_{ato}$. From the knowledge of the intensity incident on the zeroth layer, one can evaluate the energy absorbed in that layer per unit volume per unit time. One can average this quantity over one period of the radiation (λ/c), divide it by the flux density incident on it, multiply (for dimensional reasons) by λ/n_1, and denote it by W_{os}. The subscript s reminds us that the quantity is evaluated on the surface of the layer, that no averaging over the volume was performed, in order to be safe, since for the zeroth layer $d_o \ll \lambda$ need not hold.

follows. First, for given $n_{2,1}$ and R_{ato}, assume d_1 also given and select the value of ℓ_1 so that $W_1 = W_{os}$ is satisfied. This will insure that both layers in the double layer under consideration will be damaged at the same time, i.e. neither will be "weaker" than the other. Second, fix the remaining parameter, ℓ_1 so that for the chosen value g_1 be maximum (as a function of ℓ_1). This will result in maximum reduction of the intensity incidence from the left on the zeroth layer.

Having selected d_1 and ℓ_1, we turn to the second layer. Consider the assembly containing the zeroth and first layer as one layer with reflectance R_{at1}, and consider the double layer formed by this layer plus the second layer. As above, once $n_{2,2}$, d_2, ℓ_2 and R_{at1} are given, one can calculate W_2, g_2 and R_{at2}. The last of these has already been defined. The first two represent the same quantites for the second layer as W_1 and g_1 represent for the first. Again, one is free to choose d_2 and ℓ_2. We select ℓ_1 for any d_1 by requiring $W_2 = W_{os}$, and fix d_2 so that g_2 should be maximum.

Next we can turn to the third layer, repeat the same procedure, etc. until all d_m and ℓ_m are fixed. That completes the design.

Above we referred to the functions H_m, g_m and R_{atm}. These functions can be written in the form

$$
\begin{aligned}
H_m &= D_{Hm}\left(C_{Hm} + (A_{Hm}^2 + B_{Hm}^2)^{1/2}\cos(\phi_m - \phi_{Hm} - \pi)\right) \\
g_m &= D_{gm}\left(C_{gm} + (A_{gm}^2 + B_{gm}^2)^{1/2}\cos(\phi_m - \phi_{gm} - \pi)\right) \qquad (3) \\
R_m &= D_{Rm}\left(C_{Rm} + (A_{Rm}^2 + B_{Rm}^2)^{1/2}\cos(\phi_m - \phi_{Rm} - \pi)\right),
\end{aligned}
$$

where the ϕ_{im} depend linearly on ℓ_m, but all A_{im}, B_{im}, C_{im}, and D_{im} are independent[4] of ℓ_m, and $\phi_{im} = \arctan(B_{im}/A_{im})$. These equations show that, for example, H_m will go through a minimum as a function of ℓ, the location of the "magic angle" ϕ being determined by the value of ϕ_{Hm}. It is this type of behavior which permits one to increase the threshold for radiation induced damage by the appropriate choice of parameters.

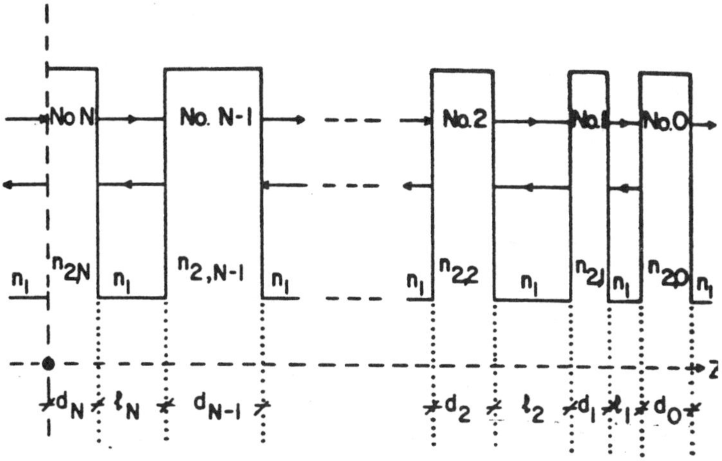

Fig. 1

Next, consider the double layer which consists of the zeroth and first layers. Once n_{21}, d_1, ℓ_1 and R_{tao} are given, one can evaluate the energy absorbed in the first layer per unit volume, and unit time, average this over one period of the elctromagnetic wave, and also over the volume of the first layer (permitted if the layer is thin enough), multiply by λ/n_1 and denote the quantity so obtained by W_1. One can write W_1 as a ratio of two functions both of which will be discussed below:

$$W_1 = H_1/g_1 \ . \tag{1}$$

The function g_1 is defined to be the ratio of the intensity incident from the left on the first layer, divided by the intensity incident from the left on the second layer.

Given $n_{2,1}$, d_1, ℓ_1 and R_{ato}, one can also evaluate the amplitude reflectance of this double layer, R_{at1}, and write it as

$$R_{at1} = \mathcal{R}_{at1}/g_1 \ , \tag{2}$$

where the function $\mathcal{R}_{at1}$ will be discussed below.

Actually once all $n_{2,m}$ are given, we are still free to choose the d_m and ℓ_m, indeed, judicious choice of these is the purpose of the design. To obtain the best result, we select d_1 and ℓ_1 as

The table shows the results of a calculation for an eleven layer mirror (N = 10), performed for free standing silver layers at $\lambda = 0.93$ μm. The following definitions are used: $R_{tm} = |R_{atm}|^2$, and $f_m = g_m^{-1}$. Clearly, $f_r = f_1 \cdot f_2 \dots f_N$.

n_r=0.04 n_i=7.795 (As in Ag when λ = 0.927 μm)

m	ℓ_m/λ	d_m/λ	R_{tm}	f_{rm}
0	-	2.000×10^{-1}	0.99741	-
1	0.4594	1.330×10^{-1}	0.99698	0.0831
2	0.4766	1.424×10^{-2}	0.99745	0.7684
3	0.4870	1.273×10^{-2}	0.99776	0.8136
4	0.4883	1.145×10^{-2}	0.99798	0.8436
5	0.4893	1.050×10^{-2}	0.99815	0.8652
6	0.4901	9.749×10^{-3}	0.99829	0.8816
7	0.4907	9.148×10^{-3}	0.99840	0.8943
8	0.4913	8.647×10^{-3}	0.99849	0.9045
9	0.4917	8.218×10^{-3}	0.99857	0.9129
10	0.4921	7.851×10^{-3}	0.99864	0.9199

$f_r = 44.0$

REFERENCES, FOOTNOTES

1. Paul L. Csonka, "Far field Laser Accelerators," in Laser Acceleration of Particles," ed. Paul J. Channell, p. 216 (1982).

2. Paul L. Csonka, "Improved Damage Threshold for Optical Elements Placed at Minima of a Radiation Field," in "Laser Acceleration of Particles," ed. Paul J. Channel, p. 248 (1982).

3. Paul L. Csonka and Hiroshi Watanabe, submitted (1985) to JOSA.

4. Expressions for A_{im}, B_{im}, C_{im} and D_{im} as a function of n_1, $n_{2,m}$, d_m and ℓ_m and $R_{at(m-1)}$, are given in Ref. 3.

ACCELERATOR TECHNOLOGY WORKING GROUP SUMMARY*

R. A. Jameson, MS H811
Los Alamos National Laboratory, Los Alamos, NM 87545

ABSTRACT

A summary is presented of workshop deliberations on basic scaling, the economic viability of laser drive power for HEP accelerators, the availability of electron beam injectors for near-term experiments, and a few very general remarks on technology issues.

INTRODUCTION

This group was set up to aid the other groups in their deliberations. The membership is listed below.

J. Fraser, Los Alamos
E. Gray, Los Alamos
R. Jameson, Los Alamos
J. Lawson, Rutherford-Appleton
A. Ruggiero, Fermilab
S. Schriber, Los Alamos
T. Smith, Stanford University
R. Sheffield, Los Alamos
J. Sheppard, SLAC
J. Simpson, Argonne
D. Sutter, DOE/OHENP
L. Teng, Fermilab
R. Williams, UCLA

In fact, the group did disperse most its efforts in interaction with the other groups. Specific contributions covered elsewhere in these proceedings include analysis by Sheppard of how to achieve better emittance using damping rings (looks favorable) or wigglers (pessimistic, because of excessive wiggler length and energy loss), and an outline by Fraser, Sheffield, and Gray on prospects for reducing electron beam source emittance using photoemitters and advanced preaccelerators.

The following comments summarize other concerns of the group relevant to questions of basic scaling, the major influence of economics on the overall viability of high-gradient accelerators, availability of electron beams for acceleration tests, and a few observations on technology issues.

SCALING RELATIONSHIPS FOR HEP REQUIREMENTS

The basic scaling equations for High-Energy Physics (HEP) colliding beam requirements (relating energy, luminosity, disruption factor, beamsstrahlung) were reviewed by Lawson. It was concluded that the set of equations is consistent and mature, and that the same approach was taken at this meeting as was used at the Frascati Workshop on the Generation of High Fields for Particle Acceleration in October 1984. There has been some refinement in

*Work supported by the US Department of Energy

the treatment of disruption factor and beamsstrahlung recently, and it should also be noted that P. Wilson pointed out in his talk that one might do better with flat, rather than round, beams because the beamsstrahlung would be less.

ECONOMIC VIABILITY OF LASER DRIVE POWER FOR HEP ACCELERATORS

Major emphasis was placed on ensuring that the significance of efficiency and economics in achieving the HEP goals for colliding beams at high energies was appreciated by the proposers of the various schemes.

In particular, it is necessary to understand that achieving very high accelerating gradients alone is a far too simplistic prescription--very high gradients hold the lure of short machines, but without very significant attendant economic advantages, very high gradients would be unaffordable in a real machine.

A short example underscores the point: As a rule of thumb, the total construction cost of a facility is about half plant, engineering, design and installation, and about half technical equipment. The technical equipment in this case is some form of linear accelerator (linac); under the usual optimization procedure[1] for today's linacs, about half the cost is for the installed drive power and half is for the linear structure cost.

$$\text{Technical Equipment Cost} \sim (\text{Unit Power Cost})(\text{Power}) + (\text{Unit Length Cost})(\text{Length})$$

The amount of power required varies inversely with length; therefore, the two terms trade directly against each other, resulting in optimum cost when the terms are roughly equal.

Putting in some numbers helps illustrate the challenges involved in the factors in this simple equation. Richter, in his opening address, gave a range of electron beam power cases; take the case of 10-MW average power. It is important to note that the requirement is for average power, because of the need for adequate luminosity.

For beam breakup and energy-spread reasons, at the present time, it is felt for HEP colliders that no more than about 10% of the power in the accelerating structure can be converted into beam power. Further, in schemes where there is an added energy conversion step between beam and drive, for example from laser drive to plasma wave to accelerated beam, another 10% efficiency (at best) is encountered. Thus, the drive power requirement is between 100 MW and 1 GW; take 100 MW.

The present cost of conventional rf average power, up to frequencies of a few GHz, is about $5/average watt. Various breakdown considerations indicate that the frequency of conventional linacs could be pushed up to a point of diminishing returns around 30 GHz, and it appears that the cost per average watt could be kept roughly the same up to that frequency. Then, the power cost term would be $500M, and one would optimize the gradient (at around

100-200 MeV/m) against length to produce an equal length cost, for a total technical equipment cost of $1B.

Laser drive power, on the other hand, is produced through a number of energy conversion steps, with resulting overall efficiencies (wall-plug to laser average power) of a few per cent at best, compared to 60-70% for rf. (The free-electron laser with energy recovery may reach a few tens of per cent, depending on the wavelength and power levels.) The laser technology group found that the present cost per average laser watt is about several hundred dollars and projected a possible reduction to about $100/average watt after extensive development. Take the cost/average laser watt to be $100. Then the cost of the power alone in our example would be $10B, clearly unfavorable and with no possibility to recoup the situation by lower length costs.

One must therefore conclude that accelerator length is not the overriding problem, and that the HEP requirements challenge even conventional unit costs and efficiencies.

An appropriate problem ranking might be the following:

- Improve drive-to-beam efficiency
- Improve electric-to-drive efficiency
- Reduce unit power costs

This very cursory system analysis further strongly suggests that a balance be maintained between longer wavelength approaches and the laser accelerator approaches. Work on longer wavelength ($\sim$1-cm) structures, transformer designs, rf breakdown, and related topics might well provide solutions appropriate to the next linear collider.

In a complementary fashion, the application of lasers to the HEP goals does address many basic physics and technology issues and is also well worth pursuing. Even while direct laser drive had to face the stringent economic realities at this meeting, new ideas on how lasers might be used were surfacing. Lasers produce peak power efficiently and might be used in that mode to provide the very strong focusing fields required, as suggested by Channell, or to provide electrical switching, as in the scheme outlined by Willis.

Also, the address of such topics (like how to achieve and use high gradients) could well be the priming influence for a new insight or breakthrough.

ELECTRON BEAM INJECTORS

Considerable interaction occurred in defining approaches and available equipment or facilities for providing the electron beams to be accelerated in near-term experiments. Table I shows the main properties of some of the prospects identified; beams could be provided at the GeV level (SLAC SLC Damping Ring, or Nuclear Physics Injector), 100 MeV from the Stanford Superconducting Accelerator, 22 MeV from ANL, and at low energies up to 10 MeV from the Los Alamos designed portable linac or commercial 9-GHz linacs from Schoenberg Radiation. One main tradeoff involved is whether one brings the linac injector to the rest of the experiment, or vice

Table I Electron beams for near-term experiments

	Energy	Np	ε (m•rad) (unnorm)	γ	σx	Δp/p,%	$
SLC Damping Ring	1 GeV	5×10^{10}	1.5×10^{-8}	2×10^{3}	3 μm	0.1%	
SLAC NPI	1 GeV	2×10^{9}	1.5×10^{-7}	2×10^{3}			
	6 GeV	2×10^{9}	3×10^{-8}	1×10^{4}			
Stanford	1 GeV	(future; seeking users, will also have storage ring)					
Stanford Super-Conducting Linac	100 MeV (20-150)	5×10^{7}	2×10^{-8}	200		∿0	Cost Recovery + $500K?
ANL	22 MeV (+ witness pulse)	6×10^{10}	2×10^{-7}	40		0.5	Cost Recovery ∿$175/hour
Los Alamos FEL							
1.3-GHz e⁻ Linac	10-20 MeV	2×10^{10}	2×10^{-6}	40	1 mm	1-2	
Optical	9-100 μm, high peak and average power						
Los Alamos 3-GHZ Linac	6-10 MeV	2×10^{8}	3×10^{-6}	20	2 mm	10	$500K
(portable--could be improved with photo-cathode)							
Schoenberg	6 MeV	2×10^{7}	1×10^{-6}		2/3 mm	10	$315K
Radiation Linac	4						$265K
9-GHz Linacs	1.5						$150K

versa. Another is the energy gain to be achieved in the experiment relative to the injection energy.

Some further particulars are listed below.

SLAC Beams

(See layout, Fig. 1.)

SLAC: CID Gun	
Charge/bunch	8 to 16 nC (5 to 10 x 10^{10} e-/pulse)
Bunch length	30 ps FWHM
Number of bunches	2
Bunch separation	arbitrary in 5.6-ns increments
Emittance	$\gamma\varepsilon \leq 15 \times 10^{-5}$ m•rad
Energy	50 MeV
$\Delta E/E$	approximately 1% FWHM
SLC: Damping Ring Beams	
Bunch separation	61.6 ns
Emittance	$\gamma\varepsilon \leq 3 \times 10^{-5}$ m•rad
Energy	1.21 GeV
Energy spread	2.0% FWHM, typical 0.2% FWHM, uncompressed
Bunch length	3.5 mm nominal (12 ps) (14 mm uncompressed)
SLC: End of Linac	
Emittance	$\gamma\varepsilon \leq 3 \times 10^{-5}$ m•rad
Energy	1 to 51 GeV
$\Delta E/E$	$\geq$0.1% FWHM
Rep rate	180 Hz
Type of beams	e^+ and e^-
NPI Gun:	
Pulse length	$\leq$ 1.6 us (4500 S-band bunches)
Charge/bunch	$\leq 2.2 \times 10^9$ e^-/bunch, max* $\leq 2.2 \times 10^8$ e^-/bunch nominal
Bunch length	5 ps FWHM
Emittance	$\gamma\varepsilon \leq 5 \times 10^{-5}$ m•rad
Energy	$\leq$ 6 GeV
$\Delta E/E$	$\geq$ 0.1% FWHM (single bunch) $\cong$ 0.7% FWHM (multibunch pulse)

*This corresponds to maximum gun output and can only be transported through the linac for pulse lengths of several nanoseconds.

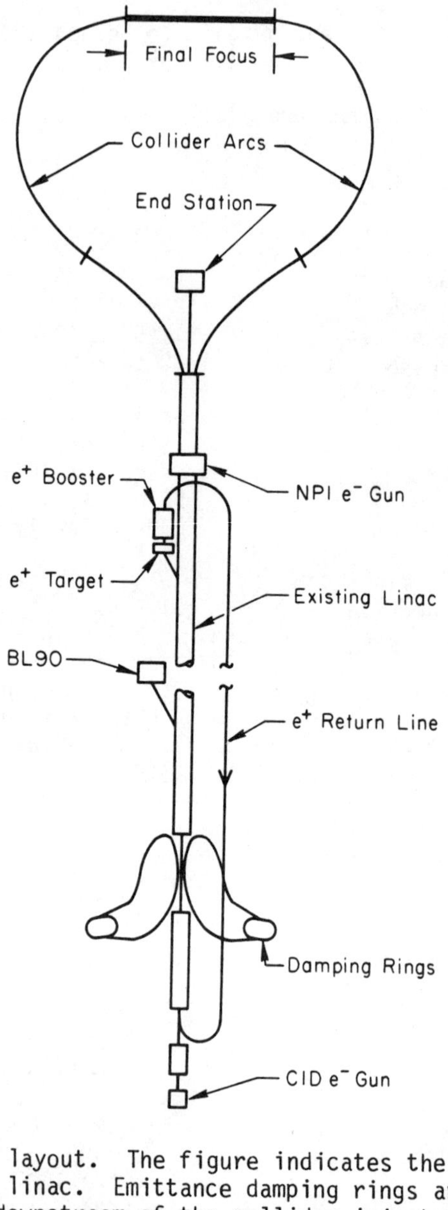

Fig. 1. Overall SLC layout. The figure indicates the general features of the SLAC linac. Emittance damping rings are located approximately 120 m downstream of the collider injector (CID). BL90 is a 14° off-axis extraction line situated about 1000 m from the gun. The dc bends associated with BL90 operate at beam energies up to 10 GeV. The Nuclear Physics Injector (NPI) is at the 2500-m point along the linac. As a point of reference, the SLC positron source is located at the 2000-m point. Overall the linac is 3 000 m long. There are three end stations at the end of the accelerator.

Stanford FEL Group

Stanford's FEL program is constructing a 1-GeV linac and storage ring, and is seeking users. Beam available in 1987.

Stanford HEPL Superconducting Linac

Average beam currents up to 400 μA are available.

Two types of pulse trains:

1) every rf bucket filled, spaced at 1/1.3 GHz = 769 ps; each pulse ∿5 ps, peak current ∿40 mA.
2) subharmonic operation - pulses at 110th subharmonic of 1.3 GHz = 85 ns, each pulse ∿7 ps, peak currents to 4 A.

Beam emittance at 50 MeV ∿ 10 mm·mrad normalized.

A cost-recovery operation is desired and, with $500K up-front, a first-rate user facility could be set up.

HEPL also has a 40-MeV infrared (1.6 μm) FEL machine that delivers about 1/4-A average current during the 10-μs pulse, with every bucket filled, and with emittance "equal to about half the Lawson-Penner criterion."

ANL Wake-Measurement Facility

Argonne has internal support to set up a multipurpose, user-oriented wake-field measurement facility around their 22-MeV high-peak-intensity linac. They hope to be able to start tests by the fall of 1985.

Linac Characteristics

Energy	22 MeV
Energy spread	±100 keV
Pulse length	30 ps*
Peak current	1 kA
Rep rate	≤ 800 Hz
ε_x, ε_y	7 x 10^{-6} m·rad (Norm)
Intensity/pulse	9 nC

*Will be ∿5-10 ps this year

L-Band, subharmonic buncher

Use at present about 5-6 shifts/wk by CHEM, PHYS, HEP. Used 25 shifts/yr for stochastic cooling pickup devices and calibration.

Beam manipulations to provide the shock pulse followed by a full or reduced energy probe pulse are being studied, as indicated in Fig. 2.

Los Alamos FEL Facility

Although programmatically committed, the Los Alamos FEL accelerator could be considered for a variety of advanced accelerator concept studies starting in 1987. Synchronized, high-quality photon and electron bunches are available.

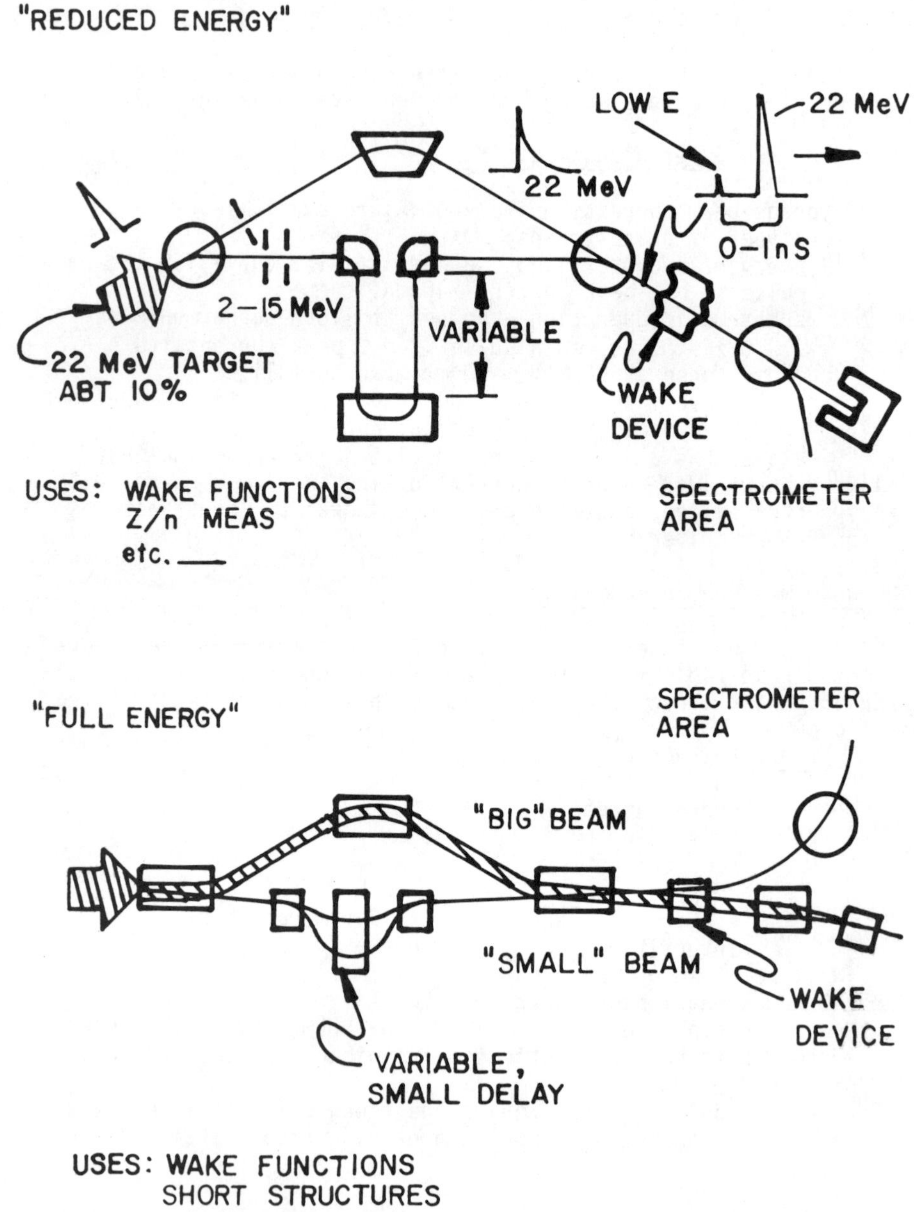

Fig. 2. Beam manipulations being studied for ANL wake measurement facility.

Electron Beam Parameters

Energy	10-20 MeV
Peak current	25-55 A
Energy spread	1-2%
Bunch length	40 ps
Pulse train	21.7 MHz for ∿100 μs
Emittance (Norm)	$8\pi \times 10^{-5}$ m•rad
(Unnorm)	$2\pi \times 10^{-6}$ m•rad
Number electrons/bunch	3×10^9 after improvements this year.

Optical Beam Parameters

Wavelength	9-35 μm (extendible to 100 μm)
Peak power	6 MW
Pulse length	30 ps
Pulse interval	46 ps
Macropulse	100 μs
Rep rate	1 Hz
Strehl ratio	0.8-0.9 (99% transmission through hole)
Raman sidebands	0.3-2.0%

With a solenoidal or periodic channel emittance filter, it might be possible to provide more than 10^7 electrons/pulse in an emittance of about 10^{-8} m•rad.

Los Alamos 3-GHz Portable Linac

This machine was packaged into seven boxes, each transportable by one or two persons, for use in extreme environments. It is not available, but the design could be used to make more machines, either portable or not. Peak current is 100 mA during the 1-μs pulse; at 3 GHz, the 30-ps micropulses occur every 330 ps.

Changing to a pulsed photoemitter cathode might improve the emittance by a factor of 10 and produce 4×10^{10} electrons/pulse, or 200-A peak current during the 30-ps pulse.

Schonberg Radiation MINAC

Inquires should be directed to Schonberg Radiation.
Address: 2560 Wyandotte Street, Mountain View, CA 94043.
Phone: (415) 964-6214.

TECHNOLOGY ISSUES

The focus of this meeting was on accelerator physics issues, rather than on accelerator technology, which will have to be addressed after some of the concepts are tested further. Thus, we do not attempt here to make lists of problems, but only summarize very briefly a few topics where preliminary discussions produced some insight that might be of general interest.

Staging/Matching

Some general considerations of the problems of staging led to the suggestion that "laser lenses" might be used to provide the extremely strong focusing needed for staging or in the final transport. Also, realization of the enormous focusing strengths inherent, for example in the plasma beat-wave scheme, led to discussion of the possibility of a continuous long channel. Staging and the interstage matching required to avoid spoiling the emittance will require much further study.

Electron Beam Emittance Reduction by Scraping (T. Himel, SLAC)

To test some accelerator concepts, one needs a very low emittance beam with a small energy spread. It has been suggested that such a beam could be obtained by scraping off all but 1 part in 10^6 of a high-quality SLAC beam. Can this be done without excessively large backgrounds from beam particles entering the edge of a collimator and scattering into the accepted region?

A study using the EGS electromagnetic shower simulation has been done by Hobey de Stabler at SLAC. Fifty-GeV electrons were incident a distance x from the edge of a copper collimator. The fraction of electrons with greater than 80% of their initial energy, which emerged from the edge of the collimator at any angle, was $0.4/(1 + x/2)^3$ where x is measured in micrometers.[3] For example, 0.4 emerged when the beam was incident at the very edge ($x = 0$) but only 0.0009 emerged when the beam was incident 5 μm from the edge. The particles emerged at an average angle of 0.2 mrad. Hence the background caused by edge scattering from a collimator, followed by a crude momentum selection, will be less than 10% if the collimator hole has a radius greater than 20 μm.

Alignment and Control

A few very general remarks to keep in mind:

The design strategy for a machine will have to find a balance between the tolerances to be achieved in the initial construction and the instrumentation and correction elements that will sense the beam and put it on the proper orbit as the machine is tuned up and run. These days, it will be found that great advantage can be had from stressing the instrumentation and correction side. Instrumentation development using the beam or perhaps structure fields will be very important.

It is usually very difficult (expensive) to make precise corrections at the high-power levels in the system. Low-power correction elements should be added, or trim modulation provided using closed-loop control. Such control will be greatly aided by thorough understanding of the machine parameters through such modeling studies as reported in these proceedings by Karl Bane on wake-field effect compensation. Models like this are crucially important; they end up being built right into the computer control system and significantly ease requirements on the hardware.

The extremely small positional and timing tolerances will certainly require closed-loop control. Sensitivity to geological disturbances might be attacked by feedback methods, but the achievable bandwidth will be low. If the rep rate is not too high, sensing and setup of components on a shot-to-shot basis might be possible. Or perhaps a probabilistic, predictive approach to feedforward should be considered. Modern control practice has produced a wealth of information on such approaches, and one can readily imagine gaining several factors in the number of successfully colliding bunches in this manner. On the timing, present projects are requiring phase and amplitude tolerances of 0.1° and 0.1% at rf frequencies--a factor of 10 better for each over existing machines. Achieving this factor of 10 is a challenge, but probably it can be done. As indicated in the laser technology summary, optical regime timing to about 1 μm pulse-to-pulse and to a few degrees phasing on the wave front appears possible. Phase-conjugation optics is also a possibility.

REFERENCE

1. R. A. Jameson, "New Linac Technology - for SSC, and Beyond?," 12th Int. Conf. on High Energy Accelerators, FNAL, August 1983, p. 497.

LINEAR ACCELERATORS FOR TeV COLLIDERS*

P. B. WILSON
Stanford Linear Accelerator Center
Stanford University, Stanford, CA 94305

I. INTRODUCTION

This paper summarizes four tutorial lectures on linear electron accelerators which were presented at this Workshop:

1. "Electron Linacs for TeV Colliders" (P. B. Wilson)
2. "Emittance and Damping Rings" (P. M. Morton)
3. "Wake Fields: Basic Concepts" (R. K. Cooper)
4. "Wake Field Effects in Linacs" (K. L. F. Bane)

The first of these lectures was intended to introduce the general requirements for electron linacs capable of delivering beams for very high energy linear colliders. Material from this lecture is presented in the next three sections. Section II introduces the basic scaling relations for important linear collider design parameters. In Sec. III some basic concepts concerning the design of accelerating structures are presented, and breakdown limitations are discussed. In Sec. IV RF power sources are considered.

The fact that two of the four lectures were concerned with wake fields and their effects emphasizes the importance of this topic for high energy collider design. Several tutorial papers which give extensive coverage to wake field concepts and wake field effects have been published recently. No attempt will be made to duplicate this material here. Some key concepts will be discussed, and some examples of wake fields for typical linac structures will be presented in Sec. V. The reader is referred to the referenced literature for further study. The importance of emittance in linear collider design is also underscored by the scaling relations in Sec. II. Some general concepts concerning emittance, and the limitations on the emittance that can be obtained from linac guns and damping rings are discussed in Sec. VI.

In connection with Lectures 3 and 4, computer generated movies were shown at the Workshop which illustrated how wake fields arise as an electron bunch moves through typical structures, and how these wake fields in turn act on the bunch to produce emittance growth. Viewing such movies greatly enhances ones physical understanding of wake fields and their effects, but unfortunately this process cannot be reproduced on the printed page.

Finally, the author takes full responsibility for the manner in which the material presented by the other three lecturers has been condensed, summarized, or rearranged, and for all omissions and errors.

---

* Work supported by the Department of Energy, contract DE–AC03–76SF00515.

II. SCALING RELATIONS FOR LINEAR COLLIDERS

A. Beam-Beam Parameters

Three parameters which characterize the interaction between two colliding bunches in a linear collider are the luminosity $\mathcal{L}$, the disruption D and the beamstrahlung δ. In the following, head-on collisions between tri-gaussian bunches are assumed. The possibility of flat bunches crossing at a slight angle in the horizontal plane will be taken into account. The expressions given here for the three beam-beam parameters in the "classical" regime are taken from Refs. 1 and 2, where a more detailed discussion and additional references can be found.

LUMINOSITY

The luminosity (in cm^{-2} sec^{-1}) times the cross section (in cm^2) gives the event rate (per second) for any physical process taking place in the colliding bunches. Along with the beam energy, it is a primary design parameter for a linear collider. Assume identical e^+e^- linacs, each with energy $E_0 = eV_0 = \gamma mc^2$, pulsed at a repetition rate f and producing trains of b bunches per linac pulse. The luminosity is given by

$$\mathcal{L} = \frac{N^2 bfH_D}{4\pi\sigma_x\sigma_y} = \frac{N^2 bf\gamma H_D}{4\pi\epsilon_n\left(\beta_x\beta_y\right)^{1/2}} \quad . \tag{1a}$$

Here N is the number of particles per bunch, σ_y and $\sigma_x = R\sigma_y$ are the bunch height and bunch width (at the interaction point, unless otherwise indicated), $\varepsilon_n = \gamma\varepsilon_x = \gamma\varepsilon_y$ is the normalized emittance (assumed equal for each dimension), and β_y and β_x are the vertical and horizontal beta functions given by the optics of the final focus system. If the disruption parameter is sufficiently large, the beams will pinch together as they pass through each other, producing an enhancement in the luminosity by a factor H_D. In practical units the luminosity is given by

$$\mathcal{L}\left(10^{32}\,\mathrm{cm}^{-2}\,\mathrm{s}^{-1}\right) = \frac{8.0\times10^{-6}}{R}\left\{\frac{b\left[N\left(10^{10}\right)\right]^2 f(Hz)\,H_D}{\left[\sigma_y(\mu m)\right]^2}\right\} \quad . \tag{1b}$$

DISRUPTION

The focussing effect produced by one beam acting on the particles in the other beam depends on the disruption parameter,

$$D = \frac{2r_0\,N\,\sigma_z}{(1+R)\,\gamma\,\sigma_y^2} \quad . \tag{2a}$$

Here σ_z is the rms bunch length and r_0 the classical electron radius. Each beam acts like a lens with focal length σ_z/D for particles near the axis in the opposing beam (if $D \lesssim 1$). For large values of D, the beams act like a plasma during the interaction, with the number of transverse plasma oscillations given[3] approximately by $(D/10)^{1/2}$. In practical units, the disruption parameter can be expressed as

$$D = \frac{.029}{1+R} \left\{ \frac{N\,(10^{10})\,\sigma_z\,(\mathrm{mm})}{E_0\,(\mathrm{TeV})\,[\sigma_y\,(\mu\mathrm{m})]^2} \right\} . \tag{2b}$$

The luminosity enhancement as a function of D must be computed by a simulation. Results from simulations made to date differ somewhat in the maximum value of the enhancement H_D that can be obtained, and in the rate of rise of H_D as a function D near $D \approx 1$. Hollebeek[3] obtains a maximum enhancement ratio in the range five to six for $D \geq 1.5$. Fawley and Lee[4] find a maximum enhancement in the range three to four at $D \geq 3$. Some representative results from these two simulations are given in Table I below. For a flat beam with large aspect ratio, the enhancement ratio is given approximately by the square root of the round beam result. For intermediate values of the aspect ratio, the enhancement can be estimated from[5]

$$H_D(R) = H_D(1)\,\frac{R}{1+(R-1)\,[H_D(1)]^{1/2}} . \tag{3}$$

Table I
Luminosity Enhancement as a Function of D

D	< 0.2	0.5	1.0	1.5	2.0	3	5
Hollebeek							
H_D (round)	1.0	1.4	3.6	5.2	5.6	5.9	6.0
H_D (flat)	1.0	1.2	1.9	2.3	2.4	2.4	2.5
Fawley & Lee							
H_D (round)	1.0	1.0	1.5	2.2	2.6	3.1	3.4
H_D (flat)	1.0	1.0	1.2	1.5	1.6	1.8	1.8

BEAMSTRAHLUNG

We turn next to a consideration of the beamstrahlung parameter δ. As the colliding bunches pass through each other, the particles in one bunch are deflected by the fields in the opposing bunch. This transverse acceleration produces synchrotron radiation, called beamstrahlung in this case. The beamstrahlung parameter δ is the average energy loss per particle, divided by the incident energy, calculated after the beams have separated. From the point of view of the physical processes occurring during the interaction between bunches, the relative energy loss in the center of mass system, $\delta/2$, is of more concern. The rms energy spread in the center of mass system may be somewhat less than $\delta/2$. Thus it is now conventional to take $\delta = 0.3$ as acceptable in collider design. However, it is well to remember that beamstrahlung is best studied by calculating the actual distribution function for the energy loss, and this can only be done by a simulation in most cases of interest. The analytic expressions for δ which follow are, however, useful for scaling.

The expression for beamstrahlung in the classical regime (this term will be defined later) for two colliding tri-gaussian bunches has been calculated by Bassetti and Gygi-Hanney:[6]

$$\delta_{c\ell} = \frac{r_0^3 N^2 \gamma}{\sigma_z \sigma_y^2} F(R) \quad , \tag{4a}$$

where $F(R)$ is a rather complicated function (see also Ref. 1) such that $F(1) = 0.22$ and $F(R >> 1) = 0.91$. Within a few percent, $F(R)$ is approximated by

$$F(R) \approx 0.22 \left(\frac{2}{1+R}\right)^2 \quad .$$

In the above calculation it was assumed that the particle trajectories do not change as the bunches collide. If the disruption parameter is large enough to cause the bunches to pinch, we would expect the beamstrahlung as well as the luminosity to be enhanced. This enhancement can be taken into account, at least approximately, by multiplying the preceding expression by H_D. A more exact beamstrahlung enhancement ratio can only be obtained by a simulation. In practical units Eq. (4a) becomes

$$\delta_{c\ell} \approx 1.0 \times 10^{-3} \left(\frac{2}{1+R}\right)^2 \left\{\frac{\left[N\,(10^{10})\right]^2\; E_0\;(\text{TeV})}{\sigma_z\;(\text{mm})\;\left[\sigma_y\;(\mu\text{m})\right]^2}\right\} H_D \quad . \tag{4b}$$

The classical synchrotron radiation spectrum for a relativistic electron moving in a uniform magnetic field B peaks up near the critical photon energy

$\hbar\omega_c = 3\hbar\gamma^2 eB/2mc$. However, when $\hbar\omega_c > \gamma mc^2$, one photon at the critical energy would have to carry more than the entire energy of the electron which emits it, and consequently the classical calculation of synchrotron radiation can no longer be valid. Define a scaling parameter Υ by

$$\Upsilon \equiv \frac{2}{3}\frac{\hbar\omega_c}{\gamma mc^2} = \gamma\frac{B}{B_c} \quad , \tag{5}$$

$$B_c \equiv \frac{m^2 c^3}{e\hbar} = \frac{e}{r_0\, \lambda\!\!\!^{-}_c} = 4.4 \times 10^{13}\ \mathrm{G} \quad .$$

Here $\lambda\!\!\!^{-}_c$ is the Compton wavelength divided by 2π. For $\Upsilon \ll 1$ the classical calculation of the energy loss by synchrotron radiation is valid, while for $\Upsilon \gtrsim 1$ quantum effects, which act to reduce the energy of the emitted photons, must be taken into account. The modification of the synchrotron radiation spectrum by these quantum effects is summarized in Refs. 7 and 8. The rate at which an electron radiates energy in the quantum regime is reduced compared to the classical radiation rate. This reduction factor, H_Υ, is plotted in Fig. 1 as a function of Υ.

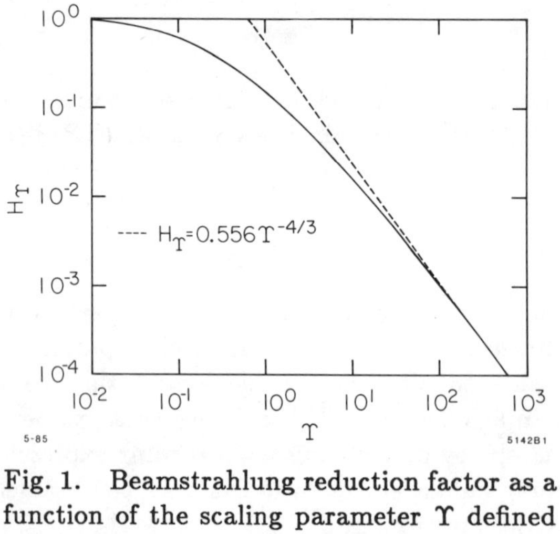

Fig. 1. Beamstrahlung reduction factor as a function of the scaling parameter Υ defined in Eq. (5).

An exact analytic calculation of the beamstrahlung parameter for gaussian bunches in the quantum regime is difficult, and in any case a simulation must be carried out if the bunches pinch significantly. However, a rough approximation for scaling purposes is useful. We first compute the average value of Υ for a flat beam. Assume the actual gaussian bunch can be modeled by a uniform particle

particle distribution of width $\sqrt{2\pi}\,\sigma_x$, half-height $\sigma_y \ll \sigma_x$ and length $\sqrt{2\pi}\,\sigma_z$. . Allowing an additional factor of 1/2 for averaging in the vertical direction, the average value of the magnetic field becomes

$$\overline{B} = \frac{eN}{2\sigma_x\sigma_z} = \frac{eN}{2R\sigma_y\sigma_z} \quad .$$

For a round beam we approximate the bunch by a uniform particle distribution of length $\sqrt{2\pi}\,\sigma_z$ and radius $2\sigma_r = 2\sigma_y$. Using a factor of 2/3 for the average field in the radial direction gives

$$\overline{B} = \frac{eN}{3\sqrt{2\pi}\,\sigma_y\sigma_z} \approx \frac{eN}{4\sigma_y\sigma_z} \quad .$$

Thus for any aspect ratio it is reasonable to approximate the average magnetic field by

$$\overline{B} \approx \frac{eN}{2(1+R)\sigma_y\sigma_z} \quad . \tag{6}$$

Substituting in Eq. (5),

$$\overline{\Upsilon} \approx \frac{r_0\,\lambda\!\!\!^{-}_c\gamma N}{2(1+R)\sigma_y\sigma_z} \quad , \tag{7a}$$

where $r_0\,\lambda\!\!\!^{-}_c = 1.09 \times 10^{-23}\ \text{cm}^2$. In practical units

$$\overline{\Upsilon} \approx \frac{1.1\times10^{-2}}{1+R}\left[\frac{N\,(10^{10})\ E_0\,(\text{TeV})}{\sigma_z\ (\text{mm})\ \sigma_y\ (\mu\text{m})}\right] \quad . \tag{7b}$$

Note from Fig. 1 that in the quantum regime H_Υ is given by $0.556\ \Upsilon^{-4/3}$. Using this together with Eqs. (2b), (4b) and (7b),

$$\delta_Q = \delta_{c\ell}\,H_\Upsilon \approx 1.4\times10^{-3}\left[\frac{DN}{1+R}\right]^{1/3} \quad , \tag{8}$$

valid in the regime $\Upsilon \gtrsim 10$. From Eqs. (2b) and (4b) we also have

$$\delta_Q \approx 5.8\left[\delta_{c\ell}\,\frac{\sigma_z^2\ (\text{mm})}{E_0^2\ (\text{TeV})}\right]^{1/3} \quad . \tag{9}$$

B. Beam Power and Wall Plug Power

The three beam-beam parameters discussed in the previous section depend only on the beam properties at the interaction point and the beam dynamics during the collision. A fourth parameter which is also independent of the accelerating linac is the beam power $P_b = bf\,N\,E_0$. For scaling purposes it is useful to introduce a normalizing voltage, current and power given by

$$\begin{aligned} V_n &= mc^2/e = e/r_0 = 0.511 \text{ MV} \\ I_n &= 4\pi V_n/Z_0 = ce/r_0 = 17.04 \text{ kA} \\ P_n &= I_n V_n = ce^2/r_0^2 = 8.71 \times 10^9 \text{ W} \quad , \end{aligned}$$

where $Z_0 = 377\Omega$ is the impedance of free space. Thus

$$P_b' = \frac{P_b}{P_n} = \left(\frac{r_0}{c}\right) bf\,N\gamma \quad . \tag{10}$$

Of direct practical interest is the total "wall plug" power P_{ac} required by each linac in a collider. The wall plug power does depend on the properties of the accelerating structure. It is related to the beam power by $P_{ac} = P_b/(\eta_{rf}\eta_s\eta_b)$, or in practical units

$$P_{ac}\,(\text{MW}) = \frac{1.6 \times 10^{-3}}{\eta_{rf}\eta_s\eta_b} \left[N\left(10^{10}\right) E_0\,(\text{TeV})\, f\,(\text{Hz})\right] \quad . \tag{11}$$

Here η_{rf} is the efficiency for the conversion of ac power into rf power, η_s is a structure efficiency which takes into account the fact that some rf energy is dissipated in the structure walls during filling, and η_b is the fraction of the energy stored in the structure which is removed per bunch. In Sec. III A it is shown that a reasonable value for the net efficiency $\eta_{rf}\eta_s$ is, assuming some future technological improvements in the production of high peak power RF, $\eta_{rf}\eta_s \approx 0.5$. The single bunch efficiency is discussed below.

C. Single Bunch Efficiency and Energy Spread

For scaling purposes, we assume that each bunch in a train b bunches long accelerated during one RF pulse has equal charge and extracts the same fraction of the energy stored in the accelerating structure. If there are, for example, eight bunches and each bunch extract 4% of the stored energy, then this approximation is rather poor unless the bunch-to-bunch beam loading is compensated by one of several possible methods. As one example, the first bunch can be injected into a traveling wave section before it is completely filled, and the spacing between bunches adjusted so that the energy added to the section between bunches just compensates for the bunch-to-bunch energy sag. In the following we assume this is done. The efficiency for energy extraction by a single bunch is

$$\eta_b = \frac{4eNk_1}{G} \quad , \tag{12}$$

where G is the accelerating gradient and k_1 is a structure constant given by

$$k_1 \equiv \frac{G^2}{4u} = \frac{C_0}{\lambda^2} \quad . \tag{13}$$

Here u is the stored energy per unit length, λ is the wavelength, and C_0 is a constant independent of wavelength which depend only on the structure geometry. For the SLAC disk-loaded structure, $C_0 = 2.1 \times 10^{11}$ V-m/C. It varies approximately as a/λ, where a is the diameter of the iris opening[9]. Substituting Eq. (13) into Eq. (12),

$$\eta_b = \frac{4eNC_0}{\lambda^2 G} = 13.4 \frac{N\,(10^{10})}{\lambda^2\,(\mathrm{cm})\;G\,(\mathrm{MV/m})} \quad . \tag{14a}$$

This can also be written in Gaussian units for a linac of length L as

$$\eta_b = \frac{4\,r_0\,C_0'\,N\,L}{\lambda^2\,\gamma_f} = 93\,\frac{r_0\,N}{\lambda^2\,G'} \quad , \tag{14b}$$

where $G' = \gamma_f/L$, $k_1' = 0.21\ \mathrm{cm}^{-2}$ and $C_0' = k_1'\lambda^2 = 23.2$ for the SLAC structure.

The single bunch energy spread is derived from the current distribution and the longitudinal wake potential, as described in Sec. V B. For a given accelerating structure and current distribution (*e.g.*, gaussian), it is a function only of σ_z/λ, η_b and θ, where θ is the angle of the center of the bunch with respect to the crest of the accelerating wave. For the SLAC disk-loaded structure, the maximum value of η_b for a 1% and 2% single bunch energy spread (defined to include 90% of the bunch current) is given[10] in Table II below for several values of σ_z/λ. The angle θ ahead of crest has been chosen to minimize the energy spread. The effective accelerating gradient is reduced with respect to the peak unloaded gradient, both because the bunch is off crest and because there is a decelerating wake field within the bunch. The reduction factor in the gradient is given in the last column. The bottom row in the table shows that a very large single bunch efficiency can be reached if the bunch length is chosen so that shape of the bunch wake is approximately the inverse of the crest of the accelerating wave, as has been proposed at Novosibirsk.[11]

Table II
Maximum η_b for 1% and 2% Energy Spread

	η_b (%)		$\theta°$		E/E_0	
σ_z/λ	1%	2%	1%	2%	1%	2%
.005	2.3	3.2	29	45	0.85	0.66
.01	3.5	6.6	14	39	0.94	0.71
.02	–	6.5	–	8	–	0.94
.04	–	≈ 25	–	6	–	0.80

D. Emittance Growth

Assuming a simple two-particle model for the bunch, several effects can cause the leading particle to drive the amplitude of the transverse oscillations of the tail particle as the bunch moves along the accelerator. To get a feel for scaling of emittance growth, consider the simplest case of a uniform machine with constant beta function, constant energy, and an offset x_0 in the leading particle at the beginning of the accelerator. From the results in Sec. VC, the amplitude of the oscillation of the tail particle at distance L, divided by the transverse size of the beam at the end of the machine, is

$$\frac{\Delta x_2}{\sigma_f} \approx \frac{1}{4} r_0 N L W_1 \left(\frac{\beta}{\gamma \epsilon_n}\right)^{1/2} x_0 \quad , \tag{15a}$$

where W_1 is the dipole wake at the tail particle due to the leading particle. In Gaussian units, $W_1 \approx 2 \times 10^5$ m^{-3} for the SLAC structure with $\sigma_z/\lambda \approx .01$. For a structure with constant geometry and fixed σ_z/λ, W_1 scales as $W_1 = C_1'/\lambda^3$, where $C_1^1 = 315$ for the SLAC structure geometry. Using this in Eq. (15a) and substituting η_b from Eq. (14b), we obtain

$$\frac{\Delta x_2}{\sigma_f} \approx \frac{C_1' \eta_b}{16 C_0'} \left(\frac{\beta \gamma}{\epsilon_n}\right)^{1/2} \left(\frac{x_0}{\lambda}\right) \quad . \tag{15b}$$

It is important to recall that the dipole wake constant C_1' depends on both the structure geometry and bunch length. Details are given in Sec. V.

A more realistic example assumes uniform acceleration from injection energy γ_0 to γ_f, with a beta function which varies as $\beta = \beta_0 \, (\gamma/\gamma_0)^{1/2}$. Assume also that the accelerator consists of M sections which are misaligned with an rms error d in transverse position. The the growth in amplitude of the tail particle in the two particle model is then given by [12]

$$\frac{\Delta x_2}{\sigma_f} = \frac{1}{2} r_0 NLW_1 \left(\frac{\beta_f}{\gamma_f \epsilon_n}\right)^{1/2} \frac{d}{M^{1/2}} \quad . \tag{16}$$

Note that, in spite of the more complex assumptions, this result is still very similar to the simple scaling leading to Eq. (15a).

A third result has been obtained assuming uniform acceleration and constant beta function. Assume also a focussing lattice with a 90° phase advance per cell, with M focussing quadrupoles which jitter in transverse position with an rms displacement d. For this case the displacement of the trailing bunch grows to an amplitude[13]

$$\frac{\Delta x_2}{\sigma_f} = \frac{r_0 NLW_1}{2\pi} \left(\frac{\beta}{\gamma_f \epsilon_n}\right)^{1/2} M^{1/2} d \quad . \tag{17a}$$

Magnet misalignment is seen to impose a stricter limitation than accelerator section misalignment. For a 90° lattice the number of magnets is $M = 4L/\pi\beta$. Introducing also the gradient $G' = d\gamma/dz$, the preceding expression becomes

$$\frac{\Delta x_2}{\sigma_f} = \frac{r_0 NLW_1 d}{\pi^{3/2} \left(\epsilon_n G'\right)^{1/2}} \quad . \tag{17b}$$

By introducing an energy spread within the bunch (Landau damping), the emittance growth due to the dipole wake can be greatly reduced.[14] In Sec. V C it is shown that, for the simple case of a uniform structure having constant energy and beta function with an initial offset x_0, the growth in the transverse oscillation amplitude of the tail particle is reduced by a factor

$$\frac{\pi\beta}{2L(\Delta p/p)} \quad . \tag{18}$$

Thus Eq. (15a) becomes

$$\frac{\Delta x_2}{\sigma_f} = \frac{\pi r_0 NW_1 \beta^{3/2} x_0}{8\left(\gamma\epsilon_n\right)^{1/2} (\Delta p/p)} \quad . \tag{19}$$

Landau damping is seen to be very effective in reducing emittance growth due to injection errors. It may be less effective in reducing the effect of alignment errors and magnet jitter, but detailed calculations remain to be done.

E. Design Strategy for Linear Colliders

Based on the relations summarized in the preceding sections, there are a number of ways to approach the design of a linear collider. We first assume that the energy, desired luminosity and allowable beamstrahlung are fixed. To carry this process further, some expressions which combine some of the preceding basic relations are useful. From Eqs. (1b) and (11),

$$\frac{\eta_{tot}\, P_{ac}\,(\mathrm{MW})}{E_0\,(\mathrm{TeV})\;[\mathcal{L}\,(10^{32})]^{1/2}} = 0.57 \left[\frac{bfR\sigma_y^2\,(\mu\mathrm{m})}{H_D}\right]^{1/2} , \tag{20}$$

where $\eta_{tot} = b\eta_b\eta_{rf}\eta_s = P_b/P_{ac}$. A reasonable upper value for η_{tot} is 0.15 (assuming $\eta_{rf}\eta_s = 0.5$, $b\eta_b \approx 0.3$). It is clear that the number of bunches, the repetition rate and the beam area $R\sigma_y^2$ should be chosen as low as possible to keep the AC power down. However, the constraint imposed by beamstrahlung must also be considered. From Eqs. (1b) and (4b),

$$\frac{E_0\,(\mathrm{TeV})\mathcal{L}\,(10^{32})}{\delta_{c\ell}} = 2.0\times10^{-3}\left(\frac{1+R}{R}\right)^2 [bfR\sigma_z\,(\mathrm{mm})] \quad . \tag{21a}$$

In order to get a high luminosity in the classical beamstrahlung regime, we see that, in contrast to the requirement set by Eq. (20) a large number of bunches and a high repetition frequency is desirable, as is a long bunch length. Since $\sigma_z = (\sigma_z/\lambda)\lambda$, this also implies a long RF wavelength. However, the aspect ratio R can be increased to allow reduced values of b, f and λ. In the quantum regime, the equivalent expression is, using Eq. (9) in Eq. (21a),

$$\frac{\mathcal{L}\,(10^{32})}{E_0\,(\mathrm{TeV})\,\delta_Q^3} = 1.0\times10^{-5}\left(\frac{1+R}{R}\right)^2\left[\frac{bfR}{\sigma_z\,(\mathrm{mm})}\right] \quad . \tag{21b}$$

It is seen that, contrary to the classical case, a short bunch length is helpful.

A final set of scaling relations is informative. Squaring Eq. (20) and dividing by Eqs. (21a) and (21b), we obtain

$$\frac{\delta_{c\ell}H_\Upsilon P_b^2\,(\mathrm{MW})}{E_0^3\,(\mathrm{TeV})\;\mathcal{L}\,(10^{32})} = 1.5\times10^2\left(\frac{R}{1+R}\right)^2\frac{\sigma_y^2\,(\mu\mathrm{m})}{H_D\,\sigma_z\,(\mathrm{mm})} , \tag{22a}$$

$$\frac{\delta_Q^3 P_b^2\,(\mathrm{MW})}{E_0\,(\mathrm{TeV})\;\mathcal{L}\,(10^{32})} = 3.0\times10^4\left(\frac{R}{1+R}\right)^2\frac{\sigma_y^2\,(\mu\mathrm{m})\,\sigma_z\,(\mathrm{mm})}{H_D} \quad . \tag{22b}$$

Suppose $E_0 = 5$ TeV, $\delta = 0.3$, $\mathcal{L} = 10^{34}$, $P_{ac} = 100$ MW, and $\eta_{tot} = 0.15$ ($P_b = 15$ MW). The two expression above then give

$$C\ell: \qquad \frac{\sigma_y^2\ (\mu\text{m})}{\sigma_z\ (\text{mm})} \approx 3.4 \times 10^{-7}\ H_D \left(\frac{1+R}{R}\right)^2$$

$$Q: \qquad \sigma_y^2\ (\mu\text{m})\ \sigma_z\ (\text{mm}) \approx 3.8 \times 10^{-9}\ H_D \left(\frac{1+R}{R}\right)^2 \quad ,$$

for the classical and quantum regimes respectively. Suppose $\sigma_z \approx 10^3\sigma_y$. Then in the classical and quantum regimes $\sigma_y \approx 3 \times 10^{-7}$ μm and 1.5×10^{-3} μm, respectively. In both cases, one is forced to extremely small bunch dimensions.

When the bunch length has been chosen, the scale of the collider design has been set. Since $\sigma_z = (\sigma_z/\lambda)\lambda$, and since σ_z/λ cannot be chosen arbitrarily, the choice of bunch length is equivalent to a choice of operating wavelength. From Eqs. (22) the transverse dimension σ_y is now fixed (we have to guess initially whether we are in the classical or quantum regime, or else iterate on H_γ in Eq. (22a). From Eq. (20) the product bfR is now fixed. It would be reasonable to choose $f = 180$ or 360. Some flexibility then remains in choosing b and R. The remaining quantities N, D, H_D, and $\overline{\Upsilon}$ are now readily calculated, and all parameters can be checked for consistency. It is left to the reader to continue this program for $E_0 = 5$ TeV, $\mathcal{L} = 10^{34}$ cm^{-2} and $P_{ac} = 100$ MW. It will be seen that for reasonable values of b, f and R, $\overline{\Upsilon} > 1$ and the parameters are pushed into the quantum regime.

III. ACCELERATING STRUCTURES

A. Structure Design

In this section we review a few basic expressions related to the design of traveling wave accelerating structures. Consider a periodic structure consisting of identical coupled cells with an RF feed at one end. For such a "constant impedance" structure, the group velocity v_g and attenuation per unit length are uniform along the length of the structure. The accelerating field is attenuated by a factor $e^{-\tau}$ along a structure of length ℓ, where

$$\tau = \left(\frac{\omega}{2Q}\right) T_f \quad , \tag{23}$$

and $T_f = \ell/v_g$ is the filling time. Thus for a given τ, the filling time varies as $\omega^{-3/2}$. The structure efficiency, η_s, is defined as the ratio $V^2(\tau)/V^2(0)$, where $V(\tau)$ is the actual voltage delivered by the structure and $V(0)$ is the voltage that would be obtained if the attenuation were zero. An input RF pulse with peak power P_0 and length T_f is assumed. Because of attenuation the energy per pulse required to reach a given accelerating gradient is increased by $1/\eta_s$. The structure efficiency is given by[15]

$$\eta_s = \left(\frac{1 - e^{-\tau}}{\tau}\right)^2 \quad . \tag{24}$$

This function is plotted in Fig. 2. The efficiency is seen to approach 100% as $\tau \to 0$. On the other hand, the peak RF power required per unit length is

$$\frac{P_0}{\ell} = \frac{G^2}{r} f(\tau) \; , \tag{25}$$

$$f(\tau) = \frac{\tau}{2} \left(1 - e^{-\tau}\right)^2 = \frac{1}{2\tau\eta_s} \; .$$

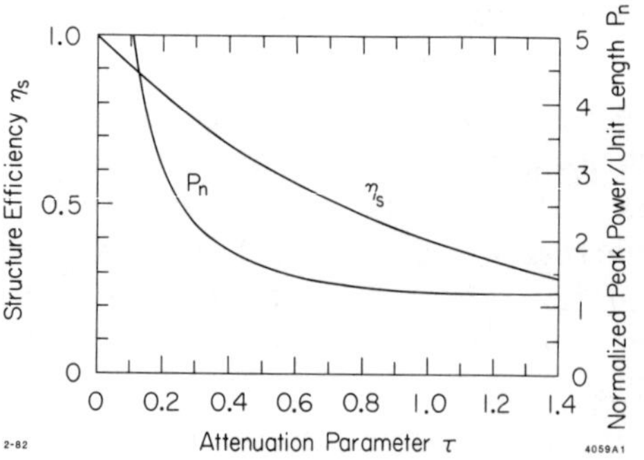

Fig. 2. Structure efficiency η_s and normalized peak power per unit length $P_n = P_0 r/G^2\ell$, as a function of the attenuation parameter τ.

Here r is the shunt impedance per unit length and $G = V/\ell$ is the average accelerating gradient. Thus 100% efficiency ($\tau \to 0$) implies both zero filling time and infinite peak power. The function $f(\tau)$ has a minimum at $\tau = 1.26$, with f(1.26) = 1.23, as shown in Fig. 2. At this minimum, however, the efficiency is only 32%. By decreasing τ to 0.5, the efficiency is increased to 62% (almost double), while the peak power is increased by only 32%. Thus $\tau \approx 0.5$ gives a reasonable compromise between efficiency and peak power requirement. For the SLAC structure $\tau = 0.57$ and $\eta_s = 58\%$.

Present-day high power pulsed klystrons operate with a conversion efficiency of 45–55%. The efficiency for conversion of AC to DC pulsed power (modulator efficiency) is 80–90%. It is difficult to predict how much these efficiencies might be improved by future technological advances. A Lasertron[16] RF source operating directly from a DC power supply might, for example, achieve an efficiency on the order of 75%. Together with a structure efficiency of 65% ($\tau = 0.45$), this gives a possible net efficiency $\eta_{rf}\eta_s = 50\%$. An additional structure parameter is the loss parameter per unit length,

$$k_1 \equiv \frac{G^2}{4u} \tag{26}$$

where u is the stored energy per unit length. The factor of four comes from the fact that the loss parameter was originally defined by $u = k_1 q^2$, where u is the energy deposited in the accelerating mode per unit length by a point charge passing through a structure originally empty of energy. For a simple pillbox cavity of length g, the parameter k_1 is given by

$$k_1 = \left[0.456 \times 10^{12} \, \frac{\Omega - m}{s}\right] \frac{T^2}{\lambda^2} \quad ,$$

$$T = \frac{\sin \frac{\pi g}{\lambda}}{\frac{\pi g}{\lambda}} \quad .$$

We see from this expression, and directly from Eq. (26), that $k_1 \sim \omega^2$. For a SLAC type disk loaded structure with $\tau = 0.5$ and period $\lambda/3$,

$$k_1 = \frac{0.20 \times 10^{12}}{\lambda^2 \text{ (m)}} \, \frac{\Omega}{s - m} \text{ or } \frac{V}{C - m} \quad . \tag{27}$$

The SLAC structure doesn't do quite as well as a chain of simple pillboxes because of the finite disk thickness and field fringing in the disk aperture.

It is important to note that the value of k_1 depends strongly on the radius $\underline{a}$ of the disk aperture. This is shown in Fig. 3 for the SLAC structure. Approximately, $k_1 \sim a^{-1}$ for $a/\lambda \approx 0.1$.

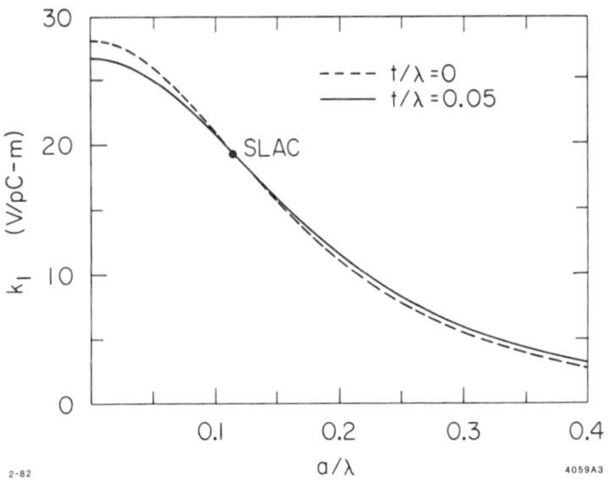

Fig. 3. Structure parameter k_1 as a function of beam aperture radius for the average cell in the SLAC disk-loaded structure ($\lambda = 10.5$ cm, $a = 1.163$ cm and $t = 0.584$ cm).

B. Peak Power Requirement

The filling time of a typical disk loaded accelerating structure with $\tau = 0.5$ will be 0.7 μs at $\lambda = 10$ cm [see Eq. (23)]. We can therefore write

$$T_f(ns) \;=\; 22\,[\lambda\ (\text{cm})]^{3/2} \quad . \tag{28a}$$

The average energy per unit length required from a power source is

$$\bar{u}\left(\frac{J}{m}\right) = \frac{G^2}{4k_1\eta_s} = \frac{\lambda^2\ (\text{cm})\ G^2\ (\text{MV/m})}{5.0\times 10^3} \tag{28b}$$

for the same structure. The peak power requirement is

$$\frac{P_0}{L}\ (\text{MW/m}) = \frac{\bar{u}}{T_f} = \frac{\lambda^{1/2}\ (\text{cm})\ G^2\ (\text{MV/m}))}{110} \quad . \tag{28c}$$

Results from Eqs. (28) are plotted in Fig. 4 for wavelengths from 1 mm to 10 cm and accelerating gradients from 50 to 500 MV/m.

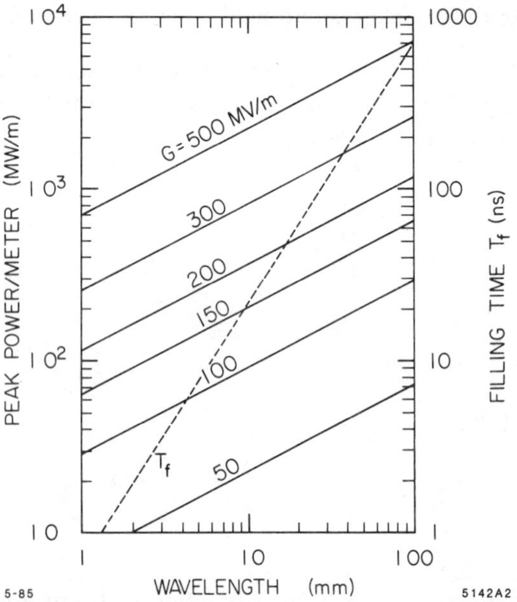

Fig. 4. Filling time and peak power per unit length as a function of wavelength for a typical disk-loaded structure with $\tau = 0.5$.

C. LIMITS ON ACCELERATING GRADIENT

Several effects can impose limitations on the RF fields in an accelerating structure. The easiest to calculate and understand in simple physical terms is surface heating. The power per unit area absorbed by a surface with surface resistance R_s is

$$P_a = \frac{1}{2} R_s \widehat{H}^2 = \frac{1}{2} G^2 \frac{R_s}{Z_s^2} , \quad (29)$$

where $\widehat{H}$ is the peak magnetic field and $Z_s = G/\widehat{H}$ is an impedance defined by the geometry of the structure. For a typical disk-loaded structure, $Z_s \approx 400\Omega$. In terms of the power per unit area and the pulse length, T_p, the temperature rise is

$$\Delta T = \frac{2P_a}{K} \left[\frac{DT_p}{\pi}\right]^{1/2} ,$$

where K is the thermal conductivity, $D = K/C_s\rho$ is the thermal diffusivity, C_s is the specific heat and ρ the density. Substituting for P_a from Eq. (29),

$$G = Z_s \left(\frac{K\Delta T}{R_s}\right)^{1/2} \left(\frac{\pi}{DT_p}\right)^{1/4} . \quad (30)$$

If T_p is set equal to the filling time (which scales are $\omega^{3/2}$), and since $R_s \sim \omega^{1/2}$, then $G \sim \omega^{1/8}$. Putting in typical numbers at $\lambda = 10$ cm for copper ($Z_s = 400\ \Omega$, $R_s = .014\ \Omega$, $K = 3.8$ W/cm–°K, $D = 1.1$ cm^2/s), and assuming also that the pulse length is equal to a typical filling time $T_p \approx 0.7\ \mu$s, then the gradient required to raise the surface to the melting point is $G \approx 1$ GeV/m. This model breaks down at $\lambda \approx 30\ \mu$m, when the diffusion distance $(DT_f)^{1/2}$ is on the order of the skin depth. Under this condition the filling time is about 5 ps and the gradient is about 2.6 GeV/m. For still shorter wavelengths, the temperature rise is determined only by the specific heat per unit volume, giving

$$\Delta T = \frac{P_a T_p}{\rho C_s \delta} = \frac{P_a T_p\, D}{K\delta} ,$$

where δ is the skin depth. Substituting for P_a from Eq. (29),

$$G = Z_s \left(\frac{2K\lambda\Delta T}{\pi D Z_0 T_p}\right)^{1/2} , \quad (31)$$

where $Z_0 = 377\Omega$. In this regime, G scales as $\omega^{1/4}$. Another limit is obviously encountered when the filling time becomes comparable to one rf cycle, $T_f \approx \lambda/c$. At this limit a velocity-of-light wave travels a distance equal to the skin depth in one filling time. Again scaling as $\lambda^{3/2}$ from $T_p = T_f \approx 0.7\ \mu s$ at $\lambda = 10$ cm, we find $\delta = cT_f$ at $\lambda \approx 0.02\ \mu m$. The gradient is on the order of 15 GeV/m, and the filling time is 6×10^{-17} sec.

The variation of gradient with wavelength due to surface heating is plotted for the different regimes in Fig. 5. A further discussion is given in the report of the Near Field Group, in these Proceedings.

The electric field limitations on gradient are less amendable to calculation. We expect the gradient limit to be a function of both frequency and pulse length. The well-known Kilpatrick criterion[17] predicts for CW or very long RF pulses,

$$E_b\ (\mathrm{MV/m}) \approx 25\ [f\ (\mathrm{GHz})]^{1/2} \quad , \tag{32}$$

for frequencies greater than a few GHz. Here E_b presumably can be taken as the peak field E_p at the surface of an accelerating structure, where typically $G \approx 0.5E_p$.

The variation in breakdown field with pulse length is also not a precisely determined function. Some data[18] at 2856 MHz on the power flow at breakdown in a resonant ring, used at SLAC to test klystron windows, are fit by[♯1]

$$E_b\,(T_p) = E_b(cw) \left\{ 1 + \frac{4.5}{[T_p\ (\mu s)]^{1/4}} \right\} \quad . \tag{33}$$

Combining Eqs. (32) and (33), for very short pulses

$$E_b \sim \omega^{1/2}\, T_p^{-1/4} \quad . \tag{34}$$

If the pulse length is equal to the filling time, and again assuming the filling time scales as $\omega^{-3/2}$, then $E_b \sim \omega^{7/8}$.

Two measurements have been made on breakdown in short resonant sections of disk-loaded structure near 3 GHz. Loew and Wang[20] at SLAC reached a peak surface field of 259 MV/m without breakdown at 2856 MHz for a pulse length of about 1.5 μs. Equations (32) and (33) predict a breakdown field of 215 MV/m. Tanabe[21] working at 2997 MHz, reports a peak field of about 240 MV/m at a pulse length of 4 μs, with some surface damage due to breakdown. If we use Eqs. (33) and (34) to scale these two results to a filling time of 0.7 μs at

♯1 The data can also be fit by a $(1+\mathrm{const}/T_p^{1/3})$ variation. This scaling with T_p is in agreement with the behavior for DC pulses.[19] However, the enhancement factor over the Kilpatrick limit at S–band is then only a factor of three, which is inconsistent with experimental measurements.[20,21] If the $T_p^{1/3}$ variation is accepted, then the electric field breakdown limit plotted in Fig. 5 varies as ω^1 instead of $\omega^{7/8}$.

3 GHz, we obtain 310 MV/m for the Loew and Wang measurement, with no breakdown, and 350 MV/m for the Tanabe measurement, with breakdown and surface damage. As a calibration point on our plot of breakdown field versus wavelength, we therefore take 160 MV/m (assuming $G \approx E_p/2$) at $\lambda = 10$ cm. This is plotted in Fig. 5.

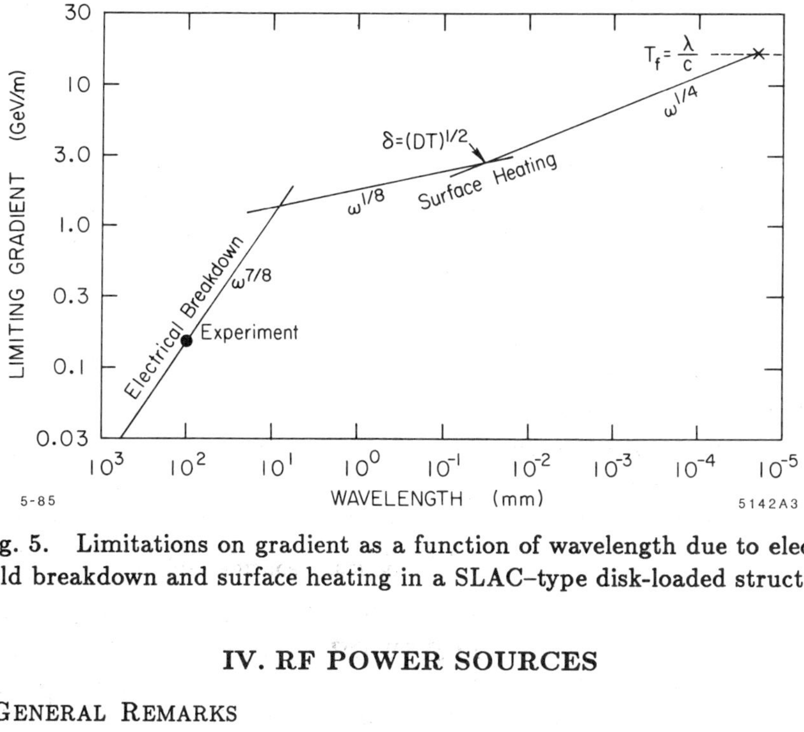

Fig. 5. Limitations on gradient as a function of wavelength due to electric field breakdown and surface heating in a SLAC–type disk-loaded structure.

IV. RF POWER SOURCES

A. General Remarks

From Fig. 4 we see that a collider linac operating at a gradient of (for example) 100 MV/m requires a peak power of 300 MW/m at $\lambda = 10$ cm and a peak power of 100 MW/m at $\lambda = 1$ cm. The pulse lengths for the two cases are about 700 ns and 20 ns, respectively. The required peak power can be generated either by external microwave tubes, or by a high current driving beam which can be external or internal to the accelerating structure. Further, the required peak power can be generated directly by the source at a pulse length equal to the filling time or alternatively at a lower peak power level and longer pulse length, followed by some pulse compression technique to raise the peak power to the required level. These alternatives are considered in the following sections.

RF sources which might be suitable for linear colliders are discussed in a recent review by Granatstein[22]. Sources which have produced peak power levels on the order of 100 MW in the wavelength range 1-10 cm are: virtual cathode oscillators (Vircators), backward wave oscillators, magnetrons, gyrotrons, klystrons and free electron lasers. Oscillators, however, are not suitable as sources to drive a collider. Many amplifiers with good phase stability, driven from a common source, will be required. This reduces the possible sources to klystrons, gyrotron amplifiers (gyroklystrons), FEL's and possibly some type of crossed field amplifier. The latter device is a dark horse and will not be considered further. FEL's and other possible two-beam accelerators are considered briefly in Sec. IV D.

B. Klystrons and Gyroklystrons

For many years, klystron have been the RF source of choice for the highest peak power at wavelengths on the order of 10 cm. In 1970 a klystron was designed at $\lambda = 9$ cm to produce a peak power of 1 GW at $T_p = 15$ ns.[23] However, the tube failed before it could be tested at full output power. Recently, a klystron has been designed at SLAC to produce 150 MW at $\lambda = 10.5$ cm at a pulse length of 1 μs. This tube has now achieved[24] the design output power with an efficiency of 55%.

We have noted that a peak power of 300 MW/m is needed to reach an interesting accelerating gradient (100 MV/m) at $\lambda = 10$ cm. Also, it would be desirable to reduce the number of sources by spacing them further apart than 1 m. Furthermore, the optimum operating wavelength for a linear collider will almost certainly be shorter than 10 cm. It is difficult to specify a precise scaling law for the variation of peak power output of a klystron with wavelength, but almost certainly it will decrease more rapidly than the $\lambda^{1/2}$ requirement given by Eq. (28c). We conclude that some form of pulse compression will be needed if klystrons are used as an RF source for a linear colliders. If so, a premium will be placed on efficiency and reliability, rather than on peak power, assuming that a power level in the range 50–100 MW can be attained at the desired wavelength.

Gyroklystron are inherently capable of operating at shorter wavelengths than klystrons. For a collider operating in the wavelength range at or below 3 cm, a gyroklystron will probably be the RF source of choice (excluding for the moment two-beam concepts). Granatstein[25] has recently reviewed the capabilities of high peak power gyroklystrons. A design calculation has been made for a gyroklystron capable of delivering 300 MW at 9 GHz. This source would power two meters of typical structure at a gradient of 100 MV/m.

C. Lasertron RF Source

In recent years a new possibility for a high efficiency RF source has been the subject of increasing interest—the Lasertron. Figure 6 shows a schematic

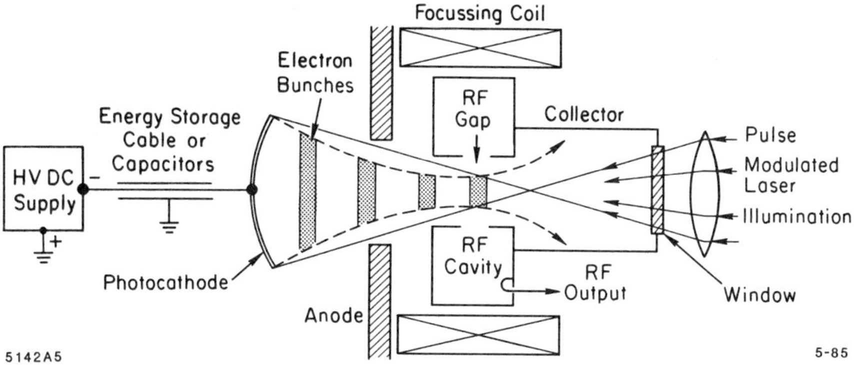

Fig. 6. Schematic diagram of a Lasertron RF power source.

diagram of this device. A laser beam, pulse modulated at the desired RF frequency, is incident on a photocathode. Electron bunches, each a small fraction of the RF period in length, are emitted by the cathode, accelerated to high voltage and passed through the gap of an RF cavity. If the RF voltage across the output gap is about equal to the DC beam voltage, each bunch is brought nearly to rest by the RF field in the gap, thereby converting DC to RF energy with very high efficiency. A further interesting feature of this device is that it can in principle by operated directly from a DC power source, eliminating the inefficiency associated with a pulse modulator. The laser-driven photocathode acts, in essence, as a switch operating at microwave frequencies, capable of the direct production of microwave power from a DC source.

Expermental work is currently underway on the Lasertron in Japan[26] and at SLAC[16]. At SLAC, a proof of principle test is underway to produce a Lasertron with a peak output power of 35 MW. Numerical simulations[27] indicate that an efficiency exceeding 70% is possible if a double output gap composed of two magnetically coupled output cavities is used. It is foreseen that peak power levels of 100 MW or more can be produced at a wavelength on the order of 10 cm. It is not so clear, however, whether this device can be scaled to produce high peak output power at substantially shorter wavelengths. Simulations are being carried out at SLAC to explore this possibility.

D. Pulse Compression

From the results of the preceding two section, it is seen that the direct generation of peak power level on the order of 300 MW/m by microwave tubes will be difficult, especially at shorter wavelengths. It should be emphasized again that it is also highly desirable to reduce the total number of RF sources by producing the required power level per meter from sources spaced at less frequent intervals. Thus some form of pulse compression and power splitting will almost certainly be required for a very long collider operating at a high gradient. Suppose we require a gradient of 150 MV/m at a wavelength of 3.5 cm. From Fig. 4, this implies a peak power of about 400 MW/m. A filling time of about 125 μs is required. Suppose power sources are available capable of generating 100 MW for 1 μs. If the peak power can be multiplied by a factor of eight and split two ways, each source is then capable of feeding two meters of structure (assuming the pulse compression can be carried out with an efficiency close to 100%).

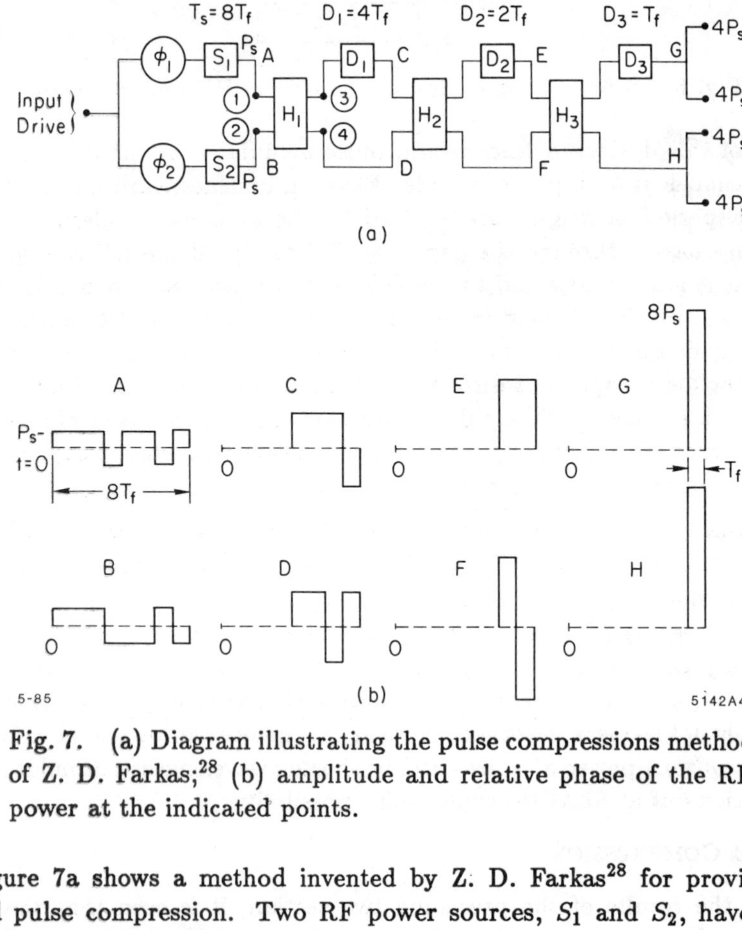

Fig. 7. (a) Diagram illustrating the pulse compressions method of Z. D. Farkas;[28] (b) amplitude and relative phase of the RF power at the indicated points.

Figure 7a shows a method invented by Z. D. Farkas[28] for providing the desired pulse compression. Two RF power sources, S_1 and S_2, have a pulse length equal to eight time the structure filling time T_f. D_1, D_2 and D_3 are delay lines having, respectively, delays of $4T_f$, $2T_f$ and T_f. H_1, H_2 and H_3 are so-called 3db hybrids. If power is applied at either input terminal of such a device (terminal 1 or terminal 2 of H_1 for example), half the power appears at each output terminal (terminals 3 and 4 of H_1). There is, however, a 90° phase shift between terminals 1 and 4 and between terminals 2 and 3. Thus, if the phase difference between the waves incident at terminals 1 and 2 is ±90°, it is readily seen by supposition that the combined power will appear at either terminal 3 or 4, depending on the sign of the phase difference. By changing the relative phase between the two input terminals by 180°, the power can therefore be switched from one output terminal to the other. By pulse coding the low power phase shifters ϕ_2 and ϕ_2 correctly in each of the eight time slots in the incident pulse, this switching is carried out in the hybrid H_1, H_2 and H_3 at increasing power levels and reduced pulse lengths. The process is illustrated in Fig. 7b, where the relative phase and power level is shown for the points indicated. The process can be extended, in principle, to any desired power multiplication by a factor of 2^n. Of course, the delay lines must not introduce significant attenuation. They can be either superconducting or over-moded room temperature copper pipes.

E. Two-Beam and Wake Field Accelerators

The energy per unit length required to produce an intense accelerating field can be produced by a variety of means other than by conventional external microwave power sources. An intense driving bunch, with appropriately shaped current distribution, can be injected on the axis of the accelerating structure ahead of the bunch to be accelerated (the collinear wake field accelerator).[29] A hollow ring-shaped driving bunch, which produces inwardly propagating wake fields in a suitable structure, can be used (the Voss-Weiland wake field accelerator).[30,31]

A low energy, high current beam moving in an external circuit parallel to the accelerating structure can be sent through a series of wigglers to generate the required RF power.[32,33] The energy lost by the driving beam is made up periodically by induction units. In addition to an FEL of this type, in which the parallel driving beam interacts with the transverse component of the RF field, a two-beam accelerator in which the driving beam interacts with a longitudinal RF field is also possible. In this device[34] a bunched beam, possibly produced by a laser-modulated photocathode, passes periodically through klystron-type cavities which extract a portion of the beam energy. The energy loss can again be made up by induction units. The disadvantage of this type of two-beam accelerator, in contrast to the FEL, is that the transverse dimensions of the RF interaction region must be comparable to the RF wavelength.

V. WAKE FIELDS

A. Delta Function Wake Potentials

The delta function wake $W(\tau)$ is the potential seen by a test charge following at a distance $c\tau$ behind a point unit charge passing through a component or structure. Both the test charge and the unit driving charge are usually assumed to be travelling on parallel paths at the speed of light. The wake potential is then causal, such that $W(\tau) \equiv 0$ for $\tau < 0$. The instantaneous forces experienced by the test charge in response to the complex pattern of "wake fields" excited in even a simple structure are not usually of interest. What matters is the integrated force, or total potential seen by the test charge on passing through an entire component, or through one period of a periodic structure. The potential may be either longitudinal or transverse.

The theory underlying the wake potential description has been extensively developed in recent years. The analytic development is somewhat complex, with many subtleties. We give here only a few results of use in scaling wake field effects for relativistic particles in typical accelerating structures. The reader is referred to Refs. 35-38 for a more complete exposition.

The longitudinal wake field for the nth mode excited by a point charge q at radius r_q and azimuthal angle $\phi = 0$ in a cylindrically symmetric periodic structure is given by

$$E_z(r, \phi, \tau) = -2qk_n \left(\frac{r}{a}\right)^m \left(\frac{r_q}{a}\right)^m \cos m\phi \, \cos \omega_n \tau \quad . \tag{35}$$

Here a is the minimum wall radius of the structure (the disk hole radius), m gives the azimuthal dependence of the mode and

$$k_n \equiv \frac{E_{on}^2}{4u_n} \quad , \tag{36}$$

where u_n is the stored energy per unit length and E_{on} is the longitudinal synchronous field component at radius $r = a$. The delta function wake potential for the nth mode is now defined as the field per unit charge and per unit offset in both r_q and r at angle $\phi = 0$. Thus

$$\begin{aligned} W_{zn}(\tau) &= \frac{2k_n}{a^{2m}} \cos \omega_n \tau \quad , \\ E_{zn} &= -q W_{zn}(\tau)\, r^m \, r_q^m \cos m\phi \quad . \end{aligned} \tag{37}$$

The wake potential can also be defined as the potential per cell of the structure, rather than per unit length. In this case E_{on} and u_n in Eq. (36) are replaced by $E_{on}p$ and $u_n p$, where p is the periodic length. Note that the longitudinal wake for azimuthally symmetric ($m = 0$) modes is independent of the radial positions of both the driving charge and the trailing test charge.

To find the total wake potential behind a point charge for a given value of m, one must in principle sum over all possible modes supported by the structure with symmetry $\cos m\phi$,

$$W_z(\tau) = \frac{2}{a^{2m}} \sum_{n=1}^{\infty} k_n \cos \omega_n \tau \quad . \tag{38}$$

In practice a finite number of modes are calculated by an appropriate computer code, and an "analytic extension" is added to take care of the modes with frequencies above the limit of the calculation. The analytic extension is based on the fact that at sufficiently high frequencies the impedance $(dk/d\omega)$ can be shown to vary as $\omega^{-3/2}$ for typical accelerating structures.

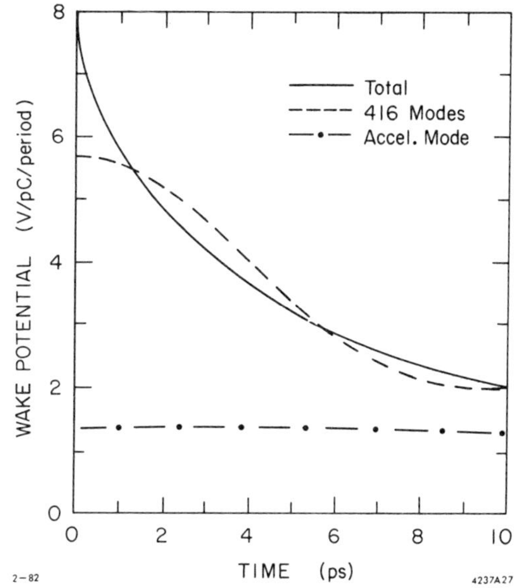

Fig. 8. Longitudinal wake potential per cell for the average cell in the SLAC disk-loaded structure in the range 0–10 ps.

The wake obtained by summing over 416 modes for the SLAC structure is shown by the dashed curve in Fig. 8 for 0–10 ps. Adding on an analytic extension gives the solid curve. The fundamental (accelerating) mode is also shown for comparison. Note that the total wake at $\tau = 0$ is about a factor of six greater than that given by the fundamental mode alone. The wake out to 300 ps is shown in Fig. 9.

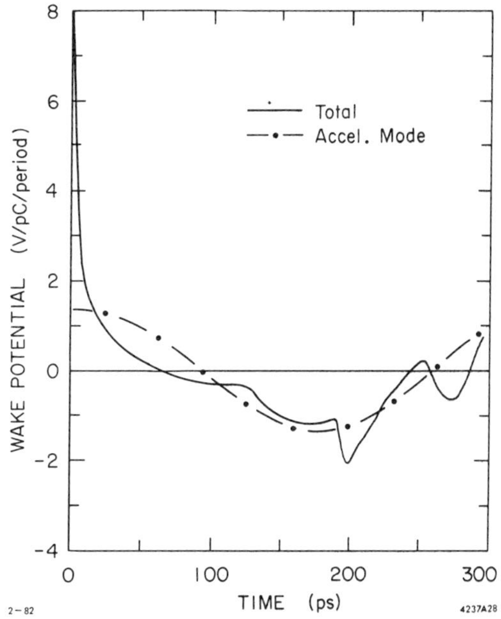

Fig. 9. Longitudinal wake potential per cell for the SLAC structure in the range 0–300 ps.

If Fig. 8 shows the wake seen by a trailing test charge, one can ask what potential is seen by the driving (point) charge itself. It is easy to show[35] from conservation of energy that the potential acting on the driving charge is just one-half of the wake potential seen by a test charge following an infinitesimal distance behind. Thus for the SLAC structure a point picocoulomb of charge experiences a retarding potential of 4 V per cell.

The transverse wake field is also of great interest for collider design. It is given, again for a single mode in a cylindrically symmetric periodic structure, by

$$\vec{E}_{\perp}(r,\phi,\tau) = 2qm\left(\frac{k_{nc}}{\omega_{na}}\right)\left(\frac{r}{a}\right)^{m}\left(\frac{r_q}{a}\right)^{m-1}\left(\hat{r}\cos m\phi - \hat{\phi}\sin m\phi\right)\sin\omega_n\tau \quad , \tag{39}$$

where $\hat{r}$ and $\hat{\phi}$ we unit vectors. Here $e\vec{E}_{\perp}$ gives the total transverse force acting on the trailing particle. The delta function wake potential is now defined by

$$W_{\perp}(\tau) = \frac{2mk_n c}{\omega_n a^{2m}}\sin\omega_n\tau \quad , \tag{40}$$

and

$$\vec{E}_{\perp} = qW_{\perp}(\tau)\,r^{m-1}\,r_q^m\left(\hat{r}\cos m\phi - \hat{\phi}\sin m\phi\right) \quad .$$

For the important case of the dipole ($m = 1$) modes, the deflection field varies linearly with the offset of the leading charge and is uniform across the entire aperture of the structure behind the leading charge. Note that the wake potentials W_z and $W_{\perp}$ are scalar function of τ only. We see also that the longitudinal and transverse wake fields are related by

$$\frac{\partial\vec{E}_{\perp}}{\partial\tau} = -c\nabla_{\perp}E_z \quad . \tag{41}$$

The total delta-function wake potential is again obtained by summing over many modes and adding an analytic extension as described in Ref. 35. Results for the dipole mode for the SLAC structure are shown in Figs. 10 and 11. Note that, in contrast to the longitudinal wake, the dipole wake (and all transverse wake potentials) starts at zero at time $\tau = 0$ and rises to a first maximum at a distance behind the driving charge which is comparable to the iris aperture radius. At long distances behind the driving charge, the total wake is given by a beating of the wakes due to the two or three lowest frequency modes. The period of the resulting semi-regular oscillation is substantially that of the lowest frequency mode.

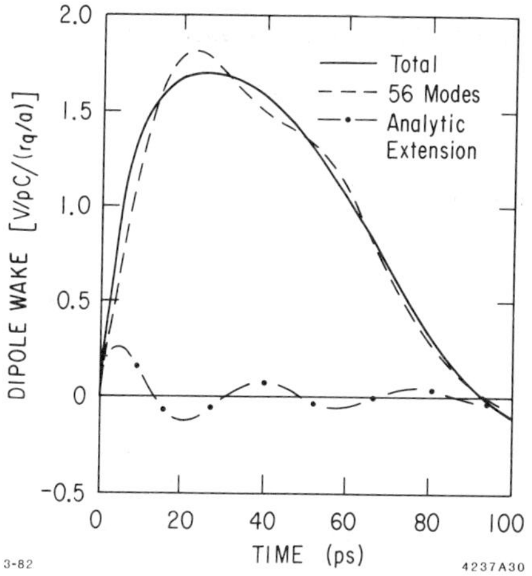

Fig. 10. Dipole wake potential per cell for the average cell in the SLAC disk-loaded structure in the range 0–100 ps.

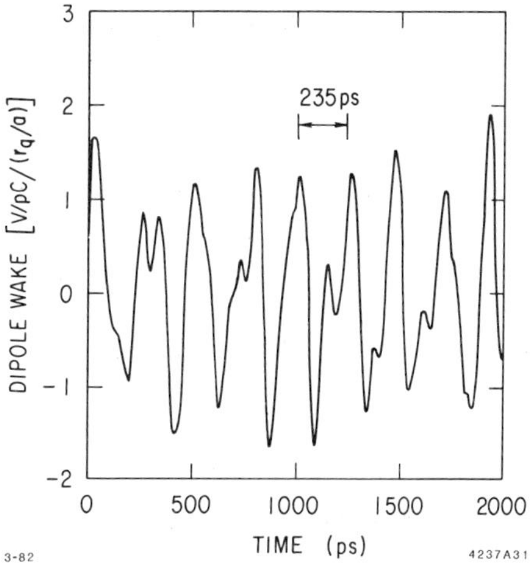

Fig. 11. Dipole wake potential per cell for the SLAC structure in the range 0–2000 ps. The period of the lowest frequency dipole mode is 235 ps.

From Eq. (36) it is seen that $k_n \sim \omega^2$. Thus from Eqs. (37) and (40) the wake potentials per unit length scale as

$$\begin{aligned} W_z^m &\sim \omega^{2m+2} , \qquad m \geq 0 \\ W_\perp^m &\sim \omega^{2m+1} , \qquad m > 0 \ . \end{aligned} \tag{42}$$

The wake potentials per cell of a periodic structure scale as one power lower than above. The horizontal time axes in Figs. 8–11 also scale, of course, in proportion to the wavelength.

The above scaling is for a constant geometry such that all dimensions vary in proportion to wavelength. The case in which only the disk hole radius is varied for a structure of fixed frequency is also of interest. This scaling has been investigated for the SLAC structure by K. Bane[39]. The intercept at $\tau = 0$ for the longitudinal wake is found to vary as

$$W_z(0) \sim a^{-1.7} \ . \tag{43a}$$

The time at which the longitudinal wake falls to one-half of its value at $\tau = 0$ is given by

$$c\tau_{1/2} \approx .09\,a \quad . \tag{43b}$$

The amplitude of the first maximum of the dipole wake is found to vary with beam aperture radius as

$$W_1(\tau_m) \sim a^{-2.25} \quad , \tag{43c}$$

where the value of τ_n varies as

$$c\tau_m \approx 0.65\,a \quad . \tag{43d}$$

The initial slope of the dipole wake for $\tau << \tau_n$ varies as

$$\frac{dW_1}{d\tau} \sim a^{-3.5} \quad . \tag{43e}$$

Finally, a note about dimensions. In the scaling relations developed in Sec. II, it was found to be convenient to use *cgs*–Gaussian units. The wake potentials in *cgs* units are readily obtained from the potential in *mks* units by multiplying by $4\pi\epsilon_0$. Thus the $m = 0$ longitudinal wake potential per unit length, with dimension $V/(C-m)$ in mks units, has dimensions $1/\text{m}^2$ (or $1/\text{cm}^2$) in Gaussian units. We have, in general, for the wake potential per unit length

		mks	Gaussian
$m = 0$	W_z	$\frac{V}{C-m}$	$\frac{1}{m^2}$
$m = 1$	$W_\perp$	$\frac{V}{C-m^2}$	$\frac{1}{m^3}$
	W_z	$\frac{V}{C-m^3}$	$\frac{1}{m^4}$
$m = 2$	$W_\perp$	$\frac{V}{C-m^4}$	$\frac{1}{m^5}$
	W_z	$\frac{V}{C-m^5}$	$\frac{1}{m^6}$

B. Wake Potentials for Charge Distributions

The delta-function wake fields or wake potentials for a point charge, discussed in the previous section, can be used as Green's functions to compute the longitudinal and transverse potentials in an arbitrary charge distribution $I(t)$. Thus for the important case of the longitudinal accelerating mode, the single bunch beam loading potential at time t within the bunch is given by

$$E_b(t) = \int_{-\infty}^{t} W_z(t-\tau)\, I(\tau)\, d\tau \quad . \tag{44}$$

If this expression is divided by the charge, the potential in the bunch per unit charge per unit length (or the potential per discrete component) is obtained. This integrated potential, or bunch potential, is sometimes also called the wake potential. It is unfortunate that the terms "wake potential" and "wake field" are used to refer to several different quantities. The reader is cautioned to check the precise meaning of these terms in each case.

The integrated wake potential for the SLAC structure is shown in Fig. 12 for three different bunch lengths. The total energy gain of a particle at time t in the distribution in than obtain by a superposition of the single bunch beam loading potential per unit length, given by Eq. (44), and the RF accelerating field produced by the external RF source:

$$E(t) = G\cos(\omega t - \theta) - E_b(t) \quad . \tag{45}$$

Here G is the unloaded peak accelerating gradient and θ is the phase angle by which the bunch center leads the crest of the accelerating wave. The total energy spread within the bunch can be minimized by adjusting θ, as described previously.

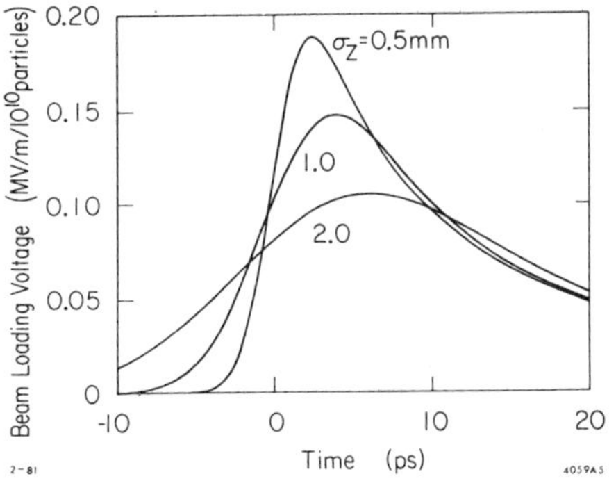

Fig. 12. Beam loading gradient within a single bunch for the SLAC disk-loaded structure. Gaussian bunches of 10^{10} particles, centered at $t = 0$, are assumed.

An additional parameter of interest is the total loss parameter, k_{tot}, given by

$$k_{tot} = \frac{1}{q^2} \int_{-\infty}^{\infty} I(t)\, E_b(t)\, dt \quad . \tag{46}$$

If $E_b(t)$ is the integrated wake due to a single normal mode, it can be shown that, for a gaussian bunch,

$$k_{tot}(n) = k_n \, e^{-\omega_n^2 \sigma^2} \quad , \tag{47}$$

where $\sigma = \sigma_z/c$. If $E_b(t)$ is the integrated wake due to all modes, then

$$k_{tot} = B(\sigma)\, k_1 \quad , \tag{48}$$

where k_1 is the loss parameter for the fundamental mode alone and $B(\sigma)$ is the beam loading enhancement factor. This function is plotted in Fig. 13 for the SLAC structure.

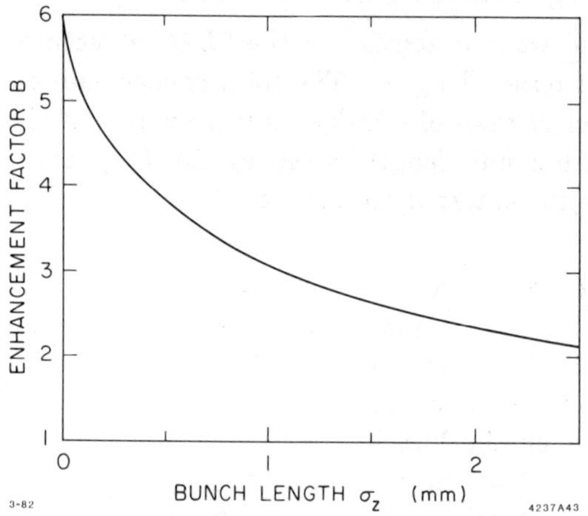

Fig. 13. Single bunch beam loading enhancement factor as a function of bunch length for the SLAC structure.

Functions similar to those defined in Eqs. (44) and (46) for the $m = 0$ case can be constructed using the transverse delta function wake potential $W_\perp(\tau)$. The integrated wake potentials for $m = 0$ and $m = 1$ are compared in Figs. 14 and 15 for the SLAC structure for various bunch lengths[40]. In these two figures, $\mathcal{W}$ is the integrated wake per cell per unit charge. The dashed and solid curves show agreement between the sum of modes method used to calculate the wake, as discussed here, and a direct time integration of Maxwell's equation computed by T. Weiland's code TBCI[41].

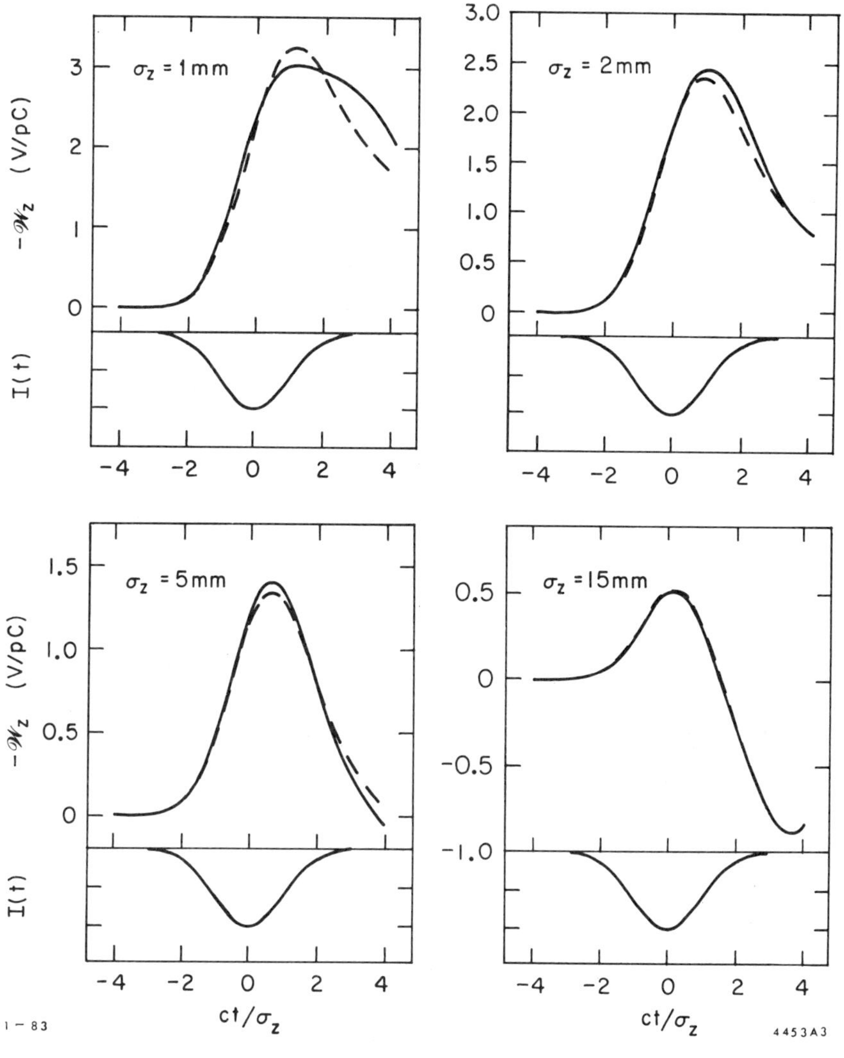

Fig. 14. Longitudinal wake potential ($m = 0$) per cell for gaussian bunches in the SLAC structure. Solid curves give TBCI results and dashed curves are results from a sum of modes.

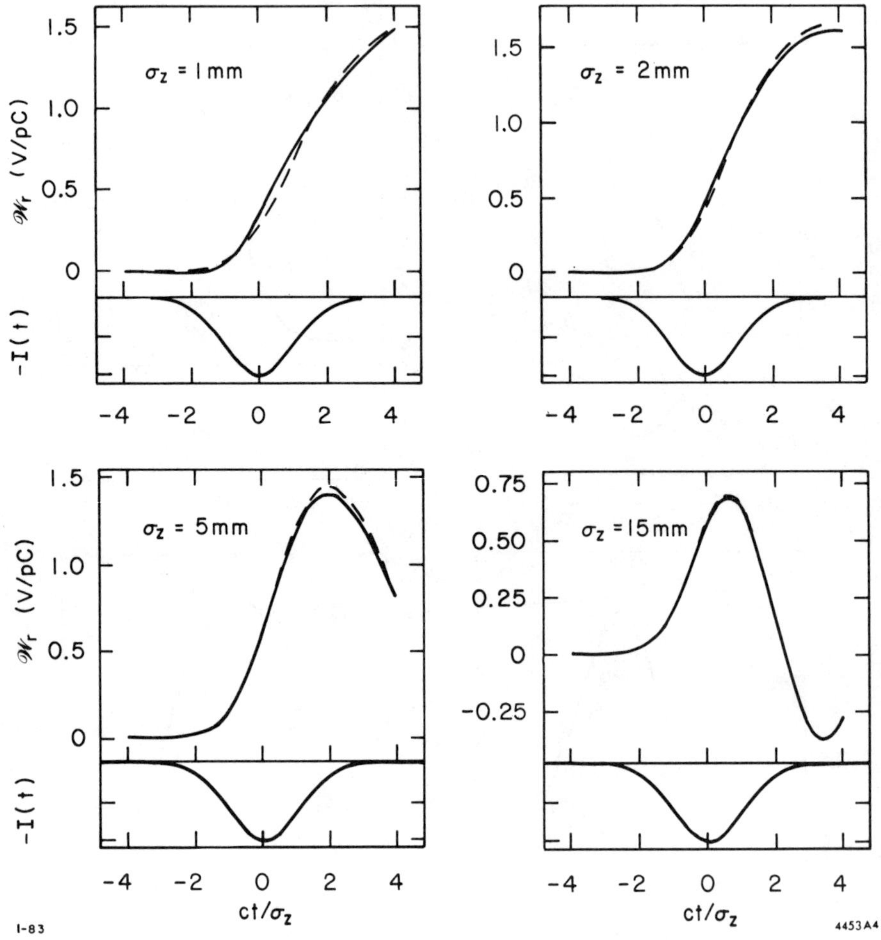

Fig. 15. Transverse wake potential ($m = 1$) per cell for gaussian bunches in the SLAC structure. Solid curves give TBCI results and dashed curves are results from a sum of modes.

C. Two Particle Model

A simple model in which the bunch current distribution is approximated by two point charges, a leading (head) particle with charge $q/2$ and a following (tail) particle with charge $q/2$, is very useful in estimating emittance growth due to dipole wake field effects in a linac. Consider the simplest case of a linac with constant energy and constant focussing strength $k = 1/\beta$, where $2\pi\beta$ is the wavelength of the betatron oscillation in the focussing lattice. The equations of motion of the two particles are

$$x_1'' + k^2\, x_1 \;=\; 0 \quad , \tag{49a}$$

$$x_2'' + (k + \Delta\, k)^2\, x_2 \;=\; C\, x_1 \quad , \tag{49b}$$

where the subscripts 1 and 2 refer to the head and tail particles respectively. Here the prime indicates dx/dz and $C = eqW_1/2E_0$, where W_1 is the dipole wake potential at the position of the trailing charge. (The separation of the two charges can be approximately by $2\sigma_z$). There may also be an energy spread in the bunch due to the longitudinal wake, or to the slope of the RF wave if the bunch is placed off crest. This energy spread can be modelled in the two particle approximation by an energy difference ΔE between the particles, leading to a difference in focussing force $\Delta k/k = \xi \Delta E/E_0$, where ξ is the chromaticity of the lattice (for a lattice with 90° phase shift per cell $\xi = -4/\pi$). For the leading particle, the solution to Eq. (49a) is

$$\widetilde{x}_1 \;=\; \widetilde{x}_{10}\, e^{ikz} \quad , \tag{50}$$

where $\tilde{x}_1$ is a complex quantity giving the amplitude and phase of the oscillation at position z, and $\tilde{x}_{10}$ is the initial value at $z = 0$. If $\Delta k = 0$, the tail particle obeys

$$\widetilde{x}_2 \;=\; \widetilde{x}_{20}\, e^{ikz} - \frac{iCz}{2k}\, \widetilde{x}_{10}\, e^{ikz} \quad . \tag{51}$$

Here the first term represents a free betatron oscillation and the second term an oscillation driven by the head particle. If $\widetilde{x}_{20} = \widetilde{x}_{10}$, the difference $\widetilde{\Delta x} = \widetilde{x}_2 - \widetilde{x}_1$, grows in amplitude as

$$\frac{\Delta x}{x_{10}} = \frac{Cz}{2k} = \frac{eqW_1 z}{4kE_0} \quad . \tag{52a}$$

In Gaussian units this becomes

$$\frac{\Delta x}{x_{10}} = \frac{r_0 N W_1 z}{4k\gamma} \quad . \tag{52b}$$

If there is an energy difference between the head and tail particles (Landau damping), the solution for x_2 is[14]

$$\widetilde{x}_2 \;=\; \widetilde{x}_{20}\, e^{i(k+\Delta k)z} - \frac{iC\widetilde{x}_{10}}{k\Delta k}\, \sin\left(\frac{z\Delta k}{2}\right)\, e^{i\left(k+\frac{\Delta k}{2}\right)z} \quad . \tag{53}$$

Again assuming $\widetilde{x}_{20} = \widetilde{x}_{10}$, the difference $\widetilde{\Delta x} = \widetilde{x}_2 - \widetilde{x}_1$, grows in amplitude as

$$\frac{\Delta x}{x_{10}} = 2\left(1 - \frac{C}{2k\Delta k}\right) \sin\left(\frac{z\Delta k}{2}\right) \quad . \tag{54}$$

As has been pointed out[14], the emittance growth is zero if either $z\Delta k = 2n\pi$ or if $C = 2k\Delta k$. This latter condition can be written, for a 90° lattice,

$$\Delta E = -\frac{\pi e q W_1}{16k^2} \quad . \tag{55}$$

However, even in the worst case the amplitude does not exceed $C/k\Delta k$, which becomes for a 90° lattice

$$\frac{\Delta x}{x_{10}} < -\frac{\pi e q W_1}{8k^2 \Delta E} \quad . \tag{56}$$

The amplitude of the wake potential W_1 can be determined from Fig. 10, at least approximately, by taking the wake potential at distance σ_z behind the bunch center. Thus for a 1 mm bunch in the SLAC structure $W_1 \approx 0.8$ V/pC/cell, or 2×10^{15} V/C-m^2. In Gaussian units this becomes $W_1 = 2 \times 10^5/\text{m}^3$.

VI. EMITTANCE

A. General Remarks

The scaling laws in Sec. II show that very small emittance beams will be required for future linear colliders operating in the energy range above 1 TeV. The normalized emittance required for a 5 TeV machine, for example, might be on the order of 10^{-7}–10^{-8} π m-rad.[42]. In this section some limitations and expectations concerning the emittance that can be obtained from linac injectors and damping rings will be briefly discussed.

Linear optics and emittance concepts in beam transport systems and in circular machines have been discussed in several tutorial papers.[43] We are concerned here with periodic linear transport systems, and for this case a word of caution is in order. In a circular machine, the lattice functions (*e.g.*, β) are truly periodic and can be defined by the characteristics of the focussing lattice alone, even where no beam is present. This is not the case for a linear transport system. Assume that the transport system consists of a finite number of identical cells. The initial beam ellipse in the phase plane at the entrance to the first cell must still be defined in order to define the initial values of the lattice functions. Alternatively, the values at the end of the last cell could also be defined by working back from the final focus. Expressed differently, the concept of a beta function in a linear collider is only maningful if the beam ellipse is defined at some location in the transport system.

B. Emittance from Guns

The emittance of a typical linac injector is for the most part determined by the emittance of the gun itself. Consider a point on the surface of a hot cathode. Electrons leving this point have a transverse momentum proportional to $(KT)^{1/2}$, where K is the Boltzmann constant. The transverse emittance is then proportional to this momentum times the radius of the cathode, or to the square root of the cathode area. For a fixed current density, the area is in turn proportional to the total current. Thus, it is reasonable to write

$$\epsilon_n = \gamma\epsilon \approx 1 \times 10^{-4} \sqrt{\hat{I}B} \;\pi \text{ m-rad} \quad , \tag{57}$$

where $\hat{I}$ is the peak current in the bunch in amperes and B is the bunching factor. This expression is also called the Lawson–Penner relation (see, for example, Ref. 44).

In addition to the transverse momentum due to the finite temperature of the cathode, other factors contribute to the beam emittance in the gun region. These factors include field fringing at grid and anode apertures, nonlinear forces in focussing lenses, and transverse RF fields which vary longitudinally over the bunch in the bunching region. A number of suggestions have been made for reducing or eliminating these deleterious effects, such as: removal of grids, very high gun voltages, tight focussing, bunching at high energy, use of a cathode in which the emission is driven by a microwave field, and photocathodes in which the emission is driven by a modulated laser beam.

The normalized emittance for typical present-day linacs at 100 A peak cathode current is the range of 1–3 $\times 10^{-4}\,\pi$ m-rad. It is expected that this can be reduced by a factor of ten or so in the case of a laser-driven photocathode. The thermal limit lies still another order of magnitude lower.

C. Emittance from Damping Rings

Low emittance storage rings are of interest as synchrotron radiation sources, as beam recirculation devices for FELs, and as injectors for linear colliders. Assume first a ring with a lattice consisting of bending magnets of length $\ell = \rho\theta$, with a waist in the β function at the center of each magnet and appropriate focussing elements between the magnets. Then it can be shown[45] that the minimum nomalized emittance is given by

$$\epsilon_n = 8.3 \times 10^{-15} \, \gamma^3 \, \theta^3 \, \pi \text{ m-rad} \quad . \tag{58a}$$

The more conventional FODO lattice produces a considerably higher minimum normalized emittance, given by[46]

$$\epsilon_n = \frac{4.8 \times 10^{-13}}{F_m} \gamma^3 \theta^3 \pi \, \text{m-rad} \quad , \tag{58b}$$

where F_m is the fraction of the ring filled by the magnets. From the standpoint of low emittance, therefore, a damping ring should consist of a large number of very short bending magnets with a large bending radius.

It is natural to ask whether there is a fundamental limit on the emittance that can be achieved using this strategy. It is clearly not productive to reduce the bending angle below the opening angle for synchrotron radiation. More precisely, it can be shown[47] that $\theta_{min} \approx 6/\gamma$. Using this in Eq. (58a),

$$\epsilon_n \approx 1.8 \times 10^{-12} \gamma^3 \theta^3 \pi \, \text{m-rad} \quad . \tag{59}$$

In addition to emittance, other factors must be considered in the design of a damping ring. For either lattice, the damping rate is given by[46]

$$[\tau \, (\text{sec})]^{-1} = 2.1 \times 10^{-3} \, F_m \, E_0^3 \, (\text{GeV}) \Big/ \rho^2 \, (\text{m}) \quad . \tag{60}$$

Thus for a fast damping rate, required for a collider with a high repetition rate, a small bending radius is desirable. This is in conflict with the requirement for low emittance and a compromise must be struck.

The Touschek effect may limit the beam lifetime is a storage ring designed for low emittance, and intrabeam scattering (multiple Touschek effect) may produce emittance growth. In a damping ring the beam lifetime needs to be only a few damping times, and the limitation on lifetime imposed by the Touschek effect is normally not of concern. Intrabeam scattering, howerver, may impose a serious limitation on emittance. It is difficult to write a precise relation giving the scaling for this effect, but the threhold current at which significant emittance growth becomes observable increases rapidly with increasing energy. Computer programs[48] are available for calculating beam lifetime limitations and emittance growth due to intrabeam scattering in storage rings designed for high brightness synchrotron radiation sources and FELs. For these applications, normalized emittances on the order of $5 \times 10^{-6} \pi$ m-rad have been achieved. This is a factor of six lower than the emittance of the SLC damping ring at SLAC ($\epsilon_n = 3 \times 10^{-5}$). For a 5 TeV collider, an emittance which is still lower by two or three orders of magnitude may be required. However, a serious effort to design damping rings capable of producing beam emittances of this order is only just beginning.

Acknowledgment

The help of P. Morton and L. Rivkin in preparing Sec. VI is gratefully acknowledged.

References

1. Ugo Amaldi, ed., Proc. of the 2nd ICFA Workshop on Possibilities and Limitations of Accelerators and Detectors, Les Diablerets, Switzerland, October 1979 (CERN, June 1980), pp. 3–20.
2. H. Wiedemann, 1981 SLAC Summer Institute on Particle Physics, SLAC–PUB–2849 (November 1981).
3. R. Hollebeek, Nucl. Instrum. Methods **184**, 333 (1981).
4. W. M. Fawley and E. P. Lee, UCID–18584, Lawrence Livermore Laboratory (1980).
5. R. Noble, private communication.
6. M. Bassetti and M. Gygi–Hanney, LEP–Note–221, CERN, Geneva (1980).
7. T. Erber and G. B. Baumgartner Jr., Proc. 12th Int. Conf. on High Energy Accelerators (Fermilab, August 1983), p. 372.
8. T. Himel and J. Siegrist, "Quantum Effects in Linear Collider Scaling Laws." These Proceedings.
9. P. B. Wilson, "High Energy Electron Linacs: Applications to Storage Ring RF Systems and Linear Colliders," in *Physics of High Energy Accelerators,* R. A. Carrigan, F. R. Huson and M. Month, eds. (AIP Conf. Proc. No. 87, New York, 1982); also SLAC–PUB–2884. (See Sec. 10.1.)
10. See Ref. 9, Sec. 12.3.
11. A. N. Skrinsky, Proc. 12th Int. Conf. on High Energy Accelerators (Fermilab, August 1983), p. 104.
12. A. Chao, SLAC Internal Note (May 1983).
13. A. Chao and L. Rivkin, SLAC Internal Note CN–263 (January 1984).
14. K. L. F Bane, "Landau Damping in the SLAC Linac," 1985 Particle Accelerator Conf. (to be published in IEEE Trans. Nucl. Sci. **NS-32**); also SLAC–PUB–3670.
15. See, for example, Ref. 9, Sec. 10.1.
16. E. L. Garwin *et al.,* "An Experimental Program to Build a Multimegawatt Lasertron for Super Linear Colliders," 1985 Particle Accelerator Conf. (to be published in IEEE Trans. Nucl. Sci. **NS–32**); also SLAC–PUB–3650.
17. W. D. Kilpatrick, Rev. Sci. Instr. **28**, 824 (1957).

18. G. Konrad, private communication.

19. R. B. Miller, *An Introduction to the Physics of Intense Charged Particle Beams* (Plenum Press, New York, 1982), p. 11.

20. G. Loew and J. Wang, "Measurement of Ultimate Accelerating Gradients in the SLAC Disk–Loaded Structure." 1985 Particle Accelerator Conf. (to be published in IEEE Trans. Nucl. Sci. **NS-32**); also SLAC–PUB–3597.

21. Eiji Tanabe, IEEE Trans. Nucl. Sci. **NS–30, No. 4**, 3551 (1983).

22. V. L. Granatstein, 1984 Summer School on High Energy Particle Accelerators, Fermilab (to be published in AIP Conf. Proc.).

23. Rome Air Development Center, Report RADC–TR–70–101 (July 1970).

24. T. G. Lee *et al.*, "The Design and Performance of a 150 MW Klystron at S–Band" (to be published in IEEE Trans. Plasma Science.); also SLAC-PUB–3619.

25. V. L. Granatstein, to be published in Int. J. Electronics **57** (1984).

26. Y. Fukushima *et al.*, "Lasertron, a Photocathode Microwave Device Switched by Laser." 1985 Particle Accelerator Conf. (to be published in IEEE Trans. Nucl. Sci. **NS-32**).

27. W. Herrmannsfeldt, SLAC–AP/41 (May 1985).

28. Z. D. Farkas, "Binary Power Multiplier." Submitted to the Second N. C. Christophilos Int. Conf. on Pulsed Power and Its Applications, Spetses, Greece, August 1985; also SLAC–PUB–3694.

29. K. L. F Bane, Pisin Chen and P. B. Wilson, "On Collinear Wake Field Acceleration," 1985 Particle Accelerator Conf. (to be published in IEEE Trans. Nucl. Sci. **NS-32**); also SLAC–PUB–3662.

30. G. Voss and T. Weiland, DESY Report 82–074 (November 1982).

31. T. Weiland, "Wake Field Work at DESY," 1985 Particle Accelerator Conf. (to be published in IEEE Trans. Nucl. Sci. **NS-32**).

32. J. Wurtele, " On Acceleration by the Transfer of Energy between Two Beams," these Proceedings.

33. A. M. Sessler, "The Free Electron Laser as a Power Source for a High Gradient Structure," in *Laser Acceleration of Particles,* P. J. Channell, ed. (AIP Conf. Proc. No. 91, New York, 1982), pp. 163–189.

34. Suggested by W. K. H. Panofsky.

35. Ref. 9, Sec. 9.

36. A. W. Chao, "Coherent Instabilities of a Relativistic Bunched Beam," in *Physics of High Energy Particle Accelerators,* M. Month, ed. (AIP Conf. Proc. No. 105, New York, 1983), pp. 353–523.

37. K. L. F. Bane, P. B. Wilson and T. Weiland, "Wake Fields and Wake Field Acceleration" in *Physics of High Energy Particle Accelerators*, M. Month, Per F. Dahl and M. Dienes, eds. (AIP Conf. Proc. No. 127, New York, 1985), pp. 875-928.

38. K. L. F. Bane and R. K. Cooper, Invited Lectures at the 1984 Summer School on High Energy Particle Accelerators (to be published in AIP Conf. Proc).

39. K. L. F. Bane, private communication.

40. K. Bane and T. Weiland, SLAC/AP-1 (January 1983).

41. T. Weiland, 11th Int. Conf. on High Energy Accelerators (Birkhäuser Verlag, Basel, 1980), pp. 570–575.

42. B. Richter, "Requirements for Very High Energy Accelerators," these Proceedings.

43. See, for example, K. L. Brown and R. V. Servranckx, "First and Second Order Charged Particle Optics," in *Physics of High Energy Particle Accelerators,* M. Month, Per F. Dahl and M. Dienes, eds. (AIP Conf. Proc. No. 127, New York, 1985), pp. 62–138.

44. L. R. Elias and G. J. Ramian, "Status Report of the UCSB FEL Experimental Program," in *Free-Electron Generators of Coherent Radiation,* C. A. Brau, S. F. Jacobs and M. O. Scully, eds., Proc. SPIE **453**, 137 (1984).

45. L. C. Teng, internal report LS–17, Argonne National Laboratory (March 1985); also internal report TM–1269, Fermilab (June 1984).

46. H. Wiedemann, 11th Int. Conf. on High Energy Accelerators (Birkhäuser Verlag, Basel, 1980), p. 693.

47. L. Rivkin, private communication

48. Computer code ZAP by M. S. Zisman, J. Bisognano and S. Chattopadhyay, Lawrence Berkeley Laboratory. Unpublished.

49. K–J. Kim *et al.,* "Storage Ring Design for a Short Wavelength FEL," 1985 Particle Accelerator Conf. (to be published in IEEE Trans. Nucl. Sci. **NS-32**).

HIGH-BRIGHTNESS PHOTOEMITTER DEVELOPMENT FOR ELECTRON ACCELERATOR INJECTORS*

J. S. Fraser, R. L. Sheffield and E. R. Gray, MS H825
Los Alamos National Laboratory, Los Alamos, NM 87545

ABSTRACT

Free-electron-laser (FEL) oscillators require a train of high-brightness bunches. Conventional subharmonic bunchers are currently used with rf linacs to generate pulse trains, but the resulting dilution of the transverse phase space and lower beam brightness are unacceptable for high-performance FELs. Recent developments suggest that photoemitters of high quantum efficiency combined with rapid acceleration can produce pulse trains of higher brightness than has been achieved before.

DISCUSSION

The prospects for development of high-brightness photoemitters are good. First, the high peak current that a FEL requires has been demonstrated. Recently, a peak current of over 200 A has been demonstrated at the Thermo Electron Corporation, using a 1-cm^2 Cs_3Sb cathode, while at SLAC a current density of 180 A/cm^2 has been produced, using a GaAs photoemitter. Secondly, the average energy of the electrons produced is less than 1.0 eV--only a little more than the thermal energy of 0.1 eV typical of dispenser cathodes. Finally, there is a variety of negative-electron-affinity photoemitters available for which the work function is close to zero. This means that there is a reasonable chance of finding a material that can be used in the demanding environment of an accelerator injector.

At the Los Alamos National Laboratory, a program is under way to develop an rf gun based on a laser-illuminated photocathode in an rf cavity. A mode-locked laser is used to generate electron bunches less than 50 ps in duration and carrying a charge of up to 10 nC. The repetition rate is about 100 MHz. The photoemitter is placed on the end wall of an rf cavity in which longitudinal electric field is about 30 MV/m. The resulting rapid acceleration reduces the time in which space-charge forces can act to degrade the emittance.

Figure 1 shows the normalized transverse emittance of typical electron linacs using subharmonic or other conventional bunching schemes plotted against the peak bunch current in amperes. The goal for photoemitter rf guns is halfway (on a log scale) between current practice with typical linacs and the thermal limit for a bright photoemitter such as GaAs.

*This work was performed under the auspices of the US Department of Energy and supported by the US Army Ballistic Missile Defense Organization.

The Los Alamos program is based on a high-quantum-yield photoemitter mounted in a 1300-MHz cavity. The vacuum pressure must be maintained at 10^{-10} torr or lower. The immediate objective of the program is to demonstrate the production of 200-A pulses at a repetition rate of 100 MHz and acceleration to an energy of 2 to 5 MeV with a transverse normalized emittance of about 10 to 30 π mm•mrad.

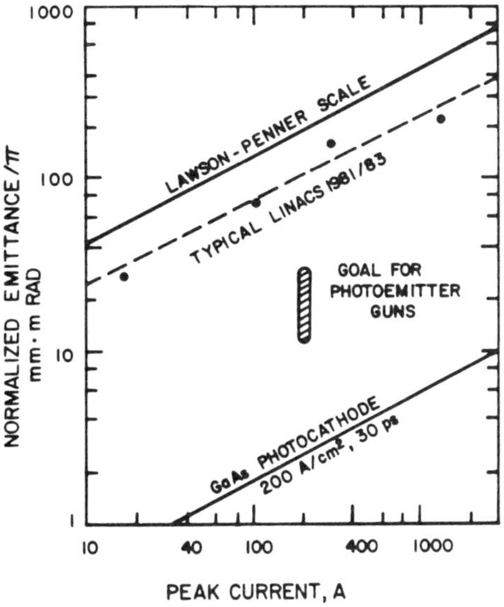

Fig. 1. Normalized transverse emittance vs peak electron current. Scope for possible improvement is indicated by the gap between the thermal limit of a bright GaAs photoemitter and the results obtained from typical linacs.

The transport and acceleration of intense electron bunches in the rf cavity is being simulated with aid of 2D, time-dependent codes. The MASK code is being used in a collaborative effort at SLAC, and the ISIS code is being used at Los Alamos. Figure 2 shows typical output from the code ISIS of the current profile near the photocathode and bunch outlines at two positions.

A conceptual design of an electron injector, an rf gun, and several high-gradient rf cavities is shown in Fig. 3. A magnet at the end of the injector linac deflects the electron beam so the pulsed laser beam can illuminate the photocathode through the linac bore hole.

The Los Alamos injector development program includes the study of a back-up scheme that retains the features of a bright emitter and rapid acceleration of the beam (Fig. 4). A LaB_6 thermionic emitter (with a current density of 200 A/cm^2) is placed in a bimodal cavity excited at 200 and 600 MHz to form pulses of about 1 ns width. The bunches are then rapidly accelerated in high-gradient 200-MHz cavities to an energy of 2 MeV or more. The required bunching to 50 ps is then accomplished with some form of magnetic compressor.

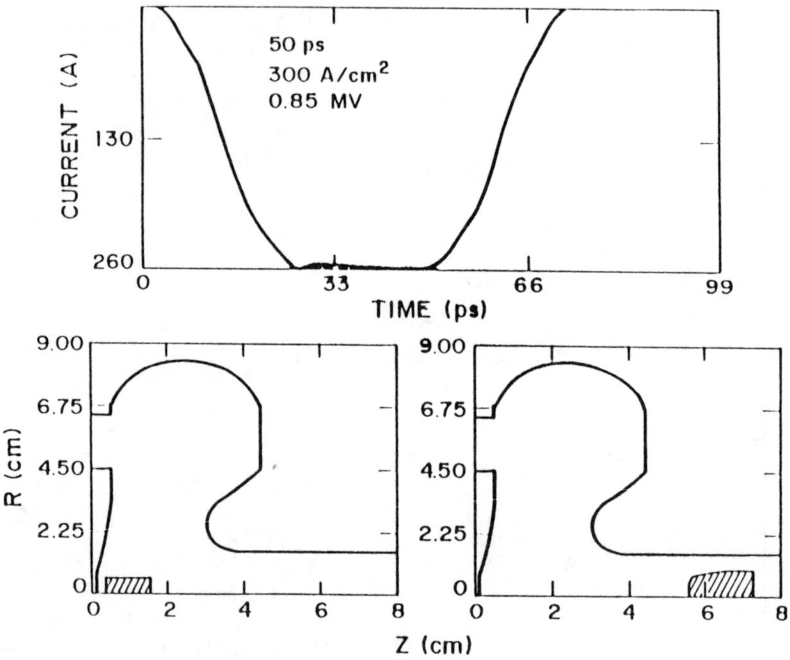

Fig. 2. Current profile and bunch profiles from the 2D, time-dependent code ISIS.

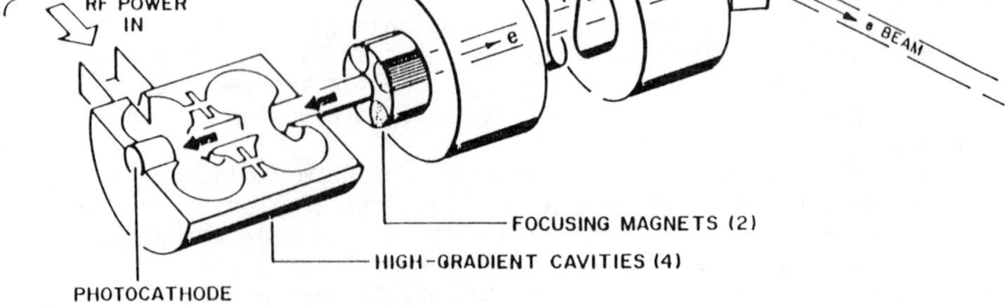

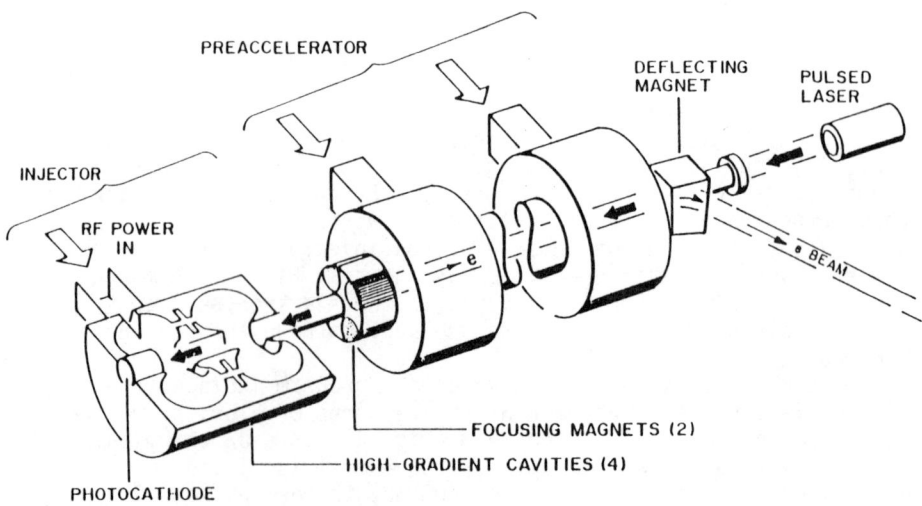

Fig. 3. Conceptual design of an electron injector based on the use of an rf-photocathode gun.

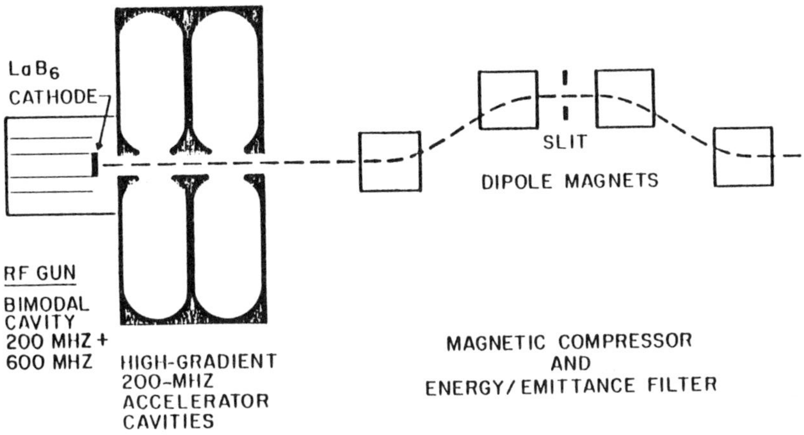

Fig. 4. Schematic diagram of back-up scheme based on a bright thermionic emitter and rapid acceleration of 1-ns bunches followed by magnetic compression.

ACKNOWLEDGMENT

The theoretical support for the time-dependent calculation is being performed by Mike Jones and Gary Rodenz at Los Alamos National Laboratory and Bill Herrmannsfeldt at Stanford Linear Accelerator Center.

QUANTUM EFFECTS IN LINEAR COLLIDER SCALING LAWS*

T. Himel
Stanford Linear Accelerator Center, Stanford, CA 94305

J. Siegrist
Lawrence Berkeley Laboratory, Berkeley, CA 94720

ABSTRACT

Compared to classical calculations, quantum corrections greatly reduce the radiation emitted when the e^+ and e^- beams collide in a linear collider. This allows a given luminosity to be obtained with much lower beam powers by making the beam size smaller.

INTRODUCTION

In the design of high energy, high luminosity e^+e^- linear colliders there are several constraints which must be satisfied by the beam parameters[1] . These beam parameters and their nominal values at SLC are:

$2\sigma_r$	:	the radius of the beam (in this paper we will assume uniform cylindrical beams)	$(2 \times 1.8\ \mu)$
$2\sqrt{3}\sigma_z$	:	the length of the bunch (The numerical factors in the definitions of σ_z and σ_r are a convention to help make formulas for cylindrical beams similar to those for Gaussian ones.)	$(2\sqrt{3} \times 1\ \text{mm})$
N	:	the number of particles per bunch	(5×10^{10})
f	:	the repetition rate of the machine	(180 Hz)
γ	:	E/mc^2 (where E is the beam energy).	(1×10^5)

From these five basic beam parameters, four other parameters can be derived. There are constraints on these four parameters placed by the needs of High Energy Physics and by beam dynamics. The derived parameters are

- The total beam power

$$P = fN\gamma mc^2 \quad (= 74 \text{ KW at SLC}) \tag{1}$$

has no strict constraints on it but small beam powers are preferred as the total AC power consumed by the accelerator and therefore its cost scales with P.

* Work supported by the Department of Energy, contract DE – AC03 – 76SF00515.

- The luminosity

$$\mathcal{L} = \frac{N^2 f}{4\pi\sigma_r^2} \quad (= 6 \times 10^{30}\text{cm}^{-2}\text{sec}^{-1} \text{ at SLC}^{[2]}) \tag{2}$$

must increase with the energy of the beams in order to keep a constant event rate. Since the interaction cross section falls as $1/\gamma^2$ we must have $\mathcal{L} = \mathcal{L}_0 \frac{\gamma^2}{\gamma_0^2}$ where e.g., $\frac{\mathcal{L}_0}{\gamma_0^2} = \frac{10^{30}}{(10^5)^2}$.

- The disruption parameter

$$D = \frac{N r_e \sigma_z}{2\gamma\sigma_r^2} \tag{3}$$

is related to how many oscillations an electron will go through as it passes through the magnetic field caused by the other beam. If it is too large there is an instability which will increase the beam size and reduce the luminosity. One must require $D < 20$.

- The beamsstrahlung parameter

$$\delta_{classical} = \frac{r_e^3}{3\sqrt{3}} \frac{N^2\gamma}{\sigma_r^2\sigma_z} \tag{4}$$

tells what fraction of a beams energy is lost due to synchrotron radiation in the magnetic field of the other beam. The subscript indicates the equation comes from a classical calculation. To do a clean high energy physics experiment, one wants the spread in the center-of-mass energy to be less than about 10%. This spread is $\delta/\sqrt{12}$, hence, one must require $\delta < 0.3$.

There are a total of 9 parameters: σ_r, σ_z, N, f, γ, P, $\mathcal{L}$, D and δ and 4 equations relating them. So one can specify 5 parameters and then solve for the other 4. For example, specifying γ, $\mathcal{L}$, f, P and D one can solve for σ_r, σ_z, N and δ. In particular

$$N = \frac{P}{fmc^2\gamma} \tag{5}$$

$$\sigma_z = \frac{DP}{2\pi mc^2 r_e \mathcal{L}} \tag{6}$$

$$\sigma_r = \frac{P}{\gamma mc^2\sqrt{4\pi f\mathcal{L}}} \tag{7}$$

$$\delta_{classical} = \frac{(4\pi)^2 r_e^4 \gamma \mathcal{L}^2 mc^2}{6\sqrt{3} fDP}. \tag{8}$$

Plugging in reasonable numbers for a 5 TeV linac[3] :

$$\begin{aligned} \gamma &= 10^7 \\ \mathcal{L} &= 10^{34}\,\mathrm{cm}^{-2}\mathrm{sec}^{-1} \\ f &= 5\,\mathrm{kHz} \\ D &= 2 \\ P &= 10\,\mathrm{MW} \end{aligned} \tag{9}$$

gives

$$\begin{aligned} N &= 2.4 \times 10^9 \\ \sigma_z &= 0.14\,\mathrm{mm} \\ \sigma_r &= 0.0049\,\mu\mathrm{m} = 49\,\text{Å} \\ \delta &= 76 \end{aligned} \tag{10}$$

It is unphysical to have each electron lose 76 times more energy than it has.

QUANTUM CORRECTIONS TO BEAMSSTRAHLUNG

The cause of this unphysical result is that $\delta_{classical}$ was derived with classical formulas for the radiation of a moving charge in a magnetic field. It turns out that we are in the quantum regime. This can be seen as follows. Consider a particle at radius r in the uniform cylindrical bunch. For the numerical examples $r = 2\sigma_r$ (a particle at the very outside of the uniform cylindrical bunch) is used. It sees a magnetic field

$$B = \frac{N_r e}{\sqrt{3} r \sigma_z} \tag{11}$$

where N_r is the number of electrons inside a circle of radius r. B is 4.8×10^7 Gauss for our example. There is an equal force coming from the electric field. The radius of curvature of the electron's trajectory in these fields is

$$\rho = \frac{\gamma mc^2}{2eB} = \frac{\sqrt{3}\gamma mc^2 r\sigma_z}{2N_r e^2} = \frac{\sqrt{3}\gamma r \sigma_z}{2N_r r_e} \tag{12}$$

This is 177 cm for our example. Now classically the synchrotron radiation spectrum peaks at $\omega \approx \frac{1}{3}\omega_c$ where $\omega_c = \frac{3c\gamma^3}{2\rho}$ [4] . The critical energy is the energy of these photons

$$E_c = \hbar\omega_c = \frac{3}{2}\hbar c\frac{\gamma^3}{\rho} = \frac{\sqrt{3}\hbar c\gamma^2 N_r r_e}{r\sigma_z}. \tag{13}$$

This is 268 ergs = 168 TeV for our example. This is much greater than the 5 TeV beam energy so clearly such photons can not be radiated.

Obviously the classical calculation is invalid and a proper quantum mechanical calculation must be used. A full quantum treatment of synchrotron radiation was done in 1952 by Sokolov, Ternov and Klepikov[4] . The results are illustrated in Fig. 1[5] . When $E/E_c \gg 1$ the classical calculation is correct. When $E/E_c \ll 1$, the power spectrum follows the classical curve and then drops exponentially at the electron's energy. Here we will use the approximation that the radiated power follows the classical curve until the photons energy equals the electron's energy where it sharply drops to zero. This approximation is always greater than the exact solution and is accurate to a few percent for $E/E_c < 0.1$.

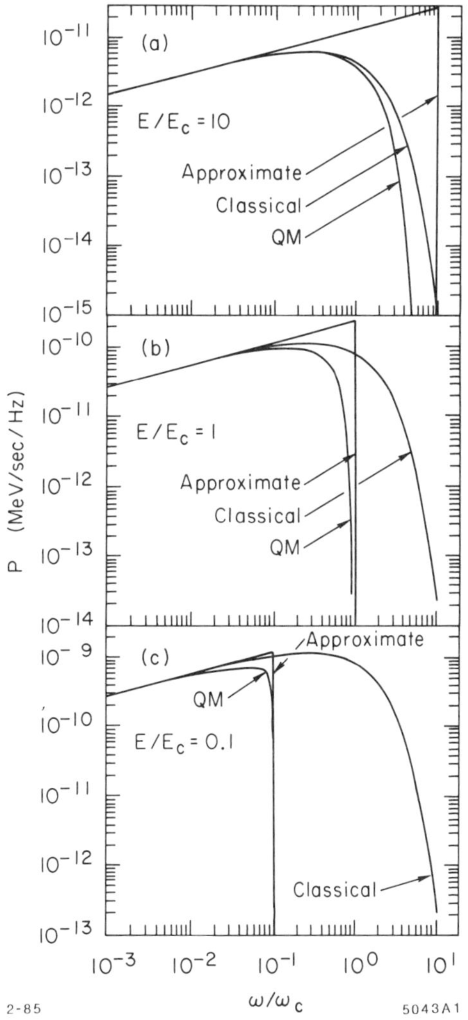

Fig. 1. Differential power spectra of the radiation emitted by 5 TeV electrons for several values of E/E_c. Parts a–c are in the classical, intermediate and quantum regimes respectively. Shown are the classical calculation, the exact quantum calculation and the approximation used here.

The classical synchrotron radiation formula for the power radiated as a function of frequency is[4]

$$P(\omega) = \frac{2r_e c\gamma^4 mc^2}{3\omega_c \rho^2} S\left(\frac{\omega}{\omega_c}\right) \tag{14}$$

where S contains the integral of a modified Bessel function. For $\omega/\omega_c \ll 1$; $S(\omega/\omega_c) \approx \frac{4}{3}(\omega/\omega_c)^{1/3}$. Hence, using this approximation, the total power radiated is

$$P_{tot} = \int_0^{E/\hbar} P(\omega)d\omega = \frac{2r_e c\gamma^4 mc^2}{3\rho^2}\left(\frac{E}{E_c}\right)^{4/3} \tag{15}$$

this power is radiated for the time it takes a particle to pass through the other bunch, namely $2\sqrt{3}\sigma_z/c$. Hence δ which is the fraction of the electrons energy which is radiated is

$$\delta_{QM} = \frac{16r_e^3\gamma N_r^2}{3\sqrt{3}r^2\sigma_z}\left(\frac{E}{E_c}\right)^{4/3} = \frac{16r_e^3\gamma N_r^2}{3\sqrt{3}r^2\sigma_z}\left(\frac{r\sigma_z mc}{\sqrt{3}\hbar\gamma N_r r_e}\right)^{4/3} \tag{16}$$

$$\text{for} \quad \frac{E}{E_c} = \left(\frac{r\sigma_z mc}{\sqrt{3}\hbar\gamma N_r r_e}\right) \ll 1. \tag{17}$$

Putting in numbers for the outside particle in the above example gives

$$\delta_{QM} = 2.7. \tag{18}$$

Note that the outside particle sees the largest magnetic field and loses the most energy. Averaging over radius for uniform cylindrical beams gives another factor of 3/4,

$$\delta_{QM} \leq \frac{r_e^3\gamma N^2}{\sqrt{3}\sigma_r^2\sigma_z}\left(\frac{2\sigma_r\sigma_z mc}{\sqrt{3}\hbar\gamma N r_e}\right)^{4/3}. \tag{19}$$

Note that at small radius the extreme quantum limit does not apply, but the approximation used here is always greater than both the exact classical and quantum formulas. Hence an upper bound on δ_{QM} has been calculated. A more accurate answer will require a computer simulation which is in progress.

EFFECTS ON SCALING LAWS OF LINEAR COLLIDERS

It is interesting to express these quantum formulas in terms of γ, $\mathcal{L}$, f, P and D instead of σ_r, σ_z, N and δ. Plugging equations (5)–(7) into (17) and (19) gives

$$\frac{E}{E_c} = \frac{P f^{1/2} D}{2\sqrt{3}(\pi \mathcal{L})^{3/2} r_e^2 \hbar c \gamma} \tag{20}$$

$$\delta_{QM} = \frac{8mc^2}{\sqrt{3}} \left(\frac{r_e}{2\sqrt{3}\hbar c}\right)^{4/3} \left(\frac{DP}{\gamma f}\right)^{1/3} . \tag{21}$$

For our example 5 TeV machine $E/E_c = 0.03$ which is truly much less than one. In fact from the form of equation (20) it is clear that for any high luminosity, high energy collider quantum effects in beamsstrahlung are important. This is because $\mathcal{L} \sim \gamma^2$ so the denominator increases like γ^4 and there are practical limits on how large P, f and D in the numerator can be made. This strong γ dependence is why at SLC quantum corrections are unimportant but at 5 TeV they are very large.

For our example 5 TeV machine the corrections were not quite large enough. We still have an average δ_{QM} of 2.1[7] which is greater than the 0.3 needed by high energy physics. Referring to equation (21) one sees that if D and P are both reduced by a factor of 20 then $\delta_{QM} = 0.29$ which is acceptable. Using equations (5)–(8) one gets the following consistent set of parameters for a 5 TeV machine

$$\begin{aligned}
\gamma &= 10^7 \\
\mathcal{L} &= 10^{34}\,\mathrm{cm}^{-2}\,\mathrm{sec}^{-1} \\
f &= 5\,\mathrm{kHz} \\
D &= 0.1 \\
P &= 0.5\,\mathrm{MW} \\
N &= 1.2 \times 10^8 \\
\sigma_z &= 0.4\,\mu\mathrm{m} \\
\sigma_r &= 2.5\,\text{Å} \\
E/E_c &= 7.5 \times 10^{-5} \quad \rightarrow \quad \text{quantum regime} \\
\delta_{classical} &= 30{,}000 \\
\delta_{QM} &= 0.29
\end{aligned} \tag{22}$$

Comparing equations (8), (20) and (21) one sees there are two possible regimes of beam parameters for a high energy (> 5 TeV), high luminosity e^+e^- linear collider.

1. One can use very high power (> 100 MW) and high repetition-rate beams. For this case the beam size is relatively large and the classical formulation of beamsstrahlung is valid. However if the accelerator is *e.g.* 3% efficient then the power consumption is 3 GW and the accelerator is obviously very expensive.
2. One can use low power beams. For this case the beam size is very small and the quantum formulation of beamsstrahlung must be used. Attaining such a small beam size will be difficult to say the least.

Note how different the scaling laws are for classical and quantum beamsstrahlung. To keep $\delta_{classical}$ small f, D and P must be large. To keep δ_{QM} small, f must be large while D and P are small.

In conclusion, quantum corrections to the beamsstrahlung are often important. Their inclusion extends the set of linear collider beam parameters which are useful for high energy physics.

REFERENCES

1. E. Keil et. al., Proceedings of the Second ICFA Workshop on Possibilities and Limitations of Accelerators and Detectors, Les Diablerets, 1979, p. 3.
2. The SLC luminosity given here includes an enhancement due to the pinching of the beams in each others magnetic field. This factor is not included in the formula.
3. B. Richter, contribution to this workshop.
4. M. Sands, Physics with Intersecting Storage Rings, Proceedings of the International School of Physics "Enrico Fermi," (Academic Press, N.Y.) p. 257.
5. A. A. Sokolov, N. P. Klepikov and I. M. Ternov, J. exp. theor. Phys. (U.S.S.R.) 23, 632 (1952).
   A. A. Sokolov and I. M. Ternov, Synchrotron Radiation (Pergamon Press, N.Y., 1968) p. 99.
6. R. Noble, private communication.
7. Particles can not lose 2.1 times their energy. We have implicitly assumed that the particles energy doesn't change much as it traverses the other beam. Actually after the electron has radiated some photons, it will have less energy and will radiate less than we calculated. Nevertheless a large δ indicates the electron lost a good fraction of its energy, presumably more than the 30% allowed.

LIST OF PARTICIPANTS

Mohammad J. Abedi	University of Oregon
Neil V. Baggett	Brookhaven National Laboratory
Russell G. Berger	University of Washington
Irving J. Bigio	Los Alamos National Laboratory
Robert Bingham	Rutherford-Appleton Laboratory
Jean Louis Bobin	University of Paris
Charles Buchanan	UCLA
Robert Byer	Stanford University
Paul Channell	Los Alamos National Laboratory
A. W. Chao	S L A C
Francis F. Chen	UCLA
K. Wendell Chen	University of Texas
Pisin Chen	UCLA
Christopher E. Clayton	UCLA
Johannes Claus	Brookhaven National Laboratory
David B. Cline	University of Wisconsin
F. T. Cole	Fermi National Accelerator Lab
Richard Cooper	Los Alamos National Laboratory
Paul B. Corkum	NRC of Canada
Paul L. Csonka	University of Oregon
Chris Darrow	UCLA
John M. Dawson	UCLA
Joseph N. D. Dodoo	University of Maryland
Shalom Eliezer	University of Texas
C. James Elliott	Los Alamos National Laboratory
Thomas Erber	UCLA
Roger G. Evans	Rutherford-Appleton Laboratory
J. J. Ewing	Spectra Technology, Inc.
Charles Fenstermacher	Los Alamos National Laboratory
Richard Fernow	Brookhaven National Laboratory
Humberto Figueroa	UCLA
Jorge R. Fontana	UC Santa Barbara
John S. Fraser	Los Alamos National Laboratory

L. Warren Funk	Chalk River Nuclear Laboratories
Suman Ganguly	Rice University
Edward R. Gray	Los Alamos National Laboratory
Phillip Goldstone	Los Alamos National Laboratory
Samuel Heifets	University of Texas
Thomas Himel	S L A C
Donald B. Hopkins	Lawrence Berkeley Laboratory
Heinrich Hora	University of New South Wales
Wendell Horton	University of Texas
Robert Huff	UCLA
Hiroyuki Ikezi	General Atomic Tech., Inc.
Jonah Jacob	Science Research Inc.
Robert A. Jameson	Los Alamos National Laboratory
Chan Joshi	UCLA
O'Dean P. Judd	Los Alamos National Laboratory
Thomas Katsouleas	UCLA
Rhon Keinigs	Los Alamos National Laboratory
Kwang-Je Kim	Lawrence Berkeley Laboratory
S. H. Kim	University of Texas
Joe Kindel	Los Alamos National Laboratory
Norman M. Kroll	UC San Deigo
William L. Kruer	Lawrence Livermore National Lab
Norman A. Kurnit	Lawrence Livermore National Lab
John Lawson	Rutherford-Appleton Laboratory
Kenneth Lee	Los Alamos National Laboratory
Kotik K. Lee	General Electric Co.
G. E. Lee-Whiting	Chalk River Nuclear Laboratories
W. T. Leland	Los Alamos National Laboratory
Erick Lindman	Los Alamos National Laboratory
Dennis D. Lowenthal	Spectra Technology, Inc.
David Luckey	M I T
Kirk T. McDonald	Princeton University
Frederick E. Mills	Fermi National Accelerator Lab
Bryan W. Montague	CERN

R. James Morgan	Itek Optical Systems
Warren B. Mori	UCLA
David L. Morse	Cornell
Phil Morton	S L A C
Norman Murray	UC Berkeley
David Neuffer	Texas Accelerator Center
Robert J. Noble	S L A C
Katsunobu Oide	KEK, Japan
Craig L. Olson	Sandia National Laboratory
J. Orthel	TRW
Robert Palmer	Brookhaven National Laboratory
Richard Pantell	Stanford University
Claudio Pellegrini	Brookhaven National Laboratory
Michael Pickup	Cornell
Melvin A. Piestrup	Adelphi Technology
Reginald Richardson	UCLA
Burton Richter	S L A C
James B. Rosenzweig	University of Wisconsin
Alessandro G. Ruggiero	Argonne National Laboratory
Ronald Ruth	S L A C
Al Saxman	Los Alamos National Laboratory
Barbara J. Schaad	DePaul University
George Schmidt	Stevens Institute of Technology
Stanley O. Schriber	Los Alamos National Laboratory
Andrew Sessler	Lawrence Berkeley Laboratory
Richard L. Sheffield	Los Alamos National Laboratory
John C. Sheppard	S L A C
J. D. Simpson	Argonne National Laboratory
Sidney Singer	Los Alamos National Laboratory
John Slater	Spectra Technology, Inc.
Todd Smith	Stanford University
John Soures	University of Rochester
Phillip Sprangle	Naval Research Laboratory
Iuliu Stumer	Brookhaven National Laboratory

Richard H. Stokes	Los Alamos National Laboratory
David Sutter	Department of Energy
Toshi Tajima	University of Texas
Kazuo Tanaka	Osaka University
Cha-Mei Tang	Naval Research Laboratory
Lee Teng	Fermi National Accelerator Lab
Maury Tigner	Cornell
Donald Umstadter	UCLA
Robert L. Warnock	Lawrence Berkeley Laboratory
David Whelan	Northrop
Ronald L. Williams	UCLA
W. Willis	CERN
Perry Wilson	S L A C
Jonathan Wurtele	M I T

AIP Conference Proceedings

		L.C. Number	ISBN
No. 1	Feedback and Dynamic Control of Plasmas – 1970	70-141596	0-88318-100-2
No. 2	Particles and Fields – 1971 (Rochester)	71-184662	0-88318-101-0
No. 3	Thermal Expansion – 1971 (Corning)	72-76970	0-88318-102-9
No. 4	Superconductivity in *d*- and *f*-Band Metals (Rochester, 1971)	74-18879	0-88318-103-7
No. 5	Magnetism and Magnetic Materials – 1971 (2 parts) (Chicago)	59-2468	0-88318-104-5
No. 6	Particle Physics (Irvine, 1971)	72-81239	0-88318-105-3
No. 7	Exploring the History of Nuclear Physics – 1972	72-81883	0-88318-106-1
No. 8	Experimental Meson Spectroscopy –1972	72-88226	0-88318-107-X
No. 9	Cyclotrons – 1972 (Vancouver)	72-92798	0-88318-108-8
No. 10	Magnetism and Magnetic Materials – 1972	72-623469	0-88318-109-6
No. 11	Transport Phenomena – 1973 (Brown University Conference)	73-80682	0-88318-110-X
No. 12	Experiments on High Energy Particle Collisions – 1973 (Vanderbilt Conference)	73-81705	0-88318-111–8
No. 13	π-π Scattering – 1973 (Tallahassee Conference)	73-81704	0-88318-112-6
No. 14	Particles and Fields – 1973 (APS/DPF Berkeley)	73-91923	0-88318-113-4
No. 15	High Energy Collisions – 1973 (Stony Brook)	73-92324	0-88318-114-2
No. 16	Causality and Physical Theories (Wayne State University, 1973)	73-93420	0-88318-115-0
No. 17	Thermal Expansion – 1973 (Lake of the Ozarks)	73-94415	0-88318-116-9
No. 18	Magnetism and Magnetic Materials – 1973 (2 parts) (Boston)	59-2468	0-88318-117-7
No. 19	Physics and the Energy Problem – 1974 (APS Chicago)	73-94416	0-88318-118-5
No. 20	Tetrahedrally Bonded Amorphous Semiconductors (Yorktown Heights, 1974)	74-80145	0-88318-119-3
No. 21	Experimental Meson Spectroscopy – 1974 (Boston)	74-82628	0-88318-120-7
No. 22	Neutrinos – 1974 (Philadelphia)	74-82413	0-88318-121-5
No. 23	Particles and Fields – 1974 (APS/DPF Williamsburg)	74-27575	0-88318-122-3
No. 24	Magnetism and Magnetic Materials – 1974 (20th Annual Conference, San Francisco)	75-2647	0-88318-123-1

No. 25	Efficient Use of Energy (The APS Studies on the Technical Aspects of the More Efficient Use of Energy)	75-18227	0-88318-124-X
No. 26	High-Energy Physics and Nuclear Structure – 1975 (Santa Fe and Los Alamos)	75-26411	0-88318-125-8
No. 27	Topics in Statistical Mechanics and Biophysics: A Memorial to Julius L. Jackson (Wayne State University, 1975)	75-36309	0-88318-126-6
No. 28	Physics and Our World: A Symposium in Honor of Victor F. Weisskopf (M.I.T., 1974)	76-7207	0-88318-127-4
No. 29	Magnetism and Magnetic Materials – 1975 (21st Annual Conference, Philadelphia)	76-10931	0-88318-128-2
No. 30	Particle Searches and Discoveries – 1976 (Vanderbilt Conference)	76-19949	0-88318-129-0
No. 31	Structure and Excitations of Amorphous Solids (Williamsburg, VA, 1976)	76-22279	0-88318-130-4
No. 32	Materials Technology – 1976 (APS New York Meeting)	76-27967	0-88318-131-2
No. 33	Meson-Nuclear Physics – 1976 (Carnegie-Mellon Conference)	76-26811	0-88318-132-0
No. 34	Magnetism and Magnetic Materials – 1976 (Joint MMM-Intermag Conference, Pittsburgh)	76-47106	0-88318-133-9
No. 35	High Energy Physics with Polarized Beams and Targets (Argonne, 1976)	76-50181	0-88318-134-7
No. 36	Momentum Wave Functions – 1976 (Indiana University)	77-82145	0-88318-135-5
No. 37	Weak Interaction Physics – 1977 (Indiana University)	77-83344	0-88318-136-3
No. 38	Workshop on New Directions in Mossbauer Spectroscopy (Argonne, 1977)	77-90635	0-88318-137-1
No. 39	Physics Careers, Employment and Education (Penn State, 1977)	77-94053	0-88318-138-X
No. 40	Electrical Transport and Optical Properties of Inhomogeneous Media (Ohio State University, 1977)	78-54319	0-88318-139-8
No. 41	Nucleon-Nucleon Interactions – 1977 (Vancouver)	78-54249	0-88318-140-1
No. 42	Higher Energy Polarized Proton Beams (Ann Arbor, 1977)	78-55682	0-88318-141-X
No. 43	Particles and Fields – 1977 (APS/DPF, Argonne)	78-55683	0-88318-142-8
No. 44	Future Trends in Superconductive Electronics (Charlottesville, 1978)	77-9240	0-88318-143-6
No. 45	New Results in High Energy Physics – 1978 (Vanderbilt Conference)	78-67196	0-88318-144-4
No. 46	Topics in Nonlinear Dynamics (La Jolla Institute)	78-057870	0-88318-145-2

No. 47	Clustering Aspects of Nuclear Structure and Nuclear Reactions (Winnepeg, 1978)	78-64942	0-88318-146-0
No. 48	Current Trends in the Theory of Fields (Tallahassee, 1978)	78-72948	0-88318-147-9
No. 49	Cosmic Rays and Particle Physics – 1978 (Bartol Conference)	79-50489	0-88318-148-7
No. 50	Laser-Solid Interactions and Laser Processing – 1978 (Boston)	79-51564	0-88318-149-5
No. 51	High Energy Physics with Polarized Beams and Polarized Targets (Argonne, 1978)	79-64565	0-88318-150-9
No. 52	Long-Distance Neutrino Detection – 1978 (C.L. Cowan Memorial Symposium)	79-52078	0-88318-151-7
No. 53	Modulated Structures – 1979 (Kailua Kona, Hawaii)	79-53846	0-88318-152-5
No. 54	Meson-Nuclear Physics – 1979 (Houston)	79-53978	0-88318-153-3
No. 55	Quantum Chromodynamics (La Jolla, 1978)	79-54969	0-88318-154-1
No. 56	Particle Acceleration Mechanisms in Astrophysics (La Jolla, 1979)	79-55844	0-88318-155-X
No. 57	Nonlinear Dynamics and the Beam-Beam Interaction (Brookhaven, 1979)	79-57341	0-88318-156-8
No. 58	Inhomogeneous Superconductors – 1979 (Berkeley Springs, W.V.)	79-57620	0-88318-157-6
No. 59	Particles and Fields – 1979 (APS/DPF Montreal)	80-66631	0-88318-158-4
No. 60	History of the ZGS (Argonne, 1979)	80-67694	0-88318-159-2
No. 61	Aspects of the Kinetics and Dynamics of Surface Reactions (La Jolla Institute, 1979)	80-68004	0-88318-160-6
No. 62	High Energy e^+e^- Interactions (Vanderbilt, 1980)	80-53377	0-88318-161-4
No. 63	Supernovae Spectra (La Jolla, 1980)	80-70019	0-88318-162-2
No. 64	Laboratory EXAFS Facilities – 1980 (Univ. of Washington)	80-70579	0-88318-163-0
No. 65	Optics in Four Dimensions – 1980 (ICO, Ensenada)	80-70771	0-88318-164-9
No. 66	Physics in the Automotive Industry – 1980 (APS/AAPT Topical Conference)	80-70987	0-88318-165-7
No. 67	Experimental Meson Spectroscopy – 1980 (Sixth International Conference, Brookhaven)	80-71123	0-88318-166-5
No. 68	High Energy Physics – 1980 (XX International Conference, Madison)	81-65032	0-88318-167-3
No. 69	Polarization Phenomena in Nuclear Physics – 1980 (Fifth International Symposium, Santa Fe)	81-65107	0-88318-168-1
No. 70	Chemistry and Physics of Coal Utilization – 1980 (APS, Morgantown)	81-65106	0-88318-169-X

No. 71	Group Theory and its Applications in Physics – 1980 (Latin American School of Physics, Mexico City)	81-66132	0-88318-170-3
No. 72	Weak Interactions as a Probe of Unification (Virginia Polytechnic Institute – 1980)	81-67184	0-88318-171-1
No. 73	Tetrahedrally Bonded Amorphous Semiconductors (Carefree, Arizona, 1981)	81-67419	0-88318-172-X
No. 74	Perturbative Quantum Chromodynamics (Tallahassee, 1981)	81-70372	0-88318-173-8
No. 75	Low Energy X-Ray Diagnostics – 1981 (Monterey)	81-69841	0-88318-174-6
No. 76	Nonlinear Properties of Internal Waves (La Jolla Institute, 1981)	81-71062	0-88318-175-4
No. 77	Gamma Ray Transients and Related Astrophysical Phenomena (La Jolla Institute, 1981)	81-71543	0-88318-176-2
No. 78	Shock Waves in Condensed Mater – 1981 (Menlo Park)	82-70014	0-88318-177-0
No. 79	Pion Production and Absorption in Nuclei – 1981 (Indiana University Cyclotron Facility)	82-70678	0-88318-178-9
No. 80	Polarized Proton Ion Sources (Ann Arbor, 1981)	82-71025	0-88318-179-7
No. 81	Particles and Fields –1981: Testing the Standard Model (APS/DPF, Santa Cruz)	82-71156	0-88318-180-0
No. 82	Interpretation of Climate and Photochemical Models, Ozone and Temperature Measurements (La Jolla Institute, 1981)	82-071345	0-88318-181-9
No. 83	The Galactic Center (Cal. Inst. of Tech., 1982)	82-071635	0-88318-182-7
No. 84	Physics in the Steel Industry (APS/AISI, Lehigh University, 1981)	82-072033	0-88318-183-5
No. 85	Proton-Antiproton Collider Physics –1981 (Madison, Wisconsin)	82-072141	0-88318-184-3
No. 86	Momentum Wave Functions – 1982 (Adelaide, Australia)	82-072375	0-88318-185-1
No. 87	Physics of High Energy Particle Accelerators (Fermilab Summer School, 1981)	82-072421	0-88318-186-X
No. 88	Mathematical Methods in Hydrodynamics and Integrability in Dynamical Systems (La Jolla Institute, 1981)	82-072462	0-88318-187-8
No. 89	Neutron Scattering – 1981 (Argonne National Laboratory)	82-073094	0-88318-188-6
No. 90	Laser Techniques for Extreme Ultraviolt Spectroscopy (Boulder, 1982)	82-073205	0-88318-189-4
No. 91	Laser Acceleration of Particles (Los Alamos, 1982)	82-073361	0-88318-190-8
No. 92	The State of Particle Accelerators and High Energy Physics (Fermilab, 1981)	82-073861	0-88318-191-6

No. 93	Novel Results in Particle Physics (Vanderbilt, 1982)	82-73954	0-88318-192-4
No. 94	X-Ray and Atomic Inner-Shell Physics – 1982 (International Conference, U. of Oregon)	82-74075	0-88318-193-2
No. 95	High Energy Spin Physics – 1982 (Brookhaven National Laboratory)	83-70154	0-88318-194-0
No. 96	Science Underground (Los Alamos, 1982)	83-70377	0-88318-195-9
No. 97	The Interaction Between Medium Energy Nucleons in Nuclei – 1982 (Indiana University)	83-70649	0-88318-196-7
No. 98	Particles and Fields – 1982 (APS/DPF University of Maryland)	83-70807	0-88318-197-5
No. 99	Neutrino Mass and Gauge Structure of Weak Interactions (Telemark, 1982)	83-71072	0-88318-198-3
No. 100	Excimer Lasers – 1983 (OSA, Lake Tahoe, Nevada)	83-71437	0-88318-199-1
No. 101	Positron-Electron Pairs in Astrophysics (Goddard Space Flight Center, 1983)	83-71926	0-88318-200-9
No. 102	Intense Medium Energy Sources of Strangeness (UC-Sant Cruz, 1983)	83-72261	0-88318-201-7
No. 103	Quantum Fluids and Solids – 1983 (Sanibel Island, Florida)	83-72440	0-88318-202-5
No. 104	Physics, Technology and the Nuclear Arms Race (APS Baltimore –1983)	83-72533	0-88318-203-3
No. 105	Physics of High Energy Particle Accelerators (SLAC Summer School, 1982)	83-72986	0-88318-304-8
No. 106	Predictability of Fluid Motions (La Jolla Institute, 1983)	83-73641	0-88318-305-6
No. 107	Physics and Chemistry of Porous Media (Schlumberger-Doll Research, 1983)	83-73640	0-88318-306-4
No. 108	The Time Projection Chamber (TRIUMF, Vancouver, 1983)	83-83445	0-88318-307-2
No. 109	Random Walks and Their Applications in the Physical and Biological Sciences (NBS/La Jolla Institute, 1982)	84-70208	0-88318-308-0
No. 110	Hadron Substructure in Nuclear Physics (Indiana University, 1983)	84-70165	0-88318-309-9
No. 111	Production and Neutralization of Negative Ions and Beams (3rd Int'l Symposium, Brookhaven, 1983)	84-70379	0-88318-310-2
No. 112	Particles and Fields – 1983 (APS/DPF, Blacksburg, VA)	84-70378	0-88318-311-0
No. 113	Experimental Meson Spectroscopy – 1983 (Seventh International Conference, Brookhaven)	84-70910	0-88318-312-9

No. 114	Low Energy Tests of Conservation Laws in Particle Physics (Blacksburg, VA, 1983)	84-71157	0-88318-313-7
No. 115	High Energy Transients in Astrophysics (Santa Cruz, CA, 1983)	84-71205	0-88318-314-5
No. 116	Problems in Unification and Supergravity (La Jolla Institute, 1983)	84-71246	0-88318-315-3
No. 117	Polarized Proton Ion Sources (TRIUMF, Vancouver, 1983)	84-71235	0-88318-316-1
No. 118	Free Electron Generation of Extreme Ultraviolet Coherent Radiation (Brookhaven/OSA, 1983)	84-71539	0-88318-317-X
No. 119	Laser Techniques in the Extreme Ultraviolet (OSA, Boulder, Colorado, 1984)	84-72128	0-88318-318-8
No. 120	Optical Effects in Amorphous Semiconductors (Snowbird, Utah, 1984)	84-72419	0-88318-319-6
No. 121	High Energy e^+e^- Interactions (Vanderbilt, 1984)	84-72632	0-88318-320-X
No. 122	The Physics of VLSI (Xerox, Palo Alto, 1984)	84-72729	0-88318-321-8
No. 123	Intersections Between Particle and Nuclear Physics (Steamboat Springs, 1984)	84-72790	0-88318-322-6
No. 124	Neutron-Nucleus Collisions – A Probe of Nuclear Structure (Burr Oak State Park - 1984)	84-73216	0-88318-323-4
No. 125	Capture Gamma-Ray Spectroscopy and Related Topics – 1984 (Internat. Symposium, Knoxville)	84-73303	0-88318-324-2
No. 126	Solar Neutrinos and Neutrino Astronomy (Homestake, 1984)	84-63143	0-88318-325-0
No. 127	Physics of High Energy Particle Accelerators (BNL/SUNY Summer School, 1983)	85-70057	0-88318-326-9
No. 128	Nuclear Physics with Stored, Cooled Beams (McCormick's Creek State Park, Indiana, 1984)	85-71167	0-88318-327-7
No. 129	Radiofrequency Plasma Heating (Sixth Topical Conference, Callaway Gardens, GA, 1985)	85-48027	0-88318-328-5